AGING

AGING
HOW AGING WORKS, HOW WE REVERSE AGING, AND PROSPECTS FOR CURING AGING DISEASES

Edited by

MICHAEL FOSSEL
Telocyte LLC, Ada, MI, United States

Academic Press is an imprint of Elsevier
125 London Wall, London EC2Y 5AS, United Kingdom
525 B Street, Suite 1650, San Diego, CA 92101, United States
50 Hampshire Street, 5th Floor, Cambridge, MA 02139, United States

Copyright © 2024 Elsevier Inc. All rights are reserved, including those for text and data mining, AI training, and similar technologies.

No part of this publication may be reproduced or transmitted in any form or by any means, electronic or mechanical, including photocopying, recording, or any information storage and retrieval system, without permission in writing from the publisher. Details on how to seek permission, further information about the Publisher's permissions policies and our arrangements with organizations such as the Copyright Clearance Center and the Copyright Licensing Agency, can be found at our website: www.elsevier.com/permissions.

This book and the individual contributions contained in it are protected under copyright by the Publisher (other than as may be noted herein).

Notices

Knowledge and best practice in this field are constantly changing. As new research and experience broaden our understanding, changes in research methods, professional practices, or medical treatment may become necessary.

Practitioners and researchers must always rely on their own experience and knowledge in evaluating and using any information, methods, compounds, or experiments described herein. In using such information or methods they should be mindful of their own safety and the safety of others, including parties for whom they have a professional responsibility.

To the fullest extent of the law, neither the Publisher nor the authors, contributors, or editors, assume any liability for any injury and/or damage to persons or property as a matter of products liability, negligence or otherwise, or from any use or operation of any methods, products, instructions, or ideas contained in the material herein.

ISBN: 978-0-443-15500-0

For information on all Academic Press publications visit our website at
https://www.elsevier.com/books-and-journals

Publisher: Mica Haley
Acquisitions Editor: Stacy Masucci
Editorial Project Manager: Billie Jean Fernandez
Production Project Manager: Fahmida Sultana
Cover Designer: Miles Hitchen
Title design by Yuri Kirillov at Age2b.com.

Typeset by TNQ Technologies

Dedication

To Cal Harley, a pioneer in cell aging and telomere biology, a friend whose Alzheimer's disease came all too quickly to a man of genius and kindness.

Contents

Contributors xi
About the editor xiii
Foreword xv
Preface xvii
Acknowledgments xix

1. **Introduction to aging and age-related disease**
   MICHAEL FOSSEL

   1. Introduction 1
   2. How did we get here? 2
   3. How to understand aging 4
   4. The limited role of biomarkers 6
   5. A unified systems model 7
   6. Caveats 13
   7. Prospects for intervention 14
   References 15

2. **Age-related disease: Central nervous system**
   MICHAEL FOSSEL

   1. Introduction 19
   2. The dementias 19
   3. Therapeutic status 21
   4. Neuroanatomy 22
   5. Cell types 22
   6. Biochemical changes 23
   7. Aging as a key process 25
   8. Unified systems model: Cell aging in the central nervous system 27
   9. Summary 30
   References 31

3. **Age-related disease: Cardiovascular system**
   PETER M. NILSSON, MICHAEL FOSSEL, JOE BEAN AND NINA KHERA

   1. Determinants of arterial stiffness 35
   2. How to define early vascular aging 36
   3. Mechanisms important for modifying arterial stiffness and vascular aging 36
   4. Factors in early life influencing arterial stiffness and early vascular aging 36
   5. Healthy and supernormal vascular aging 37
   6. Treatment of cardiovascular risk and clinical manifestations 38
   7. Emerging role of polypill for cardiovascular prevention 39
   8. Summary: Aging and cardiovascular risk—mechanisms and clinical aspects 40
   9. Role of cell aging in cardiovascular disease 40
   10. Prospects for intervention 43
   Acknowledgment 44
   References 44
   Further reading 52

4. Age-related disease: Bones
JOSHUA N. FARR AND ABHISHEK CHANDRA

1. Introduction: Bone anatomy and homeostasis ... 53
2. Skeletal maintenance and remodeling ... 54
3. Osteoporosis: An overview ... 56
4. Therapeutic options to treat osteoporosis: Current interventions ... 57
5. Cellular components of an aging bone ... 60
6. Biology of an aging bone: Mechanisms and pathways ... 62
7. Therapeutic options: Future potential interventions and emerging trials ... 65
8. Conclusion ... 65
References ... 65

5. Age-related disease: Joints
ILYAS M. KHAN

1. Cartilage structure and function: a primer ... 73
2. Aging and osteoarthritis ... 75
3. Replicative senescence in cartilage ... 76
4. Stress-induced senescence in cartilage ... 77
5. The senescence-associated secretory profile in chondrocytes ... 77
6. Autophagy: A complex interrelationship with senescence ... 78
7. The role of sirtuins in chondrosenescence ... 79
8. The emerging role of circadian clocks in senescence ... 79
9. mTOR, nutrient sensing, and senescence in cartilage ... 80
10. Stem cell exhaustion? ... 80
11. The role of the extracellular matrix in initiating and stabilising cellular senescence ... 81
12. Chondrosenescence: Can the advancing tide be reversed? ... 82
13. Conclusion ... 83
References ... 84

6. Age-related disease: Kidneys
SASWAT KUMAR MOHANTY, BHAVANA VEERABHADRAPPA, ASIT MAJHI, KITLANGKI SUCHIANG AND MADHU DYAVAIAH

1. Introduction ... 91
2. Aging kidney ... 92
3. Common alterations in cellular and molecular mechanisms of renal aging and CKD ... 97
4. Prevalence of CKD in the elderly ... 101
5. Risk factors for CKD in the elderly ... 104
6. CKD care and management for the elderly ... 107
7. Conclusion ... 109
Acknowledgment ... 110
References ... 110
Further reading ... 117

7. Age-related disease: Immune system
KARIN DE PUNDER AND ALEXANDER KARABATSIAKIS

1. Introduction ... 119
2. The aging of the immune system ... 119
3. Examples of conditions related to immune system aging ... 124
4. Biomolecular mechanisms of immunosenescence ... 126
5. Upstream risk factors ... 130
6. Biomarkers/downstream factors related to aging and immunosenescence ... 134
7. Future perspectives ... 135
8. Conclusion ... 137
References ... 137

8. **Age-related disease: Skin**
SARANYA P. WYLES, KRISHNA VYAS, J. ROSCOE WASSERBURG, RYEIM ANSAF AND JAMES L. KIRKLAND

1. Geriatric dermatology 147
2. Fountain of youth and skin health 147
3. Chronological versus biological skin aging 147
4. Clinical manifestations of skin aging 149
5. Regenerative aesthetics 150
6. Cellular senescence 151
7. Function of senescent cells 152
8. Features of senescent cells 153
9. Biochemical markers of cellular senescence 154
10. Senescence in skin aging 154
11. Targeting senescent cells in the skin 154
12. Senotherapeutic agents for skin 155
13. First- and second-generation senolytics 155
14. Translation to clinical practice 157
15. Targeting cellular senescence with senotherapies 158
16. Conclusion 159
References 159

9. **Age-related disease: Lungs**
JOSHUA O. OWUOR AND AYMAN O. SOUBANI

1. Introduction and relevant anatomy 165
2. The aging lung 166
3. Role of cell aging in age-related lung disease 167
4. Specific lung diseases correlated with advanced age 167
5. Current research, future direction, and potential interventions 171
References 172

10. **Age-related disease: Diabetes**
ALLYSON K. PALMER AND JAMES L. KIRKLAND

1. Introduction: Type 2 diabetes is an age-related disease 175
2. Cellular senescence in the pathogenesis of type 2 diabetes 176
3. The diabetic microenvironment drives senescent cell accumulation 180
4. Cellular senescence and diabetic complications 181
5. Current diabetes therapeutics: Impact on cellular senescence 182
6. Targeting cellular senescence in diabetes 184
7. Clinical trials 186
8. Challenges and future directions 186
9. Conclusions 187
References 187

11. **Age-related disease: Eyes**
COAD THOMAS DOW

1. Introduction 195
2. Presbyopia 195
3. Glaucoma 196
4. Lens 197
5. Retina 198
6. The eye as a window to other age-related diseases 200
7. Upstream opportunities for life and health extension 201
8. Summary 202
References 202

12. Age-related disease: Cancer, telomerase, and cell aging
    KURT WHITTEMORE

    1. Introduction 205
    2. A short historical perspective 205
    3. Current cancer treatments 206
    4. Relationship between cancer, aging, and telomerase 207
    5. Conclusion 212
    References 213

13. Age-related disease: Effective intervention
    MICHAEL FOSSEL AND JOHN P. COOKE

    1. Perspective 219
    2. Current approaches 221
    3. Targeting aging processes 222
    4. Targeting a dominant process in cell aging 223
    5. Where do we stand now? 228
    6. Where we are going 229
    References 229

14. Summary
    MICHAEL FOSSEL

    1. Introduction 233
    2. Is aging a disease? 235
    3. Organs 236
    References 245

Index 247

Contributors

Ryeim Ansaf Colorado State University Pueblo, Pueblo, CO, United States

Joe Bean University of Missouri School of Medicine, Kansas City, MO, United States

Abhishek Chandra Department of Physiology and Biomedical Engineering, Section of Geriatric Medicine and Gerontology, Robert and Arlene Kogod Center of Aging, Mayo Clinic, Rochester, MN, United States

John P. Cooke Houston Methodist Research Institute, Houston, TX, United States

Karin de Punder Department of Psychology, Clinical Psychology-II, University of Innsbruck, Innsbruck, Austria

Coad Thomas Dow Department of Ophthalmology and Visual Sciences, McPherson Eye Research Institute, University of Wisconsin, School of Medicine and Public Health, Madison, WI, United States

Madhu Dyavaiah Department of Biochemistry and Molecular Biology, School of Life Sciences, Pondicherry University, Kalapet, Pondicherry, India

Joshua N. Farr Department of Endocrinology, Robert and Arlene Kogod Center of Aging, Mayo Clinic, Rochester, MN, United States

Michael Fossel Telocyte LLC, Ada, MI, United States

Alexander Karabatsiakis Department of Psychology, Clinical Psychology-II, University of Innsbruck, Innsbruck, Austria

Ilyas M. Khan Swansea University Medical School, Faculty of Medicine, Health & Life Science, Swansea, UK

Nina Khera Buckingham Browne and Nichols School, Cambridge, MA, United States

James L. Kirkland Department of Biomedical Engineering, Mayo Clinic, Rochester, MN, United States; Division of General Internal Medicine, Department of Medicine, Mayo Clinic, Rochester, MN, United States

Asit Majhi Department of Biochemistry and Molecular Biology, School of Life Sciences, Pondicherry University, Kalapet, Pondicherry, India

Saswat Kumar Mohanty Department of Biochemistry and Molecular Biology, School of Life Sciences, Pondicherry University, Kalapet, Pondicherry, India

Peter M. Nilsson Clinical Sciences, Lund University, Malmö, Sweden

Joshua O. Owuor Department of Pulmonary & Critical Care Medicine, Wayne State University/Detroit Medical Center, Detroit, MI, United States

Allyson K. Palmer Robert and Arlene Kogod Center on Aging, Mayo Clinic, Rochester, MN, United States; Division of Hospital Internal Medicine, Department of Medicine, Mayo Clinic, Rochester, MN, United States

Ayman O. Soubani Department of Pulmonary & Critical Care Medicine, Wayne State University/Detroit Medical Center, Detroit, MI, United States

Kitlangki Suchiang Department of Biochemistry and Molecular Biology, School of Life Sciences, Pondicherry University, Kalapet, Pondicherry, India; Department of Biochemistry, North Eastern Hill University, Shillong, Meghalaya, India

Bhavana Veerabhadrappa Department of Biochemistry and Molecular Biology, School of Life Sciences, Pondicherry University, Kalapet, Pondicherry, India

Krishna Vyas Division of Plastic Surgery, Department of Surgery, Mayo Clinic, Rochester, MN, United States

J. Roscoe Wasserburg State University of New York, Downstate College of Medicine, Brooklyn, NY, United States

Kurt Whittemore Novo Nordisk, Watertown, MA, United States

Saranya P. Wyles Department of Dermatology, Mayo Clinic, Rochester, MN, United States

About the editor

Dr. Fossel received a joint BA and MA in Psychology at Wesleyan University then, with perfect GRE scores, won an NSF fellowship to Stanford University, where he completed his Ph.D. in Neurobiology and his MD, and taught neuroanatomy, experimental design, and other courses. He began to focus on aging, progeria, and the experimental testing of aging theories. A clinical professor of medicine for almost 3 decades, he was the executive director for the American Aging Association, editor of several journals, and a frequent speaker and chair at global conferences. He has authored more than 100 articles, chapters, and books on aging, medicine, electronic health records, and medical ethics. His invited article, A Unified Model of Dementias and Age-related Neurodegeneration, and his parallel article, A Unified Model of Age-Related Cardiovascular Disease, outline a unified systems model for age-related CNS and CV diseases and offer a novel, feasible, and potentially more effective point of intervention for age-related diseases. His textbooks and books for the general public have been translated into more than a dozen global editions and have been praised by both *The London Times* and *The Wall Street Journal* as among the best science books of the year. Dr. Fossel is the President of Telocyte, a biotech firm targeting CNS and other age-related diseases by using the telomerase gene therapy.

Foreword

There is general agreement that aging is the greatest risk factor for the leading causes of death. Yet, there is no public or private institution, in the US or any other country, with "age" or "aging" in its title that either promotes or funds research on the etiology of biological aging.

The funds available annually for research on age-associated diseases in the United States are orders of magnitude greater than what is available on alleged efforts to increase our understanding of the underlying aging process.

It is a multibillion dollar miss-understanding to believe that the resolution of any, or all, age-associated diseases will reveal information about the underlying aging process. This error is analogous to why the resolution of many childhood diseases did not increase our understanding of childhood development.

The belief that research on the etiology of aging is a futile exercise ignores the recent discovery of revolutionary methods that allow for observing the qualitative and quantitative features of the biomolecular and atomic landscape that distinguishes old from young cells in the same lineage. This knowledge would have the likelihood of discovering why the differences increase vulnerability of old cells to age-associated diseases. It could reveal that specific biomolecular configurations reveal a common cause for why the increase in pathology is age-associated.

The general belief that the leading causes of death are age-associated leads to the logical conclusion that research on the etiology of biological aging is a critical goal. This research is likely to find a significant common molecular landscape that only exists in old cells and, unlike the landscape found in young cells, increases their vulnerability to pathology. The likelihood of this finding is now great because of the enormous number of discoveries in technology that now permit the identification of changes at the molecular and atomic level that occurs in single living cells.

Leonard Hayflick

Preface

We stand on the verge of a medical revolution. With the possible exception of microbial theory and its effect upon human health over the past two centuries, the oncoming conceptual revolution in our understanding of aging and age-related disease has no precedent. In both cases—what microbial theory did for the mean human lifespan and what aging theory may well do for the maximum human lifespan—the impact resulted from a new and deeper understanding of biology and diseases. Such understandings radically change medicine and alter human society globally. The impact of a better understanding of the fundamental aging processes will not merely be upon age-related disease but, more personally and poignantly, upon the human psyche.

Ameliorating the misery, hopelessness, fear, pain, and desperation of those suffering from age-related diseases are the outcomes we may foresee if the ravages of aging are finally eased. Yet the medical and the personal implications do not stand alone: the extension of the healthy maximum human lifespan will have deep, broad, and often unforeseen consequences for human culture. For the individual, these consequences will be an almost unalloyed good: the eradication of diseases that destroy our minds and bodies. For our global culture, these consequences will be a less predictable mix of changes, and change is always difficult for those in its midst.

Yet the need and the necessity remain. We can scarcely stand by and ignore the needs of those around us: we have the necessity of compassion if we are to have any claim to being human. Compassion for the young is probably innate and almost universal among human beings; compassion for the elderly is perhaps less common. We have all been young; most are not yet aware of what it is to be old. Moreover, we tend to assume that aging and its concomitant diseases are both ubiquitous and unavoidable, yet we had much the same set of assumptions once regarding what we now call microbial diseases, which were once held to be equally ubiquitous and unavoidable. We were wrong.

To truly understand aging and age-related diseases, we must first reexamine our assumptions. We only guarantee failure if we begin by assuming that nothing can—or should—be done. While success can never be guaranteed in any human endeavor, the assumption that something is impossible surely guarantees failure.

Aging cannot only be altered, but it can be accomplished with the technical tools currently in our hands. The diseases of aging cannot only be prevented and cured, but the prospects of success are within our medical horizons. We are not adept, we are not fully cognizant of the complexities of such interventions, but we can already demonstrate an ability to intervene, if not to extent we might wish.

The Wright brothers demonstrated the reality of powered flight in 1903, but we were still decades away from the ability to transport hundreds of people from continent-to-continent safely, reliably, and affordably, a possibility few would have believed that day at Kitty Hawk. Intervention in the fundamental process of aging is already a reality, but we may still be decades away from the ability to cure and prevent age-related diseases safely, reliably, and affordably. A revolution in the making, if barely begun.

It is a medical revolution that a few are aware of. It is a revolution that we outline here in this book, but it is not driven by us alone, but by thousands of researchers, physicians, and others who see the potential, have the compassion, and are driven to the lives of those around them. These are dedicated people, less concerned about getting credit than about getting the work done. They do not work to become well-known, but to be known for making others well.

This is their book and written for all of us.

Acknowledgments

As with so many things in our lives, this textbook was not the product of my work, but was instead the product of hundreds, perhaps thousands of others who have, over the years, devoted their careers to this topic. There are endless researchers who have done the hard work, where I have merely sat and thought about what their work implied.

My thanks, admiration, and humble gratitude go out to those who mean so much to this field. I include a handful here, but could easily fill pages more with those who have my admiration and my gratitude. Leonard Hayflick, Cal Harley, Maria Blasco, Bill Andrews, John Cooke, and dozens of others spring immediately to mind. I find it impossible not to think of those others, some of them friends, some of them correspondents, some of them coauthors, some of them all but unknown to me except by their reputation for probity, hard work, perseverance, and intelligence.

To all of you: my humble thanks for the rigor and the energy of your lifelong work. In addition to those trying to understand aging, are those whose support, understanding, and shear dogged will have enabled me, as well as many others, to have moved ahead as we work to cure age-related disease: Billie Jean Fernandez and Stacy Masucci, whose efforts and faith enabled us to publish this book; Artem Iaryguine and his team at Age2B.com whose talent resulted in our cover; Peter Rayson, the friend and entrepreneur who enabled us to push onward toward human trials; our board members Michelle Hylan, Rajesh Shukla, Georgi Gospodinov, and Radomir Julina, as well as the myriad other associates who have enabled us to persevere and, hopefully, to succeed. My deepest thanks to my coauthors, for agreeing to join me when invited, and for their knowledge, understanding, and compassion for others. This book could certainly not have been written without each of you.

Thank you all—friends, associates, authors, scientists, physicians, and others—for doing where I could not, for thinking what I could not, for helping when I could not. Success does not take a mere village; it takes a globe filled with people who work harder than I do.

Thank you for your work, intelligence, and especially, your compassion.

CHAPTER 1

Introduction to aging and age-related disease

Michael Fossel

Telocyte LLC, Ada, MI, United States

1. Introduction

The moment of the rose and the moment of the yew-tree are of equal duration. —*T.S. Eliot, Little Gidding, 1942.*

The deepest mysteries are those that escape our notice.

There are times in human history when conceptual revolutions change the course of humanity. They are seldom recognized except in retrospect. The advent of microbial theory washed over us within the past 2 centuries, almost unseen and seldom appreciated by those it bore along in its powerful wake. It changed the world, but the world saw it only in the small academic eddies and the froth of clinical argument, and occasionally in a sudden change in practice. It was fought against, decried as foolishness, and ignored by the uninformed. Yet for all the resistance and confusion, our lives improved, the cost of compassionate medical care decreased, and the world became a better place. Our lives became safer, longer, and healthier than ever before.

We still suffer, albeit from different diseases than previously. Most of us will die of aging.

Until quite recently in human history, most of us died of infections, although trauma, genetic diseases, and toxic exposures have always remained steady contenders over time. Our growing understanding of microbial disease changed much of medical history, a remarkable transition that occurred over the past 200 years. Once, we died young of invisible germs that ravaged us, especially in our youth. Now, we die in old age of the ravages of age-related diseases, diseases that perhaps too often we accept as unavoidable.

In the case of infectious diseases, the problem was that despite our wishes and our wits, we simply had no concept of their microbial causes. Ironically, in the case of age-related diseases, we may suffer from much the same conceptual lacuna. Until we truly understand the aging process, we can neither understand nor effectively treat age-related diseases. We can effectively treat age-related diseases only by understanding, then reversing the fundamental processes that underlie aging itself.

What is aging? How does aging cause age-related disease? Can we reverse aging?

These are the central questions of this book, and misconceptions aside, they remain the most difficult and poorly addressed questions in modern medicine. The difficulty is inherent in the questions themselves, but our ability to answer these questions effectively and accurately also stumbles as a result of our inability to recognize our unexamined assumptions.

The problem is not merely that we don't *understand* aging, but that we think we *do*.

Our first error is the assumption that we already know what aging is: we assume that aging is wear-and-tear. As so many people would say, it just happens. We equate aging with entropy, and therefore think of it as inescapable. The naive assumption of simplicity—a belief that appears obvious but is at odds with reality and with a broader view of biology—is why we have had little success in addressing aging and age-related diseases throughout human history. As with most subjects, if we delve into aging thoroughly and rigorously, we discover unexpected complexity and find that our initial assumption of simplicity was unjustified. Aging is not simply the passive accumulation of damage, but the dynamic and complicated failure of maintenance in the face of entropy. The dynamic and complicated failure of maintenance is a process that can teach us a great deal and that offers startling and unexpected insights into potential points of intervention. To understand aging enables us to alter aging.

Our second error lies in giving lip service to the observation that aging is the single most important risk factor for all age-related diseases, but then focusing instead on the clinical findings (the hallmarks, biomarkers, and signs and symptoms of age-related disease) while ignoring the fundamental aging process. After all, if we assume aging is merely the passive accumulation of dysfunction, why delve into it further? The result has been a great deal of meticulously detailed knowledge about the onset, course, symptoms, and pathology, and a multitude of (largely ineffective) interventions in age-related diseases, with no ability to stop or reverse the fundamental processes that underlie and drive those diseases. Like a physician who confuses symptoms for the disease, we are blind to the disease process itself. We provide transient relief of the symptoms without effecting a change in the pathology and without changing the ultimately fatal outcome. Hallmarks of aging are no more aging than the symptoms of a disease are the disease. In addressing age-related disease, our current clinical interventions reflect this limitation. We can replace an osteoarthritic knee but cannot reverse the osteoarthritic process itself. We can treat coronary artery disease with stents, statins, and bypass grafts but cannot reverse the atherosclerotic process itself. Whether we use dialysis for chronic renal failure, bisphosphonates for osteoporosis, or monoclonal antibodies for dementia, the results are, as a rule, expensive, constraining, uncomfortable, and eventually ineffective. Our disappointing clinical outcomes derive from our focus on the outcomes of age-related disease rather than a straightforward consideration of how aging causes such disease. We can do better.

Our third error lies in assuming that we cannot reverse the aging process. Biologically provincial, we see aging in ourselves, our pets, and farm animals, and in a thousand other common species, and conclude that because aging occurs in species we know best, aging is universal and irreversible. A broader look at biology shows that aging is neither universal nor (in experimental settings) irreversible. Misunderstandings about the aging process are widespread among the public, but almost equally so among academics, researchers, and clinicians. We exhibit stunning insights into the hallmarks of aging but little understanding of aging as a process. What we cannot conceive, we cannot change.

2. How did we get here?

Aging occurs in cells. When we see aging at the level of the entire organism or at the level of a tissue or organ, we are actually seeing the results of aging at the cellular level. All aging is cell aging. When we see older humans become aged, or when we evaluate an age-related disease, we are seeing the outcome of the cell aging. No aging of tissue occurs independently of aging of the cells comprising that tissue.

This point—that cells are the basic setting of all aging—was first pointed out in the 1960s. Before that, there was a firmly and widely held belief that individual cells were immortal and aging was a multicellular event. The belief was understandable, although the experimental support for that belief was egregious. The error derived from the observation that all multicellular life begins from a single fertilized ovum. Because such unicellular beginnings can be traced back perhaps 3.5 billion years, since the advent of life on earth, the inference was plain. Single cells, including the germ cell lineage that produces the ovum, had not aged in all those billions of years, and yet multicellular organisms (that derive from a single fertilized cell) age. Therefore, aging had to be attributed to multicellularity. This was an apparently reasonable conclusion, but, as careful experimentation came to show, deceptive. Experimental support for this erroneous conclusion relied on the work of Alexis Carrell, a biologist who claimed to keep individual cells alive indefinitely and without aging, even while the organism itself aged and died. The implication was that aging was a gestalt effect: not cellular but multicellular. To a degree, this supported both the public idea (as seen in contemporary fiction of the time [1]) and the medical idea that aging might be ameliorated or reversed using various infusions, transfusions, or vitamins. The belief in the value of heterochronic parabiosis, for example, in which blood from a younger animal is given to an older animal, resurfaced repeatedly over the past century despite both conceptual inconsistency and experimental evidence to the contrary [2].

Alexis Carrell (1873—1944): A remarkably gifted, if credulous, French biologist with a number of accomplishments. Beginning in 1912 at the Rockefeller Institute in New York City, he grew embryonic chicken myocardial cells and apparently maintained them in serially passaged cell culture for more than 2 decades, well beyond the life span of the donor species. Despite the inability of others to replicate his work, not only it was accepted as biological dogma, it drove a number of recurring quasiscientific beliefs, such as the use of rejuvenating transfusions (e.g., parabiosis), which that continue to draw adherents and to prompt transient fashions even in the investment and biotech community.

In the early 1960s, however, Leonard Hayflick [3] convincingly proved that somatic cells taken from multicellular organisms had a replicometer. They would divide a predictable number of times and showed increasing evidence of cell aging and dysfunction, culminating in distinct morphologic changes and, finally, in the inability to divide further, an end point termed replicative senescence. This limitation in cell replication is known as the Hayflick limit and is usually expressed as the number of potential cell divisions. Human fetal fibroblasts, for example, have a Hayflick limit of approximately 50 cell divisions. The Hayflick limit may vary with the species, cell type, and state of differentiation. It applies to all somatic cells, but, as we will see, is only partially applicable to germ cells, stem cells, and cancer (or otherwise mutated) cells.

These observations have been replicated and found to be accurate in innumerable species, in varying conditions, and in countless laboratories worldwide. The Hayflick limit and its implications for aging have become a fundamental pillar of modern cell biology and have immense implications for this book and for our understanding of age-related disease.

Leonard Hayflick (1928 to current): An extraordinarily insightful and diligent researcher responsible for many key inventions, discoveries, and advances in 20th-century biology, including the WI-38 cell strain used in modern virology, the first isolation of *Mycoplasma*, and the inverted microscope. He is perhaps best known as the eponymous discoverer of the Hayflick limit, when he observed that all somatic cells maintain a memory of how many times they have divided and enforce a maximum cap for such divisions. A retired professor at the University of California, San Francisco, and Stanford, he currently lives north of San Francisco. Although perhaps among the most deserving, he has never received a Nobel Prize for his work.

Hayflick's observation of an aging replicometer prompted the obvious question of what cellular mechanism counted the replications. The observation that telomeres should shorten with each cell division, first made by Alexey Olovnikov in 1971 [4], offered an obvious candidate for a cellular replicometer. The telomere length (within, if not between species) was rapidly shown to be well-correlated with the extent of cell aging [5], but causal proof was lacking until 1998 [6,7], when researchers at Geron Corporation in California published definitive data showing that cell aging could be reset by relengthening the telomere in human fibroblasts. This was the first time that aging was reversed [8], if only in human cells and only in vitro, but although unappreciated and often unknown, these papers remain landmarks in both biology and medicine. Telomerase not only resets cell morphology, histology, and physiology in senescent cells, it resets epigenetic expression [9].

Alexey Olovnikov (1936 to current): A Russian biologist, born in Vladivostok and still living in an apartment in Moscow, he first described the implications of DNA duplication for telomeres, using an analogy based on the way the Moscow rail system would replicate rail tracks. His work was published in a Russian scientific journal, prompting many in the west to credit the observation to James Watson, who published an equivalent notion 1 year later in an English scientific journal. The first to suggest a telomere theory of aging, he remains active as a theoretical biologist and continues to offer potential insights and innovative biological theories.

The potential (and profound) clinical implications had been sketched out in a pair of papers [10,11] published in the medical literature at about the same time (as well as the first book on the topic [12]), pointing out that age-related diseases might be effectively treated if we understood the basis of cell aging and targeted the process directly. Experimental data in support of this view was soon extended from human cells to human tissues. The original work showing that this was feasible in human skin [13] was published in 2000, and this was extended to both human bone [14] and human vascular endothelial tissue [15] by 2001. Although these papers provided proof that we were able to reverse human cell and tissue aging in vitro and ex vivo, the question of reversing cell and tissue aging in vivo, in animals or humans, remained open for the next decade. Could we alter aging not merely at the cell and tissue level, but also at the level of the entire organism? What was easy to do in a Petri dish was incomparably more difficult in an adult animal or human patient.

In the early 2010s, initial published data based on the oral use of telomerase activating compounds in humans [16–18] were intriguing if not overwhelming. More compelling and technically adroit were the results of gene therapy experiments in mice, first using germ line telomerase switches in specially bred mice [19], and then using viral vector (using an AAV) delivery of the telomerase gene [mTERT] to wild-type aged mice [20]. In both cases, within the technical limitations of available gene therapy techniques at the time, the results were both confirmatory and remarkable. Not only were the hallmarks of aging reversed, the healthy life span was significantly extended. However, mice are not humans, and the technical limitations remain daunting for taking these findings into translational human trials to address clinical disease.

Yet the implications are enormous. These implications have been reviewed in medical textbooks [21] and academic articles [22–27] with the reasonable suggestion that not only can cell aging be reversed, doing so would imply reversing the aging process at the level of the organism, allowing us to intervene effectively in age-related human disease. The key question is to what extent we are technically able to reverse cell aging in human trials. Current limitations of gene therapy remain formidable, although our abilities are rapidly improving. Numerous issue remain: safety, immunogenicity, side effects, transduction rates, cell-specific delivery, and the duration of expression, as well as issues of vector choice, serotype, promoters, and so on, to say nothing of viral vectors versus lipid nanoparticles, genes versus mRNA, and a myriad other options. In short, we have the means, but we still lack the elegant expertise we would desire to prove we can use this approach to cure and prevent age-related disease with both confidence and competence.

To use a common analogy, our ability to use gene therapy effectively is much like the Wright brothers' demonstrations in 1903. Clearly, powered flight is possible, although not yet involving transatlantic flights with hundreds of passengers in safety and comfort, as well as at an acceptable cost. The use of gene therapy to effect fundamental changes in the aging process is possible, although it is not yet ready for safe, effective, and cost-effective use in our medical clinics. We have much to learn.

3. How to understand aging

Aging has been remarkably resistant to a clear understanding. The problem has been, and remains, our resistance to stepping back to move ahead: we resist an honest appraisal of our assumptions. Consider a few historical parallels.

In the Middle Ages, Vatican astronomers assumed that the sun went around the earth, not only because their religion told them so, but because quotidian experience appeared to confirm it. Observers watched the sun rise in the east and watched it set in the west, and neither saw nor felt the earth spinning beneath them. Given those starting points, they concluded that the sun went around the earth, despite the protestations of Copernicus, Kepler, Brahe, and Galileo. Rather than stepping back and reexamining their assumptions, they maintained a simple, provincial view of astronomy.

Human disease has not fared much better. A thousand years ago, Avicenna (Ibn Sina) was the world's foremost physician. His textbook, the *Canon of Medicine* remained in use for more than 700 years after his death. Happily, medical science moves a bit faster these days. An excellent observer of human disease, Avicenna's observations on bubonic plague remain unexcelled even today, but he suffered from an understandable conceptual lacuna: he could not imagine that invisible organisms (microbes) caused disease. Only in the past 2 centuries, with the advent of the microscope and the acceptance of germ theory, have we expanded our beliefs and been able to overcome an inadequate understanding of such diseases.

The problem we now face in understanding aging is similar and results in the same deficit of understanding. We lack both the wish and the wit: the wish to reexamine our assumptions (aging is entropy) and the wit to appreciate the extraordinary complexity involved in the aging process (aging just happens). The first assumption, that aging is ubiquitous, fails upon reflection. Every cell in an adult organism derives from a fertilized ovum. The maternal ovum was inherited from her mother and was already a few decades old, if unaffected by it. The provenance of that ovum can be traced back much further, a good 3.5 billion years further, and yet was fully functional on fertilization.

Retrospectively, we are all the product of an unaged cellular lineage of remarkable antiquity. Nor is the ovum the only player in this billion year–plus provenance. Many would ascribe aging to mitochondrial dysfunction, yet the mitochondria we inherit from the maternal cell has been a partner in eukaryotic cells for perhaps half as long as life has been on our planet. Although the mitochondria we find in each ovum are well more than a billion years old, they are neither dysfunctional nor aged as the human fetus begins formation. On the contrary, these antique mitochondria are entirely functional at conception, yet it is also true that they lose function progressively over the next several decades as the organism grows older.

Why is aging not evident in the germ cell line or in our inherited mitochondria? Why does aging occur in some cases and not in others? Why are germ cells and mitochondria retrospectively healthy despite being billions of years old, and yet our somatic cells and mitochondria show definite aging changes within mere decades after fertilization? We cannot invoke entropy unless we can explain why entropy holds sway in some biological contexts and not in others. The assumption that aging is purely entropy fails to offer sufficient explanation, whether in cells or in mitochondria.

Aging is a dynamic balance that includes entropy, but it also includes biological maintenance. Aging occurs when maintenance fails to keep up with entropy. Aging is held to a standstill when, as in the germ cell lineage, maintenance continuously balances entropy. The germ cell lineage avoids aging only to the degree that cell maintenance keeps up with continual entropic damage. The somatic cells age only to the extent that cell maintenance does not keep up with such continual damage. Aging is not a linear, static process of entropy, but a systems process in which the entire complex cascade of interrelated biological maintenance mechanisms is actively balanced against entropy, and those mechanisms fail as cells age.

What is needed is a unified systems model that explains the aging process. Such a model should explain why the upstream risk factors (genetic or otherwise) operate to affect age-related disease. It should explain why the downstream biomarkers occur, and why there is clinical heterogeneity among human patients and among different species.

If we point to an age-related increase in DNA damage, the immediate question becomes why DNA damage increases. If we ascribe this to slower DNA repair, the question becomes why DNA repair becomes slower. If ascribe this to downregulation of the major DNA repair enzymes, then the question becomes why they are downregulated. If we ascribe the downregulation to epigenetic changes (i.e., senescence-associated gene expression), we are once again faced with the question of what causes these epigenetic changes. Such iterative questions provide three useful insights:

(1) They remind us that biomarkers are not aging,
(2) They link us to the upstream risk factors (genetic, epigenetic, and behavioral), and
(3) They suggest novel and technically feasible points of intervention.

Upstream risk factors should likewise prompt questions, rather than (likely inaccurate and misleading) statements that a particular gene is the cause of a particular age-related disease. Whereas some diseases are purely genetic (e.g., sickle cell, hemophilia), age-related diseases play out in the setting of cell aging. A narrow emphasis on genetics will often prove frustrating if the genes are considered to be causes rather than upstream contributors to age-related diseases. This often results in the identification of druggable clinical targets that then prove unsuccessful in human trials, as in many genome-wide association studies (GWAS) [28]. Even with advanced artificial intelligence (AI) tools and large-scale patient databases, genomic (and other multiomic) approaches continue to prove frustrating when taken to human trials. When we ask the wrong questions, even the most technically adroit approach will supply the wrong answers. Success requires an intuitive leap rather than better data crunching, and neither AI nor large databases can easily provide such a leap. To hark back to an earlier analogy, even if Avicenna had had access to GWAS, AI, and large databases, he would still not have understood (or been able to cure) microbial diseases such as bubonic plague. What he needed was not technology, but insight.

The important question is not what causes aging, which implies a single, linear cause, but the more useful question of how the system works. Aging has no simple cause, but the entire, complicated process of aging can be delineated and understood if we view it as a unified system rather than a single, passive, entropic cascade. For most of us, whether as aging humans or as physicians, our wish to understand aging prompts the more practical clinical question of what the optimal point of intervention is, both clinically and financially. In short, can we do something about aging diseases? Where in the unified system, the complex cascade of processes that we simply sum up as aging, can we find the best target if we are to find cures for age-related diseases?

Consider an analogy, that of a complicated piece of modern engineering. An Airbus 380 is powered by a Rolls Royce jet engine, which is in turn powered by, for example, a Trent 900 turbofan, which (among hundreds of other parts) has a number of fan blades. What can we ask ourselves if one of those fan blades fails? The important question is not what caused the fan blade to fail", as though it were a single part operating alone. The better question is what caused the fan blade to fail in the Trent 900 turbofan, in the Rolls Royce Jet engine, in the Airbus 380, at 35,000 feet, going several hundred kilometers per hour, while operating at high rpm and a high temperature. It is a systems issue, not a single-part issue.

To return to aging and age-related disease, we should not focus on a biomarker (or a fan blade), but on the entire biological (or aeronautical) system as a complex whole. It is a unified system, not a collection of biomarkers (or a collection of individual parts). If we want to prevent fan blade failure, we should not look solely at the fan blade. If we want to prevent age-related disease, we should not look solely at biomarkers. In each case, failure occurs within an inordinately complex system, and if we are to prevent failures, we would best understand the system as a unified whole.

4. The limited role of biomarkers

Biomarkers are not aging.

The aging literature often focuses on outcomes (biomarkers of aging), which offer little or no insight into mechanisms of aging. An analogy to microbial disease makes this point: if we list the biomarkers of a COVID-19 infection, they offer no insight into the mechanism of disease (that of a viral infection), nor do they offer insight into the development of a vaccine. They may be useful diagnostic markers, but fundamental therapies require deeper insight.

Aging is much the same. Biomarkers beg the question of what causes them to occur, just as risk factors beg the question of how they contribute to age-related diseases.

At the macroscopic level of biomarkers (or hallmarks [29,30]), aging is not gray hair, wrinkles, dementia, cardiovascular disease, osteoarthritis, osteoporosis, sarcopenia, and declining glomerular filtration rate, or the myriad other common results of the aging process. We see these biomarkers in clinics, in surgeries, in those around us, and in the mirror. Although the common outcomes of aging are useful and necessary to clinical medicine, they are not the aging *process*.

At the microscopic level of biomarkers, aging is not any single cellular finding or a collection of them. Aging is not inflammation (or its markers), declining DNA repair (or increases in DNA damage), increased mutations (or the resultant increase in cancer), lipofuscin, declining ATP/ROS ratios (the ratio of Adenosine TriPhosphate to Reactive Oxygen Species, or the mitochondrial dysfunction these reflect), methylation changes, acetylation changes, Yamanaka factors, Horvath clock, senescence-associated secretory phenotype (SASP), epigenetic factors in general, telomere shortening, or the hundreds of other changes in cell function that are considered to be biomarkers of aging. Cellular biomarker of aging may be inordinately valuable, but cellular biomarkers are not the aging *process*.

At the genetic level, aging is not a particular gene or a set of genes, nor are age-related diseases defined by such genes or their alleles. In the case of Alzheimer's dementia, for example, we know of well more than 200 genes whose alleles correlate with the risk and course of Alzheimer's disease. The APOE-4 allele increases the risk of Alzheimer's, although APOE-2 alleles clearly decrease that risk, and yet neither of these alleles (nor any others) are Alzheimer's itself. They may have a role (clearly, two APOE-4 alleles have a predictably disastrous role), yet Alzheimer's is neither a single gene disease nor a simple additive combination of genes, but a complex process in which such genes are key players. Genetic biomarkers offer predictive value and potential insights into the mechanisms of age-related disease, but genetic risk factors are not the aging *process*.

At the level of species longevity, the same claims can be made: we can identify genes associated with longevity [31], but such associations do not in themselves explain the fundamental processes of either aging or age-related disease. In the case of a single species and a single gene, the observation that Hutchinson—Gilford progeria can be linked to the lamin A gene and the production of progerin protein does not tell us precisely how this gene and the protein result in the early onset of aging and age-related disease.

In all of these cases—the macroscopic level, microscopic level, and genetic level—we know an enormous amount about the specific biomarkers, yet too little about the aging process that encompasses them. We have detailed knowledge about these and other biomarkers, we have countless peer-reviewed research papers, we have foundations, institutions, and universities that tirelessly contribute to our expanding information base at all of these levels, but however carefully we consider them, biomarkers are not aging itself.

Consider what happens when we view aging as a complex process, a unified system, rather than a passive result of entropy. If we look at biomarkers of aging, at every juncture we can ask what caused the biomarker to change, prompting us to move upstream in the cascade of causation, only to find that the same question applies repeatedly as we move upstream. In short, we find we are beginning to look at a unified system rather than a collection of biomarkers. Aging is that entire system, not its downstream outcomes.

One example of a common aging biomarker is increased DNA damage. Although DNA damage increases with age in our somatic cells, this begs the question of *why* it does so. We cannot simply ascribe this to the passage of time, because the same general increase in DNA damage has not resulted in biological extinction over the past several billion years, yet we see an increased rate of DNA damage of our somatic cells as we age. Why does the germ cell lineage not show the same increase (over billions of years) that we see in the somatic cell population (over a few decades)? This prompts the conclusion that the increase in DNA damage seen in aging somatic cells occurs because repair decreases, damage increases, or both. Both occur: aging cells demonstrate a downregulation of DNA repair [32,33] as well as an increase in oxidative damage. If oxidative damage increases in somatic cells, then it begs the question of why it does so, a question generally answered by pointing to extensive evidence that mitochondria become increasingly dysfunctional in aging somatic cells [34—37]. Both of these, a downregulation of DNA repair and mitochondrial dysfunction, occur in aging somatic cells, but this brings up the point of why it occurs. A common explanation might be that aging somatic cells undergo epigenetic shifts that downregulate DNA repair and mitochondrial efficiency.

This point deserves emphasis. A century ago, some biologists (perhaps reasonably) believed that different cells had different genes. With a few interesting exceptions, we know that the differences among different cell types (neurons, podocytes, myocytes, and fibroblasts, for example) are not caused by different genes but different gene expression: they are genetically identical (again, with a few interesting exceptions), but epigenetically distinct. This same point, however, can now be made (again, with a few interesting exceptions) with regard to the difference between young cells and old cells. Younger and older cells are epigenetically distinct. The distinction between your somatic cells at age 1 and at age 100 is largely caused by epigenetic differences, not genetic damage.

With that in mind, however, we are once again faced with the question of why such epigenetic shifts occur [38,39]. Certainly, we can discuss the details of such shifts (methylation, acetylation, Yamanaka factors, the Horvath clock, and other observations), but these *whats* do not address the *whys* of the epigenetic shifts. Yet we know that epigenetic shifts occur as the result of telomere shortening, which in turn is due to cell division, which occurs as a routine part of tissue maintenance as we age and replace lost cells.

It is the aging process, not aging biomarkers that require our deeper understanding. We rely on biomarkers in medicine (signs and symptoms of disease) to distinguish between and provide fodder for clinical diagnoses, with good reason, and often with good result. We rely on biomarkers in research, defining cell aging largely based on biomarkers, whether we use histologic appearance (the fried egg appearance of aging human fibroblasts) or physiologic markers (e.g., β-galactosidase or P16 [40]). We rely on genetic biomarkers to quantify clinical risks in terms of alleles (e.g., APOE-4). These and other biomarkers are the currency by which we make judgments that define modern biology and medicine, but they are only the currency, not the economic system. We should treat biomarkers with respect, but to understand aging as a unified system we must look deeper. Can we construct a unified systems model of aging that is consistent with the upstream biomarkers (the risk factors) and the downstream biomarkers (the clinical results, for example), but that is also consistent with known laboratory and clinical data, predictively valid, and offers us a potentially more effective point of intervention, one that is feasible and can be addressed with current techniques?

Let us consider such a potential model and see where it might take us.

5. A unified systems model

We age because our cells age.

We acquire age-related diseases because our aging cells make up aging tissues and organs, which become gradually more dysfunctional, which results in age-related disease, with death as the final outcome. All age-related diseases result from the aging of multiple cells and multiple cell types. We age because our cells age.

Let us first consider what this process might look like at the cellular level in typical aging cells, and consider how the upstream inputs to the system (the genetic, epigenetic, and behavioral starting points) affect the process of cell aging and result in the downstream outputs of this system: the biomarkers and the clinical outcomes that we collectively categorize as age-related diseases.

What precisely happens as cells age? As cell aging progresses, what happens to gene expression, mitochondrial function, DNA repair, and the thousands of molecular pools that we rely on for normal cell function? First consider the most obvious starting point. Over time, cells are lost and others divide and replace those losses.

We lose cells daily, especially in tissues such as skin and the lining of the gastrointestinal tract, somewhat in tissues such as the immune systems, and seldom in cell types such as neurons or muscles. Some cells divide almost constantly, some intermittently, and some almost never. Even if cells never (or almost never) divide, however, they depend on other cells that do divide [41]. An obvious and universal example is that all of our cells depend on a vascular system that relies on vascular endothelial cells that divide and replace missing endothelial cells. Likewise, neurons rarely (or in many cases, never) divide, but whether or not they divide, those neurons depend on glial cells as well as on vascular endothelial cells. All cells depend on innumerable other cells. Whether or not a specific cell divides, every cell relies on dividing cells. Cell aging affects all cells, dividing or not. In most tissues, lost cells are replaced as other cells divide, whether locally (for example, chondrocytes of the knee) or more distantly (for example, various blood cells replaced from hematopoietic stem cells within marrow). Cell replacement requires cell division, which causes telomere shortening. In some cases, this shortening can be mitigated (or even reversed) by the expression of telomerase. When this occurs, one or both daughter cells may be able to replace at least a modicum of lost telomere length. In the case of the hematopoietic system, the (partial) replacement of the lost telomeric repeats is why even centenarians generally produce functional blood cells, despite needing to replace lost hematopoietic cells constantly for a century.

Despite exceptions within stem cell niches, the cell loss and replacement of somatic cells generally results in a loss of telomere length. As we will see, absolute telomere lengths are irrelevant, whereas relative lengths are critical because they predict the species life span [42,43] and cause subtle but pervasive changes in gene expression [44–48]. As telomeres shorten, the result is epigenetic change in ways we are only beginning to grasp. Many genes become downregulated, some are upregulated (whether directly or in reaction to changing physiologic stimuli as the cell ages), and some are altered in their responses to usual physiologic stimuli, whether the response is dampened, exaggerated, or simply different.

Epigenetic shifts are frequently viewed as independent biomarkers of cell aging, without considering the role that telomere shortening has in these epigenetic shifts. We might, for example, focus narrowly on the shifts themselves without looking upstream at the origins of such shifts or downstream at their outcomes. A narrow stress on methylation, acetylation, Yamanaka factors, the Horvath clock, or the molecular mechanics of such epigenetic shifts runs the risk of missing the point: epigenetic changes do not simply occur because time has passed or because the patient has become older. They have discernible upstream causes and critical downstream results that are far from trivial.

Perhaps the most important epigenetic change, in that it affects almost all of the central processes in aging cells, may be the downregulation of molecular turnover, which can result in the downstream biomarkers that we see as defining cell aging: mitochondrial dysfunction, slower DNA repair, increased DNA damage, increased oxidative damage, and so forth. A minor downregulation in gene expression can trigger enormous changes in cell function. Biological molecules are not static, but are in dynamic equilibrium, with balanced production and degradation. The result is that molecular damage is washed out by continual turnover; however, the rate of molecular turnover declines with age. The result is a gradual increase in the percentage of molecular damage as turnover slows.

To cite one example, the β-amyloid that we see prominently in some age-related dementias is not a static pool of extracellular molecules, but is in constant turnover, a necessity as such molecules cross-link or otherwise become dysfunctional. Microglia continually produce, secrete, bind, reinternalize, and degrade such molecules. The result is that despite constant rates of molecular damage, there is (at least with young microglia cells) little accrual of damage to the pool. In the young nervous system, such damage is washed out by a rapid recycling of such pools. In the older nervous system, however, where the microglia have a history of cell division, shortening telomeres, and an altered epigenetic pattern, such recycling slows significantly [49,50].

The results are perhaps unsurprisingly significant, but they may also serve as an example to explain one aspect of age-related neurodegenerative disease (e.g., β-amyloid accumulation in Alzheimer's disease). Specifically, as the recycling rate slows, the percentage of dysfunctional molecules climbs proportionately and results in increased amyloid microaggregates, and finally, amyloid plaques.

This is portrayed graphically in Fig. 1.1 (the numbers used here are chosen for illustrative purposes only). Assume that for any given unit time, a young cell recycles 50% of the molecular pool. Then, at equilibrium and at a constant rate of 2% molecular damage per that same unit time, we would predict that only 4% of the molecules are dysfunctional. If, however, epigenetic shifts downregulate the required enzymes responsible for recycling, such that the rate of recycling is only 10% per unit time (an arbitrary example), then at equilibrium we would expect to find that 20% of the molecules are dysfunctional. The end result is that the rate of molecular turnover is directly responsible for the percentage of molecules that are functional in any given pool. Although the numbers used here are arbitrary, the

The rate of molecular recycling determines the extent of molecular damage

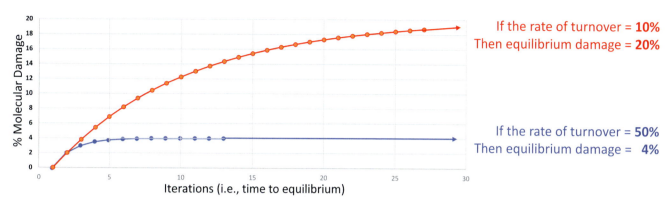

FIGURE 1.1 For any given rate of damage (e.g., here 2%/unit time), the accrued percentage of damage will be inversely proportional to the rate at which molecular turnover occurs per unit time. If turnover is rapid (as occurs in young cells), accrued damage remains low. If turnover slows (as occurs in old cells), accrued damage becomes high.

effect is a valid representation of the dynamics of molecular turnover whether we consider proteins, lipids, or other biological molecules.

That is not all. Most pools of biological molecules do not dependent on the expression and downregulation of a single enzyme. In the previous case, for example, β-amyloid depends on at least five enzyme families to provide active turnover of the pool, and this is likely a naive simplification of the recycling process. Each of these enzyme families works (to some degree) independently, so the results of downregulation of each such enzymatic step will result in an exponential rather than geometric change in function, illustrated in Fig. 1.1. Consider another example, that of DNA repair, in which at least four families of DNA repair enzymes have a role. They are responsible for tagging the base pair error (detection), removing the wrong base pair (excision), inserting the correct base pair (replacement), and rejoining the strand (ligation). Because the gene expression of each of these families of DNA repair enzymes operates (to a degree) independently, as cells age and downregulation proceeds, we might predict an exponential (rather than linear) decrease in the rate of DNA repair. Clinically, the projected outcome would be an exponential increase in the rate of mutations (and cancer) with age, which is what we see in many species. The rate of cancer shows an exponential increase with age within species, but when we compare species, we see that the exponential increase is roughly identical when adjusted for the relative species life span. Putting it simply, the increase in cancer is proportional to the occurrence of cell aging, not to the absolute duration of the life span of the animal. The increase is exponential within the life span of the species, whether that life span is 2 years or 100 years.

Curiously, this suggests that our lifetime risk of cancer is not directly related to our lifetime exposure to carcinogens, but to our decreasing ability to repair the DNA damage from such carcinogens. Were it simply a matter of accumulated exposure, we might well expect a roughly linear increase over the life span, whereas the observed exponential increase is directly predictable from the multipronged downregulation of DNA repair enzymes as cell aging progresses. To oversimplify, we develop age-related cancers not because of the occurrence of DNA damage, but because the DNA damage is repaired more slowly as cells age. Most of the increase in DNA damage observed in cell aging may well be a result, rather than a cause, of cell aging.

Aging is not just an increase in DNA and oxidative damage, a downregulation of DNA repair, an SASP, or the host of other biomarkers we associate with cell aging. These are all part of the overall process, but are in turn due to epigenetic changes that are in their turn downstream outcomes of telomere changes that occur as cells divide (Fig. 1.2).

Overall, we might consider the unified systems model of aging as encompassing a plethora of upstream risk factors (genetic, epigenetic, and life history) that occur on the complex cascade of increasing cellular dysfunction which we characterize as cell aging [51], resulting in a diverse set of downstream biomarkers and, ultimately, age-related clinical diseases. Such a model encompasses the concept of biomarkers, but also the rationale for such biomarkers in the context of cell aging. A simplified view of the overall model is illustrated by Fig. 1.3. To understand some of the complexity involved, let us take a slightly more detailed look at the upstream risk factors and the downstream outcomes.

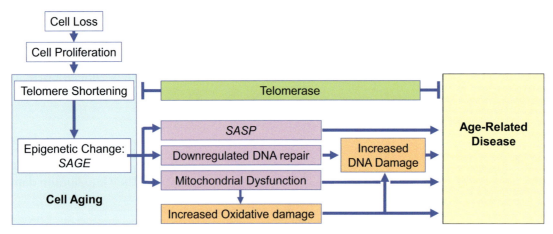

FIGURE 1.2 Age-related diseases are the downstream outcomes of common biomarkers (increased DNA and oxidative damage, mitochondrial dysfunction, downregulated DNA repair, senescence-associated secretory phenotype [SASP], etc.), which are themselves the results of alterations in gene expression, modulated by telomere loss, which occur as cells divide.

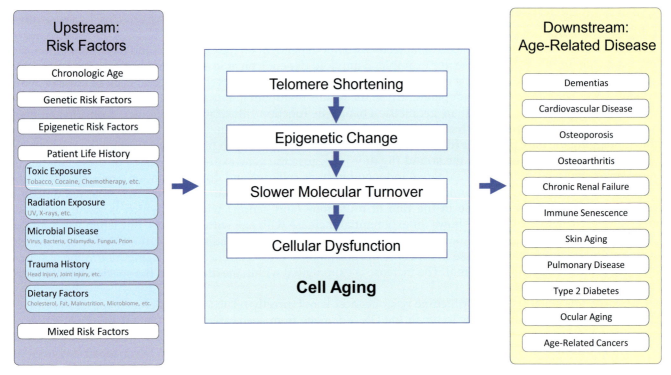

FIGURE 1.3 The aging process encompasses upstream risk factors (genetic and epigenetic causes, and life history) which have effects on cell aging, which is expressed as slower molecular turnover, increased DNA damage, mitochondrial dysfunction, senescence-associated secretory phenotype, and so on. The downstream heterogeneity of disease in individual patients hinges on individual upstream risk factors.

5.1 Upstream risk factors

We all differ in our upstream risk factors and our downstream clinical presentations. How do the upstream inputs to the system (the genetic, epigenetic, and history starting points) affect the process of cell aging and result in the downstream outputs of this system (the biomarkers and the clinical outcomes that we collectively categorize as age-related diseases)? We start with different genes (or with alleles different from those genes), different epigenetic settings, and often markedly different life histories. The latter, the behavioral or medical histories, encompass trauma, toxic exposure, drug use, infections, and a myriad other risk factors and differences. In each example (the genetic and epigenetic settings and history) enormous initial heterogeneity exists that, in the context of cell aging, culminates in the clinical heterogeneity of age-related disease. Patients differ in which age-related disease they develop, when it begins, how it manifests, and the rapidity of its course. For example, the genetic, epigenetic, and historical starting points determine whether we develop dementia and which type, the individual findings, when it appears clinically, and how rapidly it progresses. In most age-related diseases, we often see a degree of atypicality or mixed findings. For example, whereas Alzheimer's disease is classically a progressive cognitive dysfunction, some patients with Alzheimer's disease have motor dysfunction as well. Conversely, whereas Parkinson's disease is classically a progressive motor dysfunction, some patients with Parkinson's disease have cognitive dysfunction as well. If we could understand the precise upstream inputs (the genetics, epigenetics, and historical factors), we might better understand and perhaps more accurately predict the heterogeneity of clinical outcomes in age-related diseases.

Although we may categorize upstream risk factors as genetic, epigenetic, and history, the reality is more complex. One obvious omission is chronologic age per se, the most obvious risk factor for age-related disease. As an example, consider a patient who is biallelic for APOE4, as we assess their risks of Alzheimer's disease. APOE4 alleles increase the overall risk, lower the age of onset, and increase the severity, yet chronologic age itself remains the best predictor of such disease for such patients. Few patients who are biallelic for APOE4 or not, develop Alzheimer's disease in the first 2 decades of life, but the risk rises predictably with age. We ignore aging at our own risk. As Leonard Hayflick observed, "The cause of aging is ignored by the same people who argue that aging is the greatest risk factor for their favorite disease" [52,53]. That aging should be considered a risk factor is obvious, but the risk is not the passage of time; it is the aging process that occurs as time passes. Aging is not the years, but the way in which those years correlate with cell aging.

Complexity is also introduced as the upstream risk factors (genetic, epigenetic, and history) interact with one another. Hypertension, for example, is known to incorporate genetic risks, yet the patient's history (the diet, for example) can also affect risk. Interactions are probably common in the upstream causes of age-related diseases such as diabetes, dementia, lung disease, and renal failure. Such mixed risk factors are likely the rule rather than the exception and should make us leery of attributing simple causes to complex diseases.

The simplification is understandable, because some risk factors have a high correlation with the onset of age-related disease. Several common examples come to mind: UV exposure for skin aging, joint trauma for osteoarthritis, smoking for chronic obstructive pulmonary disease, APOE4 for Alzheimer's disease, and others. The mistake, however, comes in making the conceptual transition from contributory risk factor to defined cause of the disease at issue. The model discussed here would argue, and the distinction between correlation and causation would emphasize, that upstream risk factors contribute to disease not being the operative cause. Rather, they influence how they affect the timing, course, and details of the aging of cells that are central to the specific age-related disease at issue. In short, we might do well to avoid thinking of genes (or other risk factors) as causing age-related disease, while recognizing their importance in the cascade of events that result in cell aging and the expression in clinical disease.

Although the importance of genetic and history risk factors is usually accepted without argument, the epigenetic inputs are often slighted in such discussions. To put the epigenetic contribution in perspective, contrast it with the genetic contribution. Protein-expressing genes are estimated to be less than 2% of the human genome, whereas regulatory elements are estimated encompass 40% or more of the genome. Such elements define and regulate the expression of the protein-expressing genes in terms of the rate of expression as well as the stimuli and the degree to which individual genes respond to physiologic changes in the cell. Individuals start with inherited epigenetic settings, but we know little (compared with our knowledge of protein-expressing genes) about the initial epigenetic settings, let alone how these epigenetic inputs change over time and with cell aging. We continue to learn more about the epigenetic background—the regulatory settings that each of us has individually as upstream risk factors—but we would do well to be cognizant of these relatively unfamiliar inputs, despite our limited understanding.

Individual upstream risk factors (genetic and epigenetic causes, and history) are the starting vectors that, as they influence cell aging [51], determine the individual clinical outcomes. Our individual heterogeneity is precisely reflected in the heterogeneity of age-related diseases. Which diseases we experience, their age at onset, and our individual presentations are the result of our individual starting points in these upstream risk factors.

5.2 Downstream outcomes

Just as we can make the mistake of confusing any specific upstream risk factor as the cause of an age-related disease, so, too, can we confuse downstream biomarkers as the cause of such diseases. In the extreme case, this error appears so obvious as to be unlikely. For example, we would not expect anyone to confuse a cough, fever, or sore throat as being the cause of a COVID-19 infection. Even more so, we would have reservations if an effective antitussive compound were offered as a cure for COVID-19.

To a degree, however, we fall into that trap when we view many age-related diseases. In the case of age-related dementias, for example, we see the common assumption that a histologic finding (such as β-amyloid plaque) is causal. Likewise, we see pharmaceuticals targeting such plaques suggested as potential cures for Alzheimer's disease. This is not to argue that amyloid, tau tangles, or neurofilament light chains are not causal, only that we have no good data to suggest that they are not in turn caused by other processes lying upstream from such abnormalities. From a clinical perspective, the practical question is not what the cause is, but rather what the optimal point is of invention. The issue of causation may help us (and, generally does) to target the optimal point of intervention, but a full understanding of the cascade of such causation is far more likely to yield an effective cure than simply equating a downstream clinical or histologic finding with the cause.

Aging cells have a plethora of downstream biomarkers that are often suggested as causal, although in every case they raise the question of what causes those biomarkers in turn. In medicine, we seldom confuse symptoms and signs of a disease with the underlying pathology, yet in our limited understanding of aging, this is a common refrain and should be resisted. As noted earlier, biomarkers such as SASP, DNA damage, epigenetic changes, and increased oxidative damage are clearly present in aging cells, but it would be unwise to construe them as the primary movers in aging.

Clinically, the downstream outcomes are countless and vary among patients. We see enormous heterogeneity among patients of a single species (i.e., humans), and this is even more true among species. Rats, mice, dogs, Rhesus monkeys, and other species vary in age-related diseases. Most species demonstrate a decline in behavioral function (cognitive function), but species differ in the specifics of the behavior decline as well as the histologic underpinnings. Although behavioral decline may be universal in mammals, prominent amyloid plaques, for example, are not. Whether in a comparison of humans or species, it becomes a truism to attribute the heterogeneity to genes; to an extent, this is reasonable. Humans and mice, for example, share perhaps 75% of protein-expressing genes. The other 25% (and the regulatory elements mentioned earlier) are assumed to underlie the differences in clinical aging between the species. In a similar vein, we tend to attribute the clinical heterogeneity of human age-related disease to genetic differences while allowing for the important effects of epigenetic starting points and the life history of each patient.

If we were to delineate the individual clinical starting points fully for each patient, we might better predict the intermediate effects on cell aging in relevant tissues and reveal a clearer ability to foresee individual clinical outcomes. To an extent, potential risk factors are often, and often erroneously, viewed as the direct cause of such clinical outcomes. For example, some toxins (e.g., paraquat) have been viewed as causing specific age-related diseases (e.g., Parkinson's disease), although such data are notoriously difficult to assess and careful analyses often undermine such assertions [54]. Nonetheless, the idea of addressing upstream risk factors to at least postpone or ameliorate age-related disease has been an almost universal working hypothesis. Minimizing trauma, infections, tobacco use, poor dietary practices, and so on, are the usual preventative targets in modern medicine, with good reason.

In all cases, however (whether genetic, epigenetic, or history), the effects are likely far more complex and difficult to tease apart than we might assume. As a relatively simple example, consider the effects of a known risk factor, such as APOE4 alleles for Alzheimer's disease, as it influences aging cells. Consider how a genetic risk may interact with cell aging to cause disease. Whereas different genetic risk factors (e.g., APOE2 vs. APOE4 in the case of Alzheimer's dementia) may have little clinical significance in a young adult with a high rate of protein recycling, the effects become magnified in middle age as cell aging progresses and protein recycling slows (Fig. 1.4). In the case of amyloid, we know that the turnover of the pool of extracellular amyloid molecules slows with cell aging. Production, secretion, binding, internalization, and degradation all show deceleration, increasing the time during which molecular degradation may occur. This may have little detrimental effect if the amyloid molecules are less prone to forming aggregates (as in patients with APOE2 alleles), but these effects are multiplied in patients who produce amyloid molecules that are more prone to forming such aggregates (patients with two APOE4 alleles). As shown in Fig. 1.4 (numbers are for illustrative purposes only), instead of a baseline rate of damage of 2%/unit time, for example, patients with APOE4 alleles might have a baseline rate of damage that was several times as large, which may accelerate the deposition of amyloid, increasing microaggregate and plaque formation, and triggering clinical disease. This model, encompassing both the effects of cell aging on molecular turnover and the effects of genetic risks, predicts that whereas the overall process may be the same whether the patient carries APOE2 or APOE4 genes, patients who are biallelic for APOE4 will display earlier and more progressive disease, which correctly predicts the clinical experience. The model suggests that much the same process pertains to tau tangles, neurofilament light chains, and other biomarkers of age-related dementia.

Epigenetic risk factors might well follow a similar pattern, because the epigenetic settings in young patients have an increasingly crucial role as cell aging progresses. For example, the effects of a slightly lower than average rate of

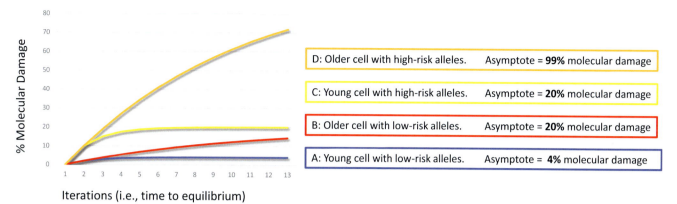

FIGURE 1.4 In the case of high-risk alleles, even young cells (C) rapidly accrue damage similar to that of older cells with lower-risk alleles. The graphs illustrate expected risks in young (A) and older (B) cells with low-risk alleles (C) versus young (C) and older cells (D) with high-risk alleles.

gene expression, as modulated by inherited regulatory elements, may be magnified as cell aging, and its other multiple changes in gene expression, progresses.

Historical factors can likewise be expected to become more relevant as cell aging progresses. Thus, injury, toxin exposure, radiation, or infection all may be expected to increase cell loss, cell division, and cell aging. The observed clinical effects of head injury in dementia, joint injury in osteoarthritis, UV damage in skin aging, or microbial damage in affected tissues align well with predictions of the unified systems model of aging.

6. Caveats

The unified systems model of age-related disease outlined here is simply that: a model. It has the advantage of being consistent with both research data and clinical experience, as well as being predictively valid. Consistency and explanatory value are an initial requirement of any model; the definitive tests are interventional trials. The first caveat is that the unified systems model of aging is provisional until taken to clinical testing. *Caveat medicus*.

We would also do well to avoid the most common pitfall in evaluating the unified systems model of aging: that of confusing aging with entropy. Aging is a failure of cell maintenance in the face of entropy. Oddly, even those agreeing with this point may occasionally fall into the error of confusing entropy with aging. *Caveat physicus*.

Confusing terminology is also a hazard in the relevant literature (Fig. 1.5). The phrase cell senescence may at times refer to the spectrum of cell aging (the entire gamut of cell aging from young to old cells) and at other times refer to replicative senescence (the end point of cell aging). Unless we clarify to which of these (the spectrum or the end point) we are referring, data may be misinterpreted and may mislead readers. In this text, we will try to distinguish the process of cell aging from the end point of replicative senescence.

Another confusing term is the telomere length. Data often fail to specify whether they refer to the mean telomere length of a cell or, as is more relevant to cell function, the shortest telomeres within a cell (or cell population). Whereas a mean length is often used or assumed, careful researchers often specify the shortest decile or other indicator of telomere variance. Moreover, we might ask whether the telomere measurements are from individual cells or cell populations. A population of cells may well have relatively long telomeres, but a small number of aging cells may have significant effects on the larger cell population, such as when SASP has a role in the tissue pathology.

In addition, in discussions of the role of telomere length in relationship to disease, particularly when comparing species, the telomere length is not the operative variable, because multiple studies show that the important factor (in species life span, for example) is not the length of the telomeres per se, but the rate of telomere shortening. In short, the absolute telomere length is not a crucial parameter, whereas the relative telomere length (e.g., compared with the initial telomere length) is predictive of the species life span as well as the extent of cell aging. The epigenetic changes that occur with cell aging are modulated by relative, not absolute, telomere lengths and should therefore be the appropriate parameter in assessing the role of telomere changes as they relate to age-related disease, particularly when making cross-species comparisons.

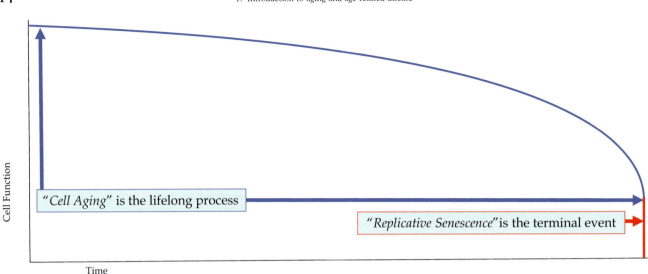

FIGURE 1.5 "The term cell senescence" might be used in either of two senses. "Cell aging" encompasses a gradual spectrum of epigenetic and functional changes over the organism's lifetime. "Replicative senescence" is the terminal end point, a biologically narrower event with specific and critical changes including the inability to divide further.

Finally, there is an inappropriate tendency to use peripheral leukocyte telomeres in assessing organism aging, as well as in making correlative conclusions with regard to telomere lengths (see the earlier discussion) and their role in age-related disease in non—leukocyte related tissues [55—64]. For example, the appropriate cells to measure in considering central nervous system (CNS) disease would be neurons, glial cells, and the vascular endothelial cells lining the CNS vascular system, rather than circulating leukocytes. The appropriate cells to measure in evaluating cardiovascular disease would be the vascular endothelial cells as well as cardiac myocytes and fibroblasts, and so on, and not leukocytes. To study possible correlations between age-related CNS or CV (cardiovascular) disease and circulating leukocyte telomeres is disingenuous [65—67]. Leukocytes are readily accessed compared with almost any other cell types that influence age-related disease, but the practice and its rationale are dubious because they promote confusion and incorrect conclusions regarding cell aging and age-related disease. Compounding this fault, peripheral leukocytes (and the rate of telomere loss) are more responsive to stress, infection, and other factors than are their hematopoietic stem cells. Stem cell populations may remain relatively unaffected, whereas peripheral expansion will accelerate the apparent loss of telomeres. Equally, the resolution of stress, infection, and other factors may result in an apparent increase in peripheral telomere lengths (because circulating cells are less likely to divide) whereas the misleading increase in telomere lengths is not reflected in hematopoietic stem cells. Any apparent increase in peripheral leukocyte telomere lengths caused by interventions in lifestyle, for example, is not a reversal of cell aging, but merely the outcome of a decreased turnover in circulating leukocytes.

These and other caveats notwithstanding, the unified systems model of aging offers a logical framework for our understanding of aging [51] as well as potential benefits in the indicated therapeutic targets.

7. Prospects for intervention

In any clinical discussion of aging and disease, the key question is not about causation, but the more practical question of intervention. A clear and realistic understanding of the aging process may be necessary to identify the optimal point of intervention, but the end goal is not a better *theory*, but a better *therapy*. Poor assumptions lead to ineffective therapies. One problem with confusing biomarkers, hallmarks, or histologic findings with the aging process is the predictable frustration when human trials fail to show a significant improvement in clinical outcomes. Aging is complex, but we need to understand the complexity to intervene effectively.

Ironically, if we regard aging as a simple, passive, linear process of wear-and-tear, experimental interventions become complex, difficult, multiple, and largely ineffective when applied to human trials. In contrast, when we regard aging as a complex, active, systems process, prospective points of interventions begin to appear relatively simple, easy, singular, and potentially effective (Fig. 1.6).

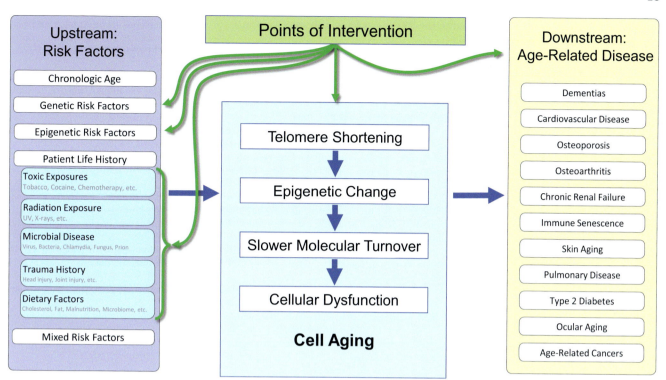

FIGURE 1.6 Age-related diseases offer many potential points of intervention. Historically, we have emphasized the importance of history-based risk factors to lower the risk of disease and be fully engaged in treating the clinical complications of such diseases. However, interventions targeting the process of cell aging itself might be far more effective.

Consider the history of interventions in age-related disease. Upstream points of intervention (Fig. 1.5) have focused on alterations in human behavior (e.g., smoking cessation), prevention (e.g., trauma), and small-molecular approaches (e.g., blood pressure and glucose control). The clinical results have been encouraging in the propensity to delay or slow age-related diseases, but those age-related diseases have not been stopped, let alone reversed by addressing upstream risk factors [51]. Downstream points of intervention have focused on biomarkers (e.g., the use of monoclonal antibodies in Alzheimer's disease), replacement (e.g., the use of artificial joints), or cosmetic concerns (e.g., the use of botulinum toxin for wrinkles), but again the downstream effects have not stopped, let alone reversed the aging process at either the cellular or diagnostic levels. In both approaches, upstream risks and downstream outcomes, there are significant benefits to patients, as assessed by patient satisfaction, morbidity, and mortality.

The unified systems model suggests that if we can directly address the cell aging process, the benefits may be remarkable. In the rest of this book, we will look at the role of cell aging in each of several organs systems and assess the potential for more effective interventions in each of those systems. Ultimately, the goal is to address age-related disease more effectively by directly addressing the cell aging process.

References

[1] Heinlein RA. Methuselah's children. Gnome Press; 1958. p. 181.
[2] Yankova T, Dubiley T, Shytikov D, Pishel I. Three month heterochronic parabiosis has a deleterious effect on the lifespan of young animals, without a positive effect for old animals. Rejuvenation Res 2022;25(4):191—9. https://doi.org/10.1089/rej.2022.0029.
[3] Hayflick L, Moorhead PS. The serial cultivation of human diploid strains. Exp Cell Res 1961;25:585—621.
[4] Olovnikov AM. Principle of marginotomy in template synthesis of polynucleotides. Dokl Akad Nauk SSSR 1971;201(6):1496—9 [Article in Russian].
[5] Harley CB, Futcher AB, Greider CW. Telomeres shorten during ageing of human fibroblasts. Nature 1990;345(6274):458—60. https://doi.org/10.1038/345458a0.
[6] Vaziri H, Benchimol S. Reconstruction of telomerase activity in normal cells leads to elongation of telomeres and extended replicative life span. Curr Biol 1998;8:279—82. https://doi.org/10.1016/s0960-9822(98)70109-5.
[7] Bodnar AG, Chiu C, Frolkis M, Harley CB, Holt SE, Lichtsteiner S, et al. Extension of life-span by introduction of telomerase into normal human cells. Science 1998;279:349—52. https://doi.org/10.1126/science.279.5349.349.

[8] Vaziri H. Extension of life span in normal human cells by telomerase activation: a revolution in cultural senescence. J Anti Aging Med 1998;1: 125—30. https://doi.org/10.1089/rej.1.1998.1.125.
[9] Shelton DN, Chang E, Whittier PS, Choi D, Funk W. Microarray analysis of replicative senescence. Curr Biol 1999;9:939—45. https://doi.org/10.1016/S0960-9822(99)80420-5.
[10] Fossel M. Telomerase and the aging cell: implications for human health. JAMA 1998;279:1732—5. https://doi.org/10.1001/jama.279.21.1732.
[11] Banks D, Fossel M. Telomeres, cancer, and aging: altering the human lifespan. JAMA 1997;278:1345—8. https://doi.org/10.1001/jama.1997.03550160065040.
[12] Fossel M. Reversing human aging. New York: William Morrow & Co.; 1996.
[13] Funk WD, Wang CK, Shelton DN, Harley CB, Pagon GD, Hoeffler WK. Telomerase expression restores dermal integrity to in vitro-aged fibroblasts in a reconstituted skin model. Exp Cell Res 2000;258:270—8. https://doi.org/10.1006/excr.2000.4945.
[14] Yudoh K, Matsuno H, Nakazawa F, Katayama R, Kimura T. Reconstituting telomerase activity using the telomerase catalytic subunit prevents the telomere shorting and replicative senescence in human osteoblasts. J Bone Miner Res 2001;16:1453—64. https://doi.org/10.1359/jbmr.2001.16.8.1453.
[15] Matsushita H, Chang E, Glassford AJ, Cooke JP, Chiu CP, Tsao PS. eNOS activity is reduced in senescent human endothelial cells. Preservation by hTERT immortalization. Circ Res 2001;89:793—8. https://doi.org/10.1161/hh2101.098443.
[16] Harley CB, Liu WM, Blasco M, Vera E, Andrews WH, Briggs LA, et al. A natural product telomerase activator as part of a health maintenance program. Rejuvenation Res 2011;14:45—56. https://doi.org/10.1089/rej.2010.1085.
[17] de Jesus BB, Schneeberger K, Vera ME, Tejera A, Harley CB, Blasco MA. The telomerase activator TA-65 elongates short telomeres and increases health span of adult old mice without increasing cancer incidence. Aging Cell 2011;10:604—21. https://doi.org/10.1111/j.1474-9726.2011.00700.x.
[18] Harley CB, Liu W, Flom PL, Raffaele JM. A natural product telomerase activator as part of a health maintenance program: metabolic and cardiovascular response. Rejuvenation Res 2013;16:386—95. https://doi.org/10.1089/rej.2013.1430.
[19] Jaskelioff M, Muller FL, Paik JH, Thomas E, Jiang S, Adams A, et al. Telomerase reactivation reverses tissue degeneration in aged telomerase deficient mice. Nature 2011;469:102—6. https://doi.org/10.1038/nature09603.
[20] de Jesus BB, Vera E, Schneeberger K, Tejera AM, Ayuso E, Bosch F, et al. Telomerase gene therapy in adult and old mice delays aging and increases longevity without increasing cancer. EMBO Mol Med 2012;4:691—704. https://doi.org/10.1002/emmm.201200245.
[21] Fossel M. Cells, aging, and human disease. Oxford University Press; 2004.
[22] Fossel M. Cell senescence in human aging: a review of the theory. In Vivo 2000;14(1):29—34. PMID: 10757058.
[23] Fossel M. The role of cell senescence in human aging. In: Bercu B, Walker R, editors. Endocrinology of aging. New York: Springer-Verlag; 2000.
[24] Fossel M. The role of cell senescence in human aging. Adv Cell Aging Gerontol 2001;7:163—204. https://doi.org/10.1016/S1566-3124(01)07019-5.
[25] Fossel M. Cell senescence in human aging and disease. Annals NY Acad Sci 2002;959:14—23. https://doi.org/10.1111/j.1749-6632.2002.tb02078.x.
[26] Fossel M. Role of cell senescence in human aging. J Anti Aging Med 2009;3(1). https://doi.org/10.1089/rej.1.2000.3.91.
[27] Libertini G, Corbi G, Conti V, Shubernetskaya O, Ferrara N. Evolutionary gerontology and geriatrics. Springer; 2021.
[28] Fang J, Zhang P, Wang Q, Chiang CW, Zhou Y, Hou Y, et al. Artificial intelligence framework identifies candidate targets for drug repurposing in Alzheimer's disease. Alz Res Therapy 2022;14:7. https://doi.org/10.1186/s13195-021-00951-z.
[29] Schmauck-Medina T, Molière A, Lautrup S, Zhang J, Chlopicki S, Madsen HB, et al. New hallmarks of ageing: a 2022 Copenhagen ageing meeting summary. Aging 2022;14(16):6829—39. https://doi.org/10.18632/aging.204248.
[30] Keshavarz M, Xie K, Schaaf K, Bano D, Ehninger D. Targeting the "hallmarks of aging" to slow aging and treat agerelated disease: fact or fiction? Mol Psychiatry July 15, 2022. https://doi.org/10.1038/s41380-022-01680-x.
[31] Sleiman MB, Roy S, Gao AW, Sadler MC, von Alvensleben GVG, Li H, et al. Sex- and age-dependent genetics of longevity in a heterogeneous mouse population. Science 2022;377(6614):eabo3191. https://doi.org/10.1126/science.abo3191.
[32] Galderisi U, Helmbold H, Squillaro T, Allessio N, Komm N, Khadang B, et al. In vitro senescence of rat mesenchymal stem cells is accompanied by downregulation of stemness-related and DNA damage repair genes. Stem Cell Dev 2009;18(7). https://doi.org/10.1089/scd.2008.0324.
[33] Toufektchan E, Toledo F. The guardian of the genome revisited: p53 downregulates genes required for telomere maintenance, DNA repair, and centromere structure. Cancers 2018;10(5):135. https://doi.org/10.3390/cancers10050135.
[34] Ojaimi J, Byrne E. Mitochondrial function and Alzheimer's disease. Biol Signals Recept 2001;10:254—62. https://doi.org/10.1159/000046890.
[35] Trounce I, Byrne E, Marzuki S. Decline in skeletal muscle mitochondrial respiratory chain function: possible factor in ageing. Lancet 1989;333: 637—9. https://doi.org/10.1016/S0140-6736(89)92143-0.
[36] Ojaimi J, Masters CL, McLean C, Opeskin K, McKelvie P, Byrne E. Irregular distribution of cytochrome c oxidase protein subunits in aging and Alzheimer's disease. Ann Neurol 2001;46:656—60. https://doi.org/10.1002/1531-8249(199910)46:4<656::AID-ANA16>3.0.CO;2-Q.
[37] Gao X, Yu X, Zhang C, Wang Y, Sun Y, Sun H, et al. Telomeres and mitochondrial metabolism: implications for cellular senescence and age-related diseases. Stem Cell Rev Rep April 23, 2022. https://doi.org/10.1007/s12015-02210370-8.
[38] Libertini G, Shubernetskaya O, Corbi G, Ferrara N. Is evidence supporting the subtelomere-telomere theory of aging? Biochem (Mosc). 2021; 86(12—13):1526—39. https://doi.org/10.1134/S0006297921120026.
[39] Libertini G, Corbi G, Ferrara N. Importance and meaning of TERRA sequences for aging mechanisms. Biochem (Mosc). 2020;85(12—13): 1505—17. https://doi.org/10.1134/S0006297920120044.
[40] Safwan-Zaiter H, Wagner N, Wagner KD. P16INK4A-more than a senescence marker. Life 2022;12(9):1332. https://doi.org/10.3390/life12091332.
[41] Libertini G, Ferrara N. Aging of perennial cells and organ parts according to the programmed aging paradigm. Age (Dordr) 2016;38(2):1—13. https://doi.org/10.1007/s11357-016-9895-0.
[42] Whittemore K, Vera E, Martínez-Nevado E, Sanpera C, Blasco MA. Telomere shortening rate predicts species life span. Proc Natl Acad Sci USA 2019;116(30):15122—7. https://doi.org/10.1073/pnas.1902452116.

References

[43] Vera E, DeJesus BB, Foronda M, Flores JM, Blasco M. The rate of increase of short telomeres predicts longevity in mammals. Cell Rep 2012. https://doi.org/10.1016/j.celrep.2012.08.023.

[44] Baur JA, Zou Y, Shay JW, Wright WE. Telomere position effect in human cells. Science 2001;292(5524):2075−7. https://doi.org/10.1126/science.1062329.

[45] Whittemore K, Martinez-Nevado E, Blasco MA. Slower rates of accumulation of DNA damage in leukocytes correlate with longer lifespans across several species of birds and mammals. Aging 2019. https://doi.org/10.18632/aging.102430.

[46] Robin JD, Ludlow AT, Batten K, Magdinier F, Stadler G, Wagner KR, et al. Telomere position effect: regulation of gene expression with progressive telomere shortening over long distances. Genes Dev 2014;28(22):2464−76. https://doi.org/10.1101/gad.251041.114.

[47] Kim W, Shay JW. Long-range telomere regulation of gene expression: telomere looping and telomere position effect over long distances (TPE-OLD). Differentiation 2018;99:1−9. https://doi.org/10.1016/j.diff.2017.11.005.

[48] Mukherjee AK, Sharma S, Sengupta S, Saha D, Kumar P, Hussain T, et al. Telomere length-dependent transcription and epigenetic modifications in promoters remote from telomere ends. PLoS Genet 2018;14(11):e1007782. https://doi.org/10.1371/journal.pgen.1007782.

[49] Flanary BE, Sammons NW, Nguyen C, Walker D, Streit WJ. Evidence that aging and amyloid promote microglial cell senescence. Rejuvenation Res 2007;10:61−74. https://doi.org/10.1089/rej.2006.9096.

[50] Flanary B. The role of microglial cellular senescence in the aging and Alzheimer diseased brain. Rejuv Res 2005;8:82−5. https://doi.org/10.1089/rej.2005.8.82.

[51] Libertini G. Chapter 5: Programmed aging paradigm and aging of perennial neurons. In: Ahmad SI, editor. Aging. Exploring a complex phenomenon. CRC Press; 2017.

[52] Fossel M. A unified model of dementias and age-related neurodegeneration. Alzheimer's Dementia J Alzheimer's Assoc 2020;16:365−83. https://doi.org/10.1002/alz.12012.

[53] Hayflick L. The greatest risk factor for the leading cause of death is ignored. Biogerontology 2020;22(1):133−41. https://doi.org/10.1007/s10522-020-09901-y.

[54] Weed DL. Does paraquat cause Parkinson's disease? a review of reviews. Neurotoxicology 2021;86:180−4. https://doi.org/10.1016/j.neuro.2021.08.006.

[55] Bhattacharyya J, Mihara K, Bhattacharjee D, Mukherjee M. Telomere length as a potential biomarker of coronary artery disease. Indian J Med Res 2017;145:730−7. 10.4103/0971-5916.216974.

[56] Schneider CV, Schneider KM, Teumer A, Rudolph KL, Hartmann D, Rader DJ, et al. Association of telomere length with risk of disease and mortality. JAMA Intern Med 2022. https://doi.org/10.1001/jamainternmed.2021.7804.

[57] De Meyer T, Van Daele CM, De Buyzere ML, Denil S, De Bacquer D, Segers P, et al. No shorter telomeres in subjects with a family history of cardiovascular disease in the Asklepios study. Arterioscler Thromb Vasc Biol 2012;32:3076−81. https://doi.org/10.1161/ATVBAHA.112.300341.

[58] Doroschuk NA, Postnov AY, Doroschuk AD, Ryzhkova AI, Sinyov VV, Sazonova MD, et al. An original biomarker for the risk of developing cardiovascular diseases and their complications: telomere length. Toxical Rep 2021;8:499−504. https://doi.org/10.1016/j.toxrep.2021.02.024.

[59] Codd V, Denniff M, Swinfield C, et al. Measurement and initial characterization of leukocyte telomere length in 474,074 participants in UK Biobank. Nature Aging 2022;2:170−9. https://doi.org/10.1038/s43587-021-00166-9.

[60] Vecoli C, Basta G, Borghini A, Gaggini M, Del Turco S, Mercuri A, et al. Advanced glycation end products, leukocyte telomere length, and mitochondrial DNA copy number in patients with coronary artery disease and alterations of glucose homeostasis: from the GENOCOR study. Nutr Metab Cardiovasc Dis 2022;32(5):1236−44. https://doi.org/10.1016/j.numecd.2022.01.021.

[61] Nordfjäll K, Eliasson M, Stegmayr B, Melander O, Nilsson P, Roos G. Telomere length is associated with obesity parameters but with a gender difference. Obesity September 6, 2012. https://doi.org/10.2337/diabetes.54.1.8.

[62] Fyhrquist F, Saijonmaa O, Strandberg T. The roles of senescence and telomere shortening in cardiovascular disease. Nat Rev Cardiol 2013;10:274−83. https://doi.org/10.1038/nrcardio.2013.30.

[63] Zimnitskaya OV, Petrova MM, Lareva NV, Cherniaeva MS, Al-Zamil M, Ivanova AE, et al. Leukocyte telomere length as a molecular biomarker of coronary heart disease. Genes 2022;13:1234. https://doi.org/10.3390/genes13071234.

[64] Aviv H, Khan MY, Skurnick J, Okuda K, Kimura M, Gardner J, et al. Age dependent aneuploidy and telomere length of the human vascular endothelium. Atherosclerosis 2001;159:281−7. https://doi.org/10.1016/S0021-9150(01)00506-8.

[65] Fossel M. Use of telomere length as a biomarker for aging and age-related disease. Curr Transl Geriatr Exp Gerontol Rep 2012;1:121−7. https://link.springer.com/article/10.1007/s13670-0120013-6.

[66] Semeraro MD, Almer G, Renner W, Gruber HJ, Herrmann M. Telomere length in leucocytes and solid tissues of young and aged rats. Aging (Albany NY) February 27, 2022. https://doi.org/10.18632/aging.203922.

[67] Daios S, Anogeianaki A, Kaiafa G, Kontana A, Veneti S, Gogou C, et al. Telomere length as a marker of biological aging: a critical review of recent literature. Curr Med Chem July 13, 2022. https://doi.org/10.2174/0929867329666220713123750.

CHAPTER 2

Age-related disease: Central nervous system

Michael Fossel

Telocyte LLC, Ada, MI, United States

1. Introduction

O the mind, mind has mountains; cliffs of fall
Frightful, sheer, no-man-fathomed. Hold them cheap
May who ne'er hung there. **Gerard Manley Hopkins**, *1885.*

Few of us reading this chapter have hung on the "cliffs of fall/Frightful, sheer, no-man-fathomed" that are the essence of the dementias, but all of us know—and all too well—people who have been on those same sheer cliffs. Age-related neurodegenerative diseases, often simply lumped as the dementias, are frightfully common, hard to fathom and, more importantly, difficult to cure. All of us have parents, friends, neighbors, family, or patients who have experienced dementia. The epitome of the dementias, Alzheimer's disease, is "the disease that steals our souls." Whether or not we have hung on those metaphorical cliffs, few can underestimate the horror evoked by these diseases.

Although the dementias can be carefully defined and the clinical diagnoses are far too real, the reality is slipperier when we see enough patients to appreciate the clinical variation displayed in life. Alzheimer's disease, for example, is typified by cognitive dysfunction: memory loss, impaired executive function, linguistic deficits, and other behavioral changes. Yet even within this rubric, patients may differ significantly in presenting symptoms. Nor does the clinical variance stop there: the age of onset, rapidity of course, emotional lability, and other features vary in clinical reality, which is well-defined in our neurology texts but fuzzy in our neurology clinics.

Ironically, we might observe that at least to some degree, the atypical is nearly typical. Alzheimer's disease is usually defined by cognitive dysfunction, and Parkinson's disease by motor dysfunction, yet a significant subset of patients with Alzheimer's disease may demonstrate motor problems, whereas a similar subset of patients with Parkinson's disease may demonstrate cognitive problems. This is not to denigrate the reality or the utility of our diagnostic categories: they are too tangible for any to gainsay. Our diagnoses are useful conceptual attractors, allowing us to identify, understand, and (one hopes) treat the dementias. Yet when faced with hundreds of patients, we find that our diagnostic concepts have fuzzy edges that become all the more apparent when we look at the plethora of other age-related neurodegenerative diseases. Like weeds, acronyms that we use to define these diseases multiply every year as we dive more deeply into the risk factors and biomarkers of these diseases. Might the atypicality, the mixed presentations, and the expanding number of diagnostic categories for age-related central nervous system (CNS) disease hide a deeper, more fundamental commonality of these diseases? Perhaps our growing recognition of the atypical and the mixed presentations (the trees) allows us to appreciate a common [1,2], underlying process (the forest) in age-related neurodegenerative disease.

That underlying process is aging itself.

2. The dementias

In 1906, Alois Alzheimer presented (and published in 1907) his classic description of what we now call Alzheimer's disease [3], but the history of the dementias goes back much further than a dozen decades. Age-related neurodegenerative disorders, under various terms, have been described repeatedly over the past several thousand years and throughout the globe, including by the ancient Egyptians as well as by Pythagoras, Hippocrates, Plato, Aretaeus [4], Avicenna [5], Chinese physicians [6], Indian physicians [7], and countless others around the world and throughout human existence.

The word dementia (from *de*, without and *mens*, mind) has been used (and perhaps overused [8]) repeatedly and with varying definitions, medical and otherwise, for centuries. More recently, at least in medical use, it has been replaced by what are intended to be more specific terms, usually with the use of disease rather than dementia, as in Parkinson's disease or Lewy body disease. Regardless, for better or worse, the term dementia has remained in general use among the public. Whereas we might prefer the more technical age-related neurodegenerative diseases in a medical context, the sheer pithiness of the generic term dementia makes the word sufficiently convenient and effective for our discussion.

Loosely speaking, the dementias share certain classic features. They are age-related and attributed to dysfunction of the nervous system per se, rather than resulting from the failure of other organ systems, such as renal, cardiac, or pulmonary dysfunction. The case of vascular dementia is, in some sense, a borderline example, in which vascular pathology occurs in the cerebral circulation. Although vascular comorbidity is common, pure vascular dementia may be relatively rare [9], and it will not be addressed directly in this chapter. Although other borderline or diagnostically confusing presentations (such as adult-onset acquired neuroinflammatory disorders [10]) might included, our focus will be on more common dementias. Most medical sources specify that dementias (as a general category) are defined by broad sets of symptoms, including impairments in memory, intellect, language, personality, the ability to carry on daily tasks, and so on. In some cases, such as Parkinson's disease, the presenting symptoms are predominantly motor rather than cognitive, but they may still be usefully described and discussed in the same context, as resulting from an age-related primary dysfunction of the nervous system.

Despite the attribution of shared features, modern diagnostic use has tended toward specific findings, both clinical and histologic, in defining the disease categories of what we might otherwise lump together as the dementias. Thus we have bodies of literature devoted not only to Alzheimer's disease and Parkinson's disease, but to an entire litany of age-related neurodegenerative diseases, including frontotemporal dementia (FTD), Lewy body dementia, Huntington's disease, corticobasal syndrome (CBS), medial temporal lobe atrophy, familial Alzheimer's disease, posterior cortical atrophy, limbic-predominant age-related TDP-43 encephalopathy (LATE) [11], primary progressive aphasia, progressive supranuclear palsy, autosomal dominant Alzheimer's disease, and late-onset Alzheimer's disease. Because we are cognizant of the genetic correlates that serve as diagnostic indicators (such as APOE4 and its variants [12,13], presenilins, the Paisa mutation [14], amyloid-β precursor protein (APP), TREMS2, trisomy 21 [15,16], and so forth [17–21]), we might also define dementias by such risk factors.

Adding to this fractionating set of diagnostic categories is the realization that many patients show atypical or mixed findings [22–24]. Patients vary not only in the age of onset, rapidity of course, and predominance of specific symptom with the diagnostic category, but sometimes have findings more typical of other diagnostic categories. As noted earlier, patients with Alzheimer's disease can have motor dysfunction more typical of Parkinson's disease, and vice versa. The diagnostic categories are all too real, but there is also a certain amount of variance both within and between such categories, which is typical of many disease categories in other organs and in non–age related diseases. What is remarkable is that this is not remarkable: disease seldom remains in the neat categories into which we attempt to force it. Regardless, however, the conceptual utility of such categories remains obvious and is generally on a secure footing.

Regardless of our tendency to narrow the diagnostic focus, a key observation remains cogent and accurate: the single most predictive variable for age-related dementias is age [25,26]. Although age is often blithely given lip service, most current research, both basic and clinical, has delved into increasingly narrow limits, ignoring the aging process and focusing instead on more specific targets, perhaps in the belief that such targets may be more amenable to intervention than is the aging process itself. That such specific targets have thus far proven ineffective in therapeutic trials has only increased the frustration of those performing the research, to say nothing of the physicians, patients, investors, insurers, and government agencies involved.

To be fair, much has improved.

Until perhaps the past few decades, diagnosis was almost exclusively clinical (unless we include postmortem findings) and had a practical focus on eliminating other, sometimes more treatable diagnoses, with the final diagnosis largely one of clinical suspicion relying on exclusion. Diagnoses based on laboratory [27] or imaging [28] results have now largely supplanted clinical impressions, although clinical concern remains the impetus for instituting the more modern and more accurate laboratory and imaging tests. This transformation of available diagnostic information has prompted new, and still evolving, scoring systems [29] that are becoming the standard in both clinical and experimental settings as we attempt to find (and be able to prove) more effective points of intervention.

3. Therapeutic status

Although diagnostic efficacy has improved, therapeutic efficacy has not. Reference to the most widely used global registry for clinical trials [30] shows a large and growing list of interventional trials for the dementias. In late 2022, for example, Parkinson's disease had more than 2500 interventional trials, whereas Alzheimer's disease had more than 2400 interventional trials. The results of such trials, however, is not commensurate with the urgency of the need for effective interventions. Despite the poignant need (and occasionally the strident claims) for a cure, there is no current intervention for the dementias that can prevent, stop, or reverse the clinical course. In almost all cases, current treatments are symptomatic rather than targeting the basic disease process. Even in the few cases, such as Parkinson's disease, for which we attempt to ameliorate specific biochemical abnormalities, such as the effects of the lower availability of dopamine in the striatum and substantia nigra, the benefits of our targeted medications, such as L-DOPA (levodopa), are limited and wane over time.

Some dementia interventions may arguably slow the disease course, although the clinical value of these interventions is often in dispute, the costs may be high, and the consensus is often depressing. For example, despite current therapeutic advances [31], no pharmacologic therapies can prevent or delay Parkinson disease progression [32]. Much the same is true for Alzheimer's disease, for which even the value of monoclonal antibody techniques and tau antibody techniques have proven questionable [33,34] if not frankly disappointing [14,35,36]. Investors view Alzheimer's disease as the graveyard of biotech, whereas physicians stress increasingly palliative [37], rather than curative, targets in age-related neurologic disease. Few believe that we can, at best, do more than slow disease progression, and visions of an actual cure are rarely even entertained.

Attainable or not, the ideal goal is clear. The desired target is a disease-modifying therapy. Optimally, such a therapy would not merely stop neurologic dysfunction, but would reverse it. The current reality, however, is that even the best interventional results claim to slow the rate of decline, and even those claims generate disagreement. In many current trials for Alzheimer's disease, monoclonal antibodies target amyloid plaque. The belief is that such plaques have a central role in the pathology [38], but that if we can reduce the plaque burden [39] we can slow the disease, particularly if therapy is instituted early in the disease course [40]. Although the reference standard would be cognitive improvement rather than a decrease in the rate of cognitive decline, such partial delays in progression have been reported in several human trials. The perceived benefit would be to extend the time during which patients are capable of, for example, self-care, driving, and other activities of daily living. It is depressing to consider that we are, at best, confining ourselves to the hope that we can prolong rather than cure such diseases (Fig. 2.1).

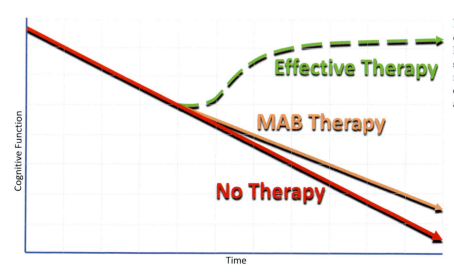

FIGURE 2.1 The progressively downhill clinical course of a typical dementia (e.g., Alzheimer's disease) over time may at best be **slowed** using current interventions such as monoclonal antibodies (MABs), whereas an effective intervention would **reverse** the loss and **stabilize** cognitive function.

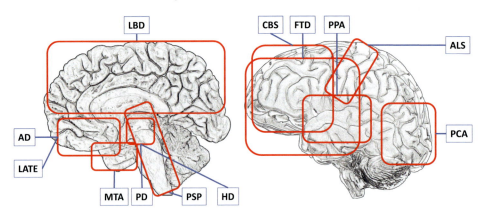

FIGURE 2.2 Initial neurodegeneration is often imprecisely located, becoming more widespread as diseases progress. These images are illustrative only, to suggest the correlation of the central nervous system location with symptoms. *AD*, Alzheimer's disease; *ALS*, amyotrophic lateral sclerosis (also motor neurone disease); *CBS*, corticobasal syndrome; *FTD*, frontotemporal dementia; *HD*, Huntington's disease; *LATE*, limbic-predominant age-related TDP-43 encephalopathy; *LBD*, Lewy body disease; *MTA*, medial temporal lobe atrophy; *PCA*, posterior cortical atrophy; *PD*, Parkinson's disease; *PPA*, primary progressive aphasia; *PSP*, progressive supranuclear palsy.

Whereas clinical classification schemes have changed over the years, current consensus has settled on the ATN (Amyloid, Tau, Neurodegeneration) classification system [29], which forms a useful [28] and practical view of available biomarkers, particularly in therapeutic trials. With regard to the underlying, fundamental cell processes that mediate and result in such biomarkers, however, we may want consider alternative parameters. When we consider the clinical foundations of the dementias, we are at least initially struck by three potential axes of findings: (1) the neuroanatomic location of the dysfunctional cells, (2) the specific types of the dysfunctional cells, and (3) the characteristic biochemical abnormalities of the dysfunctional cells. Do any of these axes provide insight into the fundamental disease mechanisms?

4. Neuroanatomy

If we focus on the neuroanatomic locations of the pathology, we find a correlation between the location of the (initial or most characteristic) pathology and the clinical function (or dysfunction) that we associate with any specific dementia (Fig. 2.2). For example, in the case of FTD (and allowing for a degree of oversimplification), we expect and observe that the loss of frontal and temporal lobe neurons is associated with progressive language dysfunction (e.g., progressive semantic and nonfluent aphasia) as the relevant neuroanatomic structures (e.g., Broca's area, Wernicke's area) are affected. Similarly, primary progressive aphasia affecting the temporal lobe has the predictable clinical outcome consistent with its name. The loss of the relevant neurons in CBS results in the expected apraxias and extrapyramidal symptoms and the loss of occipital neurons in posterior cortical atrophy result in visual processing deficits. Progressive supranuclear palsy has a wider ambit, with equally variable but predictable clinical findings with regard to balance, motor, and ocular function. In the case of LATE, we expect memory problems in regions that the affected limbic structures support.

Whereas some dementias are named based on the initial or predominant CNS structures involved, such terms are only of clinical convenience. Terminology based on neuroanatomic location has a clear-cut value in correlating clinical findings with abnormalities seen on imaging (or postmortem) studies, but it offers little or no insight into the fundamental mechanisms involved. Categorization by location has conceptual limits. To switch organs systems, we could, for example and with equal justification, categorize osteoarthritis based on the specific joint involved. In either case, whether CNS or joint disease, location-based anatomic terminology may be clinically convenient, but it provides little conceptual traction if we are to understand the fundamental process that underlies and drives these diseases.

5. Cell types

Moving from location to cell type, we advance only marginally in our bid to clarify the mechanisms of neuropathology. Cell types, including neurons, microglia, and astrocytes (see Fig. 2.3), show a remarkable degree of

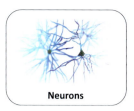 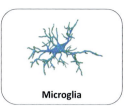 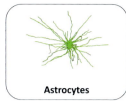

FIGURE 2.3 Although neuron pathology is the focus of most histologic descriptions and deserves a common emphasis in understanding clinical pathology, microglial and astrocyte pathology remains an almost ubiquitous characteristic in most age-related neurodegenerative diseases.

concordance as we compare the range of age-related neurodegenerative diseases. Moreover, not only do different dementias resemble one another by the cell type affected, a broad range of cell types are apparently affected to varying degrees in all of the dementias. For example, in the case of amyotrophic lateral sclerosis (ALS), we see prominent pathology in the pyramidal neurons (Betz cells) of the frontal motor cortex, but microglia and astrocytes also show significant changes. The same broad cellular effects (neurons, microglia, astrocytes, etc.) are characteristic of Alzheimer's disease, Parkinson's disease, and other dementias. Cellular pathology is not strictly neuronal; it encompasses various glial cell types in the age-related neurodegenerative diseases. A predilection for specific cell types is not an obvious defining characteristic for the age-related neurodegenerative diseases. All cell types demonstrate pathology and all dementias have a remarkable similarity in the spectrum of cell types that are affected.

Although the literature gives precedence to neuron degeneration (hence neurodegenerative disease), the reality is that age-related dementias are actually neuromicroglial-astrocytic-degenerative diseases. There is, for example, growing awareness that glial cells have a central role in age-related dementias in both animals [41] and humans [42], in which we see glial cell activation [43,44], inadequate amyloid turnover [41,45–48], tau protein abnormalities [49,50], methylation [51], homeostatic changes [52], and so forth. Although much remains unclear, glial cells are clearly key players in age-related neurodegenerative disease. The fundamental processes involved in what we call age-related neurodegenerative diseases are catholic in their effects. The crucial inference is that neither location (save as a clinical aid) nor cell type (save as a more apparent primary finding in histological surveys) offers significant insights as we struggle to understand the fundamental mechanisms that underlie the dementias. The key to the fundamental pathology underlying the dementias is not the specific neuroanatomic location (although it explains the clinical findings) or the specific cell types (although they explain the local findings); it lies at a deeper level. If we wish to understand how the dementias occur, let us move to the level of biochemistry.

6. Biochemical changes

Upon first consideration, biochemical abnormalities suggest more potential in elucidating the underlying processes of age-related neurodegenerative diseases. There is a plethora of such abnormalities. Within broad limits, these vary from one type of dementia to another. For example, the findings of amyloid plaques and tau tangles are characteristic of Alzheimer's disease [38] and have been well-described (particularly in the case of amyloid plaques) for much more than a century. Not all cases of dementia need involve amyloid, however, although controversy remains regarding the role (or necessity) of amyloid in such pathology [53]. In the past few decades, the list of known biochemical markers has increased enormously and, to some extent, allowed us to categorize the dementias in another and (as many hope) more useful manner. At the highest level, we categorize these abnormalities as proteopathies, tauopathies, synucleinopathies, and so on, and then subdivide these into more specific findings based on, for example, the specific proteins [54] involved.

Known biochemical abnormalities found in the human dementias encompass β-amyloid (extracellular microaggregates and plaques), tauopathies (intraneuronal neurofibrillary tangles) [55], TDP-43 proteins [11] (or perhaps TMEM106B) [56], α-synuclein proteins (intraneuronal Lewy bodies), neurofilament light chains (NfL), other misfolded proteins (as in prion diseases) [57], and dozens of others [54], including not only proteins but also lipids and other metabolites [58]. As we inquire into the specifics of each of these abnormalities, we note a commonality: pathology occurs when molecules usually adopt abnormal forms that then accumulate.

β-Amyloid, for example, causes problems when a fragment of the transmembrane APP, necessary for normal neuronal function, forms fibrils, then microaggregates, and finally plaques. The initial problem occurs when an abnormally folded amyloid β-protein forms. Although such abnormal patterns of folding can occur in any individual, the probability of such abnormal folding is far higher for some individuals than it is for others. The increased accumulation of abnormally folded molecules correlates well with age, but the increased propensity for abnormal folding itself depends on genetics. Those with two APOE4 alleles, for example, have an increased risk, earlier onset, and more rapid clinical disease course than those with two APOE4 alleles. In both cases, amyloid aggregates rise with age, but the vector is steeper in those with APOE4 alleles. This observation deserves emphasis: the risks of an amyloid-based neurodegenerative disease have at least two evident predictive variables: the genetic starting point of the patient [59] and the age of the patient. Both genes and aging are operative variables in the amyloid-based dementias, but the same is true for other age-related neurodegenerative diseases in which the major risks may be grouped under two distinct rubrics: age and genetics.

Tau proteins produce a problem parallel to that of β-amyloid. Normally phosphorylated tau proteins, which modulate the stability of axonal microtubules and are necessary to normal axonal function, can become hyperphosphorylated, allowing them to link to other, similar protein threads, forming neurofibrillary tangles. Whereas the insoluble tau aggregates may differ in their degree of phosphorylation and the specific tau isoforms involved, the clinical outcome is roughly the same, because they interfere with normal neuronal function. As in the case of amyloid, both aging and genetic risk factors have a role [60], but aging remains the major predictive risk factor.

NfL, like most of the molecules listed here, are a necessary part of normal neuronal function. In this case, they function in the cytoskeleton, especially in axons, and when found in the cerebrospinal fluid, they are considered a useful marker of neurologic injury and disease, particularly age-related neurodegenerative disease. Once again, both genetics [61] and aging itself are risk factors, with aging the most predictive single variable.

α-Synuclein is a protein involved in presynaptic function, but it can aggregate into insoluble inclusion bodies (synucleinopathy). Specific genetic elements at the locus have been identified [62]. Here again, aging remains the major predictive risk factor for the synucleinopathies.

Huntingtin protein is the defining finding of Huntington's disease (note the difference in spelling of the disease and the protein) and has a clear genetic underpinning [63,64] especially the *huntingtin* (HTT) gene itself. The HTT gene results in a trinucleotide repeat expansion of the affected protein that is prone to misfolding and aggregation and results in an architecture similar to that seen in other age-related neurodegenerative proteinopathies. As the disease progresses, the aggregates form inclusions and cause increasing and increasingly pervasive damage to the brain. Not surprisingly, whereas the HTT gene has a high (though variable) penetrance [65], age is again the major predictive risk factor [66].

Transactive response DNA binding protein 43 (TDP-43 protein) is encoded by the TARDBP gene [11,67] and regulates transcription and RNA processing. Mutations of the gene have a known role in FTD, ALS, and some forms of Alzheimer's disease. TDP-43 forms a hyperphosphorylated protein that goes on to impair cell function. As with other forms of protein abnormality, the effects of TDP-43 (or perhaps TM106B) [56] are age-related. Genetics are the defining risk factor, but aging is the more predictive risk factor.

A number of prion diseases also demonstrate misfolded proteins that multiply by enforcing a similarly misfolded shape on other normal versions of the protein. Such diseases include Creutzfeldt Jakob disease (CJD), fatal familial insomnia, and kuru. Although the mean onset for CJD is age 60 years, these diseases are not necessarily age-related in the sense discussed here, and the mechanisms discussed in this chapter may not be relevant to such prion diseases.

We have certainly not presented a complete list of diagnostic terms used to parse the clinical presentations of age-related neurodegenerative diseases, nor have we presented a comprehensive summary of the biochemical abnormalities associated with these diseases. To do so would be infeasible here for two reasons: (1) a catalog of the known abnormalities would be too long to fit into this chapter; and (2) the literature on these abnormalities is expanding rapidly, and it is extremely unlikely that we are even roughly aware of the full magnitude of such changes that could (and will) be found in future research.

Nevertheless, even the abbreviated thumbnail summary given earlier suggests the formidable complexity involved when these diseases are viewed from a purely biochemical point of view. The sheer number of the biochemical abnormalities associated with the age-related neurodegenerative diseases is daunting and should make us pause, step back, and reassess our chosen perspective. When we find this many (and doubtless more to come) seemingly independent biochemical events involved in the etiology of a group of otherwise clinically related diseases (age-related dementias), we might well ask ourselves whether we might gain a clearer understanding of these diseases by looking at a more fundamental process that plays into the clinical expression of all of these biochemical abnormalities (Fig. 2.4).

To use an analogy, if we were ignorant of the cause of a common microbial disease such as COVID-19, we might still be cognizant of the clinical and laboratory findings. Having an in-depth compendium of such findings (fever, cough, fatigue, pain, pharyngitis, diarrhea, conjunctivitis, headache, anosmia, rash, dyspnea, etc.) would not

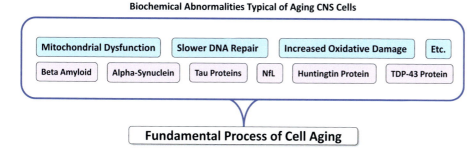

FIGURE 2.4 Myriad biochemical abnormalities (only a few are listed here) are observed in age-related neurodegenerative diseases, which might be simplified if we can gain a deeper understanding of why they occur and why cell aging is the common denominator, appearing to drive all of the abnormalities. *CNS*, central nervous system; *NfL*, neurofilament light chain.

necessarily identify the underlying cause: infection with the COVID-19 virus. A similar conceptual lacuna operates when we attempt to understand age-related neurodegenerative disease by focusing solely on biochemical biomarkers or other downstream findings. Rather than delving into the biochemical details of each dementia, we might wonder whether there is a shared, fundamental process driving the disparate findings that we innocently diagnose as independent neurologic diseases. Could we identify a unified process that results in the disparate diagnoses that we collectively call the dementias? That there are a number of apparently independent biochemical abnormalities is clear, but might these different presentations result from different genetic starting points, then play out on a common stage, a shared process that permits the veiled genetic risk factors to reach a diagnostic threshold?

That aging is the greatest predictive risk factor in age-related disease is a truism, but it may be a critical observation. Cell aging may be precisely the common stage, the shared process that permits a variety of upstream risk factors, including genetic risk factors, to reach a biologically critical state in which cell function becomes compromised and results in clinical disease. Rather than viewing aging as a passive entropic outcome, a more careful look at cell aging reveals it to be a dynamic systems process in which the genetic risks discussed earlier become operative as a result of cell aging. This observation clarifies the mechanisms of pathology and offers a potentially more effective point of intervention. A systems approach suggests an optimal target, one central to the cascading pathology, whereas downstream findings (often used as current interventional targets) are merely symptomatic results of the fundamental upstream process.

7. Aging as a key process

Why is aging, rather than other identified risk factors, the best predictive risk factor in all of these diseases?

If we were given a new patient, and knowing *nothing whatsoever* about our hypothetical patient, we were asked to predict the risk of the patient receiving the diagnosis of Alzheimer's disease in the next year, the most useful predictive indicator would be age. Regardless of APOE status, the family history, the history of head trauma or encephalitis, alcohol use, or any of dozens of other potential risk factors, age alone accounts for most of the variance in assessing the current Alzheimer's risk for a random human.

We often discount this risk factor for (what are perhaps) plausible reasons.

The first reason is that age is so easy to assess that we seldom need to use medical records, imaging, or laboratory tests, at least regarding the rough age of the hypothetical patient. At a glance, we know the patient is a child, an adult, middle-aged, old, or very old. That glance does not tell us whether the patient is aged 70 or 71 years, but we instantly know whether he or she is aged 7 or 70 years. Knowing the approximate decade is far more predictive of disease risk than is the specific year within any decade. We do not require a birth certificate; we rely on a simple glance to judge the rough biological age, a glance that accounts for the most useful data in assessing risk. Bluntly, we are far less likely to see dementia presenting in a 7-year-old (even if the patient is at high risk with two APOE4 alleles) than in a 70-year-old (even if the patient is at low risk with two APOE2 alleles). This observation is not merely true of Alzheimer's disease: age alone is the major single risk factor for all age-related neurodegenerative diseases. Even with regard to genes known to carry a high degree of risk (such as the presenilins), age is the more predictive risk factor. The major predictive risk factor for the dementias is age, but it is so transparently obvious that we disregard it.

The second reason that we tend to discount age as a risk factor is that we assume that we cannot alter aging or the aging process (you cannot change your age), so there is little point in giving it specific attention in a clinical context.

Although we might make the same point with regard to genetic risk (you cannot change your genes), we aggressively research the mechanisms of the genetic risks precisely on the assumption that even if we cannot change the offending gene (and we are beginning to do precisely that), if we can sufficiently understand how the gene results in clinical disease, we may well be able to intervene downstream and develop an effective clinical intervention for the products of that gene.

Much the same reasoning pertains to other upstream risk factors, many of which are more easily mutable. We routinely try to minimize infections, trauma, radiation, toxic exposures, or the raft of other health risks that increase the likelihood of an age-related dementia. All of these approaches are laudable, and although many of them may delay or ameliorate the course of age-related dementias, none have shown promise in stopping (let alone reversing) the course of such diseases. We tend to ignore aging as being immutable, but that assumption likely prevents us from trying to understand the aging process fully at its most fundamental cellular, genetic, and epigenetic levels. We suspect, and are beginning to demonstrate, that we can change either our genes or the effects of those genes, but the same possibility may be true for the fundamental aging process. The major impediment may not be an inability to alter the aging process, but an inability to reconsider our assumptions and a far too simplistic view of how aging works.

There is a deeply rooted tendency for people to regard aging diseases as simply accumulative rather than resulting from a shift in a dynamic equilibrium. Put simply, many regard these diseases as occurring merely as the result of the gradual and incremental accumulation of damaged proteins (or other molecules) over time. In this view, aging is simply a passive, entropic process. The consequence of that assumption is that although we give lip service to the defining truism that aging is the major risk factor in all age-related neurodegenerative diseases, aging serves only as a passive temporal measure of what we take to be the ongoing, simple accumulation of pathologic protein (or other) aggregates in a passive downhill clinical course. In this view, disease is the linear result of the passive increase in the percentage of such aggregates.

However, this view neglects the active recycling of molecules, which is true for all biological pools. Molecular pools (proteins, for example) are continually being turned over with the result that at equilibrium, there is a relatively stable (and generally small) percentage of damaged or dysfunctional molecules. As pointed out in Chapter 1, however, the point of equilibrium changes with age as cell aging progresses and the rate of molecular turnover decreases [68,69] (Fig. 2.5).

Compare this with genetic diseases of the young, in which the rate of molecular turnover has not had time to decelerate. In these cases, the genetic error is often so substantive that it manifests almost immediately in severe clinical disease, even in the young patient. In such cases age is not the most significant risk factor. The disease is one of accumulative damage without regard to the deceleration of molecular turnover that occurs as cell aging becomes the prominent driver of pathology. In the case of age-related diseases, however, the genetic risks remain relatively hidden at young ages, but manifest clinically not through mere accumulation of dysfunctional protein, for example, but through the decrease in the rates of synthesis and degradation (the rate of molecular turnover), with the result that molecular abnormalities increase as a percentage of the available pool [70].

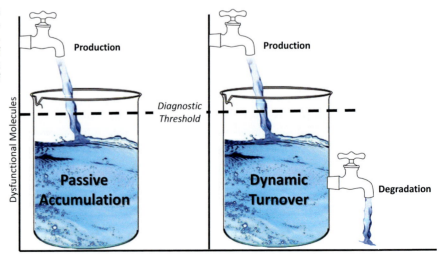

FIGURE 2.5 The percentage of damage is a product of production and degradation, a recycling process that slows as cells age. The diagnostic threshold is reached not because of passive accumulation of damage, but because a slower rate of turnover allows the percentage of damaged molecules to increase.

In age-related disease, the problem is not that abnormal complexes occur, but that the rate molecular turnover and the clearance of such complexes fall [71,72] and that this failure becomes increasingly significant as cell aging progresses. In normally functional, young cells, the quality control mechanisms serve to maintain normal synthesis and the elimination of defective molecules. This mechanism becomes insufficient under any of three conditions [72] (or when a combination of such conditions pertain):

(1) specific genetic mutations decreasing such quality control
(2) abnormal cellular stress inhibiting or overwhelming quality control
(3) cell aging that slows the rate at which quality control occurs

The result of an insufficient rate of molecular turnover is a rise in the percentage of defective molecules that then accumulate as an increasing percentage of the molecular pool, forming inclusion bodies or aggregates, and decreasing cell function. The percentage of damaged or dysfunctional molecules increases with cellular aging not as a result of chronological age per se, but of cell senescence, which correlates with, but is distinct from, chronological age [70]. Cell senescence results in age-related damage, not the other way around. The cellular equilibrium between aggregation and degradation can lead to age-related neurodegenerative diseases, particularly when the genetic predilection increases the risk of abnormal molecular folding, leading to an earlier onset or more accelerated disease course (Fig. 2.6).

8. Unified systems model: Cell aging in the central nervous system

To date, we have failed to cure, prevent, or significantly slow any of the age-related neurodegenerative diseases. This is despite a substantial pipeline of candidate drugs [73], thousands of registered trials [74], tens of thousands of patients in clinical trials [75], billions of dollars in both US federal and pharmaceutical company investment [76], more than a century of clinical expertise, thousands of professionals, and dozens of pharmaceutical and biotechnology firms in dedicated attempts to prevent, slow, or alter the course of the dementias.

We fail because we aim at the wrong targets. The fault lies not with a lack of targets (for example, amyloid and tau proteins) or with a lack to technical capability (for example, the use of monoclonal antibodies), but with the lack of a unified model that can convincingly explain how such diseases occur [77]. Such a model needs to account for the disparate and large numbers of upstream risk factors and explain how such risk factors result in the variety of downstream diseases. Moreover, it should account for the heterogeneity of clinical findings within our species (for example, the various clinical diagnoses discussed earlier, as well as their various mixed and atypical presentations), but also the heterogeneity that we observe among species (for example, differences in age-related cognitive decline seen in humans versus mice, rats, cats, dogs, Rhesus monkeys, etc.).

The model must explain the role of aging itself (i.e., cell aging) in the dementias and provide a systems view of the multiple cascades of pathology seen in these diseases. Here, the emphasis is not on the individual item of pathology

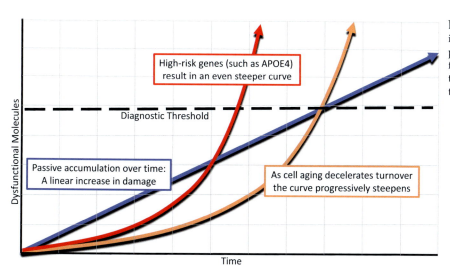

FIGURE 2.6 Depending on the genetic impact, age-related diseases result not from the passive accumulation of damage over time, but from a gradual decrease in molecular turnover that results in a roughly exponential increase in the percentage of dysfunctional molecules.

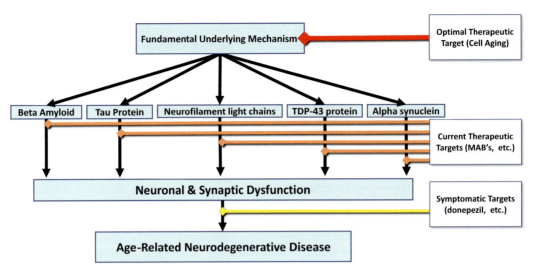

FIGURE 2.7 The current therapeutic focus is on biochemical findings associated with age-related dementias (such as amyloid) or on symptomatic therapies, rather than addressing fundamental age-related changes in the cell that drive downstream biomarkers and disease. *MAB's*, monoclonal antibodies. *Adapted from Fossel [70].*

per se, but the way in which that individual aspect of pathology functions within the entire cascade of events from risk factors to clinical symptoms that is, our emphasis is on a systems rather than component view. Too often [78], we focus too narrowly: geneticists focus on genes, pathologists focus on pathology, and chemists focus on chemistry, each taking a component approach rather than a systems approach [79,80]. The result of this narrow component view has been a waste of investment funds and an almost universal failure of clinical trials. A model that relies solely on the role of β-amyloid in Alzheimer's disease is not merely insufficient, it directly results in the predictable clinical frustration we find in human trials. A more reasonable model must also account for all of the other biochemical abnormalities discussed earlier, as well as the parallel findings that occur in aging cells, including inflammation, mitochondrial dysfunction, slower DNA repair, and increases in oxidative damage [81]. In short, if we are to treat the dementias effectively, we should not be targeting symptoms, nor should we target the biochemical outcomes of aging. We must use a more comprehensive model to target the fundamental underlying mechanisms of these diseases (Fig. 2.7).

The model must needs adopt a systems view, encompassing all of these into a broad exegesis of age-related CNS disease. This point (the lack of a unified systems model and the requirements needed in an effective model) has been emphatically and cogently made by several authors, pointing out that our persistent clinical failures to intervene suggest the need to "... reassess current ideas on biological underpinnings ... [and it] ... reinforces the necessity for a reexamination of the neurobiological premises for therapy development" [82]. These authors suggested that any unified model should address 10 key points: (1) how the aging process itself has a central role; (2) operative effects of known risk factors; (3) disease patterns; (4) the clinical sequence; (5) the anatomic specificity of lesions; (6) pathologic interactions; (7) the occurrence of mixed pathologies [22] and comorbidities [83], (8) clinical heterogeneity; (9) relationships between biomarkers; and, perhaps most important, (10) innovative and potentially more effective points of intervention.

The need for a unified model has been addressed in articles [70,84] that sketched out such a unified systems model and attempted to address exactly the 10 points that such a model requires. This model posits the central role of cell aging as the common stage on which upstream risk factors play out and result in the downstream findings discussed earlier. Although the central stage (cell aging) is the shared and central process for all age-related neurodegenerative diseases, the heterogeneity of starting points provides the heterogeneity of clinical end points. All patients undergo cell aging (the driving force in age-related disease), but those with different genetic starting points will end with different clinical presentations. Those with similar genetic starting points will end with similar clinical presentations. Similarly, those with different behavioral histories (trauma, toxic exposures, infections, gum disease, sleep disturbances, education, other comorbidities [83,85], etc.) will also end with predictably different clinical outcomes.

To use a specific example, both patients with APOE2 alleles and patients with APOE4 alleles will undergo cell aging, but the outcomes will be markedly different because the genetic starting points differ. As pointed out in Chapter 1, APOE4 alleles increase the rate of molecular damage (i.e., cross-linking of amyloid molecules [86]) that causes pathology even early in the cell aging process. Patients with APOE2 alleles, on the other hand, have

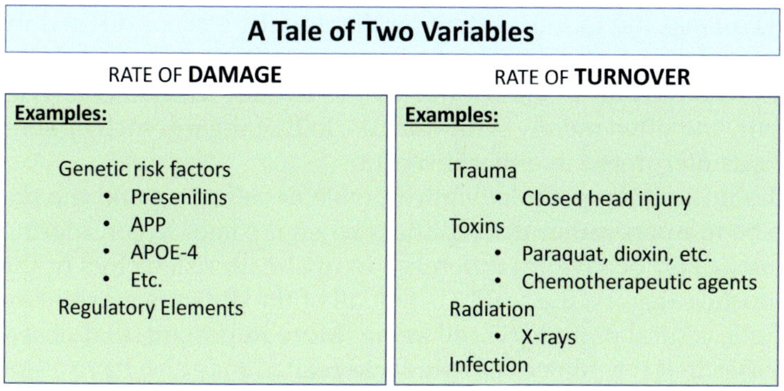

A Tale of Two Variables

RATE OF DAMAGE

Examples:

Genetic risk factors
- Presenilins
- APP
- APOE-4
- Etc.

Regulatory Elements

RATE OF TURNOVER

Examples:

Trauma
- Closed head injury

Toxins
- Paraquat, dioxin, etc.
- Chemotherapeutic agents

Radiation
- X-rays

Infection

FIGURE 2.8 The equilibrium percentage of damage is the joint result of two variables: (1) the rate of damage (determined by genetic and epigenetic starting points), and (2) the rate of molecular turnover (determined by cell aging). The rate of damage is determined by specific (e.g., high-risk) genes and regulatory elements. The rate of molecular turnover is determined by the degree of cell aging, which is affected by a host of variables that result in cell loss, including trauma, infection, and toxins. *APP*, amyloid-β precursor protein.

lower rates of molecular damage and become at risk for a similar degree of pathology only very late in the cell aging process. The outcome is that the onset and pace of neurologic disease will vary with the genetic starting point, even if the rate of cell aging is identical. Much the same is true for any of the upstream variables: they may affect either the rate of molecular damage (e.g., APOE4 alleles) or the degree of cell aging (i.e., the rate of molecular turnover), which slows molecular replacement (Fig. 2.8).

Overall, we are dealing with two major varieties of variables: the rate of molecular damage and the rate of molecular turnover. The former type of variable is determined by the genetic and epigenetic starting point of each patient. These variables might be thought of as the substrate on which cell aging works. In young cells, only highly abnormal genes will result in high rates of damaged molecules, and hence in reaching a diagnostic threshold for pathology. Relatively abnormal genes will result in significant rates of damage even if only a moderate degree of cell aging has occurred. Those with the lowest-risk genes will achieve similar rates of damage (if they do so at all) only when cell aging is advanced. To use the APOE example again, those who are biallelic for APOE4 will demonstrate early and rapid pathology with only minimal degrees of cell aging, whereas those who are biallelic for APOE2 will demonstrate late and more leisurely pathology and with only extensive degrees of cell aging.

Although we have emphasized the roll of genes as one of the two major variables, the other variable should also be discussed (i.e., variables affecting the rate of cell aging itself). Any process or event that results in cell loss is likely to result in cell replacement, and thus in cell aging. As cells are lost, the remaining neighboring cells (in some cases) or the stem cell niche (in other cases) are prompted to divide. This results in telomere shortening, which results in epigenetic alterations [87–89], which (among other outcomes) results in the downregulation of gene expression and slowing of molecular turnover in crucial molecular pools. These include biochemical candidates identified as, and likely responsible for, the cascade of pathology in age-related neurodegenerative diseases such as β-amyloid. Independent of the genetic background, the slowing of molecular turnover results in a gradually increasing percentage of dysfunctional molecules and consequent dysfunction of the involved cells. Variables increasing the risk of cell loss are legion. The more obvious examples for neurodegenerative diseases include trauma (variously designated as closed head injury or traumatic brain injury) [90–92], infection (including encephalitides from any number of microorganisms [93,94], as well as COVID-19 infections [95,96]), toxins [97–99] (alcohol is a typical example), chemotherapy [100], and radiation (e.g., radiation therapy). Metabolic risks can also result in cell loss, such as hypoxia and hypoglycemia. A host of potential vulnerabilities, generally under the rubric of medical history, can be expected to accelerate cell aging in the CNS.

In many cases, the risk factors may overlap or be difficult to tease apart. For example, regulatory elements, which make up a much greater portion of the human genome than do the protein-encoding genes, may regulate (in this case, downregulate) the rate of expression for a specific gene (increasing the rate of damage) or may affect the likelihood of cell survival (increasing the rate of cell aging). In all likelihood, both of these occur, sometimes simultaneously. In addition, there are cases in which patients with both genetic and behavioral history (or comorbidities [83]) have obvious risks, and it becomes all but impossible (and perhaps moot) to tease apart the etiology and even to defend a specific neurologic diagnosis. A classic example of this would be a patient with clear presentation for Parkinson's disease, but who also has a long history of repetitive head injury (as occurred in the case of Mohammad Ali [101]). The same diagnostic complication occurs in patients with both APOE-based risk and a history of head trauma [102].

Practically speaking, viewing cell aging as the linchpin for age-related neurodegenerative disease suggests that such diagnostic distinctions have only circumscribed value. A specific clinical diagnosis may be valuable for prognostic reasons (to help the patient and others make plans) and to some extent for therapeutic reasons (to identify a potential benefit, such as from L-DOPA in the case of Parkinson's diseases, for example). Overall, however, the prognosis for most age-related neurodegenerative diseases remains dismal, if not always equally dismal. Likewise, therapeutic options remain limited, frequently transient, and often merely symptomatic. In that regard, such options have a clear if restricted value, which should not be misinterpreted in either direction.

From the broader point of view, looking at both the underlying pathology attributable to cell aging and the potential for effective intervention, we might do well to be lumpers rather than splitters when it comes to considering neurodegenerative diseases. Without ignoring the genetic and behavioral difference in upstream risk factors or the downstream clinical characteristics of a specific diagnostic category, the reality is that all of the diseases we think of as age-related neurodegenerative diseases have a single, central pathology: cell aging. More important, that observation suggests that all of the neurodegenerative diseases that we think of as being age-related may also have a single, effective point of therapeutic intervention, which may prove far more effective than any target we have yet pursued.

The implications of cell aging for clinical disease, and the potential for a far more effective point of clinical intervention, was first pointed out in the medical literature several decades ago [103,104] and has been repeatedly raised since then [2,105,106]. The efficacy of any specific therapeutic intervention, however, critically depends on its target. To invoke an example from the microbial world, if we exclusively target a narrow clinical outcome (for example, a bleeding diathesis) in a generally fatal viral infection (for example, Ebola virus), we might marginally lower the mortality, but we are not targeting a fundamental process. To offer another example from the microbial world, an effective antitussive pharmaceutical may offer a degree of symptomatic relief to those with COVID-19 infections but is unlikely to lower the overall morbidity or mortality of the disease. Similarly, when we exclusively target a dementia biomarker, such as amyloid plaque in the case of Alzheimer's disease, we cannot expect (nor do the data support) evidence that we can stop the course of the disease, let alone reverse it.

In targeting downstream outcomes, no matter how conscientious we are and no matter how unbounded the funding, we face two intrinsic problems: (1) a host of such outcomes may require multiple independent therapeutic targeting; and (2) we are not targeting the underlying disease itself, which progresses regardless of our efficacy in targeting only downstream clinical outcomes. Although some might suggest that greater efficacy might be obtained if we use a shotgun approach (targeting amyloid, tau, NfL, mitochondrial dysfunction, and dozens of other biomarkers simultaneously and en masse), that these are outcomes rather than the underlying disease process suggests that the result would be much like targeting the outcomes of Ebola virus infection (fever, pain, bleeding, vomiting, diarrhea, etc.) using a shotgun approach (antipyretics, antiinflammatories, blood products, antiemetics, antidiarrheals, etc.) rather than targeting the underlying disease (with an Ebola vaccine). No matter how many downstream outcomes (biomarkers, symptoms, signs, etc.) we target, the disease process continues unabated and drives the clinical pathology.

Much the same problem applies to targeting upstream risk factors. Current standards of preventative medicine, for example, recommend that we can, should, and have attempted to lower behavioral risks by avoiding head trauma, infections, toxic exposures, and so on, as well as treating comorbid [83] conditions in age-related dementias. These and many similar recommendations may ameliorate the onset or rate of progress of age-related neurodegenerative diseases, but there is no evidence that adherence to these recommendations can fully prevent disease, let alone reverse it. Although gene therapy is only beginning to demonstrate efficacy in altering the upstream genetic risks of such diseases, it is equally unlikely that gene therapy will be able to prevent or cure age-related neurodegenerative diseases fully. In addressing upstream risk factors, whether behavioral or genetic, the same two basic problems occur with downstream risk factors: (1) a host of upstream risk factors may require multiple independent targeting; and (2) we are not targeting the disease process itself, which progresses regardless of our efficacy in targeting only upstream clinical outcomes.

Regardless of how effectively we target either downstream or upstream aspects of age-related neurodegenerative diseases, the fundamental process of cell aging will proceed apace and even under the best of circumstances will proceed to neurologic disease as the patient undergoes sufficient aging. The only point at which we might effectively intervene to stop and reverse age-related neurodegenerative disease is cell aging itself.

9. Summary

If we are to intervene effectively in age-related neurodegenerative disease, we must understand the full complexity of the pathology involved in these diseases. An outline of this understanding is thus::

(1) Age-related neurodegenerative diseases are fundamentally driven by cell aging.
(2) Upstream risk factors and downstream biomarkers have a role in the disease process but are insufficiently effective as optimal points of intervention.
(3) Cell aging itself is the optimal point of intervention if we intend to cure or prevent age-related neurologic diseases.

The implications of such an approach is that we may demonstrate the ability to stop and (to an undetermined extent) reverse age-related neurodegenerative diseases, including the dementias. To paraphrase the opening lines of this chapter, we must learn to fathom such diseases if we are to render them less frightful. The approach outlined here, if proven effective in human trials, promises to lower both the human and financial costs of these diseases.

References

[1] Libertini G. Programmed aging paradigm and aging of perennial neurons, ch. 5. In: Ahmad SI, editor. Aging. Exploring a complex phenomenon. CRC Press; 2017.
[2] Fossel M. Cells, aging, and human disease. New York: Oxford University Press; 2004.
[3] Alzheimer A. Uber eine eigenartige Erkrankung der Hirnrinde. Allgemeine Zeitschrift fur Psychiatrie und phychish-Gerichtliche Medizin 1907;64:146—8.
[4] Yang HD, Kim DH, Lee SB, Young LD. History of Alzheimer's disease. Dement Neurocogn Disord 2016;15(4):115—21. https://doi.org/10.12779/dnd.2016.15.4.115.
[5] Taheri-Targhu S, Gjedde A, Araj-Khodaei M, Rikhtegar R, Parsian Z, Zarrintan S, et al. Avicenna (980—1037 CE) and his early description and classification of dementia. J Alzheim Dis 2019;71(4):1093—8. https://doi.org/10.3233/JAD-190345.
[6] Li S, Wu Z, Le W. Traditional Chinese medicine for dementia. Alzheimers Dement. 2021;17(6):1066—71. https://doi.org/10.1002/alz.12258.
[7] Manyam BV. Dementia in ayurveda. J Altern Complement Med 1999;5(1):81—8. https://doi.org/10.1089/acm.1999.5.81.
[8] Jellinger KA. Should the word 'dementia' be forgotten? J Cell Mol Med 2010;14(10):2415—6. https://doi.org/10.1111/j.1582-4934.2010.01159.x.
[9] Oveisgharan S, Dawe RJ, Yu L, Kapasi A, Arfanakis K, Hachinski V, et al. Frequency and underlying pathology of pure vascular cognitive impairment. JAMA Neurol October 24, 2022;79(12):1277—86. https://doi.org/10.1001/jamaneurol.2022.3472.
[10] Ayrignac X, Carra-Dallière C, Marelli C, Taïeb G, Labauge P. Adult-onset genetic central nervous system disorders masquerading as acquired neuroinflammatory disorders: a review. JAMA Neurol 2022;79(10):1069—78. https://doi.org/10.1001/jamaneurol.2022.2141.
[11] Nelson PT, Dickson DW, Trojanowski JQ, Jack CR, Boyle PA, Arfanakis K, et al. Limbic-predominant age-related TDP-43 encephalopathy (LATE): consensus working group report. Brain 2019;142(6):1503—27. https://doi.org/10.1093/brain/awz099.
[12] Le Guen Y, Belloy ME, Grenier-Boley B, de Rojas I, Castillo-Morales A, Jansen I, et al. Association of rare APOE missense variants V236E and R251G with risk of alzheimer disease. JAMA Neurol 2022;79(7):652—63. https://doi.org/10.1001/jamaneurol.2022.1166.
[13] Rabinovici GD, Dubal DB. Rare APOE missense variants—can we overcome APOE ε4 and alzheimer disease risk? JAMA Neurol 2022;79(7): 649—51. https://doi.org/10.1001/jamaneurol.2022.0854.
[14] Rubin R. Much anticipated alzheimer disease prevention trial finds no clinical benefit from drug targeting amyloid; highlights need to consider other approaches. JAMA 2022;328(10):907—10. https://doi.org/10.1001/jama.2022.11490.
[15] Moni F, Petersen ME, Zhang F, Lao PJ, Zimmerman ME, Gu Y, et al. Probing the proteome to explore potential correlates of increased Alzheimer's-related cerebrovascular disease in adults with down syndrome. Alzheimers Dement. 2022;18:1744—53. https://doi.org/10.1002/alz.12627.
[16] Janelidze S, Christian BT, Price J, Laymon C, Schupf N, Klunk WE, et al. Detection of brain tau pathology in down syndrome using plasma biomarkers. JAMA Neurol 2022;79(8):797—807. https://doi.org/10.1001/jamaneurol.2022.1740.
[17] Ming C, Wang M, Wang Q, Neff R, Wang E, Shen Q, et al. Whole genome sequencing—based copy number variations reveal novel pathways and targets in Alzheimer's disease. Alzheimers Dement. 2022;18:1846—67. https://doi.org/10.1002/alz.12507.
[18] Reyes-Dumeyer D, Faber K, Vardarajan B, Goate A, Chao M, Boeve B, et al. The national institute on aging late-onset Alzheimer's disease family based study: a resource for genetic discovery. Alzheimers Dement. 2022;18:889—1897. https://doi.org/10.1002/alz.12514.
[19] Gao Y, Felsky D, Reyes-Dumeyer D, Sariya S, Rentería MA, Ma Y, et al. Integration of GWAS and brain transcriptomic analyses in a multi-ethnic sample of 35,245 older adults identifies DCDC2 gene as predictor of episodic memory maintenance. Alzheimers Dement 2022;18: 1797—811. https://doi.org/10.1002/alz.12524.
[20] Schramm C, Wallon D, Nicolas G, Charbonnier C. What contribution can genetics make to predict the risk of Alzheimer's disease? Rev Neurol 2022;178(5):414—21. https://doi.org/10.1016/j.neurol.2022.03.005.
[21] Tang YP, Gershon E. Genetic studies in Alzheimer's disease. Dialogues Clin Neurosci 2022;5(1):17—26. https://doi.org/10.31887/DCNS.2003.5.1/yptang.
[22] Toledo JB, Abdelnour C, Weil RS, Ferreira D, Rodriguez-Porcel F, Pilotto A, et al. Dementia with Lewy bodies: impact of co-pathologies and implications for clinical trial design. Alzheimers Dement October 14, 2022. https://doi.org/10.1002/alz.12814.
[23] Harris E. Large autopsy study estimates prevalence of "LATE" neuropathologic change. JAMA 2022;328(9):815—6. https://doi.org/10.1001/jama.2022.11513.
[24] Abbasi J. Debate sparks over LATE, a recently recognized dementia. JAMA 2019;322(10):914—6. https://doi.org/10.1001/jama.2019.12232.
[25] Lindsay J, Laurin D, Verreault R, Hebert R, Helliwell B, Hill GB, et al. Risk factors for Alzheimer's disease: a prospective analysis from the Canadian study of health and aging. Am J Epidemiol 2002;156:445—53. https://doi.org/10.1093/aje/kwf074.
[26] Riedel BC, Thompson PM, Brinton RD. Age, APOE and sex: triad of risk of Alzheimer's disease. J Steroid Biochem Mol Biol 2016;160:134147. https://doi.org/10.1016/j.jsbmb.2016.03.012.

[27] Pontecorvo MJ, Lu M, Burnham SC, Schade AE, Dage JL, Shcherbinin S, et al. Association of donanemab treatment With exploratory plasma biomarkers in early symptomatic Alzheimer disease - a secondary analysis of the TRAILBLAZER-ALZ randomized clinical trial. JAMA Neurol October 17, 2022;79(12):1250–9. https://doi.org/10.1001/jamaneurol.2022.3392.

[28] Strikwerda-Brown C, Hobbs DA, Gonneaud J, St-Onge F, Binette AP, Ozlen H, et al. Association of elevated amyloid and tau positron emission tomography signal with near-term development of alzheimer disease symptoms in older adults without cognitive impairment. JAMA Neurol 2022;79(10):975–85. https://doi.org/10.1001/jamaneurol.2022.2379.

[29] Jack CR, Bennett DA, Blennow K, Carrillo MC, Feldman HH, Frisoni GB, et al. A/T/N: an unbiased descriptive classification scheme for Alzheimer disease biomarkers. Neurology 2016;87(5):539–47. https://doi.org/10.1212/WNL.0000000000002923.

[30] https://clinicaltrials.gov.

[31] Armstrong MJ, Okun MS. Diagnosis and treatment of Parkinson disease - a review. JAMA 2020;323(6):548–60. https://doi.org/10.1001/jama.2019.22360.

[32] Fox SH, Katzenschlager R, Lim SY, Barton B, de Bie RMA, Seppi K, et al. Movement disorder society evidence-based medicine committee. International parkinson and movement disorder society evidence-based medicine review: update on treatments for the motor symptoms of Parkinson's disease. Mov Disord 2018;33(8):1248–66. https://doi.org/10.1002/mds.27372.

[33] Ross EL, Weinberg MS, Arnold SE. Effectiveness—essential for cost-effectiveness—reply. JAMA Neurol October 03, 2022;79(11):1205–6. https://doi.org/10.1001/jamaneurol.2022.3107.

[34] Mattke S. Cost-effectiveness of aducanumab and donanemab for early Alzheimer disease—estimating the true value. JAMA Neurol October 03, 2022;79(11):1204. https://doi.org/10.1001/jamaneurol.2022.3104.

[35] Teng E, Manser PT, Pickthorn K, Brunstein f, Blendstrup M, Bohorquez SS, et al. Safety and efficacy of semorinemab in individuals with prodromal to mild alzheimer disease: a randomized clinical trial. JAMA Neurol 2022;79(8):758–67. https://doi.org/10.1001/jamaneurol.2022.1375.

[36] Ostrowitzki S, Bittner T, Sink KM, Mackey H, Rabe C, Honig LS, et al. Evaluating the safety and efficacy of crenezumab vs placebo in adults with early Alzheimer disease: two phase 3 randomized placebo-controlled trials. JAMA Neurol September 19, 2022;79(11):1113–21. https://doi.org/10.1001/jamaneurol.2022.2909.

[37] Holloway RG, Kramer NM. Advancing the neuropalliative care approach—a call to action. JAMA Neurol October 17, 2022;80(1):7–8. https://doi.org/10.1001/jamaneurol.2022.3418.

[38] Jack CR, Knopman DS, Jagust WJ, Shaw LM, Aisen PS, Weiner MW, et al. Hypothetical model of dynamic biomarkers of the Alzheimer's pathological cascade. Lancet Neurol 2010;9(1):119–28. https://doi.org/10.1016/S1474-4422(09)70299-6.

[39] Shcherbinin S, Evans CD, Lu M, Anderson SW, Pontecorvo MJ, Willis BA, et al. Association of amyloid reduction after donanemab treatment with tau pathology and clinical outcomes: the TRAILBLAZER-ALZ randomized clinical trial. JAMA Neurol 2022;79(10):1015–24. https://doi.org/10.1001/jamaneurol.2022.2793.

[40] Ozlen H, Pichet Binette A, Köbe T, Meyer PF, Gonneaud J, St-Onge F, et al. Spatial extent of amyloid-β levels and associations with tau-PET and cognition. JAMA Neurol 2022;79(10):1025–35. https://doi.org/10.1001/jamaneurol.2022.2442.

[41] Njie EG, Boelen E, Stassen FR, Steinbusch HWM, Borchelt DR, Streit WJ. Ex vivo cultures of microglia from young and aged rodent brain reveal age-related changes in microglial function. Neurobiol Aging 2012;33:195.e1–195.e12. https://doi.org/10.1016/j.neurobiolaging.2010.05.008.

[42] Langston RG, Beilina A, Reed X, Kaganovich A, Singleton AB, Bleauwendraat C, et al. Association of a common genetic variant with Parkinson's disease is mediated by microglia. Sci Transl Med 2022;14:655. https://doi.org/10.1126/scitranslmed.abp8869.

[43] Edison P, Donat CK, Sastre M. In vivo imaging of Glial activation in Alzheimer's disease. Front Neurol 2018;9:625. https://doi.org/10.3389/fneur.2018.00625.

[44] Nordengen K, Kirsebom BE, Henjum K, Selnes P, Gísladóttir B, Wettergreen M, et al. Glial activation and inflammation along the Alzheimer's disease continuum. J Neuroinflammation 2019;16:46. https://doi.org/10.1186/s12974-019-1399-2.

[45] Hickman SE, Allison EK, Khoury JE. Microglial dysfunction and defective beta-amyloid clearance pathways in aging Alzheimer's disease mice. J Neurosci 2008;28:8354–60. https://doi.org/10.1523/JNEUROSCI.0616-08.2008.

[46] Takata K, Kitamura Y, Yanagisawa D, Morikawa S, Morita M, Inubushi T, et al. Microglial transplantation increases amyloid-β clearance in Alzheimer model rats. FEBS Lett 2007;581:343–580. https://doi.org/10.1016/j.febslet.2007.01.009.

[47] Ennerfelt H, Frost EL, Shapiro DA, Holliay C, Zengeler KE, Voithofer G, et al. SYK coordinates neuroprotective microglial responses in neurodegenerative disease. Cell 2022;185:1–18. https://doi.org/10.1016/j.cell.2022.09.030.

[48] Mawuenyega KG, Sigurdson W, Ovod V, Ling M, Kasten T, Morris JC, et al. Decreased clearance of CNS β-amyloid in Alzheimer's disease. Science 2010;330:1774. https://doi.org/10.1126/science.1197623.

[49] Takatori S, Wang W, Iguchi A, Tomita T. Genetic risk factors for Alzheimer disease: emerging roles of microglia in disease pathomechanisms. In: Guest P, editor. Reviews on biomarker studies in psychiatric and neurodegenerative disorders. Advances in experimental medicine and biology, vol 1118. Cham: Springer; 2019. https://doi.org/10.1007/978-3-030-05542-4_5.

[50] Hernandez I, Luna G, Rauch JN, Reis SA, Giroux M, Karch CM, et al. A farnesyltransferase inhibitor activates lysosomes and reduces tau pathology in mice with tauopathy. Sci Transl Med 2019;11(485):eaat3005. https://doi.org/10.1126/scitranslmed.aat3005.

[51] Gasparoni G, Bultmann S, Lutsik P, Kraus TFJ, Sordon S, Vlcek J, et al. DNA methylation analysis on purified neurons and glia dissects age and Alzheimer's disease-specific changes in the human cortex. Epigenet Chromatin 2018;11:41. https://doi.org/10.1186/s13072-018-0211-3.

[52] Pluvinage JV, Haney MS, Smith BAH, Sun J, Iran T, Bonanno L, et al. CD22 blockade restores homeostatic microglial phagocytosis in ageing brains. Nature 2019;568:187–92. https://doi.org/10.1038/s41586-019-1088-4.

[53] Jack C, Knopman D, Chételat G, Dickson D, Fagan AM, Frisoni GB, et al. Suspected non-Alzheimer disease pathophysiology – concept and controversy. Nat Rev Neurol 2016;12:117–24. https://doi.org/10.1038/nrneurol.2015.251.

[54] Nilsson J, Cousins KAQ, Gobom J, Portelius E, Chen-Plotkin A, Shaw LM, et al. Cerebrospinal fluid biomarker panel of synaptic dysfunction in Alzheimer's disease and other neurodegenerative disorders. Alzheimers Dement October 14, 2022. https://doi.org/10.1002/alz.12809.

[55] Crary JF, Trojanowski JQ, Schneider JA, Abisambra JF, Abner EL, Alafuzoff I, et al. Primary age-related tauopathy (PART): a common pathology associated with human aging. Acta Neuropathol 2014;128:755–66. https://doi.org/10.1007/s00401-014-1349-0.

References

[56] Jiang YX, Cao Q, Sawaya MR, Abskharon R, Ge P, DeTure M, et al. Amyloid fibrils in FTLD-TDP are composed of TMEM106B and not TDP-43. Nature 2022;605:304—9. https://doi.org/10.1038/s41586-022-04670-9.

[57] Majd S, Power JH, Grantham HJM. Neuronal response in Alzheimer's and Parkinson's disease: the effect of toxic proteins on intracellular pathways. BMC Neurosci 2015;16:69. https://doi.org/10.1186/s12868-015-0211-1.

[58] Novotny BC, Fernandez MV, Wang C, Budde JP, Bergmann K, Eteleeb AM, et al. Metabolomic and lipidomic signatures in autosomal dominant and late-onset Alzheimer's disease brains. Alzheimers Dement October 17, 2022. https://doi.org/10.1002/alz.12800.

[59] Pimenova AA, Raj T, Goate AM. Untangling genetic risk for Alzheimer's disease. Biol Psychiatr 2018;83(4):300—10. https://doi.org/10.1016/j.biopsych.2017.05.014.

[60] Lee VM, Goedert M, Trojanowski JQ. Neurodegenarative tauopathies. Ann Rev Neurosci 2001;24:1121. https://doi.org/10.1146/annurev.neuro.24.1.1121.

[61] Herrera-Rivero M, Hofer E, Maceski A, Leppert D, Benkert P, Kuhle J, et al. Evidence of polygenic regulation of the physiological presence of neurofilament light chain in human serum. Research Square 2022. https://doi.org/10.21203/rs.3.rs-422221/v1. preprint.

[62] Prahl J, Coetzee GA. Genetic elements at the alpha-synuclein locus. Front Neurosci 2022;16:889802. https://doi.org/10.3389/fnins.2022.889802.

[63] Gusella JF, MacDonald ME. Huntington's disease: the case for genetic modifiers. Genome Med 2009;1:80. https://doi.org/10.1186/gm80.

[64] Jurcau A, Jurcau MC. Therapeutic strategies in Huntington's disease: from genetic defect to gene therapy. Biomedicines 2022;10(8):1895. https://doi.org/10.3390/biomedicines1008185.

[65] Nopoulos PC. Huntington disease: a single-gene degenerative disorder of the striatum. Dialogues Clin Neurosci 2022;18(1):91—8. https://doi.org/10.31887/DCNS.2016.18.1/pnopoulos.

[66] Oh YM, Lee SW, Kim WK, Chen S, Church VA, Cates K, et al. Age-related Huntington's disease progression modeled in directly reprogrammed patient-derived striatal neurons highlights impaired autophagy. Nat Neurosci 2022. https://doi.org/10.1038/s41593-022-01185-4.

[67] Ou SH, Wu F, Harrich D, García-Martínez LF, Gaynor RB. Cloning and characterization of a novel cellular protein, TDP-43, that binds to human immunodeficiency virus type 1 TAR DNA sequence motifs. J Virol 1995;69(6):3584—96. https://doi.org/10.1128/JVI.69.6.3584-3596.

[68] Rattan SI, Derventzi A, Clark BF. Protein synthesis, posttranslational modifications, and aging. Ann N Y Acad Sci 1992;663:48—62. https://doi.org/10.1111/j.1749-6632.1992.tb38648.x.

[69] Fossel MB. See chapter 3: Cell senescence in aging. In: Cells, aging, and human disease. New York: Oxford University Press; 2004.

[70] Fossel M. A unified model of dementias and age-related neurodegeneration. Alzheimer's Dementia J Alzheimer's Assoc 2020;16:365—83. https://doi.org/10.1002/alz.12012.

[71] Zakariya SM, Zehra A, Khan RH. Biophysical insight into protein folding, aggregate formation and its inhibition strategies. Protein Pept Lett 2022;29(2):22—36. https://doi.org/10.2174/0929866528666211125114421.

[72] Upadhyay A, Sundaria N, Dhiman R, Prajapati VK, Prasad A, Mishra A. Complex inclusion bodies and defective proteome hubs in neurodegenerative disease: new clues, new challenges. Neuroscientist 2022;28(3). https://doi.org/10.1177/1073858421989582.

[73] Cummings J, Lee G, Zhong K, Fonseca J, Taghva K. Alzheimer's disease drug development pipeline: 2021. Alzheimers Dement (N Y). 2021; 7(1):e12179. https://doi.org/10.1002/trc2.12179. 25.

[74] https://clinicaltrials.gov/ct2/results?cond=Dementia&term=&type=Intr&rslt=&age_v=&gndr=&intr=&titles=&outc=&spons=&lead=&id=&cntry=&state=&city=&dist=&locn=&rsub=&strd_s=&strd_e=&prcd_s=&prcd_e=&sfpd_s=&sfpd_e=&rfpd_s=&rfpd_e=&lupd_s=&lupd_e=&sort=.

[75] Watson JL, Ryan L, Silverberg N, Cahan V, Bernard MA. Obstacles and opportunities in Alzheimer's clinical trial recruitment. Health Aff 2014;33(4):574—9. https://doi.org/10.1377/hlthaff.2013.1314.

[76] Cummings J, Reiber C, Kumar P. The price of progress: funding and financing Alzheimer's disease drug development. Alzheimers Dement 2018;4:330—43. https://doi.org/10.1016/j.trci.2018.04.008.

[77] Khachaturian AS, Hayden KM, Mielke MM, Tang Y, Lutz MW, Gold M, et al. New thinking about thinking, part two. Theoretical articles for Alzheimer's & Dementia. Alzheimers Dement 2018;14:703—6. https://doi.org/10.1016/j.jalz.2018.05.002.

[78] Khachaturian ZS. Editor-in-Chief of Alzheimer's & Dementia, Personal communication. November 2018.

[79] Khachaturian ZS. Perspectives on Alzheimer's disease: past, present and future. In: Carrillo MC, Hampel H, editors. Alzheimer's disease — modernizing concept, biological diagnosis and therapy. Advances in biological psychiatry, vol 28. Basel: Karger; 2012. p. 179—88.

[80] Nurse P. Life, logic and information. Nature 2008;454:424—6. https://doi.org/10.1038/454424a.

[81] Miller MB, Huang AY, Kim J, Zhou Z, Kirkham SL, Maury EA, et al. Somatic genomic changes in single Alzheimer's disease neurons. Nature 2022;604:714—22. https://doi.org/10.1038/s41586-022-04640-1.

[82] Khachaturian ZS, Mesulam MM, Khachaturian AS, Mohs RC. Editorial: the special topics section of Alzheimer's & dementia. Alzheimers Dement 2015;11:1261—4. https://doi.org/10.1016/j.jalz.2015.10.002.

[83] Calvin CM, Conroy MC, Moore SF, Kuźma E, Littlejohns TJ. Association of multimorbidity, disease clusters, and modification by genetic factors with risk of dementia. JAMA Netw Open 2022;5(9):e2232124. https://doi.org/10.1001/jamanetworkopen.2022.32124.

[84] Fossel M. A unified model of age-related disease. OBM Geriatrics January 2020. https://doi.org/10.21926/obm.geriatr.2001100.

[85] Aisen PS, Cummings J, Jack Jr CR, Morris JC, Sperling R, Frölich L, et al. On the path to 2025: understanding the Alzheimer's disease continuum. Alzheimers Res Ther 2017;9(1):60. https://doi.org/10.1186/s13195-017-0283-5.

[86] Urbanc B. Cross-linked amyloid β-protein oligomers: a missing link in Alzheimer's disease pathology? J Phys Chem B 2021;125(5):1307—16. https://doi.org/10.1021/acs.jpcb.0c07716.

[87] Ramdhani S, Navarro E, Udine E, Schilder BM, Parks M, Raj T. Tensor decomposition of stimulated monocyte and macrophage gene expression profiles identifies neurodegenerative disease-specific trans-eQTLs. BioRxiv 2018. https://doi.org/10.1101/499509.

[88] Dyer M, Phipps AJ, Mitew S, Taberlay PC, Woodhouse A. Age, but not amyloidosis, induced changes in global levels of histone modifications in susceptible and disease-resistant neurons in Alzheimer's disease model mice. Front Aging Neurosci 2019;11:68. https://doi.org/10.3389/fnagi.2019.00068.

[89] Katsumata Y, Nelson PT, Estus S. The Alzheimer's Disease Neuroimaging Initiative (ADNI), Fardo DW. Translating Alzheimer's disease–associated polymorphisms into functional candidates: a survey of IGAP genes and SNPs. Neurobiol Aging 2019;74:135—46. https://doi.org/10.1016/j.jalz.2018.05.002.

[90] Villapol S. Neuropathology of traumatic brain injury and its role in the development of Alzheimer's disease. In: Amyloidosis. IntechOpen; 2018. https://doi.org/10.5772/intechopen.81945.

[91] Becker RE, Kapogiannis D, Greig NH. Does traumatic brain injury hold the key to the Alzheimer's disease puzzle? Alzheimers Dement 2018; 14:431–43. https://doi.org/10.1016/j.jalz.2017.11.007.

[92] Stern RA, Adler CH, Chen K, Navitsky M, Luo J, Dodick DW, et al. Tau positron-emission tomography in former National Football League players. N Engl J Med 2019;380:1716–25. https://doi.org/10.1056/NEJMoa1900757.

[93] Itzhaki RF, Lathe R, Balin BJ, Ball MJ, Bearer EL, Braak H, et al. Microbes and Alzheimer's disease. J Alzheimers Dis. 2016;51:979–84. https://doi.org/10.3233/JAD-160152.

[94] Dominy SS, Lynch C, Ermini F, Benedyk M, Marczyk A, Konradi AW, et al. Porphyromonas gingivalis in Alzheimer's disease brains: evidence for disease causation and treatment with small-molecule inhibitors. Sci Adv 2019;5. https://doi.org/10.1126/sciadv.aau3333.

[95] Douaud G, Lee S, Alfaro-Almagro F, Arthofer C, Wang C, McCarthy P, et al. SARS-CoV-2 is associated with changes in brain structure in UK biobank. Nature 2022;604:697–707. https://doi.org/10.1038/s41586-022-04569-5.

[96] Levine KS, Leonard HL, Blauwendraat C, Iwaki H, Johnson N, Bandres-Ciga S, et al. Virus exposure and neurodegenerative disease risk across national biobanks. Neuron January 19, 2023. https://doi.org/10.1016/j.neuron.2022.12.029.

[97] Geraghty AC, Gibson EM, Ghanem RA, Greene JJ, Ocampo A, Goldstein AK, et al. Loss of adaptive myelination contributes to methotrexate chemotherapy-related cognitive impairment. Neuron 2019;103(2):250–265.e8. https://doi.org/10.1016/j.neuron.2019.04.032.

[98] de la Monte SM, Tong M, Wands JR. The 20-year voyage aboard the journal of Alzheimer's disease: docking at 'Type 3 Diabetes', environmental/exposure factors, pathogenic mechanisms, and potential treatments. J Alzheimers Dis. 2018;62:1381–90. https://doi.org/10.3233/JAD-170829.

[99] Sarkar S, Rokad D, Malovic E, Luo J, Harischandra DS, Jin H, et al. Manganese activates NLRP3 inflammasome signaling and propagates exosomal release of ASC in microglial cells. Sci Signal 2019;12(563):eaat9900. https://doi.org/10.1126/scisignal.aat9900.

[100] Kesler SR, Rao V, Ray WJ, Rao A, Alzheimer's Disease Neuroimaging Initiative. Probability of Alzheimer's disease in breast cancer survivors based on gray-matter structural network efficiency. Neuroimaging 2017;9:67–75. https://doi.org/10.1016/j.dadm.2017.10.002.

[101] Okun MS, Mayberg HS, DeLong MR. Muhammad Ali and young-onset idiopathic parkinson disease—the missing evidence. JAMA Neurol 2023;80(1):5–6. https://doi.org/10.1001/jamaneurol.2022.3584.

[102] Atherton K, Han X, Chung J, Cherry JD, Baucom Z, Saltiel N, et al. Association of APOE genotypes and chronic traumatic encephalopathy. JAMA Neurol 2022;79(8):787–96. https://doi.org/10.1001/jamaneurol.2022.1634.

[103] Banks D, Fossel M. Telomeres, cancer, and aging: altering the human lifespan. JAMA 1997;278:1345–8. https://doi.org/10.1001/jama.1997.03550160065040.

[104] Fossel M. Telomerase and the aging cell: implications for human health. JAMA 1998;279:1732–5. https://doi.org/10.1001/jama.279.21.1732.

[105] Bhatia-Dey N, Kanherkar RR, Stair SE, Makarev EO, Csoka AB. Cellular senescence as the causal nexus of aging. Front Genet 2016;7:13. https://doi.org/10.3389/fgene.2016.00013.

[106] Fossel M. The telomerase revolution. Dallas: BenBella Books; 2015.

CHAPTER 3

Age-related disease: Cardiovascular system

Peter M. Nilsson[1], Michael Fossel[2], Joe Bean[4] and Nina Khera[3]

[1]Clinical Sciences, Lund University, Malmö, Sweden [2]Telocyte LLC, Ada, MI, United States [3]Buckingham Browne and Nichols School, Cambridge, MA, United States [4]University of Missouri School of Medicine, Kansas City, MO, United States

The cardiovascular system is shaped by evolution [1] and supplies the physiologic demands of oxygen and nutrients. It also adapts to physical work, upright standing, and strains imposed by hypovolemic shock during bleeding, thirst, and fluid loss, for example. These hemodynamic reflexes are vitally important during pregnancy and labor, because pregnant women represent the most important link in the evolutionary chain for the survival and continuation of humanity. During aging, some adaptations occur within normal physiology, such as stiffening of the arteries, less active hemodynamic reflexes, and changes within cardiac and renal functions of importance for the cardiovascular system as a whole [2]. When these changes become more pronounced, clinical consequences and complications may develop, such as cardiovascular disease (CVD). A common phenomenon is the development of essential hypertension with linearly increasing systolic blood pressure over decades, but with increasing diastolic blood pressure only until about age 60–70 years, and then a decrease, increasing the pulse pressure as a marker of arterial aging and stiffness [3], in itself a risk marker for future CVD [4]. Moreover, changes occur in the arterial wall of both large elastic arteries (aorta, carotids, and femoral) and smaller arteries (e.g., in the coronaries). *Atherosclerosis* starts in the arterial intima with lipid deposits, chronic inflammation, and plaque formation [5], but in the arterial media another process starts, so-called *arteriosclerosis* (arterial stiffness) [6]. Both processes are independent risk markers for CVD but both contribute to a more general pathologic change of the arterial wall with increasing age. Several studies have described the process of atherosclerosis, its risk factors, and possible treatment, but increased interest has focused on the determinants and consequences of arteriosclerosis (arterial stiffness), often referred to as early vascular aging (EVA), first described in 2008 [7,8]. This has been followed by a variety of studies exploring cardiovascular aging as a fruitful concept to look for new mechanisms and treatment targets [9]. The core component of EVA is supposed to be arterial stiffness, as measured by an elevated carotid-femoral pulse wave velocity (c-f PWV) along the aorta. It is possible to measure c-f PWV directly with some technical devices [10,11]. Aortic distensibility can also be determined using carotid ultrasound and magnetic resonance imaging. Other indirect methods try to provide an estimate of pulse wave velocity (PWV), but inherent technical shortcomings may preclude a correct assessment obtained by direct methods. In a study in patients undergoing coronary angiography who had central hemodynamics measured at the same time, it was shown that most devices providing a direct measurement of central hemodynamic and aortic c-f PWV were reliable, but not so for all devices providing indirect measurements (for example, using age and systolic blood pressure alone in an algorithm for an indirect estimation of aortic PWV) [12].

1. Determinants of arterial stiffness

Population-based studies have shown that close correlations exist between the level of blood pressure and the degree of arterial stiffness (aortic PWV). The higher the brachial or central blood pressure is, the higher is the aortic PWV. Because hemodynamic factors have an important role in the morphology and functionality of the arterial wall, hypertension is a crucial factor in determining the level of arterial stiffness c-f PWV [13]. However, some studies also documented that several nonhemodynamic factors linked to glucose metabolism, chronic inflammation, and renal impairment are also important for PWV [14,15]. Prospective studies indicated that increased blood pressure predicts future arterial stiffness (PWV), but also that arterial stiffness can predict incident hypertension [16,17] as well as incident type 2 diabetes [18]. This shows the close association between these entities. In fact, genetic studies documented that a genetic risk score for hyperglycemia, as a phenotypic trait in the normal, nondiabetic, elderly population, is independently associated with arterial stiffness (c-f PWV) [19]. If a true causal mechanism exists, this implies that a more focused treatment directed toward these mechanisms, such as using newer antidiabetic drugs, could also alleviate arterial stiffness beyond the effect of blood pressure lowering per se.

2. How to define early vascular aging

The definition of EVA has been discussed [20], but to date, no established definition is available. However, because increased c-f PWV is the core feature of EVA, one can try to define EVA as the upper 10%, 20%, or 25% of the c-f PWV distribution in relation to a background population. For example, a reference population in Europe with a PWV measurement exists for comparison [21]. Several studies and meta-analyses documented that PWV is predictive of fatal and nonfatal CVD events, but also of total mortality [22–24]. It seems that the predictive power of PWV is more pronounced in middle-aged subjects compared with the elderly population, for whom selective survival bias may influence the observational findings [23]. Differential aging in general, and of the vasculature in particular, may be more visible in middle-aged subjects, and this is linked to risk. European guidelines therefore mention the usefulness of determining arterial stiffness (PWV), although with a lower level of evidence than, for example, the measurement of blood pressure, an established risk factor for which intervention studies show clear benefits [25]. Because impaired glucose metabolism is closely associated with vascular aging when measured by PWV [14,15] it makes sense to offer the measurement of fasting glucose, HbA_{1c}, or even an oral glucose tolerance test (75 g glucose) to subjects at risk, such as after a myocardial infarction. Even if chronic inflammation seems to be important for EVA, there is no consensus regarding whether inflammatory biomarkers should be measured in the clinic, including fibrinogen, high-sensitive C-reactive protein, or interleukins.

3. Mechanisms important for modifying arterial stiffness and vascular aging

Several studies documented the association of chronic inflammation with increased arterial stiffness and PWV as measures of vascular aging, such as in rheumatoid arthritis (RA) and inflammatory bowel disease (i.e., ulcerous colitis and morbus Crohn's disease) [26]. It is suggested that chronic inflammation and increased oxidative stress will negatively affect the proteins and structure of the arterial wall, but also impair vasodilation. A review documented the importance of vascular smooth muscle cell (VSMC) changes in relation to arterial stiffness [27]. Those authors state that the first most important components that contribute to arterial stiffening are extracellular matrix (ECM) proteins that support the mechanical load, whereas the second most important components are VSMCs, which regulate actomyosin interactions for contraction but also mediate mechanotransduction in cell–ECM homeostasis. VSMC plasticity and signaling in both conductance and resistance arteries are highly relevant to the physiology of normal aging and EVA [27]. This process also involves the architecture of cytoskeletal proteins and focal adhesion, the large and small artery cross-talk that results in target organ damage and inflammatory pathways leading to calcification or atherosclerosis [27].

4. Factors in early life influencing arterial stiffness and early vascular aging

Another aspect of EVA is the hypothesis that early life factors such as fetal growth, birth weight adjusted for gestational age, prematurity, and postnatal growth patterns may influence both adult blood pressure regulation and arterial stiffness, as measured by PWV [28] or the augmentation index (Aix), another more complex marker of stiffness and central hemodynamics, as well as total peripheral resistance [29]. In a study from Austria, these early life factors were predictive of EVA in adolescents (mean age, 16 years) [30]. An ongoing debate aims to resolve the controversy regarding whether genetic factors (nature) link the parental (maternal) trait of hypertension with the same trait in the offspring, when low birth weight is just a marker of the trait [31], or whether environmental factors (nurture) have the most important role (i.e., maternal diet [32] calorie intake, and lifestyle [smoking and alcohol] for these associations). Probably genetic factors form a background structure, whereas environmental factors can have a modifying role (epigenetics) for the phenotypic outcome.

4.1 Vascular aging and the brain

Hypertension is a well-documented risk factor for stroke and other cerebrovascular disease manifestations such as microangiopathy and white matter lesions, which often affect elderly people. Also, arterial stiffness can contribute to these pathologies in different ways [33,34], because stiffness means that the pulse wave energy is not dampened but instead transmitted to the microcirculation of the brain, causing microangiopathy. One result is impaired cognition and an increased risk of vascular dementia in these patients. Thus, hypertension and EVA may cause more damage in the population at large than is visible solely from the hospital statistics of stroke. If impaired cognition and dementia occur several years earlier than in people without these risk conditions, this will substantially affect activities of daily living capacities and the independence of care in aging populations.

4.2 Treatment of vascular aging and its consequences

The treatment of EVA and hypertension is based on an improved healthy lifestyle and the treatment of conventional risk factors, based on current best evidence, as shown in both European [25] and US guidelines [35]. Several observational studies indicated that antihypertensive treatment may reduce arterial stiffness (PWV) beyond reducing blood pressure reduction, when blockers of the renin-angiotensin-aldosterone system in particular seem to be greatly valuable [36]. In the SPRINT, study an estimated PWV (ePWV) was lowered relatively more than the intensive blood pressure control implicated in the intervention arm [37]. These results suggest that in this trial, ePWV predicted outcomes independent of the Framingham Risk Score (which includes blood pressure), indicating an increased role of markers of aortic stiffness on cardiovascular risk. The authors concluded that the better survival of individuals whose ePWV responded to antihypertensive treatment independently of reduced systolic blood pressure suggests a role for markers of aortic stiffness as effective treatment targets in individuals with hypertension [37].

Evidence for the usefulness of PWV for risk stratification and as a target for therapy came with the SPARTE study in France, where individuals at risk were randomized to a treatment strategy aimed at lowering PWV or conventional treatment for multiple risk factor intervention and control based on guidelines [38]. The results of the study indicated that a PWV-based strategy was more effective in controlling arterial stiffness over time than the conventional treatment strategy, but SPARTE was underpowered to show effects on clinical end points [38].

It was shown that newer antidiabetes drugs such as the SGLT-2 inhibitor empagliflozin may lower both office and ambulatory blood pressure [39] and have beneficial effects on Aix (a marker of aortic dysfunction) in patients with type 2 diabetes [40].

5. Healthy and supernormal vascular aging

A novel approach to EVA is to turn it around and look for factors that protect from EVA and are associated with healthy vascular aging [41–43], or even supernormal vascular aging [44]. If such protection from vascular aging could be better defined and understood based on advanced phenotyping using genetics and omics, there is the potential to find new drug targets of protection. Other models of protection exist but are poorly understood, including an astonishing lack of major cardiovascular or renal complications in a few patients with type 1 diabetes who had the disease for more than 40–50 year [45–47]. It would be interesting to examine vascular function and the escape from hemodynamic aging in these fortunate subjects [48]. Another example consists of subjects with obesity who were not hospitalized for decades in midlife despite risk factors and drug treatment [49–51]. Some seem to be fat and fit as a way to cope with obesity and its risks. Even if obesity is a strong risk factor for developing type 2 diabetes, many of these subjects with metabolically healthy obesity (HMO) escape diabetes. Although they are not thought to be protected from complications overall, such HMO subjects may benefit from postponing complications for a substantial time if they are able to maintain a nonsedentary lifestyle [52].

6. Treatment of cardiovascular risk and clinical manifestations

The current approach to treating cardiovascular risk after a detailed medical history and risk assessment based on a clinical examination and laboratory measurements is an improvement in lifestyle, followed by drug treatment based on guidelines [53–55]. For more advanced disturbances in blood flow or cardiac function, technical solutions can be considered, including percutaneous coronary interventions, coronary bypass graft surgery [56], and other selected interventions, such as renal denervation of sympathetic nerves along the renal artery for resistant hypertension [57] (Table 3.1).

Among the most recommended lifestyle interventions are physical activity, weight control, smoking cessation, and a heart-friendly diet such as the Mediterranean diet or its equivalents (Dietary Approaches to Stop Hypertension, Nordic Diet, etc.) [58–60]. This traditional lifestyle message must be updated based on evidence and tailored to the right patient, so-called precision prevention or precision nutrition within the framework of personalized medicine [61].

To treat risk factors with drugs, we have accumulated evidence from many randomized clinical trials and national as well as international guidelines regarding, for example, the treatment of *hypertension*. Blood pressure control in general is more important than the way it is achieved, based on the combination of several synergistic drugs sometimes used in fixed drug combinations recommended in guidelines [20]. Among such useful combinations are blockers of the renin-angiotensin-aldosterone system, such as angiotensin converting enzyme inhibitors and angiotensin receptor blockers, thiazide or thiazide-like diuretics (chlorthalidone), as well as calcium antagonists and modern selective β-receptor blockers. Thiazides and β-receptor blockers should be used at lower dosages to avoid metabolic disturbances and weight gain (β-receptor blockers). The recommended blood pressure goal has been much debated, but it should aiming for less than 140/90 mmHg, and for many even lower, less than 130/80 mmHg if well-tolerated [20]. For some patients who are at high risk, this lower and more ambitious blood pressure goal should be the norm, especially for younger patients with diabetes or nephropathy with varying degrees of proteinuria.

For *hyperlipidemia*, we rely on using statins at increasing dosages corresponding to the increasing cardiovascular risk, sometimes combined with cholesterol uptake inhibitors in the gastrointestinal system (resins or ezetimibe), or less often with fibrates. Because there is a linear relationship more or less between the low-density lipoprotein (LDL) cholesterol and cardiovascular risk achieved based on evidence from both clinical studies and genetic analyses (mendelian randomization) [62], it becomes important to aim for more rigorous therapy goals of lowering LDL cholesterol according to how high the cardiovascular risk is [63]. This means a goal of less than 2.5 mmol/L in most, but lower than 1.8 in patients at medium to high risk, and lower than 1.2 mmol/L in patients at very high risk, such as after a myocardial infarction, sometimes combined with diabetes mellitus. Another therapeutic alternative is the use of monoclonal antibody therapy (PCSK-9 inhibitors) [64] or small interfering RNA (siRNA) molecules (inclisiran) to block mechanisms of cholesterol metabolism and achieve impressive reductions in LDL cholesterol [65]. However, the cost of these drugs is relatively high and the safety and effectiveness for siRNA drugs are not fully documented. Nevertheless, several large studies are ongoing for these and effective classes of lipid-lowering drugs.

TABLE 3.1 Features of cardiovascular aging and its clinical consequences: from normal variation to pathophysiology.

- Cellular senescence and telomere attrition
- Age-related blood pressure elevation, and metabolic and renal changes
- Atherosclerosis of the arterial intima larger and smaller arteries
- Arterial remodeling
- Arteriosclerosis of the arterial media in large elastic arteries
- Established hypertension, diabetes mellitus, hyperlipidemia, and nephropathy
- Cardiovascular disease (coronary heart disease, stroke, and peripheral artery disease
- Congestive failure
- Vascular dementia
- Renal failure caused by nephrosclerosis.

For the treatment of *diabetes,* we have several drug alternatives besides insulin for patients with type 1 diabetes as well as for patients with longstanding type 2 diabetes with complications. Modern treatment algorithms for type 2 diabetes, as shown in both European [66] and American [67] guidelines, advocate the increasing use of antidiabetes drugs such as SGLT-2 inhibitors and GLP-1 receptor agonists and analogs (CLP-1 RA), in addition to more traditional alternatives such as metformin, sulphonylurea, thiazolidinediones (glitazones), and glinides. DPP-4 inhibitors also belong to the more modern incretin-promoting drugs like GLP-1 receptor agonists but have not been shown to be more effective than placebo for cardiovascular prevention, the major complication of type 2 diabetes. The reason for this change in drug use favoring SGL-2 inhibitors and GLP-1 receptor agonists is the growing evidence base for cardiovascular and renal prevention associated with the newer drug classes [68,69], but which is mostly lacking for the more traditional drugs. Even if metformin is the mainstay for drug treatment of type 2 diabetes when lifestyle is no longer sufficient to control hyperglycemia, evidence for cardiovascular prevention is at best weak [70]. Nevertheless, metformin can be used in combination treatment with most of the other drugs, provide that renal and hepatic function is within relatively normal limits. Side effects occur associated with all of these drugs, but they seem to be easier to tolerate with the modern drugs in general, even if occasionally problematic or even serious adverse effects can occur with SGLT-2 inhibitors (e.g., urinary tract infection, normoglycemic diabetic ketoacidosis) and GLP-1 receptor agonists (e.g., nausea, vomiting). Development continues to find even more sophisticated combination drugs to influence glucose metabolism favorably, combining incretion action (GLP-1 receptor agonists and GIP receptor agonists) with tirzepatide [71], and glucagon modification, triple therapy. Evidence for cardiovascular prevention is lacking, but large-scale intervention studies are under way.

Another perspective for treatment is to control *chronic inflammation* that often encompasses CVD development. Observational studies in patients with RA who are treated with tumor necrosis factor-α inhibitors to control inflammation reported a lower risk of CVD than expected in many of these patients [72], who were otherwise at increased cardiovascular vascular risk owing to inflammation and limited possibilities for movement and physical activity. A few randomized, controlled studies showed cardiovascular benefits with the use of other antiinflammatory drugs such as canakinumab [73], a monoclonal antibody drug-blocking interleukin 1β with a reduction in the risk of new infarcts in patients already experiencing a first myocardial infarction. Another example is colchicine, a traditional drug to treat hyperuricemia and gout, but also used in patients with RA. Both canakinumab and colchicine have been tested versus placebo and have been efficacious for cardiovascular protection [73,74]. However, that was not the case for a third antiinflammatory drug, methotrexate.

Finally, drug alternatives exist to support *smoking cessation,* besides nicotine replacement therapy (NRT). Some drugs acting on the central nervous system (varenicline and bupropion) can support patients in quitting their smoking habit when NRT and individual- or group-based support is not enough [75]. Adverse effects include depression and even suicide when brain reward systems are blocked and nicotine abstinence becomes severe.

For secondary prevention after coronary artery disease (CAD) and stroke, it is important to prevent thromboembolic events using various antiplatelet and antithrombotic drugs, including aspirin, warfarin, or a direct oral anticoagulant. These oral medications specifically inhibit factor IIa or Xa. They are also known as new oral anticoagulants or target-specific oral anticoagulants. Use of these drugs has been shown to be effective for preventing manifestations of myocardial infarction or stroke [76].

7. Emerging role of polypill for cardiovascular prevention

A trend in cardiovascular primary prevention is to use a multiple-combination drug approach, the polypill strategy. This was introduced a few decades ago by Wald et al. [77] but has attracted increased interest because a few large-scale randomized trials have shown benefits in both primary and secondary prevention [78–80]. Such polypill tablets or capsules (polycaps) include one or two antihypertensive drugs together with a lipid-lowering statin, and sometimes other components such as aspirin, folate, or vitamin D. According to leading cardiologists including Professor Salim Yusuf (Hamilton, Canada) (the Primary Investigator of some of these trials), the use of polypills or polycaps is the only realistic strategy to address CVD on a global scale, especially in many low-resource settings in low- and middle-income countries [80]. Time will tell whether this strategy could also be implemented in some western high-income countries to support the compliance of drug intake, simplification of therapy, and use of synergistic pharmacologic effects.

A general problem concerning all drug therapies is how to reach the right patient or person at risk with a cost-effective and well-tolerated drug therapy to reduce cardiovascular risk by halting or reversing vascular aging with its cardiac, cerebral, and renal complications. Structural and educational aspects have an important role (i.e., how health care services are organized and reimbursed, as well as the staffing of health care and how the education of patients and their relatives is provided). Even the best drug therapies and strategies for prevention will fail if the structure, financing, educational services, and follow-up systems do not work in optimally. When personalized medicine with individual therapy tailored to individual biology (biomarkers, genetics, and technical findings) is regarded as a more sophisticated way forward, one should not ignore the other side of the issue: a simplified strategy (polypill/polycaps and lifestyle interventions) for a larger group of patients and many individuals at risk in the general population [81]. The solution to this dilemma may be that personalized (advanced) medicine could be the choice for patients at higher risk and secondary prevention, whereas the simplified strategy could be offered for the much larger segment of the general population at low or medium cardiovascular risk. If not, the health care system and primary healthcare workers risk becoming overburdened by too many patients or individuals at low risk for an overwhelming number of health examinations, treatment, and follow-up visits—a situation that no health care system is able to sustain.

8. Summary: Aging and cardiovascular risk—mechanisms and clinical aspects

The aging of the cardiovascular system reflects the aging in general of human biology, but it might lead to clinical complications as influenced by genetics, biomarkers, risk factors, and an unhealthy lifestyle when social, cultural, and ethnic factors could also be important. The EVA concept has enriched translational research activities to find new mechanisms of CVD risk [81] and inspired researchers to find treatment targets to lower the CVD risk beyond what can be achieved by conventional risk factor control alone. Screening of EVA has been tried based on measurements at pharmacies offered to the public [82] and seems to be feasible. The remaining problem is to reach individuals who are at risk and combine this approach with public health and societal measures to control CVD. New treatment strategies are needed, ranging from simplified treatment based on polypill/polycaps to more sophisticated individualized treatment to control risk factors and counteract cardiovascular aging.

9. Role of cell aging in cardiovascular disease

At the most fundamental level, all aging is cell aging. Whereas tissue-level (i.e., histologic) changes are both prominent and historically well-documented, cell-level changes have come to be recognized as the key venue of age-related pathology, increasingly so over the past few decades as our knowledge of cell aging, genetics, and epigenetics has given us insight into what appear to be more fundamental mechanisms of age-related disease. This observation holds equally well in the case of age-related CVD. Regarding atherosclerosis, for example, tissue-level changes (e.g., plaque) are obvious on clinical workup or at autopsy, in which the complex etiology deriving from changes in vascular endothelial cells (as emphasized over monocytes, smooth muscle cells, etc.) have come forward. This observation—that cell changes underlie and drive tissue-level changes—is equally apt whether applied to arterial (e.g., atherosclerosis, aneurysms), venous (e.g., venous status, varicosities), or capillary (age-related pruning, etc.) portions of the cardiovascular system.

Although much of the clinical pathology has been attributed to a progressive loss of arterial competence with age, fine vessel disease is also clinically significant and may go unsuspected, undiagnosed, and untreated [83]. Arterial and myocardial disease are the most obvious (and often the most abruptly apparent) clinical findings, yet peripheral arterial disease and capillary pruning have a ubiquitous pathology affecting all organs and becoming progressively more common with age. Such peripheral disease is more difficult to assess than is, for example, coronary artery, aortic, or carotid disease [84]. With regard to myocardial disease, the clinical focus is usually on CAD, but coronary capillary changes likely have a role here as well. As with myocardial aging, so, too, with many other end organs. It can be difficult to know whether age-related organ disease is primary to the organ's aging (and the aging of its intrinsic parenchymal cells) or result from the aging of its vasculature (and the aging of, for example, endothelial cells lining that vasculature) [85]. Thus, age-related neurodegenerative disease [86–90], renal disease [91,92], sarcopenia, age-related skin changes, and other age-related changes may well have dual or interactive origins with both intrinsic (end organ cell aging) and extrinsic (e.g., vascular cell aging) causes.

Even in the heart, where, understandably, we tend to focus on secondary disease resulting from arterial insufficiency (e.g., CAD), there has been increasing interest in age-related (and possibly primary) changes in the myocytes and fibroblasts [93,94] of the myocardium. In the cases of cardiac hypertrophy and cardiac failure, it remains unclear to what extent the pathology is driven by intrinsic cell aging of the myocytes and fibroblasts, as opposed to the more obvious role of extrinsic loss of arterial supply, although increasing evidence suggests that telomere shortening has a central role [95—97] and offers an effective point of intervention [98]. Even if the preponderance of cardiac disease were to be secondary (i.e., owing to coronary artery insufficiency) intrinsic cardiac aging may have a role in certain clinical presentations [99]. In heart failure, for example, myocyte failure is typified by changes in gene expression [100,101], as is endothelial disease in general [100,102,103], with concomitant inflammatory changes [104,105] and an increasing loss of mitochondrial function [106], often related to cell senescence [107] and telomere shortening [108—112]. Changes in the degree of cell phospholipid saturation, cardiac myocyte hydrogen peroxide, and oxygen-radical scavenger enzyme levels [113] might be the direct result of cell aging with consequent changes in gene expression, decelerated molecular turnover, and a gradual loss of metabolic efficiency [114]. Independent of other contributory factors, aging cardiomyocytes display a decrease in contractile proteins and a parallel increase in connective tissue protein [115]. Whatever the primary etiology, there are notable age-related changes, including increased ventricular wall thickness (correlated with increased systemic blood pressure) and increased accumulation of lipofuscin pigment [116—118]. Although the degree to which these findings in cardiac myocytes and cardiac fibroblasts can be ascribed to intrinsic cell aging or result from arterial insufficiency remains arguable, it has become clear that such cells show age-related changes and that they fit the biomarker definition of cell aging.

A number of cells that are central to age-related CVD (notable examples being vascular endothelial cells and cardiac fibroblasts) (Fig. 3.1) are known to undergo cell division over the human life span, demonstrating changes typical of cell aging, but this is less clear for other cells such as cardiac myocytes themselves. Classically, myocardial cells were considered to be postmitotic, and hence unlikely to undergo cell aging, but more recent data on the contribution of myocardial satellite cells and myocyte telomere shortening in heart failure [98] altered our views on the potential importance of myocardial cell senescence [107,119—123].

With regard to age-related arterial pathology, the issue is clearer. The vascular endothelial cells have long been regarded as the initial site in the cascade of age-related atherosclerotic disease [124—127], and there is no question that these and neighboring cells demonstrate cell aging in patients with CVD. Such cells demonstrate obvious morphologic changes, becoming more irregular and thinner, losing mitochondria, with changes in the basement membranes [128—132], and become generally less functional [85,133]. The role of endothelial cells is crucial and does not merely encompass the role of a passive arterial cell lining. Endothelial cells actively participate in regulating blood pressure [134] and modulate elasticity and compliance in the subendothelial layer. As we age, the endothelial response to vasodilators decreases [135] and the turnover of vascular NO decreases [136]. Such changes may be especially critical to capillary function, in which endothelial cells comprise the entire vascular wall and display age-related deterioration [128].

Subendothelial layers, however, have received more attention, in which classic histologic changes of atherosclerosis consist of patchy, irregular thickening with intracellular and extracellular deposition of lipids (fatty streaks) and calcification (hardening of the arteries). Smooth muscle cells clearly proliferate and display altered morphology and function [137], may migrate into the inner layers of the vessel, and are accompanied by monocyte-derived foam cells, matrix changes [138], and inflammation [139—141], starting with macrophages and progressing to lymphocyte infiltration, changes that correlate with plaque formation [142], which may be the result of cell aging [143].

In atherosclerosis, for example, the initial formation of a fatty streak, composed of lipid-laden macrophages (foam cells) within the intima, is followed by VSMC proliferation, fibrous plaques, cholesterol deposits, and inflammatory cell recruitment. The observable fatty streak formation is, however, preceded by a less obvious inflammatory response within the endothelium that has been linked to a number of insults, including the shear stress of hypertension, diabetes, tobacco use, and the inflammatory secretory pattern directly associated with cell aging. This results in the recruitment of inflammatory cells (e.g., monocytes) and the loss of adhesion in endothelial cell junctions. The failure of cell junction integrity causes the leakage of plasma components, including LDL cholesterol, into the

FIGURE 3.1 Although vascular endothelial (and subendothelial) cells are the focus of most histologic descriptions and mechanisms of clinical pathology, fibroblasts and myocardial cells are central to some age-related cardiovascular pathology.

subendothelial compartment. Inflammation further contributes to endothelial damage and predisposes to LDL phagocytosis. Recruited monocytes transform to macrophages, phagocytosing oxidized LDL and forming atherosclerotic foam cells. They also secrete inflammatory factors, induce smooth muscle cell proliferation, and often form fibrotic tissue.

The earliest fatty streaks, in children and adolescents, lack this fibrous component and other pathologic features that characterize atherosclerosis in elderly patients, in whom cell aging is prominent. Pathology can occur absent senescent cells, such as in the presence of significant physical or oxidative damage to endothelium, as evidenced by arterial fatty streaks in many patients by age 20, yet the arterial remodeling typical of the young population is slower or absent as cell aging progresses and other changes, such as mitochondrial dysfunction, occur.

As pathology advances, the vessel wall thickens and intrudes on the arterial lumen, often accompanied by hemorrhage and cell necrosis, further accelerating the pathology. Thickening of the elastic and smooth muscle layers occurs at the expense of the lumen rather than the external diameter. As the lumen narrows and the fibrous cap covering the atherosclerotic lesion thins, obstruction owing to thrombus (often resulting from fibrous cap rupture [144,145]) becomes increasingly probable, with a clinical outcome that is usually sudden and often fatal.

Despite the histology and the historical emphasis on subendothelial layers, the classic pathology appears to be initiated within the endothelial layers. This is perhaps to be expected, given the propensity for risk factors to operate from the lumen. The vascular endothelial cells lie between the lumen and the subendothelial pathology and may be expected to mediate the cascade of pathology from the one to the other. It is the endothelial cells that withstand the worst of the insults owing to serum cholesterol abnormalities, hypertension, hyperglycemia, tobacco derivatives, infectious agents, and so on. Although many of these pathologic insults are linked to atherogenesis, they may operate not directly, but via their effects on cell aging. One clue to this causal chain is evidenced by Hutchinson Gilford progeria, in which we find severe atherosclerosis (and premature cell aging) [146,147] and early death [148,149] without the presence of other classic risk factors [150–152] such as hypercholesterolemia [153], tobacco use, hypertension, or diabetes [154–156]. This is not to discount the correlation between such risk factors and atherosclerosis, only to suggest that cell aging may be the pathway through which they operate. Moreover, studies suggest that therapy targeting senescence and telomeric changes may be effective in reversing the cellular abnormalities seen in progeric cells [157].

Another perspective is provided by the use of agents that block cell division, slowing atherogenesis even in the face of abnormal serum cholesterol. In vascular grafts, restenosis can be blocked using antisense oligonucleotides, arresting cell division without disrupting normal endothelial histology and function [158–160]. In such cases, even the presence of ischemia [161] or high-lipid diets [162] do not lead to macrophage invasion, foam cell deposition, and so forth. The classic risk factors clearly have a clinical role, but not it is indirect (i.e., through their effect on cell aging). Cell aging itself is not only the necessary and shared pathway of atherogenic pathology; it may in itself present a more effective target for clinical intervention that do those same risk factors.

A myriad of risk factors may function via their effects on cell aging, including hypertension [163–165] (which may result from as well as accelerate endothelial aging [133,166–168] and telomere loss [169,170]), proton pump inhibitors [171], hypercholesterolemia [172], hyperglycemia [173], tobacco use, diet, alcohol, obesity, lack of exercise, high homocysteine [174,175] (which accelerates cell senescence) [176], oxidation (which also accelerates cell senescence) [177–179], individual cholesterol fractions and ratios, apolipoprotein E4 [180], estrogen levels, tocopherol levels [181,182], prothrombotic mutations [183], C-reactive protein [140,184,185], myeloperoxidase [186], stress [187], dental infections [188,189], and arterial wall DNA and RNA viral (e.g., herpesvirus, cytomegalovirus, Coxsackie virus) or bacterial (e.g., *Chlamydia, Helicobacter*, etc.) infection [190–193], and inflammatory biomarkers [194]. In general, we might divide these upstream risk factors into genetic (and epigenetic) risk factors and historical (e.g., medical and behavioral history) risk factors. The former include broadly genetic risks (often multiply determined), under which we may subsume diabetes, immune dysfunction, hypertension, lipid metabolism disorders, and so on. The latter include a number of risk factors alluded to earlier, such as tobacco use, alcohol use, lack of exercise, and poor diet (Fig. 3.2).

The overall model is clear: aging is itself the primary cardiovascular risk factor, and cell senescence is the central mechanism. Endothelial injury [195] (e.g., owing to circulating toxins, hypertension [169], dyslipidemia, hyperglycemia) [196], for example, is followed by cell loss and cell replacement, telomere shortening, and changes in gene expression [197]. This, in turn, prompts pathology in the intermediate layers, followed by the loss of luminal area and decreased endothelial adhesion, with the increasing likelihood of thrombus and embolism. Cell senescence and telomere shortening herald the onset of, and are more conspicuous in, areas of arterial disease [198–200].

Within the endothelial layer cell senescence has a pivotal and compelling role in atherogenesis and offers a rational target for effective intervention [201,202]. Endothelial cell turnover rates are higher than cells elsewhere

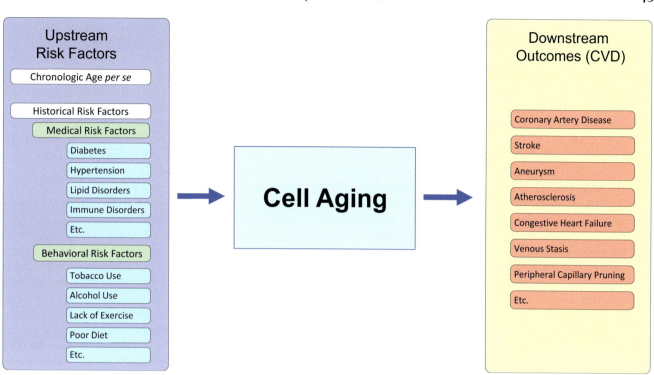

FIGURE 3.2 The downstream disease outcomes are the result of cell aging, but depend on the specific upstream "starting points" (i.e., various medical risk factors, both medical and behavioral, which are often multiply determined by a patient's underlying genetic or epigenetic risks). Clinical heterogeneity is the result of risk factor heterogeneity. CVD, cardiovascular disease.

in the vascular system (e.g., venous walls) or than subendothelial cells [195]. Telomere lengths are specifically shorter in areas of high vascular stress and atheroma formation than in areas of low stress, where pathology is less common [203,204]. A similar, though less important correlation is frequently found between shortened telomeres in circulating leukocytes and CVD [205–208]. That this correlation exists is accepted; that it is no more than a correlation and not causal has been pointed out repeatedly [209,210] and is likely misleading, conceptually akin to demonstrating that patients with gray hair are more likely to have heart disease. Key measures are telomere lengths in the cardiovascular cells in vivo, not those in circulating leukocytes, and when such cells are measured, results are clearer [198,199,211–213].

The central feature of cell aging is the loss of telomere length, causing a shift in gene expression, which causes slower molecular turnover, and which results in an increased proportion of dysfunctional molecules, cellular dysfunction, and ultimately, tissue-level pathology. Although cell aging is ubiquitous in the cardiovascular system, the specific CVDs depend on the genetic, epigenetic, and behavioral starting points of each patient. The downstream clinical outcomes—the specific cardiovascular pathology—hinge on those starting points, but in all cases, they develop during the central process of cell aging.

10. Prospects for intervention

Although component interventions, such as those targeting diet, lifestyle, hypertension, hyperglycemia, cholesterol, or discrete surgical abnormalities (e.g., arterial insufficiency, aneurysms) can offer clinical benefits, particularly when initiated early, such approaches neither stop nor reverse age-related pathology [150,153,214,215]. Even the most aggressive lifestyle interventions often prove inadequate in the face of continued cell aging of and prior damage to the arterial walls [153,202,216–220]. A more thorough and efficient approach to age-related CVD would take a systems approach, targeting the fundamental and shared pathology evidenced in cell aging and its cascade of pathology [114,221,222], as first suggested a few decades ago [223,224].

Although aging is usually regarded as a nonmodifiable risk factor [225], the optimal target for effective clinical intervention is likely to be cell aging per se [90,112,114,222,226,227]. The classic statement is that we are only as old as our arteries [228], but a more accurate statement is that we are only as old as the cells constituting the cardiovascular system. Hayflick aptly observed that "the cause of aging is ignored by the same people who argue that aging is the greatest risk factor for their favorite disease" [229]. Our best target, then, might be aging itself. The gamut of abnormalities we find in senescent endothelial cells are ameliorated (or reversed) by human telomerase reverse transcriptase (hTERT) treatment in vitro. Senescent human aortic endothelial cells treated with hTERT revert to the pattern of gene expression seen in normal young endothelial cells [201,202,230,231]. Research suggests that telomeric repeat-binding factor 2 can restore cell size and sarcomere density in cardiac myocytes [98]. Effective intervention in age-related CVD requires that we not merely expound on and target known risk factors and biomarkers [232], but that we achieve a broader understanding of aging itself, specifically cell aging. We require a unified systems model of age-related pathology rather than a component view that focuses narrowly on upstream risk factors or downstream outcomes. Such a unified systems model [114] suggests that age-related CVDs are driven by cell aging and that effective intervention in this central mechanism is feasible.

Some interest remains in approaches that, rather than reversing cell aging, would ablate senescent cells (i.e., using senolytics) [226,233,234]. This has suggestive merit, but the approach does not address the underlying issue of cell aging and may result in initial improvement followed by accelerated age-related pathology as remaining cells divide and senesce more rapidly [235]. Removing aged cells is not the same as reversing aging in those cells. Senomorphic approaches [236,237], in which specific biomarkers of cell aging are targeted, also does not address or intervene in the fundamental process of cell aging. Targeting one or a few biomarkers of cell aging does not target the multiple downstream effects of cell aging, let alone cell aging itself.

Reversing cell aging has been accomplished in human cells [238] and tissues [239,240], including human vascular endothelial tissue [230], as well as in animal trials [241–243]. The suggestion that telomeres are an optimal clinical target [244,245] in cardiovascular patients [246–249] is well-warranted. Indeed, the use of telomerase reverse transcriptase (TERT) gene therapy in mice with acute myocardial infarction improves ventricular function and reduces infarct scars [250]. Rigorous human trials using telomerase as a therapeutic intervention, using either messenger RNA [251,252] or viral vectors [114], would be attractive, and are planned.

An optimal intervention should have three key characteristics. It should: (1) be curative and preventative, (2) be effective in targeting a wide range of age-related CVDs, and (3) lower medical costs compared with available therapies. Cell aging promises to be the optimal point of intervention for all three points. Beyond the theoretical clinical benefits, targeting cell aging, as a more efficient and effective target than those in use, may offer increased financial efficiency as well. The low costs of single-treatment therapy and the potentially unparalleled efficacy may lower the costs of medical care globally, nationally, and personally.

Acknowledgment

Dr. Nilsson's work was supported by grants from the Swedish Research Council (2013-2756) and Heart and Lung Foundation (2015-0427) to PMN. Dr. Fossel is the founder and president of a gene therapy company targeting age-related diseases.

References

[1] Monahan-Earley R, Dvorak AM, Aird WC. Evolutionary origins of the blood vascular system and endothelium. J Thromb Haemostasis 2013; 11 Suppl 1(Suppl. 1):46–66.
[2] Pietri P, Stefanadis C. Cardiovascular aging and longevity: JACC state-of-the-art review. J Am Coll Cardiol 2021;77(2):189–204.
[3] Franklin SS, Gustin 4th W, Wong ND, Larson MG, Weber MA, Kannel WB, et al. Hemodynamic patterns of age-related changes in blood pressure. The Framingham Heart Study. Circulation 1997;96(1):308–15.
[4] Franklin SS, Larson MG, Khan SA, Wong ND, Leip EP, Kannel WB, et al. Does the relation of blood pressure to coronary heart disease risk change with aging? The Framingham Heart Study. Circulation 2001;103(9):1245–9.
[5] Hansson GK. Inflammation, atherosclerosis, and coronary artery disease. N Engl J Med 2005;352(16):1685–95.
[6] Laurent S, Boutouyrie P. Recent advances in arterial stiffness and wave reflection in human hypertension. Hypertension 2007;49(6):1202–6.
[7] Nilsson PM. Early vascular aging (EVA): consequences and prevention. Vasc Health Risk Manag 2008;4(3):547–52.
[8] Nilsson PM, Lurbe E, Laurent S. The early life origins of vascular ageing and cardiovascular risk: the EVA syndrome. J Hypertens 2008;26: 1049–57.
[9] O'Rourke MF, Safar ME, Dzau V. The cardiovascular continuum extended: aging effects on the aorta and microvasculature. Vasc Med 2010; 15(6):461–8.
[10] Vlachopoulos C, Xaplanteris P, Aboyans V, Brodmann M, Cífková R, Cosentino F, et al. The role of vascular biomarkers for primary and secondary prevention. A position paper from the European society of cardiology working group on peripheral circulation: endorsed by the association for research into arterial structure and physiology (artery) society. Atherosclerosis 2015;241(2):507–32.
[11] Climie RE, Alastruey J, Mayer CC, Schwarz A, Laucyte-Cibulskiene A, Voicehovska J, et al. Vascular ageing - moving from bench towards bedside. Eur J Prev Cardiol February 4, 2023:zwad028. https://doi.org/10.1093/eurjpc/zwad028. Epub ahead of print. PMID: 36738307.

References

[12] Salvi P, Scalise F, Rovina M, Moretti F, Salvi L, Grillo A, et al. Noninvasive estimation of aortic stiffness through different approaches. Hypertension 2019;74(1):117–29.

[13] Nilsson PM, Boutouyrie P, Laurent S. Vascular aging: a tale of EVA and ADAM in cardiovascular risk assessment and prevention. Hypertension 2009;54:3–10.

[14] Gottsäter M, Östling G, Persson M, Engström G, Melander O, Nilsson PM. Non-hemodynamic predictors of arterial stiffness after 17 years of follow-up: the Malmö Diet and Cancer study. J Hypertens 2015;33(5):957–65.

[15] Guzik TJ, Touyz RM. Oxidative stress, inflammation, and vascular aging in hypertension. Hypertension 2017;70(4):660–7.

[16] Kaess BM, Rong J, Larson MG, Hamburg NM, Vita JA, Levy D, et al. Aortic stiffness, blood pressure progression, and incident hypertension. JAMA 2012;308(9):875–81.

[17] AlGhatrif M, Strait JB, Morrell CH, Canepa M, Wright J, Elango P, et al. Longitudinal trajectories of arterial stiffness and the role of blood pressure: the Baltimore Longitudinal Study of Aging. Hypertension 2013;62:934–41.

[18] Muhammad IF, Borné Y, Östling G, Kennbäck C, Gottsäter M, Persson M, et al. Arterial stiffness and incidence of diabetes: a population-based cohort study. Diabetes Care 2017;40(12):1739–45.

[19] Gottsäter M, Hindy G, Orho-Melander M, Nilsson PM, Melander O. A genetic risk score for fasting plasma glucose is independently associated with arterial stiffness: a Mendelian randomization study. J Hypertens 2018;36(4):809–14.

[20] Cunha PG, Boutouyrie P, Nilsson PM, Laurent S. Early vascular ageing (EVA): definitions and clinical applicability. Curr Hypertens Rev 2017;13(1):8–15.

[21] Reference Values for Arterial Stiffness' Collaboration. Determinants of pulse wave velocity in healthy people and in the presence of cardiovascular risk factors: 'establishing normal and reference values'. Eur Heart J 2010;31:2338–50.

[22] Vlachopoulos C, Aznaouridis K, Stefanadis C. Prediction of cardiovascular events and all-cause mortality with arterial stiffness: a systematic review and meta-analysis. J Am Coll Cardiol 2010;55:1318–27.

[23] Ben-Shlomo Y, Spears M, Boustred C, May M, Anderson SG, Benjamin EJ, et al. Aortic pulse wave velocity improves cardiovascular event prediction: an individual participant meta-analysis of prospective observational data from 17,635 subjects. J Am Coll Cardiol 2014;63: 636–46.

[24] Zhong Q, Hu MJ, Cui YJ, Liang L, Zhou MM, Yang YW, et al. Carotid-femoral pulse wave velocity in the prediction of cardiovascular events and mortality: an updated systematic review and meta-analysis. Angiology 2018;69(7):617–29.

[25] Williams B, Mancia G, Spiering W, Agabiti Rosei E, Azizi M, Burnier M, et al. 2018 ESC/ESH guidelines for the management of arterial hypertension: the task force for the management of arterial hypertension of the European society of cardiology and the European society of hypertension: the task force for the management of arterial hypertension of the European society of cardiology and the European society of hypertension. J Hypertens 2018;36(10):1953–2041.

[26] Zanoli L, Rastelli S, Granata A, Inserra G, Empana JP, Boutouyrie P, et al. Arterial stiffness in inflammatory bowel disease: a systematic review and meta-analysis. J Hypertens 2016;34(5):822–9.

[27] Lacolley P, Regnault V, Segers P, Laurent S. Vascular smooth muscle cells and arterial stiffening: relevance in development, aging, and disease. Physiol Rev 2017;97(4):1555–617.

[28] Visentin S, Grumolato F, Nardelli GB, Di Camillo B, Grisan E, Cosmi E. Early origins of adult disease: low birth weight and vascular remodeling. Atherosclerosis 2014;237(2):391–9.

[29] Sperling J, Nilsson PM. Does early life programming influence arterial stiffness and central hemodynamics in adulthood? J Hypertens 2020; 38(3):481–8.

[30] Stock K, Schmid A, Griesmaier E, Gande N, Hochmayr C, Knoflach M, et al. Early vascular aging (EVA) study group. The impact of being born preterm or small for gestational age on early vascular aging in adolescents. J Pediatr 2018;201:49–54.

[31] Warrington NM, Beaumont RN, Horikoshi M, Day FR, Helgeland Ø, Laurin C, et al. Maternal and fetal genetic effects on birth weight and their relevance to cardio-metabolic risk factors. Nat Genet 2019;51(5):804–14.

[32] Symonds ME, Stephenson T, Budge H. Early determinants of cardiovascular disease: the role of early diet in later blood pressure control. Am J Clin Nutr 2009;89(5):1518S–22S.

[33] Henskens LH, Kroon AA, van Oostenbrugge RJ, Gronenschild EH, Fuss-Lejeune MM, Hofman PA, et al. Increased aortic pulse wave velocity is associated with silent cerebral small-vessel disease in hypertensive patients. Hypertension 2008;52:1120–6.

[34] Savoia C, Battistoni A, Calvez V, Cesario V, Montefusco G, Filippini A. Microvascular alterations in hypertension and vascular aging. Curr Hypertens Rev 2017;13(1):16–23.

[35] Arnett DK, Blumenthal RS, Albert MA, Buroker AB, Goldberger ZD, Hahn EJ, et al. 2019 ACC/AHA guideline on the primary prevention of cardiovascular disease: executive summary: a report of the American college of cardiology/American heart association task force on clinical practice guidelines. Circulation 2019;140(11):e563–95.

[36] Ong KT, Delerme S, Pannier B, Safar ME, Benetos A, Laurent S, et al., Investigators. Aortic stiffness is reduced beyond blood pressure lowering by short-term and long-term antihypertensive treatment: a meta-analysis of individual data in 294 patients. J Hypertens 2011; 29:1034–42.

[37] Vlachopoulos C, Terentes-Printzios D, Laurent S, Nilsson PM, Protogerou AD, Aznaouridis K, et al. Association of estimated pulse wave velocity with survival: a secondary analysis of SPRINT. JAMA Netw Open 2019;2(10):e1912831.

[38] Laurent S, Schlaich M, Esler M. New drugs, procedures, and devices for hypertension. Lancet 2012;380(9841):591–600.

[39] Mancia G, Cannon CP, Tikkanen I, Zeller C, Ley L, Woerle HJ, et al. Impact of empagliflozin on blood pressure in patients with type 2 diabetes mellitus and hypertension by background antihypertensive medication. Hypertension 2016;68(6):1355–64.

[40] Chilton R, Tikkanen I, Cannon CP, Crowe S, Woerle HJ, Broedl UC, et al. Effects of empagliflozin on blood pressure and markers of arterial stiffness and vascular resistance in patients with type 2 diabetes. Diabetes Obes Metab 2015;17(12):1180–93.

[41] Niiranen TJ, Lyass A, Larson MG, Hamburg NM, Benjamin EJ, Mitchell GF, et al. Prevalence, correlates, and prognosis of healthy vascular aging in a western community-dwelling cohort: the Framingham heart study. Hypertension 2017;70(2):267–74.

[42] Nilsson PM, Laurent S, Cunha PG, Olsen MH, Rietzschel E, Franco OH, et al. Metabolic syndrome, Arteries REsearch (MARE) Consortium. Characteristics of healthy vascular ageing in pooled population-based cohort studies: the global metabolic syndrome and Artery REsearch Consortium. J Hypertens 2018;36(12):2340–9.

[43] Ji H, Teliewubai J, Lu Y, Xiong J, Yu S, Chi C, et al. Vascular aging and preclinical target organ damage in community-dwelling elderly: the Northern Shanghai Study. J Hypertens 2018;36(6):1391–8.

[44] Laurent S, Boutouyrie P, Cunha PG, Lacolley P, Nilsson PM. Concept of extremes in vascular aging. Hypertension 2019;74(2):218–28.

[45] Bain SC, Gill GV, Dyer PH, Jones AF, Murphy M, Jones KE, et al. Characteristics of type 1 diabetes of over 50 year's duration (the golden years cohort). Diabet Med 2003;20:808–11.

[46] Sun JK, Keenan HA, Cavallerano JD, Asztalos BF, Schaefer EJ, Sell DR, et al. Protection from retinopathy and other complications in patients with type 1 diabetes of extreme duration: the Joslin 50-year medalist study. Diabetes Care 2011;34:968–74.

[47] Adamsson Eryd S, Svensson AM, Franzén S, Eliasson B, Nilsson PM, Gudbjörnsdottir S. Risk of future microvascular and macrovascular disease in people with type 1 diabetes of very long duration: a national study with 10-year follow-up. Diabet Med 2017;34:411–8.

[48] Nilsson PM. Hemodynamic aging as the consequence of structural changes associated with early vascular aging (EVA). Aging Dis 2014;5:109–13.

[49] Tremmel M, Lyssenko V, Zöller B, Engström G, Magnusson M, Melander O, et al. Characteristics and prognosis of healthy severe obesity (HSO) subjects – the Malmo Preventive Project. Obesity Med 2018;11:6–12.

[50] Korduner J, Bachus E, Jujic A, Magnusson M, Nilsson PM. Metabolically healthy obesity (MHO) in the Malmö diet cancer study - epidemiology and prospective risks. Obes Res Clin Pract 2019;13(6):548–54.

[51] Nilsson PM, Korduner J, Magnusson M. Metabolically healthy obesity (MHO)-new research directions for personalised medicine in cardiovascular prevention. Curr Hypertens Rep 2020;22(2):18.

[52] Zhou Z, Macpherson J, Gray SR, Gill JMR, Welsh P, Celis-Morales C, et al. Are people with metabolically healthy obesity really healthy? A prospective cohort study of 381,363 UK Biobank participants. Diabetologia 2021;64(9):1963–72.

[53] Authors/task force members; ESC committee for practice guidelines (CPG); ESC national cardiac societies. 2019 ESC/EAS guidelines for the management of dyslipidaemias: lipid modification to reduce cardiovascular risk. Atherosclerosis 2019;290:140–205. Erratum in: Atherosclerosis. 2020; 292:160-162. Erratum in: Atherosclerosis. 2020; 294:80-82. PMID: 31591002.

[54] Visseren FLJ, Mach F, Smulders YM, Carballo D, Koskinas KC, Bäck M, et al., ESC National Cardiac Societies; ESC Scientific Document Group. 2021 ESC Guidelines on cardiovascular disease prevention in clinical practice. Eur Heart J 2021;42(34):3227–337. 54.

[55] Arnett DK, Blumenthal RS, Albert MA, Buroker AB, Goldberger ZD, Hahn EJ, et al. 2019 ACC/AHA guideline on the primary prevention of cardiovascular disease: a report of the American college of cardiology/American heart association task force on clinical practice guidelines. Circulation 2019;140(11):e596–646. Erratum in: Circulation. 2019; 140(11):e649-e650. Erratum in: Circulation. 2020; 141(4):e60. Erratum in: Circulation. 2020; 141(16):e774. 56.

[56] Writing Committee Members, Lawton JS, Tamis-Holland JE, Bangalore S, Bates ER, Beckie TM, Bischoff JM, et al. 2021 ACC/AHA/SCAI guideline for coronary artery revascularization: executive summary: a report of the American college of cardiology/American heart association joint committee on clinical practice guidelines. J Am Coll Cardiol 2022;79(2):197–215. Erratum in: J Am Coll Cardiol. 2022; 79(15): 1547.

[57] Barbato E, Azizi M, Schmieder RE, Lauder L, Böhm M, Brouwers S, et al. Renal denervation in the management of hypertension in adults. A clinical consensus statement of the ESC Council on Hypertension and the European Association of Percutaneous Cardiovascular Interventions (EAPCI). Eur Heart J 2023;44(15):1313–30.

[58] Estruch R, Ros E, Salas-Salvadó J, Covas MI, Corella D, Arós F, et al., PREDIMED Study Investigators. Primary prevention of cardiovascular disease with a mediterranean diet supplemented with extra-virgin olive oil or nuts. N Engl J Med 2018;378(25):e34–58.

[59] Tang C, Wang X, Qin LQ, Dong JY. Mediterranean diet and mortality in people with cardiovascular disease: a meta-analysis of prospective cohort studies. Nutrients 2021;13(8):2623.

[60] Risérus U. Healthy Nordic diet and cardiovascular disease. J Intern Med 2015;278(5):542–4.

[61] Currie G, Delles C. Precision medicine and personalized medicine in cardiovascular disease. Adv Exp Med Biol 2018;1065:589–605.

[62] Holmes MV, Asselbergs FW, Palmer TM, Drenos F, Lanktree MB, Nelson CP, et al., UCLEB Consortium, Doevendans PA, Balmforth AJ, Hall AS, North KE, Almoguera B, Hoogeveen RC, et al. Mendelian randomization of blood lipids for coronary heart disease. Eur Heart J 2015;36(9):539–50.

[63] Ferraro RA, Leucker T, Martin SS, Banach M, Jones SR, Toth PP. Contemporary management of dyslipidemia. Drugs 2022;82(5):559–76.

[64] Khan SU, Yedlapati SH, Lone AN, Hao Q, Guyatt G, Delvaux N, et al. PCSK9 inhibitors and ezetimibe with or without statin therapy for cardiovascular risk reduction: a systematic review and network meta-analysis. BMJ 2022;377:e069116.

[65] Hardy J, Niman S, Pereira E, Lewis T, Reid J, Choksi R, et al. A critical review of the efficacy and safety of inclisiran. Am J Cardiovasc Drugs 2021;21(6):629–42.

[66] Davies MJ, Aroda VR, Collins BS, Gabbay RA, Green J, Maruthur NM, et al. Management of hyperglycaemia in type 2 diabetes, 2022. A consensus report by the American diabetes association (ADA) and the European association for the study of diabetes (EASD). Diabetologia 2022;65(12):1925–66.

[67] ElSayed NA, Aleppo G, Aroda VR, Bannuru RR, Brown FM, Bruemmer D, et al. On behalf of the American diabetes association. 9. Pharmacologic approaches to Glycemic treatment: standards of care in diabetes-2023. Diabetes Care 2023;46(Suppl. 1):S140–57.

[68] Palmer SC, Tendal B, Mustafa RA, Vandvik PO, Li S, Hao Q, et al. Sodium-glucose cotransporter protein-2 (SGLT-2) inhibitors and glucagon-like peptide-1 (GLP-1) receptor agonists for type 2 diabetes: systematic review and network meta-analysis of randomised controlled trials. BMJ 2021;372:m4573. Erratum in: BMJ. 2022; 376:o109.

[69] Kawai Y, Uneda K, Yamada T, Kinguchi S, Kobayashi K, Azushima K, et al. Comparison of effects of SGLT-2 inhibitors and GLP-1 receptor agonists on cardiovascular and renal outcomes in type 2 diabetes mellitus patients with/without albuminuria: a systematic review and network meta-analysis. Diabetes Res Clin Pract 2022;183:109146.

[70] Li X, Celotto S, Pizzol D, Gasevic D, Ji MM, Barnini T, et al. Metformin and health outcomes: an umbrella review of systematic reviews with meta-analyses. Eur J Clin Invest 2021;51(7):e13536.

[71] Frías JP, Davies MJ, Rosenstock J, Pérez Manghi FC, Fernández Landó L, Bergman BK, et al. SURPASS-2 investigators. Tirzepatide versus semaglutide once weekly in patients with type 2 diabetes. N Engl J Med 2021;385(6):503–15.

[72] Sattin M, Towheed T. The effect of TNFα-inhibitors on cardiovascular events in patients with rheumatoid arthritis: an updated systematic review of the literature. Curr Rheumatol Rev 2016;12(3):208–22.

References

[73] Ridker PM, Everett BM, Thuren T, MacFadyen JG, Chang WH, Ballantyne C, et al., CANTOS Trial Group. Antiinflammatory therapy with canakinumab for atherosclerotic disease. N Engl J Med 2017;377(12):1119–31.

[74] Deftereos SG, Beerkens FJ, Shah B, Giannopoulos G, Vrachatis DA, Giotaki SG, et al. Colchicine in cardiovascular disease: in-depth review. Circulation 2022;145(1):61–78.

[75] Thomas KH, Dalili MN, López-López JA, Keeney E, Phillippo D, Munafò MR, et al. Smoking cessation medicines and e-cigarettes: a systematic review, network meta-analysis and cost-effectiveness analysis. Health Technol Assess 2021;25(59):1–224.

[76] Chen A, Stecker E, Warden BA. Direct oral anticoagulant use: a practical guide to common clinical challenges. J Am Heart Assoc 2020;9(13): e017559.

[77] Wald NJ, Law MR. A strategy to reduce cardiovascular disease by more than 80%. BMJ 2003;326(7404):1419.

[78] Roshandel G, Khoshnia M, Poustchi H, Hemming K, Kamangar F, Gharavi A, et al. Effectiveness of polypill for primary and secondary prevention of cardiovascular diseases (PolyIran): a pragmatic, cluster-randomised trial. Lancet 2019;394(10199):672–83.

[79] Yusuf S, Joseph P, Dans A, Gao P, Teo K, Xavier D, et al. International polycap study 3 investigators. Polypill with or without aspirin in persons without cardiovascular disease. N Engl J Med 2021;384(3):216–28.

[80] Castellano JM, Pocock SJ, Bhatt DL, Quesada AJ, Owen R, Fernandez-Ortiz A, et al. SECURE investigators. Polypill strategy in secondary cardiovascular prevention. N Engl J Med 2022;387(11):967–77.

[81] Nilsson PM, Boutouyrie P, Cunha P, Kotsis V, Narkiewicz K, Parati G, et al. Early vascular ageing in translation: from laboratory investigations to clinical applications in cardiovascular prevention. J Hypertens 2013;31(8):1517–26.

[82] Danninger K, Hafez A, Binder RK, Aichberger M, Hametner B, Wassertheurer S, et al. High prevalence of hypertension and early vascular aging: a screening program in pharmacies in Upper Austria. J Hum Hypertens 2020;34(4):326–34.

[83] Hirsch AT, Criqui MH, Treat-Jacobson D, Regensteiner JG, Creager MA, Olin JW, et al. Peripheral arterial disease detection, awareness, and treatment in primary care. JAMA 2001;286:1317–24. https://doi.org/10.1001/jama.286.11.1317.

[84] Newman AB. Peripheral arterial disease: insights from population studies of older adults. J Am Geriatr Soc 2000;48:1157–62.

[85] Britten M, Schachinger V. The role of endothelial function for ischemic manifestations of coronary atherosclerosis. Herz 1998;23:97–105.

[86] Graves SI, Baker DJ. Implicating endothelial cell senescence to dysfunction in the ageing and diseased brain. Basic Clin Pharmacol Toxicol 2020;127:102–10. https://doi.org/10.1111/bcpt.13403.

[87] Eitan E, Hutchison ER, Mattson MP. Telomere shortening in neurological disorders: an abundance of unanswered questions. Trends Neurosci 2014;37:256–63. https://doi.org/10.1016/j.tins.2014.02.010.

[88] Kota LN, Bharath S, Purushottam M, Moily NS, Sivakumar PT, Varghes M, et al. Reduced telomere length in neurodegenerative disorders may suggest shared biology. J Neuropsychiatry Clin Neurosci 2015;27:e92–6. https://doi.org/10.1176/appi.neuropsych.13100240.

[89] Liu MY, Neme A, Zhou QG. The emerging roles for telomerase in the central nervous system. Front Mol Neurosci 2018;11:160. https://doi.org/10.3389/fnmol.2018.00160.

[90] Fossel MB. Unified model of dementias and age-related neurodegeneration. Alzheimers Dement 2020;16:365–83. https://doi.org/10.1002/alz.12012.

[91] Fang Y, Gong AY, Haller ST, Dworkin LD, Liu Z, Gong R. The ageing kidney: molecular mechanisms and clinical implications. Ageing Res Rev 2020;63:101151. https://doi.org/10.1016/j.arr.2020.101151.

[92] Long DA, Mu W, Price KL, Johnson RJ. Blood vessels and the aging kidney. Nephron Exp Nephrol 2005;101:e95–9. https://doi.org/10.1159/000087146.

[93] Travers JG, Kamal FA, Robbins J, Yutzey KE, Blaxall BC. Cardiac fibrosis: the fibroblast awakens. Circ Res 2016;118:1021–40. https://doi.org/10.1161/CIRCRESAHA.115.306565.

[94] Hu C, Zhang X, Teng T, Ma ZG, Tang QZ. Cellular senescence in cardiovascular diseases: a systematic review. Aging Dis 2022;13:103–28. https://doi.org/10.14336/AD.2021.0927.

[95] Gevaert AB, Shakeri H, Leloup AJ, Van Hove CE, De Meyer GR, Vrints CJ, et al. Endothelial senescence contributes to heart failure with preserved ejection fraction in an aging mouse model. Circ Heart Fail 2017;10:e003806. https://doi.org/10.1161/CIRCHEARTFAILURE.116.003806.

[96] Serrano AL, Andres V. Telomeres and cardiovascular disease. Does size matter? Circ Res 2004;94:575–84. https://doi.org/10.1161/01.RES.0000122141.18795.9C.

[97] Edo MD, Andres V. Aging, telomeres, and atherosclerosis. Cardiovasc Res 2005;66:213–21. https://doi.org/10.1016/j.cardiores.2004.09.007.

[98] Eguchi A, Gonzalez AF, Torres-Bigio SI, Koleckar K, Birnbaum F, Zhang JZ, et al. TRF2 rescues telomere attrition and prolongs cell survival in Duchenne muscular dystrophy cardiomyocytes derived from human iPSCs. Proc Natl Acad Sci USA February 7, 2023;120(6):e2209967120. https://doi.org/10.1073/pnas.2209967120.

[99] Mehdizadeh M, Aguilar M, Thorin E, Ferbeyre G, Nattel S. The role of cellular senescence in cardiac disease: basic biology and clinical relevance. Nat Rev Cardiol 2022;19:250–64. https://doi.org/10.1038/s41569-021-00624-2.

[100] Hwang JJ, Dzau VJ, Liew CC. Genomics and the pathophysiology of heart failure. Curr Cardiol Rep 2001;3:198–207. https://doi.org/10.1007/s11886-001-0023-z.

[101] Kim JW, Baek BS, Kim YK, Herlihy JT, Ikeno Y, Yu BP, et al. Gene expression of cyclooxygenase in the aging heart. J Gerontol A Biol Sci Med Sci 2001;56A:B350–5. https://doi.org/10.1093/gerona/56.8.B350.

[102] Cooper LT, Cooke JP, Dzau VJ. The vasculopathy of aging. J Gerontol 1994;49:191–6. https://doi.org/10.1093/geronj/49.5.B191.

[103] Monajemi H, Arkenbout EK, Pannekoek H. Gene expression in atherogenesis. Thromb Haemostasis 2001;86:404–12. https://doi.org/10.1055/s-0037-1616238.

[104] Ramos GC, van den Berg A, Nunes-Silva V, Weirather J, Peters L, Burkard M, et al. Myocardial aging as a T-cell–mediated phenomenon. Proc Natl Acad Sci USA 2017;114:E2420–9. https://doi.org/10.1073/pnas.162104711.

[105] Akishita M, Horiuchi M, Yamada H, Zhang L, Shirakami G, Tamura K, et al. Inflammation influences vascular remodeling through AT2 receptor expression and signaling. Physiol Genom 2000;2:13–20. https://doi.org/10.1152/physiolgenomics.2000.2.1.13.

[106] Guertl B, Noehammer C, Hoefler G. Metabolic cardiomyopathies. Int J Exp Pathol 2000;81:349–72. https://doi.org/10.1046/j.1365-2613.2000.00186.x.

[107] Tang X, Li PH, Chen HZ. Cardiomyocyte senescence and cellular communications within myocardial microenvironments. Front Endocrinol 2020;11:280. https://doi.org/10.3389/fendo.2020.00280. Published 2020 May 21.
[108] Moslehi J, DePinho RA, Sahin E. Telomeres and mitochondria in the aging heart. Circ Res 2012;110:1226−37. https://doi.org/10.1161/CIRCRESAHA.111.246868.
[109] Vecoli C, Borghini A, Andreassi MG. The molecular biomarkers of vascular aging and atherosclerosis: telomere length and mitochondrial DNA 4977 common deletion. Mutat Res 2020;784:108309. https://doi.org/10.1016/j.mrrev.2020.108309.
[110] Kajstura J, Pertoldi B, Leri A, Beltrami CA, Deptala A, Darzynkiewicz Z, et al. Telomere shortening is an in vivo marker of myocyte replication and aging. Am J Pathol 2000;156:813−9.
[111] Borges A, Liew CC. Telomerase activity during cardiac development. J Mol Cell Cardiol 1997;29:2717−24.
[112] Anderson R, Lagnado A, Maggiorani D, Walaszczyk A, Dookun E, Chapman J, et al. Length-independent telomere damage drives postmitotic cardiomyocyte senescence. EMBO J 2019;38:e100492. https://doi.org/10.15252/embj.2018100492.
[113] Muscari C, Giaccari A, Giordano E, Clo C, Guarnieri C, Caldarera CM. Role of reactive oxygen species in cardiovascular aging. Mol Cell Biochem 1996;161:159−66. https://doi.org/10.1007/978-1-4613-1279-6_22.
[114] Fossel M, Bean J, Khera N, Kolonin MG. A unified model of age-related cardiovascular disease. Biology 2022;11:1768. https://doi.org/10.3390/biology11121768.
[115] Lakatta EG, Boluyt MO. Age-associated changes in the cardiovascular system in the absence of cardiovascular disease. Chapter eight. In: Hosenpud JD, Greenberg BH, editors. Congestive heart failure. Philadelphia: Lippincott Williams & Wilkins; 2000.
[116] Roffe C. Ageing of the heart. Br J Biomed Sci 1998;55:136−48.
[117] Terman A, Brunk UT. On the degradability and exocytosis of ceroid/lipofuscin in cultured rat cardiac myocytes. Mech Ageing Dev 1998;100: 145−56. https://doi.org/10.1016/S0047-6374(97)00129-2.
[118] Helenius M, Hanninen M, Lehtinen SK, Salminen A. Aging induced up regulation of nuclear binding activities of oxidative stress responsive NF kB transcription factor in mouse cardiac muscle. J Mol Cell Cardiol 1998;28:487−98. https://doi.org/10.1006/jmcc.1996.0045.
[119] Yin H, Price F, Rudnicki MA. Satellite cells and the muscle stem cell niche. Physiol Rev 2013;93:23−67. https://doi.org/10.1152/physrev.00043.2011.
[120] Mercola M, Ruiz-Lozano P, Schneider MD. Cardiac muscle regeneration: lessons from development. Genes Dev 2011;25:299−309. http://www.genesdev.org/cgi/doi/10.1101/gad.2018411.
[121] Laflamme MA, Murry CE. Heart regeneration. Nature 2011;473:326−35. https://doi.org/10.1038/nature10147.
[122] Raman SV, Cooke GE, Binkley PF. Evidence that human cardiac myocytes divide after myocardial infarction. N Engl J Med 2002;345:1130−1.
[123] Quaini F, Urbanek K, Beltrami AP, Finato N, Beltrami CA, Nadal-Ginard B, et al. Chimerism of the transplanted heart. N Engl J Med 2002; 346:5−15.
[124] De Caterina R. Endothelial dysfunctions: common denominators in vascular disease. Curr Opin Lipidol 2000;11:9−23.
[125] Muller MM, Griesmacher A. Markers of endothelial dysfunction. Clin Chem Lab Med 2000;38:77−85. https://doi.org/10.1515/CCLM.2000.013.
[126] Forgione MA, Leopold JA, Loscalzo J. Roles of endothelial dysfunction in coronary artery disease. Curr Opin Cardiol 2000;5:409−15.
[127] Herrmann J, Lerman A. The endothelium: dysfunction and beyond. J Nucl Cardiol 2001;8:197−206.
[128] Kalaria RN. Cerebral vessels in ageing and Alzheimer's disease. Pharmacol Ther 1996;72:193−214. https://doi.org/10.1016/S0163-7258(96)00116-7.
[129] De Jong GI, De Vos RA, Steur EN, Luiten PG. Cerebrovascular hypoperfusion: a risk factor for Alzheimer's disease? Animal model and postmortem human studies. Ann N Y Acad Sci 1997;826:56−74. https://doi.org/10.1111/j.1749-6632.1997.tb48461.x.
[130] Shah GN, Mooradian AD. Age related changes in the blood brain barrier. Exp Gerontol 1997;32:501−19. https://doi.org/10.1016/S0531-5565(96)00158-1.
[131] Degens H. Age related changes in the microcirculation of skeletal muscle. Adv Exp Med Biol 1998;454:343−8. https://doi.org/10.1007/978-1-4615-4863-8_40.
[132] Robert AM, Schaeverbeke M, Schaeverbeke J, Robert L. Aging and brain circulation. Role of the extracellular matrix of brain microvessels. C R Seances Soc Biol Fil 1997;191:253−60.
[133] Kimura Y, Matsumoto M, Den YB, Iwai K, Munehira J, Hattori H, et al. Impaired endothelial function in hypertensive elderly patients evaluated by high resolution ultrasonography. Can J Cardiol 1999;15:563−8.
[134] Dzau VJ, Gibbons GH, Morishita R, Pratt RE. New perspectives in hypertension research. Potentials of vascular biology. Hypertension 1994; 23:1132−40. https://doi.org/10.1161/01.HYP.23.6.1132.
[135] Andrawis N, Jones DS, Abernethy DR. Aging is associated with endothelial dysfunction in the human forearm vasculature. J Am Geriatr Soc 2000;48:193−8. https://doi.org/10.1111/j.1532-5415.2000.tb03911.x.
[136] Cooke JP, Dzau VJ. Nitric oxide synthase: role in the genesis of vascular disease. Annu Rev Med 1997;48:489−509. https://doi.org/10.1146/annurev.med.48.1.489.
[137] Lindop GB, Boyle JJ, McEwan P, Kenyon CJ. Vascular structure, smooth muscle cell phenotype and growth in hypertension. J Hum Hypertens 1995;9:475−8.
[138] Robert L, Robert AM, Jacotot B. Elastin elastase atherosclerosis revisited. Atherosclerosis 1998;140:281−95. https://doi.org/10.1016/S0021-9150(98)00171-3.
[139] Danesh J. Smoldering arteries? Low grade inflammation and coronary heart disease. JAMA 1999;282:2169−71. https://doi.org/10.1001/jama.282.22.2169.
[140] Koenig W. Inflammation and coronary heart disease: an overview. Cardiol Rev 2001;9:31−5.
[141] Selzman CH, Miller SA, Harken AH. Therapeutic implications of inflammation in atherosclerotic cardiovascular disease. Ann Thorac Surg 2001;71:2066−74. https://doi.org/10.1016/S0003-4975(00)02597-2.
[142] Okimoto T, Imazu M, Hayashi Y, Fujiwara H, Ueda H, Kohno N. Atherosclerotic plaque characterization by quantitative analysis using intravascular ultrasound correlation with histological and immunohistochemical findings. Circ J 2002;66:173−7. https://doi.org/10.1253/circj.66.173.

References

[143] Katsuumi G, Shimizu I, Yoshida Y, Minamino T. Vascular senescence in cardiovascular and metabolic diseases. Front. Cardiovasc Med. 2018; 5:18. https://doi.org/10.3389/fcvm.2018.00018.

[144] Ambrose JA, Tannenbaum MA, Alexopoulos D, Hjemdahl-Monsen CE, Leavy J, Weiss M, et al. Angiographic progression of coronary artery disease and the development of myocardial infarction. J Am Coll Cardiol 1988;12:56−62. https://doi.org/10.1016/0735-1097(88)90356-7.

[145] Bentzon JF, Otsuka F, Virmani R, Falk E. Mechanisms of plaque formation and rupture. Circ Res 2014;114:1852−66. https://doi.org/10.1161/CIRCRESAHA.114.302721.

[146] Olive M, Harten I, Mitchell R, Beers JK, Djabali K, Cao K, et al. Cardiovascular pathology in Hutchinson-Gilford progeria: correlation with the vascular pathology of aging. Arterioscler Thromb Vasc Biol 2010;30:2301−9. https://doi.org/10.1161/ATVBAHA.110.209460.

[147] Matrone G, Thandavarayan RA, Walther BK, Meng S, Mojiri A, Cooke JP. Dysfunction of iPSC-derived endothelial cells in human Hutchinson−Gilford progeria syndrome. Cell Cycle 2019;18:2495−508. https://doi.org/10.1080/15384101.2019.1651587.

[148] Fossel M. In: Robinson R, editor. Accelerated aging: progeria. In *genetics*. New York, NY, USA: Macmillan; 2003.

[149] Fossel M. Human aging and progeria. J Pediatr Endocrinol Metab 2000;13:1477−81.

[150] Parsons C, Agasthi P, Mookadam F, Arsanjani R. Reversal of coronary atherosclerosis: role of lifestyle and medical management. Trends Cardiovasc Med 2018;28:524−31. https://doi.org/10.1016/j.tcm.2018.05.002.

[151] Tuzcu EM, Kapadia SR, Tutar E, Ziada KM, Hobbs RE, McCarthy PM, et al. High prevalence of coronary atherosclerosis in asymptomatic teenagers and young adults: evidence from intravascular ultrasound. Circulation 2001;103:2705−10. https://doi.org/10.1161/01.cir.103.22.2705.

[152] Li Y, Gilbert TR, Matsumoto AH, Shi W. Effect of aging on fatty streak formation in a diet-induced mouse model of atherosclerosis. J Vasc Res 2008;45:205−10. https://doi.org/10.1159/000112133.

[153] Björkegren JLM, Hägg S, Talukdar HA, Asl HF, Jain RK, Cedergren C, et al. Plasma cholesterol-induced lesion networks activated before regression of early, mature, and advanced atherosclerosis. PLoS Genet 2014;10:e1004201. https://doi.org/10.1371/journal.pgen.1004201.

[154] Vapaatalo H, Mervaala E. Clinically important factors influencing endothelial function. Med Sci Mon Int Med J Exp Clin Res 2001;7:1075−85.

[155] Cosentino F, Luscher TF. Effects of blood pressure and glucose on endothelial function. Curr Hypertens Rep 2001;3:79−88.

[156] Counter CM, Gupta J, Harley CB, Leber B, Bacchetti S. Telomerase activity in normal leukocytes and in hematologic malignancies. Blood 1995;85:2315−20.

[157] Xu Q, Mojiri A, Boulahouache L, Morales E, Walther BK, Cooke JP. Vascular senescence in progeria: role of endothelial dysfunction. Eur Heart J Open 2022;2:oeac047. https://doi.org/10.1093/ehjopen/oeac047.

[158] von der Leyen HE, Mann MJ, Dzau VJ. Gene inhibition and gene augmentation for the treatment of vascular proliferative disorders. Semin Intervent Cardiol 1996;1:209−14.

[159] Mann MJ, Gibbons GH, Tsao PS, von der Leyen HE, Cooke JP, Buitrago R, et al. Cell cycle inhibition preserves endothelial function in genetically engineered rabbit vein grafts. J Clin Invest 1997;99:1295−301. https://doi.org/10.1172/JCI119288.

[160] Mann MJ, Dzau VJ. Genetic manipulation of vein grafts. Curr Opin Cardiol 1997;12:522−7. https://journals.lww.com/co-cardiology/toc/1997/11000.

[161] Poston RS, Tran KP, Mann MJ, Hoyt EG, Dzau VJ, Robbins RC. Prevention of ischemically induced neointimal hyperplasia using ex vivo antisense oligodeoxynucleotides. J Heart Lung Transplant 1998;17:349−55.

[162] Dzau VJ, Mann MJ, Morishita R, Kaneda Y. Fusigenic viral liposome for gene therapy in cardiovascular diseases. Proc Natl Acad Sci U S A 1996;93:11421−5. https://doi.org/10.1073/pnas.93.21.1142.

[163] Roger I, Milara J, Belhadj N, Cortijo J. Senescence alterations in pulmonary hypertension. Cells 2021;10:3456. https://doi.org/10.3390/cells10123456.

[164] McCarthy CG, Wenceslau CF, Webb RC, Joe B. Novel contributors and mechanisms of cellular senescence in hypertension-associated premature vascular aging. Am J Hypertens 2019;32:709−19. https://doi.org/10.1093/ajh/hpz052.

[165] Westhoff JH, Hilgers KF, Steinbach MP, Hartner A, Klanke B, Amann K, et al. Hypertension induces somatic cellular senescence in rats and humans by induction of cell cycle inhibitor p16INK4a. Hypertension 2008;52:123−9. https://doi.org/10.1161/HYPERTENSIONAHA.107.099432.

[166] Raitakari OT, Celermajer DS. Testing for endothelial dysfunction. Ann Med 2000;32:293−304.

[167] Taddei S, Virdis A, Ghiadoni L, Salvetti G, Salvetti A. Endothelial dysfunction in hypertension. J Nephrol 2000;13:205−10.

[168] Aviv A. Chronology versus biology: telomeres, essential hypertension, and vascular aging. Hypertension 2002;40:229−32.

[169] Aviv A, Aviv H. Telomeres and essential hypertension. Am J Hypertens 1999;12(4 Pt 1):427−32. https://doi.org/10.1016/S0895-7061(00)00202-7.

[170] Aviv A, Aviv H. Reflections on telomeres, growth, aging, and essential hypertension. Hypertension 1997;29:1067−72. https://doi.org/10.1161/01.HYP.29.5.1067.

[171] Yepuri G, Sukhovershin R, Nazari-Shafti TZ, Petrascheck M, Ghebre YT, Cooke JP. Proton pump inhibitors accelerate endothelial senescence. Circ Res 2016;118:e36−42. https://doi.org/10.1161/CIRCRESAHA.116.308807.

[172] Shi Q, Hubbard GB, Kushwaha RS, Rainwater D, Thomas CA, Leland MM, et al. Endothelial senescence after high-cholesterol, high-fat diet challenge in baboons. Am J Physiol Heart Circ Physiol 2007;292:H2913−20. https://doi.org/10.1152/ajpheart.01405.2006.

[173] Maeda M, Hayashi T, Mizuno N, Hattori Y, Kuzuya M. Intermittent high glucose implements stress-induced senescence in human vascular endothelial cells: role of superoxide production by NADPH oxidase. PLoS One 2015;10:e0123169. https://doi.org/10.1371/journal.pone.0123169.

[174] The Homocysteine Studies Collaboration. Homocysteine and risk of ischemic heart disease and stroke. JAMA 2002;288:2015−22.

[175] Wilson PWF. Homocysteine and coronary heart disease: how great is the hazard? JAMA 2002;288:2042−3.

[176] Xu D, Neville R, Finkel T. Homocysteine accelerates endothelial cell senescence. FEBS Lett 2000;470:20−4.

[177] von Zglinicki T, Pilger R, Sitte N. Accumulation of single-strand breaks is the major cause of telomere shortening in human fibroblasts. Free Radic Biol Med 2000;28:64−74.

[178] Voghel G, Thorin-Trescasesa N, Farhata N, Nguyen A, Villeneuvea L, Mamarbachi AM, et al. Cellular senescence in endothelial cells from atherosclerotic patients is accelerated by oxidative stress associated with cardiovascular risk factors. Mech Ageing Dev 2007;128:662−71. https://doi.org/10.1016/j.mad.2007.09.006.

[179] Aviv A. Telomeres, sex, reactive oxygen species, and human cardiovascular aging. J Mol Med 2002;80:689−95.

[180] Hazzard WR. What heterogeneity among centenarians can teach us about genetics, aging, and longevity. J Am Geriatr Soc 2001;49:1568–9. https://doi.org/10.1046/j.1532-5415.2001.4911256.x.

[181] Mezzetti A, Zuliani G, Romano F, Costantini F, Pierdomenico SD, Cuccurullo F, et al. Vitamin E and lipid peroxide plasma levels predict the risk of cardiovascular events in a group of healthy very old people. J Am Geriatr Soc 2001;49:533–7. https://doi.org/10.1046/j.1532-5415.2001.49110.x.

[182] Cherubini A, Zuliani G, Costantini F, Pierdomenico SD, Volpato S, Mezzetti A, et al. High vitamin E plasma levels and low low-density lipoprotein oxidation are associated with the absence of atherosclerosis in octogenarians. J Am Geriatr Soc 2001;49:651–4. https://doi.org/10.1046/j.1532-5415.2001.49128.x.

[183] Psaty BM, Smith NL, Lemaitre RN, Vos HL, Heckbert SR, LaCroix AZ, et al. Hormone replacement therapy, prothrombotic mutations, and the risk of incident nonfatal myocardial infarction in postmenopausal women. JAMA 2001;285:906–13. https://doi.org/10.1001/jama.285.7.906.

[184] Ridker PM, Stampfer MJ, Rifai N. Novel risk factors for systemic atherosclerosis: a comparison of C-reactive protein, fibrinogen, homocysteine, lipoprotein(a), and standard cholesterol screening as predictors of peripheral arterial disease. JAMA 2001;285:2481–5. https://doi.org/10.1001/jama.285.19.2481.

[185] Patel VB, Robbins MA, Topol EJ. C reactive protein: a 'golden marker' for inflammation and coronary artery disease. Cleve Clin J Med 2001; 68:521–4. 527 534.

[186] Zhang R, Brennan M-L, Fu X, Aviles RJ, Pearce GL, Penn MS, et al. Association between myeloperoxidase levels and risk of coronary artery disease. JAMA 2001;286:2136–42. https://doi.org/10.1001/jama.286.17.2136.

[187] Nordstrom CK, Dwyer KM, Merz CN, Shircore A, Dwyer JH. Work-related stress and early atherosclerosis. Epidemiology 2001;12:180–5. https://www.jstor.org/stable/3703620.

[188] Valtonen VV. Role of infections in atherosclerosis. Am Heart J 1999;138:S431. https://doi.org/10.1016/S0002-8703(99)70269-3.

[189] Slavkin HC, Baum BJ. Relationship of dental and oral pathology to systemic illness. JAMA 2000;284:1215–7. https://doi.org/10.1001/jama.284.10.1215.

[190] Mattila KJ, Valtonen VV, Nieminen MS, Asikainen S. Role of infection as a risk factor for atherosclerosis, myocardial infarction, and stroke. Clin Infect Dis 1998;26:719–34. https://doi.org/10.1086/514570.

[191] Muhlestein JB. Bacterial infections and atherosclerosis. J Invest Med 1998;46:396–402.

[192] Shah PK. Plaque disruption and thrombosis: potential role of inflammation and infection. Cardiol Rev 2000;8:31–9.

[193] Movahed MR. Infection with Chlamydia pneumoniae and atherosclerosis: a review. J S C Med Assoc 1999;95:303–8.

[194] Pradhan AD, Manson JE, Rossouw JE, Siscovick DS, Mouton CP, Rifai N, et al. Inflammatory biomarkers, hormone replacement therapy, and incident coronary heart disease. JAMA 2003;288:980–7. https://doi.org/10.1001/jama.288.8.980.

[195] Chang E, Harley CB. Telomere length and replicative aging in human vascular tissues. Proc Nat Acad Sci USA 1995;92:1190–11194.

[196] Lefkowitz RJ, Willerson JT. Prospects for cardiovascular research. JAMA 2001;285:581–93.

[197] Shelton DN, Chang E, Whittier PS, Choi D, Funk WD. Microarray analysis of replicative senescence. Curr Biol 1999;9:939–45.

[198] Zglinicki T, Martin-Ruiz CM. Telomeres as biomarkers for ageing and age-related diseases. Curr Mol Med 2005;5:197–203. https://doi.org/10.1152/ajpheart.00332.2004.

[199] Ogami M, Ikura Y, Ohsawa M, Matsuo T, Kayo S, Yoshimi N, et al. Telomere shortening in human coronary artery diseases. Arterioscler Thromb Vasc Biol 2004;24:546–50. https://doi.org/10.1161/01.ATV.0000117200.46938.e7.

[200] Okuda K, Khan MY, Skurnick J, Kimura M, Aviv H, Aviv A. Telomere attrition of the human abdominal aorta: relationships with age and atherosclerosis. Atherosclerosis 2000;152:391–8. https://doi.org/10.1016/S0021-915000482-7.

[201] Minamino T, Komuro I. Role of telomere in endothelial dysfunction in atherosclerosis. Curr Opin Lipidol 2002;13:537–43.

[202] Minamino T, Miyauchi H, Yoshida T, Ishida Y, Yoshida H, Komuro I. Endothelial cell senescence in human atherosclerosis: role of telomere in endothelial dysfunction. Circulation 2002;105:1541–4. https://doi.org/10.1161/01.CIR.0000013836.85741.17.

[203] Aviv H, Yusuf Khan M, Skurnick J, Okuda K, Kimura M, Gardner J, et al. Age dependent aneuploidy and telomere length of the human vascular endothelium. Atherosclerosis 2002;159:281–7.

[204] Cooke JP. Flow, NO, and atherogenesis. Proc Natl Acad Sci USA 2003;100:768–70. https://doi.org/10.1073/pnas.0430082100.

[205] Samani NJ, Boultby R, Butler R, Thompson JR, Goodall AH. Telomere shortening in atherosclerosis. Lancet 2001;358:472–3.

[206] Nowak R, Siwicki JK, Chechlinska M, Markowicz S. Telomere shortening and atherosclerosis. Lancet 2002;359:976. discussion 976-977.

[207] Cawthon RM, Smith KR, O'Briend E, Sivatchenkoc A, Kerberc RA. Association between telomere length in blood and mortality in people aged 60 Years or older. Lancet 2003;361:393–5.

[208] Jeanclos E, Schork NJ, Kyvik KO, Kimura M, Skurnick JH, Aviv A. Telomere length inversely correlates with pulse pressure and is highly familial. Hypertension 2000;36:195–2000.

[209] Fossel M. Use of telomere lengths as a biomarker for aging and age-related disease. Curr Translat Geriatr Experiment Gerontol Rep 2012;1: 121–7. https://doi.org/10.1007/s13670-012-0013-6.

[210] Semeraro MD, Almer G, Renner W, Gruber HJ, Herrmann M. Telomere length in leucocytes and solid tissues of young and aged rats. Aging 2022;14:1713–28. https://doi.org/10.18632/aging.203922.

[211] Khan S, Naidoo DP, Chuturgoon AA. Telomeres and atherosclerosis: review. S Afr J Diabetes Vasc Dis. 2015;12. Available online: https://hdl.handle.net/10520/EJC181963. [Accessed 30 March 2023].

[212] den Buijs JO, Musters M, Verrips T, Post JP, Braam B, van Riel N. Mathematical modeling of vascular endothelial layer maintenance: the role of endothelial cell division, progenitor cell homing, and telomere shortening. Am J Physiol 2004;287:H2651–8. https://doi.org/10.1152/ajpheart.00332.2004.

[213] Xie Y, Lou D, Zhang D. Melatonin alleviates age-associated endothelial injury of atherosclerosis via regulating telomere function. J Inflamm Res 2021;14:6799–812. https://doi.org/10.2147/JIR.S329020.

[214] Schade DS, Gonzales K, Eaton RP. Stop stenting; start reversing atherosclerosis. Am J Med 2021;134:301–3. https://doi.org/10.1016/j.amjmed.2020.10.009.

[215] Nissen SE, Tuzcu EM, Schoenhagen P, Brown BG, Ganz P, Vogel RA, et al. Effect of intensive compared with moderate lipid-lowering therapy on progression of coronary atherosclerosis: a randomized controlled trial. JAMA 2004;291:1071−80. https://doi.org/10.1001/jama.291.9.1071.

[216] Wu CM, Zheng L, Wang Q, Hu YW. The emerging role of cell senescence in atherosclerosis. Clin Chem Lab Med 2020;59:27−38. https://doi.org/10.1515/cclm-2020-0601.

[217] Buja LM, et al. Vascular pathobiology: atherosclerosis and large vessel disease. In: Cardiovascular pathology. 4th ed. New York, NY, USA: Academic Press; 2015. p. 87−9.

[218] Wang J, Uryga AK, Reinhold J, Figg N, Baker L, Finigan AJ, et al. Vascular smooth muscle cell senescence promotes atherosclerosis and features of plaque vulnerability. Circulation 2015;132:1909−19. https://doi.org/10.1161/CIRCULATIONAHA.115.016457.

[219] Honda S, Ikeda K, Urata R, Yamazaki E, Emoto N, Matoba S. Cellular senescence promotes endothelial activation through epigenetic alteration, and consequently accelerates atherosclerosis. Sci Rep 2021;11:14608. https://doi.org/10.1038/s41598-021-94097-5.

[220] Dominic A, Banerjee P, Hamilton DJ, Le NT, Abe J. Time-dependent replicative senescence vs. disturbed flow-induced pre-mature aging in atherosclerosis. Redox Biol 2020;37:101614. https://doi.org/10.1016/j.redox.2020.101614.

[221] Martinez P, Blasco MA. An enzyme to cure age-related diseases. Nat Catal 2021;4:738−9. https://doi.org/10.1038/s41929-021-00677-z.

[222] Fossel M. Cells, aging, and human disease. Oxford, UK: Oxford University Press; 2004.

[223] Fossel M. Telomerase and the aging cell: implications for human health. JAMA 1998;279:1732−5. https://doi.org/10.1001/jama.279.21.1732.

[224] Banks D, Fossel M. Telomeres, cancer, and aging: altering the human lifespan. JAMA 1997;278:1345−8. https://doi.org/10.1001/jama.1997.03550160065040.

[225] Wald NJ, Simmonds M, Morris JK. Screening for future cardiovascular disease using age alone compared with multiple risk factors and age. PLoS One 2011;6:e18742. https://doi.org/10.1371/journal.pone.0018742.

[226] Childs BG, Li H, van Deursen JM. Senescent cells: a therapeutic target for cardiovascular disease. J Clin Invest 2018;128:1217−28. https://doi.org/10.1172/JCI95146.

[227] Evangelou K, Vasileiou PVS, Papaspyropoulos A, Hazapis O, Petty R, Demaria M, et al. Cellular senescence and cardiovascular diseases: moving to the "heart" of the problem. Physiol Rev 2022;103:609−47. https://doi.org/10.1152/physrev.00007.2022.

[228] Weber T, Mayer CC. "Man is as old as his arteries" taken literally: in search of the best metric. Hypertension 2020;76:1425−7. https://doi.org/10.1161/HYPERTENSIONAHA.120.16128.

[229] Hayflick L. The greatest risk factor for the leading cause of death is ignored. Biogerontology 2020;22:133−41. https://doi.org/10.1007/s10522-020-09901-y.

[230] Matsushita H, Chang E, Glassford AJ, Cooke JP, Chiu CP, Tsao PS. eNOS activity is reduced in senescent human endothelial cells: preservation by hTERT immortalization. Circ Res 2001;89:793−8.

[231] Cooke JP. The endothelium: a new target for therapy. Vasc Med 2000;5:49−53.

[232] Booth LK, Redgrave RE, Tual-Chalot S, Spyridopoulos I, Phillips HM, Richardson GD. Heart disease and ageing: the roles of senescence, mitochondria, and telomerase in cardiovascular disease. Subcell Biochem 2023;103:45−78. https://doi.org/10.1007/978-3-031-26576-1_4.

[233] Bloom SI, Islam MT, Lesniewski LA, Donato AJ. Mechanisms and consequences of endothelial cell senescence. Nat Rev Cardiol 2022;19:1−4. https://doi.org/10.1038/s41569-022-00739-0.

[234] Murakami T, Inagaki N, Kondoh H. Cellular senescence in diabetes mellitus: distinct senotherapeutic strategies for adipose tissue and pancreatic β cells. Front Endocrinol 2022;13:869414. https://doi.org/10.3389/fendo.2022.869414.

[235] Fossel M. Cell senescence, telomerase, and senolytic therapy. OBM Geriatr 2019;3:1−14. https://doi.org/10.21926/obm.geriatr.1901034.

[236] Lagoumtzia SM, Chondrogiannia N. Senolytics and senomorphics: natural and synthetic therapeutics in the treatment of aging and chronic diseases. Free Radic Biol Med 2021;171:169−219. https://doi.org/10.1016/j.freeradbiomed.2021.05.003.

[237] Stojanović SD, Fiedler J, Bauersachs J, Thum T, Sedding DG. Senescence-induced inflammation: an important player and key therapeutic target in atherosclerosis. Eur Heart J 2020;41:2983−96. https://doi.org/10.1093/eurheartj/ehz919.

[238] Bodnar AG, Ouellette M, Frolkis M, Holt SE, Chiu CP, Morin GB, et al. Extension of life-span by introduction of telomerase into normal human cells. Science 1998;279:349−52. https://doi.org/10.1126/science.279.5349.349.

[239] Funk WD, Wang CK, Shelton DN, Harley CB, Pagon GB, Hoeffler WK. Telomerase expression restores dermal integrity to in vitro-aged fibroblasts in a reconstituted skin model. Exp Cell Res 2000;258:270−8. https://doi.org/10.1006/excr.2000.4945.

[240] Yudoh K, Matsuno H, Nakazawa F, Katayama R, Kimura T. Reconstituting telomerase activity using the telomerase catalytic subunit prevents the telomere shorting and replicative senescence in human osteoblasts. J Bone Miner Res 2001;16:1453−64. https://doi.org/10.1359/jbmr.2001.16.8.1453.

[241] Mojiri A, Walther BK, Jiang C, Matrone G, Holgate R, Xu Q, et al. Telomerase therapy reverses vascular senescence and extends lifespan in progeria mice. Eur Heart J 2021;42:4352−69. https://doi.org/10.1093/eurheartj/ehab547.

[242] Jaskelioff M, Muller FL, Paik JH, Thomas E, Jiang S, Adams A, et al. Telomerase reactivation reverses tissue degeneration in aged telomerase deficient mice. Nature 2011;469:102−6. https://doi.org/10.1038/nature09603.

[243] Shim HS, Horner JW, Wu CJ, Li J, Lan ZD, Jiang S, et al. Telomerase reverse transcriptase preserves neuron survival and cognition in Alzheimer's disease models. Nat Aging 2021;1:1162−74. https://doi.org/10.1038/s43587-021-00146-z.

[244] Yeh JK, Lin MH, Wang CY. Telomeres as therapeutic targets in heart disease. JACC (J Am Coll Cardiol) 2019;4:855−65. https://doi.org/10.1016/j.jacbts.2019.05.009.

[245] Ramunas J, Yakubov E, Brady JJ, Corbel SY, Holbrook C, Brandt M, et al. Transient delivery of modified mRNA encoding TERT rapidly extends telomeres in human cells. Faseb J 2015;29:1930. https://doi.org/10.1096/fj.14-259531.

[246] Maier R, Bawamia B, Bennaceur K, Dunn S, Marsay L, Amoah R, et al. Telomerase activation to reverse immunosenescence in elderly patients with acute coronary syndrome: protocol for a randomized pilot trial. JMIR Res Protoc 2020;9:e19456. https://doi.org/10.2196/19456.

[247] Nazari-Shafti TZ, Cooke JP. Telomerase therapy to reverse cardiovascular senescence. Methodist DeBakey Cardiovasc J 2015;11:172. https://doi.org/10.14797/mdcj-11-3-172.

[248] Hoffmann J, Richardson G, Haendeler J, Altschmied J, Andrés V, Spyridopoulos I. Telomerase as a therapeutic target in cardiovascular disease. Arterioscler Thromb Vasc Biol 2021;41:1047−61. https://doi.org/10.1161/ATVBAHA.120.315695.

[249] Cooke JP. Mechanisms of atherosclerosis: new insights and novel therapeutic approaches. Methodist Debakey Cardiovasc J 2015;11:154. https://doi.org/10.14797/mdcj-11-3-154.

[250] Bär C, de Jesus BB, Serrano RM, Tejera AM, Ayuso E, Jimenez V, et al. Telomerase expression confers cardioprotection in the adult mouse heart after acute myocardial infarction. Nat Commun 2014;5:5863. https://doi.org/10.1038/ncomms6863.

[251] Chanda PK, Sukhovershin R, Cooke JP. mRNA-enhanced cell therapy and cardiovascular regeneration. Cells 2021;10:187. https://doi.org/10.3390/cells10010187.

[252] Cooke JP, Youker KA. Future impact of mRNA therapy on cardiovascular diseases. Methodist DeBakey Cardiovasc J 2022;18(5). https://doi.org/10.14797/mdcvj.1169.

Further reading

[1] Yusuf S, Pinto FJ. The polypill: from concept and evidence to implementation. Lancet 2022;400(10364):1661−3.

CHAPTER 4

Age-related disease: Bones

Joshua N. Farr[1] and Abhishek Chandra[2]

[1]Department of Endocrinology, Robert and Arlene Kogod Center of Aging, Mayo Clinic, Rochester, MN, United States
[2]Department of Physiology and Biomedical Engineering, Section of Geriatric Medicine and Gerontology, Robert and Arlene Kogod Center of Aging, Mayo Clinic, Rochester, MN, United States

1. Introduction: Bone anatomy and homeostasis

The skeletal system consists of approximately 206 bones that provide a framework for supporting and protecting the softer tissues and other organ systems. In addition, the skeletal system also fulfills several unique tasks, including hematopoiesis, storage of several mineral and metabolic entities, endocrine signaling, and providing support for tendons and ligaments. The skeleton of mammals comprises two types of bone: cortical (or compact) and trabecular (or cancellous) bone. The cortical bone, with an outer layer (periosteum) and an inner surface (endosteum) [1], comprises ~80% of the adult skeleton, with the remaining comprised of trabecular bone, which is spongy in nature (Fig. 4.1). Cortical bone forms the shaft of long bones, located at the ends of joints and in the vertebrae. Based on the microarchitecture, bone can be characterized as woven or lamellar bone [2]. Woven bone consists of osteocytes and type I collagen with no structural organization and is seen mostly in cases of fracture or trauma. Woven bone can appear de novo, without the presence of previous bone or cartilage [3]. Lamellar bone consists of a more organized mature bone with collagen fibers arranged in parallel sheaths on flat surfaces.

Trabecular bone makes up most of vertebral bodies and fills the medullary cavity of long bones. Trabecular bone, while being porous, has a much larger surface area than cortical bone [4]. It also forms much more intricate connections with soft tissue such as bone marrow, vascular, and connective tissue. Trabecular bone is susceptible to changes in metabolic processes and conditions subjecting to bone loss, which affect trabecular bone at a higher rate than cortical bone [5,6]. Age shifts the balance of remodeling toward bone resorption over formation which leads to thinning of trabecular bone over time [5].

The bony matrix comprises of type I collagen fibers and noncollagenous proteins. $Ca_{10}(PO_4)_6(OH)_2$, hydroxyapatite, forms ~95% of its mineral weight, with carbonate and other small impurities. Type I collagen is approximately 90% of the total protein in bone, while absorbed plasma proteins and proteins synthesized by bone-forming cells form the noncollagenous component. Bony matrix glycoproteins and proteoglycans together stabilize the mineral crystal [7]. Calcium and phosphorus form most of the mineral content in bone, with magnesium, sodium, and bicarbonate forming a smaller subset of these minerals. Bone serves as a reservoir for these minerals [8], which are critical for normal bone health and maintenance of skeletal integrity, while regulating functions in other tissues and organs. These endocrine and exocrine regulators form a complex interplay in maintaining perfect homeostatic regulation of circulating levels of these minerals. At the cellular level, the skeleton is generally maintained by the tight regulatory function of the bone-forming osteoblasts and the bone-resorbing osteoclasts, a process better known as bone remodeling, which continues long after the longitudinal growth of bones has ceased (Fig. 4.1). Mesenchymal stem cells or marrow stromal fibroblasts give rise to osteoblasts and are influenced by local growth factors including BMPs, IGFs, and TGF-β [9–11]. Osteoblasts deposit mineral and form collagen-enriched bone matrix, which embeds the osteocytes, the most abundant cell type in bone with a vast interconnecting network referred to as the canalicular network. Osteocytes are known to closely regulate bone homeostasis, controlling the function of both osteoblasts and osteoclasts using signals which travel through the canalicular network. Osteoclasts, on the other hand, are large multinucleate cells of hematopoietic origin, that are responsible for resorbing the bone matrix, with tight regulation from

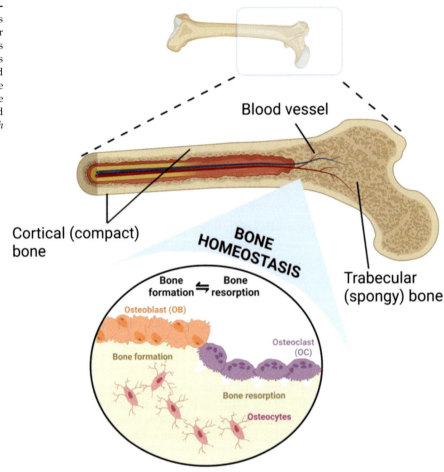

FIGURE 4.1 **Bone anatomy and homeostasis.** Structural organization of bone includes compact or cortical bone, which forms the outer shell, while trabecular or spongy bone forms the inner framework. Intercellular interactions between bone-forming osteoblasts (OBs) and bone-resorbing osteoclasts, together with the regulation from osteocytes, maintains the bone homeostasis, where both the formation and resorption processes are balanced. *Created with BioRender.com.*

osteoblasts and osteocytes, while receiving constant support from progenitors such as bone marrow monocytes or macrophage precursors [12–14]. The resorption of bone by osteoclasts creates a surface to be laid by new bone orchestrated by osteoblasts (Fig. 4.2).

2. Skeletal maintenance and remodeling

As mentioned above, skeletal remodeling is a highly orchestrated and dynamic process, with rejuvenation occurring in a continuum, with bone resorption preceding bone formation. Any imbalance to this orchestration causes uncoupling between the osteoblasts and osteoclasts, as seen in age-related bone loss. This process of bone formation and resorption is carried out by unique clusters of osteoblasts and osteoclasts often referred to as basic multicellular units (BMUs) (also known as the bone-remodeling units). The process of remodeling can be divided into four phases: activation, resorption, reversal, and formation (Fig. 4.2).

2.1 Activation (phase 1)

Activation is the process of recruitment of mononucleated osteoclast precursors in circulation to the bone surface where they fuse to form multinucleated pre-osteoclasts, and is often initiated due to structural stresses, damage, or hormonal changes (e.g., estrogen or PTH) [15]. Osteoblasts and marrow stromal cells (and possibly T cells) provide cues that help recruit pre-osteoclasts in order to initiate a bone remodeling unit [16]. The bone remodeling

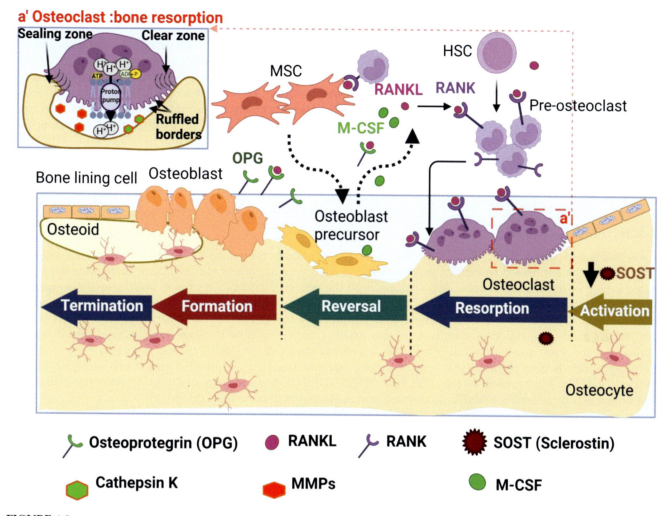

FIGURE 4.2 **Bone remodeling: an overview.** The process of bone remodeling is depicted here with a sequence of events, *activation*, *resorption*, *reversal*, and *formation*, presented graphically. The termination step which is not an event by itself but the end point of bone formation or mineralization, results in a newly formed bone, with embedded osteocytes. The activation step is a trigger from environmental stresses, hormones, or injury, where cells on the bone surface secrete factors that active the mononuclear hematopoietic stem precursor cells (HSCs) to fuse and form the pre-osteoclast and eventually bone resorbing osteoclasts. The osteoclasts (a′) create an acidic environment with a milieu of proteins and enzymes that help in the resorption process. The resorption phase involving the multinucleated osteoclasts lasts for few weeks, and ends up in a newly laid down surface for the newly reversed osteoblast precursors to lay down bone in the formation phase. The bone formation phase takes weeks to months and results in several layers of bone as the process of remodeling continues. *Created with BioRender.com.*

compartment provides a perfect microenvironment to facilitate coupling between osteoblasts and osteoclasts. Factors such as growth hormone, parathyroid hormone, vitamin D and local factors including IL-1, IL-6, RANKL, and colony-stimulating factors, specifically M-CSF, regulate the interaction between osteoblasts and osteoclasts [5,15–17]. Apart from M-CSF, RANKL is another key component responsible for the recruitment and differentiation of osteoclasts. RANKL is produced by osteoblasts and mesenchymal stem cells or osteoprogenitors, which then allow interaction with the RANK receptor on the pre-osteoclasts, which then facilitates differentiation and maturation of osteoclasts making them capable of resorbing bone. Osteoprotegerin (OPG) belongs to the TNF receptor super-family, and is mainly produced by osteoblasts [18]. OPG is found in a free-floating form, and serves as a soluble decoy receptor to RANKL, thereby inhibiting osteoclastogenesis and bone resorption [18]. OPG levels in serum increase with old age and are probably in compensation to the enhanced bone resorption seen during estrogen deprivation or in old age [19]. Vitamin D_3, parathyroid hormone, prostaglandin E2, and IL-1, IL-4, IL-6, IL-11, IL-17, TNF-α have been shown to induce osteoclastogenesis by inhibiting OPG while simultaneously stimulating the production of RANKL [18,20]. Estrogen has been shown to inhibit RANKL and RANKL-stimulated osteoclastogenesis [20].

2.2 Resorption (phase 2)

The resorption phase involves an erosion cavity which is generated across the bone surface. Mononuclear precursors are recruited to the bone surface, combine, and form larger, activated, multinucleated osteoclasts [1]. Hydrogen ions, which provide the acidic environment required for resorption, are transferred in the resorption cavity through the proton pumps (Fig. 4.2) [5,21]. Structurally, osteoclasts have fold-like structures called "ruffled border," which contain the lysosome enzymes to digest mineral matrix [21]. The attachment to the bone surface is termed the "sealing zone" which is supported by an actin-filament enriched region of "clear zone" [21]. Deficiency in the presence of ruffled border or clear zone has been shown to causes osteopetrosis [22]. The secretion of cathepsin K, which is a protease, enriched with serine, cystine or aspartic acid residues, tartrate-resistant acid phosphatase (TRACP), and matrix metalloproteinases (MMPs), forms a milieu of enzymes that resorb the mineral matrix [5,23,24]. Following resorption, the osteoclasts undergo apoptosis and pave the way for reversal.

2.3 Reversal (phase 3)

As mentioned above, the reversal phase (last 4–5 weeks) initiates following the resorption step. This stage continues to see the unfinished resorption process and preparation of the newly exposed bone surface for the orchestrated laying of the osteoblast cells. This requires both the action of catabolic agents as well as anabolic agents to smoothen the surface for laying of new osteoblasts. Mononuclear cells in the reversal zone are unique in nature, with the majority of them expressing Runx2 and without any monocytic or osteoclastic markers. Osteoblasts in the process of mineralization synthesize a protein matrix comprising type I collagen-rich osteoid which lays the foundation for subsequent mineralization. This recruitment of osteoblasts to fill the resorbed surface defines coupling which is also regulated by several factors released from the resorbed matrix, such as IGF-I and -II, BMPs, PDGF, fibroblast growth factors, and TGF-β [5,25–27]. TGF-β is a known inhibitor for osteoclast differentiation which acts by reducing RANKL production and suppression of bone resorption [28].

2.4 Formation (phase 4)

Bone formation and bone resorption do not happen in synchrony, as resorption sites and formation sites occur at different places. Bone formation begins with the assembly of osteoblasts which lay down the osteoid that allows a framework for mineralization. During this process a subset of osteoblasts get embedded in the mineralized bone and form osteocytes. Due to still partially understood mechanisms, interactions between osteocytes and cells on the bone surface regulate the overall mineralization process, a step which can last for months and eventually provides strength and density to the new bone [29]. Once the remodeling of the bone is complete, the osteoblasts involved in the process follow different fates [5], where some become bone-lining cells, and others undergo apoptosis. Bone-lining cells regulate the calcium ions and other ionic flux into and out of the bone, modulate local bone formation and remodeling, and also have the capability to reactivate from a quiescent to active state in the setting of mechanical loading [30,31]. While remodeling regulates both cortical and trabecular regions, during aging an imbalance in the remodeling process on the periosteal and endosteal surfaces leads to a decrease in cortical thickness and thinning of the trabecular plates [5]. Bone loss to trabecular bones often outpaces resorption of the cortical bone. This causes the trabecular bones to be more prone to fracture [32]. Uncoupling between the resorption and formation process can lead to loss of trabeculae and increased porosity in bone [33], which results in osteoporosis.

3. Osteoporosis: An overview

Osteoporosis is a single major metabolic bone disease characterized by reduced bone mass with increased risk of fragility fractures, comorbidities, and mortality, which makes it one of the most physically debilitating diseases associated with aging. Among the aged worldwide population, only 31%–36% of those older than 70 years of age have normal bone mass [9]. The lifetime fracture risk in men is generally lower than that of women who are predisposed to osteoporosis due to loss of estrogen levels following the menopause [7,34,35].

The loss of estrogen in post-menopausal women of all ages and later in aged men leads to a substantial burden of osteoporosis and osteoporotic fractures [36]. A similar decrease in BMD is observed with patients with declining androgen levels [37]. However, age-related osteoporosis and post-menopausal osteoporosis are caused by distinct underlying mechanisms. As discussed later in this chapter, it is evident now that cellular senescence and apoptosis drive age-related bone loss, however post-menopausal bone loss is largely independent of these mechanisms of aging [38]. Both post-menopausal- and age-related osteoporosis result in an imbalance in bone remodeling, where bone resorption outpaces bone formation and elevated levels of markers of bone turnover are seen in older individuals. For example, the EPIDOS trial reported that elevated levels of osteocalcin, N-telopeptide, C-telopeptide, and bone-specific alkaline phosphatase were observed in elderly women with the lowest BMD [39]. By contrast, bone formation markers, such as procollagen peptide, do not increase [40]. Several co-morbidities of aging such as diabetes, chronic kidney disease, drug interactions, and immobility further affect bone metabolism and may further exacerbate osteoporosis and the risk of fracture [41–48].

4. Therapeutic options to treat osteoporosis: Current interventions

Serum calcium is regulated within a narrow physiologic range, and an imbalance in calcium levels triggers the activation of osteoclasts which resorb bone to release calcium, thereby maintaining calcium homeostasis. Vitamin D (cholecalciferol) is responsible for calcium resorption in the gastrointestinal (GI) tract; hence any loss of vitamin D also leads to a loss of calcium which is excreted through the GI, urine, sweat, or skin. Chronic glucocorticoid excess, hyperthyroidism, and vitamin D deficiency lead to conditions of negative mineral balance [49]. Aging exacerbates the efflux of circulating minerals from bone [49]. Thus, an early intervention to treat bone health is dietary supplementation of calcium and vitamin D, but while these supplements may help slow down some level of bone loss, often the imbalance in osteoclast and osteoblast activity is too extreme to overcome these mineral imbalances. Drugs targeting osteoclasts are termed "antiresorptives" and activation of bone formation as a therapy is often referred as "anabolic" or "osteoanabolic" therapy. Whether a patient is prescribed an anabolic or antiresorptive often depends on their T-score, a measure of the BMD by bone densitometry representing standard deviation from peak bone mass of an average 30-year-old. Defined by a World Health Organization (WHO) report, T-score was used to define a normal BMD (T-score ≥ -1), low bone mass or osteopenia (T-score of < -1 and > -2.5), osteoporosis (T-score ≤ -2.5), and severe osteoporosis (T-score ≤ -2.5) accompanied by fragility fractures [50,51]. Typically, antiresorptives are favored over anabolics between the T-score of < -1 and > -2.5, while they are less effective for severe osteoporosis. Bone-forming anabolics are prescribed to higher risk individuals over antiresorptives when the T-score is < -2.5 and the patient has a history of fragility fracture (Fig. 4.3).

4.1 Antiresorptive and osteoclast-targeting therapeutics

Because osteoclasts respond to a negative calcium balance in the body by releasing calcium through bone resorption, these cells represent a logical target to treat osteoporosis. Thus, the strategy to eliminate osteoclasts and/or suppress their functions, has led to several approved therapeutics as well as others which are in the pipeline. Bisphosphonates form the major class of antiresorptive drugs that are characterized by a pyrophosphate-like chemical structure which facilitates a strong binding affinity to calcium. The nitrogen-containing bisphosphonates, etidronate, clodronate, risedronate, alendronate, olpadronate, ibandronate, and zoledronate [52,53] all target farnesyl diphosphate synthase (FPP synthase), a mevalonate pathway enzyme, to thereby inhibit osteoclast function. In several placebo-controlled trials, bisphosphonates have been shown to reduce the risk of hip and femur fractures, although long-term use of bisphosphonates can increase atypical femoral fracture risk as well as the risk of osteonecrosis of the jaw; nevertheless, the benefit of risk reduction far outweighs the risk of these uncommon serious adverse events [54].

Receptor activator of nuclear factor kappa-B (RANK) ligand (RANKL), a protein produced by osteoblasts or marrow stromal cells, is responsible for differentiation of osteoclasts after binding to the RANK receptor on the pre-osteoclast surface [9,10]. Denosumab, formerly known as AMG162, is a fully human monoclonal antibody, that acts by blocking the binding of RANKL to RANK, thereby mimicking osteoprotegerin (OPG), an endogenous decoy receptor for RANKL, to thereby prevent osteoclast differentiation (Fig. 4.3). This permits Denosumab to suppress osteoclast function and function as an effective antiresorptive therapy for patients with osteoporosis where it improves bone parameters and reduces the risk of vertebral, nonvertebral, and hip fractures in men and women, and has been associated with fewer adverse events as compared to other antiresorptives [55–58]. Furthermore, Denosumab was also found to be effective in suppressing bone turnover in patients with breast cancer-related bone metastases [59].

Since loss of estrogen is a major trigger for osteoclast-based bone loss, interest in using estrogen as a target has remained since the mid-20th century. The efficacies of estrogen and selective estrogen receptor modulators (SERMs)

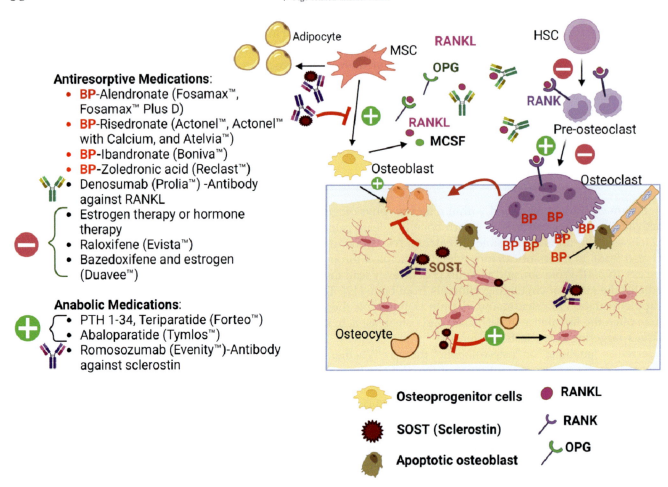

FIGURE 4.3 **Therapeutics for osteoporosis.** This figure depicts various FDA-approved treatments for osteoporosis. While the antiresorptives work by suppressing osteoclast function, the anabolic drugs work by promoting bone formation by activating osteoprogenitor transition to osteoblasts, which then mineralize to form new bone. *BP*, bisphosphonates; *PTH*, parathyroid hormone. A negative sign indicates negative regulation of osteoclast function, mainly showing estrogen-related therapies, while a positive sign indicates the presence of anabolic drugs, which have a role both in promoting osteoblast and osteoclast functions. Sost or sclerostin is a negative Wnt inhibitor, and blocking sclerostin by a neutralizing antibody promotes Wnt-based positive regulation of bone formation and suppression of age- or disease-related bone marrow adiposity. *Created with BioRender.com.*

have been explored as a treatment for post-menopausal osteoporosis. Raloxifene is one such SERM which has been shown to prevent ovariectomy-related bone resorption in rodents [60–63] and suppress post-menopausal osteoporosis in women, by not only suppressing osteoclast function and resorption, but having positive induction of the bone-forming osteoblasts [64]. From the early days of development of Raloxifene, the focus was to reduce the significant side effects caused by these SERMs mainly in reproductive organs, and Raloxifene achieved that at that time [65]. In direct comparisons of SERMs with nitrogenous bisphosphonates, SERMs were found to be more efficacious, as bisphosphonates also have a negative effect on bone accrual and quality of bone [66]. However, as compared to estrogen therapy, the SERMs were always found to be less efficacious [67]. To find a better alternative, several studies showed that an analog of Raloxifene, Raloxifene HCL, was more effective than estrogen or Raloxifene [68]. In a randomized control trial, Raloxifene not only improved bone mineral density, but also reduced fracture risk [69]. Bazedoxifene, another SERM, has been tested alone and in combination with estrogen therapy to suppress osteoclast function and hence post-menopausal osteoporosis [70–73].

As illustrated in Fig. 4.4, most classical antiresorptive drugs will suppress bone resorption, often kill the osteoclast, hence affecting osteoclast–osteoblast bidirectional coupling, and also affecting osteoblast viability and eventually bone formation [74,75]. However, the risk of osteonecrosis is considered low and is outweighed by the benefits as a treatment for osteoporosis.

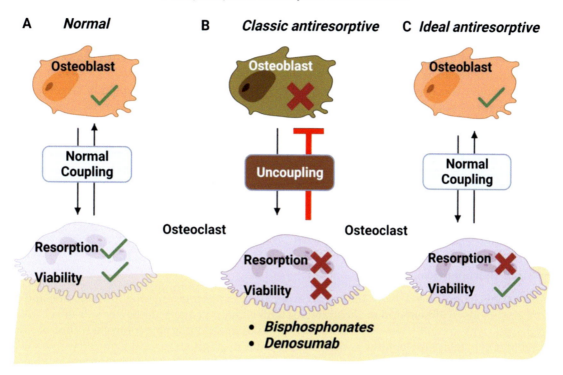

FIGURE 4.4 **Theoretical take on an ideal antiresorptive.** Based on studies done in vitro, preclinical, and clinical studies, it is now understood that the release of factors from osteoblasts maintains osteoclast function, while osteoclasts release proteins know as coupling factors, that maintain osteoblast proliferation, differentiation and cell death. (A) A physiological state where this relationship between osteoblast and osteoclast is bidirectional and often in balance. (B) Classic antiresorptive drugs reduce osteoclast viability, hence reducing the release of coupling factors, which affects osteoblast function and bone formation. Bisphosphonates and denosumab both fall in this classic antiresorptives class of drugs. (C) An ideal antiresorptive includes those drugs that can block bone resorption, but maintain osteoclast viability, which then allows normal release of coupling factors supporting or maintaining bone formation. *Created with BioRender.com.*

Cathepsin K, secreted by lysosomes in the osteoclasts, is one of the most potent cysteine proteases, with the major function to resorb bone. Cathepsin K is the most abundant cathepsin that is expressed in the bone [76]. The cathepsin K inhibitor, Odanacatib, falls into another class of drugs which have the properties of an ideal antiresorptive, where the resorption function of the osteoclasts is suppressed without affecting their viability, thereby maintaining the osteoblast function through coupling (Fig. 4.4). Unfortunately, Odanacatib caused several adverse events, such as adverse cardiovascular events (stroke, atrial fibrillation, and atrial flutter), and morphea-like skin reactions. These adverse events did not allow the clinical use of Odanacatib, but a future avenue was opened for the ideal antiresorptive.

4.2 Bone anabolics

Bone anabolics or osteo-anabolics function by activating quiescent bone-lining cells or osteoblasts on the bone surface, sometimes because of bone remodeling, with osteoblasts being responsible for filling in the remodeling cavities. The osteoanabolic drugs tilt the balance of modeling-based bone formation over remodeling-based bone formation, resulting in a net gain in bone accrual (Fig. 4.3).

Parathyroid hormone (PTH) is made by the parathyroid glands, with its major function being as a regulator of calcium levels in the blood. PTH was one of the first anabolic molecules approved by FDA and the European Union as a biosynthetic peptide, PTH 1-34 (Teriparatide; Forteo) for the treatment of osteoporosis [77–79]. The intact PTH (1-84) was approved for treatment in many other countries outside the United States. An analog of PTH-related peptide (PTHrP), known as Abaloparatide (marketed as TYMLOS), has been shown to increase BMD at the lumbar spine, femoral neck, and hip [80,81]. Abaloparatide may serve as a superior anabolic treatment option when compared to Teriparatide [82]. However, because of the increased occurrence of osteosarcoma in animals receiving prolonged treatment of Teriparatide or Abaloparatide, both these modalities have a 2-year limitation for use [83].

Despite these black box warnings and a 2 year limitation, human studies have failed to show any incidence of osteosarcoma in patients receiving Teriparatide treatment [84].

We have also reported the efficacy of teriparatide to promote DNA repair by activation of the Wnt pathway [85,86], thereby suggesting a role of Teriparatide in suppressing cellular senescence.

Another osteoanabolic therapeutic was developed in the form of a monoclonal antibody targeting sclerostin, an inhibitor of the canonical WNT signaling. Sclerostin became a central focus with the identification of two rare autosomal recessive syndromes: sclerosteosis and van Buchem disease [87,88], which are both high bone mass phenotypes resulting from genetic deletion of the sclerostin gene, *SOST* [89,90]. A similar targeted deletion of the *Sost* gene in transgenic mice phenocopied the human high bone mass phenotype [91] and enhanced fracture healing [92]. Further studies using a neutralizing antibody against sclerostin demonstrated its efficacy for improving bone formation and bone strength in a rat model of post-menopausal osteoporosis [93]. This led to multiple clinical trials [94–97] and an FDA approval to use Romosozumab as a combination anabolic and antiresorptive treatment for severe osteoporosis [98].

4.3 Prospective therapeutics

In addition to the FDA-approved drugs for osteoporosis, there is rigorous ongoing investigation and new clinical trials for more efficacious treatment options or strategies with better outcomes and fewer side effects. A denosumab biosimilar, QL1206, was recently tested in a phase III randomized, placebo-controlled clinical trial showing improved lumbar spine, total hip, femoral neck, and trochanter in postmenopausal Chinese women with osteoporosis [99]. Combinatorial therapies using an antiresorptive after the osteoanabolic treatments are also being explored. One such study assessed the transition to denosumab after romosozumab treatment [100]. The combination of teriparatide and denosumab increased BMD more than either compound alone [101]. Phytomedicinal compounds such as *Zingiber officinale* (ginger) and *Curcuma longa* L. (turmeric) have recently been tested in clinical trials demonstrating their potential benefits for bone health [102].

5. Cellular components of an aging bone

5.1 Osteoblasts

Osteoblasts are cells of mesenchymal origin that are regulated by several signaling molecules such as BMPs, IGFs, and TGF-β [9–11]. Several transcriptional regulators, including the homeodomain proteins (e.g., Msx-2, Dlx-2, Dlx-5, and BAPX1), steroid receptors, as well as the helix-loop-helix (HLH) proteins Id, Twist, and Dermo, maintain the osteoprogenitor nature of mesenchymal cells [103]. Osteoblasts regulate the bone remodeling process through the action of several signaling molecules including PTH, sex hormones, 1,25-dihydroxyvitamin D, glucocorticoids, growth hormone, thyroids hormone, interleukins, TNF α, prostaglandins, IGFs, TGF-β, BMPs, WNTs, fibroblast growth factors (FGFs), platelet-derived growth factors (PDGFs), vascular endothelial growth factors (VEGFs), and interferon γ (IF γ) [104–106].

Several factors secreted from osteoblasts regulate the osteoclast function, thereby controlling the bone remodeling process. M-CSF secreted by osteoblasts promotes the proliferation and survival of osteoclast precursor [107], while RANKL, which is expressed on the osteoblast surfaces, interacts with RANK receptor to promote osteoclastogenesis. OPG (also known as Tumor necrosis factor receptor super family 11B [TNFRSF11B]), is a peptide produced by the osteoblasts, that acts as a nonsignaling decoy receptor for RANKL, thereby suppressing osteoclast function and bone resorption [18]. OPG is elevated in the serum of elderly people which may serve as a compensatory response to age-related changes in estrogen deficiency and bone loss [19]. The correlation between age, menopause, and disease with levels of RANKL and OPG often decides the BMD status [108], with RANKL levels reported to increase with age, while no change or decline in OPG levels is seen with age [108]. In addition to physiological aging, the RANKL and OPG ratio determines bone health in several disease conditions [109–112].

BMPs and the Wnt signaling pathway positively regulate bone formation, via a key transcriptional regulator, Runx2 [113]. The runt-related transcription factor 2/core binding factor α1 (Runx2/Cbfa1) commits the fate of mesenchymal progenitors to the osteoblast lineage, thereby regulating osteoblast differentiation [114–116]. Osterix (Osx), a downstream regulator of Runx2, controls the terminal stages of osteoblast differentiation [117]. Based on these studies, both Runx2 and Osx have important roles in the formation of functional osteoblasts [117].

Bone formation occur in two steps, where osteoblasts create a layer of osteoid which lays the foundation to promote mineralization (Fig. 4.2). The resulting lacunae and canaliculi are filled with calcium and phosphorus with the inhibition of mineral-degrading enzymes [118]. A subset of newly formed osteoblasts gets embedded in the bone matrix and become osteocytes, while some others form the lining cells. These bone-lining cells are quiescent in nature and have been shown to be activated by bone anabolic signals [119], thus serving important functions for future bone formation.

A subset of these osteoblasts undergoes apoptosis following completion of the mineralization process. Several growth factors, such as insulin, IGF, Wnt's, and PTH, have been shown to regulate osteoblast survival [85,86,120−122]. Multiple agents have a negative impact on osteoblast survival [123−126]. As discussed later in this chapter, cellular senescence is another fate which the majority of osteoblasts undergo both in physiological [127,128] and pathological aging of bone [129,130].

5.2 Osteocytes

Osteoblasts that get embedded in the bony matrix differentiate into osteocytes, a cell type which has extended cytoplasmic processes called canaliculi that connect to create a network responsible for cell-to-cell communication. Osteocytes interact with other osteocytes, and with cells on the bone surface through these canalicular networks through secretory signals often carried by microvesicular structures, which travel through canaliculi releasing factors in the bone marrow environment [131,132]. An adult human skeleton has 23 million osteocytic canalicular connections between osteocytes and with the bone cells on the bone surface [133]. Osteocytes are the most abundant cell types in the bone matrix, and recent studies on these transduced signals from these osteocytes have shown the importance of these previously considered inactive cell type as important mechano-sensors communicating the signals for mechanical loading [134,135]. Osteocytes serve as positive or negative regulators of bone remodeling [136], but become mostly negative during aging when the lacunar density declines and connectivity between these lacunae is disrupted [137,138]. Longitudinal studies in mice have shown that the number of osteocytes also declines with age and has an inverse relationship with the empty lacunae which increase with old age [139,140]. Following the identification of gap junction proteins such as Connexin 43(Cx43) [141,142], their presence in an osteocyte cell line was also detected [143]. Cx43 was found to be essential for bone material property, as deletion of the *Cx43* gene in osteocytes resulted in defective mineralization [144]. Cx43 is a key component in relaying mechanosensory signals from and to the osteocytes [145] via prostaglandin E2 [146,147]. Overexpression of Cx43 promotes osteocyte survival during physiological aging [148] and also protects against oxidative stress [149].

5.3 Osteoclasts

Osteoclasts are multinucleated cells of hematopoietic origin, which are formed by the coalescing of the mononuclear precursors on the bone surface [1]. This process, which is seemingly random, is highly orchestrated under the influence of systemic factors such as growth hormone, parathyroid hormone, vitamin D, and local factors including IL-1, IL-6, RANKL, and colony-stimulating factors, specifically M-CSF which stimulates the interaction between osteoblast and osteoclast [5,15−17]. These interactions promote the differentiation and maturation of osteoclast precursors which coalesce and mature into bone-resorbing osteoclasts [1]. These activated osteoclasts propagate protons in the resorption pit, which allows the lowering of pH with the release of hydrogen ions via proton pumps [5,21]. The mature osteoclast have a peculiar and visible "ruffled border," with unique projections containing the bone-resorbing lysosomal enzymes [21]. The osteoclasts adhere tightly to the bone surface forming a "clear zone" which engulfs the ruffled border and is composed of actin-like filaments (Fig. 4.2) [21]. The secreted proteases dissolve the matrix and mineral within the erosion cavity [5]. A lack of ruffled border and a clear zone were key characteristics of osteopetrotic rats, thus showing the importance of these morphological features of the osteoclast in its function [22]. The other components of the resorption mix include the enzymes TRACP, cathepsin K, and matrix metalloproteinase MMP-9, which assist in the resorption of the collagen and noncollagen matrix components [5,23,24]. Aging, menopause, and low-dose radiation are some of the conditions which activate osteoclastogenesis and promote bone resorption [150−152]. However, higher doses of radiation have been shown to inhibit osteoclast formation and activity [153]. There is a direct correlation between osteoclasts and RANKL levels, while an inverse correlation exists with OPG with advancing age [154].

5.4 Adipocytes

The mesenchymal stem cells which derive both osteoblast and adipocytic cell lineages, maintain a crucial balance during physiological aging. Bone marrow adipose tissue (BMAT) has an inverse relationship with BMD with advancing age [10]. Estrogen deficiency, immobilization, glucocorticoid treatment, and radiotherapy are some other conditions which skew the fate of mesenchymal stem cells to adipocytes at the expense of bone volume [10]. With aging, mesenchymal stem cells tend to form more adipocytes than osteoblasts with increases in both the number and size of these adipocytes [155]. An increase in marrow adiposity and inverse reduction in bone quality with elevated risks of fractures are currently used as predictors of osteoporosis [156,157]. Several animal studies have shown that loss of function of BMAT-related genes can improve bone mass [158,159]; however, the actual causes of BMAT are still being investigated. Cellular senescence correlates with the occurrence of BMAT, both in physiological bone aging, as well as with pathological bone aging [160]. Based on studies performed in transgenic mouse models that permit inducible clearance of senescent cells, it was shown that $p21^{Cip1}$ could be a key senescence-related regulator of BMAT and pathological osteoporosis [161], with $p16^{Ink4a}$ playing a larger role during age-related osteoporosis and related BMAT [162].

5.5 Endothelial cells

Endothelial cells are a key component of bone vasculature, and while they contribute to the formation of blood vessels, they have other functions as well. Alterations in skeletal vasculature with aging were recently reported together with the identification of the key skeletal specific endothelial cells which were type H ($CD31^{hi}$ and Endomucin^{hi}) and type L ($CD31^{lo}$ and Endomucin^{lo}) [163]. These type H cells were found in the proximity of bone progenitor cells which were osterix positive. With aging, the type H endothelial cells decline, while the type L remained constant [163,164]. Impairments in wound healing abilities with old age [165] are a major health burden.

6. Biology of an aging bone: Mechanisms and pathways

6.1 Biology of an aging bone: Cellular senescence

It was first shown by Hayflick and Moorehead that cells can undergo replicative senescence [166], a term which is often used interchangeably with cellular aging. Together with replicative senescence, which may mimic physiological aging to some extent, cytotoxic stress-induced premature senescence (SIPS) is also triggered by oncogene activation [167], free radical generation and accumulation, ROS, DNA damage, telomeric DNA damage, proteostasis, mitochondrial dysfunction, activation of pro-survival pathways, etc. [168,169]. Senescent cells have a peculiar cellular morphology characterized by flattened and enlarged size, with ruffled cellular surfaces [170], and often with an altered chromatin, also known as senescence-associated heterochromatin foci (SAHF) [171]. Genomic alteration to the nuclear component marker lamin B1 (LMNB1), together with a reduction in its expression levels, triggers the spatial relocalization of the perinuclear H3K9me3-positive heterochromatin, thus triggering SAHF formation [172]. Histone modification has been used to detect SAHF as a biomarker of senescent cells [173].

Several biomarkers of senescence are now well identified, but there is no universal biomarker for senescence, and the actual trigger of senescence during physiological aging remains unclear, and while it is a gradual process, sometimes the generation of senescent cells is accelerated during pathological conditions. The healthspan in old mice has been improved following clearance of senescent cells using pharmacological approaches and genetic mouse models of senolysis [174,175]. The pharmacological clearance of senescent cells is currently being explored as a therapeutic for healthy aging in several ongoing clinical trials.

Cellular senescence is dependent on the cytotoxic triggers but is mostly regulated by two pathways, the ataxia telangiectasia mutated (ATM)/p53/$p21^{Cip1}$ and $p16^{INK4a}$/RB pathways. The $p21^{Cip1}$ and $p16^{INK4a}$ pathways may drive cellular senescence both independently [176,177] and depending on each other [178], which is decided by the initial cytotoxic insult. *$p16^{INK4a}$* interacts directly with and inhibits the catalytic activity of cyclin-dependent kinase (CDK) 4 [179] and CDK6 [180].

The DNA damage response (DDR) stabilizes the tumor suppressor protein, p53, thereby activating the CDK inhibitor (CDKi), *$p21^{Cip1}$*, and thus causing growth arrest [181,182]. The activation of CDKi's blocks the CDKs, which is followed by the activation of the tumor suppressor protein, retinoblastoma (pRB). Apart from the $p21^{Cip1}$ and p16^{INK4a} pathways, p27^{Kip1} is also a CDKi which is known to block the function of CDK2/4 [183,184]. Loss of both p21^{Cip1} and p27^{Kip1} results in tumorigenesis in mice, suggesting a synergistic effects of these CDKi's [185].

SAM-R/3 and SAM-P/6 mice were among the earliest attempts to study the role of cellular senescence in bone deterioration [186]. These mice and many related strains shared several characteristics of age-associated co-morbidities [187]. The seminal studies which identified senescent cells and biomarkers related to senescence and the senescence-associated secretory phenotype (SASP) were detected in aged bone cells from mice and humans [127] and in radiated bones in a model of pathological bone aging [129]. Pharmacological and genetic models of clearance of senescent cells established the importance of senescent cells in causing physiological and pathological bone aging [129,161,162]. Furthermore, targeted clearance of senescent cells using the senolytic drug cocktail, Dasatinib plus Quercetin (D + Q), was shown to enhance fracture healing in a transient manner [188]. Using two independent genetic mouse models where senescent cells expressing either *p21^{Cip1}* or *p16^{Ink4a}* were cleared, it was shown that radiation-induced skeletal aging was largely a p21^{Cip1}-driven process [161].

One key feature of senescent cells is their proinflammatory secretome, termed the SASP, which is now known to be dependent on cell type and the inducer of senescence [189]. Recent studies have clarified that SASP may also vary based on the expression of either *p21^{Cip1}* or *p16^{Ink4a}* [161]. Pharmacological blocking of the SASP could alleviate age-related frailty [190] and osteoporosis in mice [162].

6.2 Biology of an aging bone: DNA damage and genomic instability

One of the major causes of cellular senescence and apparently in aging is the accumulation of multiple mutations, and failure to maintain the genomic integrity, which is more prominent due to additional cytotoxic insults. With advancing age these mutations continue to increase, and with exogenous exposures to chemical, physical, and biological agents, the chances of chromatin modifications become more likely. The endogenous conditions that cause these mutations may include error-prone DNA replication, ROS, and hydrolytic reactions [191]. ROS induces a DDR response that is tightly controlled by several pathways depending on the kind of DNA damage. Among several kinds of DNA damage and the DDRs, the most crucial are: (i) fixing of alkylated bases resulting in an error-free repair, (ii) a base excision repair (BER) fixes oxidative, deamination, alkylation, and abasic single base damage, (iii) nucleotide excision repair (NER) controls bulky base repairs, (iv) mismatch repair (MMR) resolves replicative fidelity, (v) inter-cross-link repair (ICL) resolves covalent linkages of adjacent DNA strands, and (vi) the DNA break repair including single-strand break (SSB) repair and double-strand break (DSB) repair. Some key DDR components include ATM kinase, ataxia telangiectasia and Rad3-related (ATR) kinase, PARP1, and three DSB repair pathways [classical nonhomologous end joining (c-NHEJ), alternative (alt)-NHEJ, and homology-directed repair (HDR)]. Genomic instability can also result from replication stress and high speed of fork progression [192]. Deletion of topoisomerase I can result in genomic instability and neurodegeneration, a disease prevalent among the elderly [193].

Hutchinson—Gilford progeria syndrome (HGPS) and progeroid laminopathies are severe forms of rare early-onset premature aging diseases that are triggered by mutations in the lamin genes [194] resulting in severe age-related morbidities, including osteolysis, craniofacial disproportion, short and sculptured nose, delayed dentition, and scoliosis [195]. Mutations in the Xeroderma pigmentosum (XP)-type D (XPD) gene causes the human disorder trichothiodystrophy (TTD). Premature aging and osteoporosis were among the key features in TTD mice [196]. Deficiency in the ATM kinase gene in mice resulted in bone loss with increased osteoclastogenesis and disrupted osteoblastogenesis [197].

6.3 Biology of an aging bone: Telomere dysfunction and shortening

Telomeres consist of repeat DNA sequences of TTAGGG located at the ends of chromosomes. They act as a molecular clock and regulate the replicative ability of a cell. Intact telomeres and their importance in genomic stability were identified by several groups in the past century [198]. Telomere length is maintained by *telomerase*, which consists of an RNA subunit and a catalytic protein subunit called telomerase reverse transcriptase (TERT). Telomerase expression can increase the life span of cells [199], improve longevity in adult and old mice [200], and reverse age-related tissue degeneration in telomerase-deficient mice [201]. Interestingly, telomerase gene therapy can also improve bone mineral density in both 12-month-old and 24-month-old mice [200].

Senescence in human cells is associated with shortening of the telomeres, which is driven by an ATM, p53, and p21^{Cip1}-based mechanism. The telomere sequence is generally shielded by the shelterin complex, which consists of six distinct proteins: TRF1, TRF2, Rap1, TIN2, TPP1, and POT1 [202]. Without the presence of shelterin, the telomeric site is detected as a DNA damage site thereby targeted by several DDR pathways, including ATM and ATR kinases, PARP1, c-NHEJ, (alt)-NHEJ, and HDR [202,203]. The shelterin subunits involved with this process vary with different DDR pathways. Werner's syndrome (WS) and dyskeratosis congenita are two accelerated models of aging in mice in which WS helicase and telomerase have been genetically deleted [204]. Further studies showed that single mutation in the telomerase gene, *Terc* also showed age-related bone loss [205].

Telomere dysfunction, as detected by the co-localization of DNA damage at sites of telomeric DNA [206], is now used as a robust marker of cellular senescence that has been used in several studies to identify senescent osteoblasts, osteocytes, and bone marrow cells [127,129,160,161,207].

6.4 Biology of an aging bone: Mitochondrial dysfunction

Mitochondria are considered the cellular powerhouse, requiring enormous amounts of intracellular oxygen which is accompanied by the production of energy, and generation of ROS as a byproduct. Oxidative stress and ROS due to hyperoxia cause cells to undergo DNA damage and cellular senescence [208]. In the absence of mitochondria, senescent cells fail to produce enough ROS to generate chromatin aberrations and SASP [209]. Levels of ROS can regulate the balance of function and the cross-talk between osteoblasts and osteoclasts [210]. Elevated levels of ROS cause osteoblast and osteocyte apoptosis and deterioration in bone architecture [211]. An elevated ROS also activates osteoclasts while suppressing osteoblast function, thereby disturbing the bone homeostasis, resulting in bone loss [210,212,213]. Osteoclasts resorb bone in an environment requiring high levels of energy, generated by the mitochondria. This energy production is required for acid production used for resorbing bone. Mutation in mtDNA polymerase gamma (Polg), a mitochondrial DNA polymerase, results in elevated osteoclast activity and reduced osteogenesis, thus causing an accelerated osteoporosis phenotype [214]. Another study reported subchondral bone loss including the articular cartilage in the joints of prematurely aging $Polg^{D275A}$ mutant mice [215]. Primary mitochondrial diseases involving the dysfunctional mitochondrial respiratory chain, are linked to increased risk of fractures [216]. Another antioxidant enzyme countering H_2O_2, catalase, when overexpressed, has been shown to both extend the lifespan and reduce age-associated pathologies in mice [217,218]. The role of catalase in age-related osteoporosis is still unclear, but there are some studies which suggest that catalase may regulate skeletal tissue remodeling [219]. PTEN-induced kinase1 (PINK1), which is a serine/threonine kinase that regulates mitochondrial quality control was found to be indispensable for osteoblast differentiation [220]. Sirtuin 3 (Sirt3), a protein deacetylase, was also shown to play important roles in mitochondrial biogenesis, dynamics, and mitophagy. In addition, Sirt3 has been reported to have important roles in bone loss caused by aging or estrogen-deficiency [221–223], which conflicts with reports that Sirt3 is important for osteoblast function and bone formation [224].

6.5 Biology of an aging bone: Epigenetic regulations

Epigenetic regulators have important roles in skeletal aging of bone cells. Indeed, epigenetic changes have a direct correlation with age-related bone loss, including changes in methylation of CpG islands [225,226]. In patients with osteoporotic fractures, *SOST* gene expression was elevated as compared to normal patients, however the patients with osteoporotic fractures had a higher demethylation pattern in the *SOST* gene, suggesting demethylation of *SOST gene* could be a risk factor for osteoporosis [227].

Histone modification posttranslationally can occur either by acetylation, methylation, ubiquitylation, or phosphorylation. Acetylation of histones regulates the transcription of several genes involved in osteogenic differentiation [228–230]. Histone deacetylase 3 (Hdac3), a nuclear enzyme which removes acetyl groups from lysine residues in histones, is key for bone formation during aging [231]. Hdac8 was found to suppress osteogenic differentiation by suppressing the transcriptional activity of Runx2 [229]. Histone methylation at lysine sites, such as H3K27 by EZH2, a histone methyltransferase, regulates osteogenesis [232–235]. A genome-wide study identified unique methylation sites in patients with osteoporosis and osteoarthritis [236]. However, blood is not a good source to detect DNA methylation as a predictor for age-related osteoporosis [237].

6.6 Biology of an aging bone: Loss of proteostasis

The cellular machinery maintains homeostasis by regulating the balance between protein synthesis, folding, and degradation. Impaired proteostasis may result in the aggregation of damaged proteins or misfolded proteins which are now known to cause multiple age-related diseases, including Parkinson's and Alzheimer's [238–241].

The maintenance of protein homeostasis requires the disposal of unwanted proteins, handled by either the 26S-proteasome or through the autophagy pathway. The 26S proteasome activity is reduced in senescent cells, causing accumulation of cross-linked and damaged proteins [242–244]. Inhibition of proteasome has been shown to induce senescence [245], however suppression of proteasome suppresses osteoclast differentiation [246,247], promotes osteoblast differentiation in vitro and bone formation in vivo, both in normal conditions [248,249] as well as in multiple myeloma patients [246,250]. Proteasome inhibition was also shown to stabilize DNA repair proteins, Ku70 and DNA-PKC, in radiated osteoblasts, promoting DNA repair and survival in osteoblasts and thus preventing radiation-induced bone damage in mice [251]. These results suggest that proteostasis may have a larger function than is seen in other organ systems and may be cell type specific. Due to this ambiguity and side effects in other organs, none of the proteasome inhibitors which were tested preclinically, it failed to reach clinical trials as a treatment for osteoporosis.

7. Therapeutic options: Future potential interventions and emerging trials

Thousands of clinical trials have been conducted to treat osteoporosis, even using unique behavioral modifications such as yoga [252] and dietary supplements [253]. More studies are attempting to treat osteoporosis using methods other than targeting bone resorption pathways or promoting bone formation. A randomized controlled trial looking at the effects of selective beta blockers on bone turnover markers identified the role of the sympathetic nervous system in bone metabolism via β2-adrenergic receptors [254]. A glucagon-like peptide-1 (GLP-1) receptor agonist is also being tested to improve trabecular bone score in postmenopausal women with T2D [255].

Identification of novel therapeutic drugs based on the concept of targeting mechanisms of cellular aging led to the preclinical and clinical advent of senotherapeutics [129,162,256]. Eliminating senescent cells as an option to improve bone health in older women is currently being tested in a randomized clinical trial (NCT04313634). Other potential senolytic compounds such as Navitoclax (ABT263) [257], BCL-XL inhibitors [258], HSP90 inhibitors [256], Piperlongumine [259], RG7112 [260], O-Vanillin [260], ABT-737 [261], CD153 vaccine [262], and aspirin [263,264], as recently reviewed by Robbins et al. [265], many of which are in preclinical stages, may serve as therapeutic options to treat age-related osteoporosis. Another approach which does not target senescent cells, but rather blocks their SASP, using a senomorphic drug (Ruxolitinib), showed promising results in preclinical studies to treat age-related osteoporosis in old mice [162]. Other senomorphic drugs such as Rapamycin and Rapalogs (analogs of Rapamycin) [266,267], have been shown to improve bone mass in mice [268,269], however, whether these strategies are effective in humans will require future trials.

8. Conclusion

Aging is a complex process which often results in tissue dysfunction. Mechanisms underlying age-related skeletal dysfunction are being identified and therapeutics targeting these mechanisms are emerging as novel approaches to treat age-related diseases, such as osteoporosis. There is still reliance on antiresorptive and bone anabolic drugs as the only available source of treatment, which have limited efficacy, potential side effects, and do not benefit other comorbidities of aging. The concept of targeting fundamental mechanisms of old age to thereby delay the onset of multiple co-morbidities, including osteoporosis, as a group, is currently being tested in humans and may become a viable option to slow or delay the onset of age-related diseases in the future.

References

[1] Jilka RL. Biology of the basic multicellular unit and the pathophysiology of osteoporosis. Med Pediatr Oncol 2003;41(3):182–5.
[2] Martin RB, Burr DB, Sharkey NA. Skeletal biology. In: Skeletal tissue mechanics. Springer Nature; 1998. p. 29–78.
[3] Boyde A. The real response of bone to exercise. J Anat 2003;203(2):173–89.
[4] Weiss L. Cell and tissue biology. In: Histology. Springer Nature; 1983. p. 1–87.
[5] Dempster D. Anatomy and functions of the adult skeleton. In: Favus M, editor. Primer on the metabolic bone diseases and disorders of the mineral metabolism. 6th ed. Washington, DC: The American Society for Bone and Mineral Research; 2006. p. 9.
[6] Silva MJ, Gibson LJ. Modeling the mechanical behavior of vertebral trabecular bone: effects of age-related changes in microstructure. Bone 1997;21(2):191–9.
[7] McKee MD, Addison WN, Kaartinen MT. Hierarchies of extracellular matrix and mineral organization in bone of the craniofacial complex and skeleton. Cells Tissues Organs 2006;181(3–4):176–88.

[8] Anderson JJB, Garner SC. Calcium and phosphorus in health and diseasevol. 11(7); 2009. p. 1040.
[9] Srouji S, Livne E. Bone marrow stem cells and biological scaffold for bone repair in aging and disease. Mech Ageing Dev 2005;126(2):281–7.
[10] Nuttall ME, et al. Human trabecular bone cells are able to express both osteoblastic and adipocytic phenotype: implications for osteopenic disorders. J Bone Miner Res 1998;13(3):371–82.
[11] Canalis E. Skeletal growth factors. In: Osteoporosis. Philadelphia: Lippincott Williams & Wilkins; 2000. p. 391–410.
[12] Yasuda H, et al. Identity of osteoclastogenesis inhibitory factor (OCIF) and osteoprotegerin (OPG): a mechanism by which OPG/OCIF inhibits osteoclastogenesis in vitro. Endocrinology 1998;139(3):1329–37.
[13] Kong YY, et al. OPGL is a key regulator of osteoclastogenesis, lymphocyte development and lymph-node organogenesis. Nature 1999; 397(6717):315–23.
[14] Xiong J, et al. Matrix-embedded cells control osteoclast formation. Nat Med 2011;17(10):1235–41.
[15] Roodman GD. Cell biology of the osteoclast. Exp Hematol 1999;27(8):1229–41.
[16] Boyle WJ, Simonet WS, Lacey DL. Osteoclast differentiation and activation. Nature 2003;423(6937):337–42.
[17] Troen BR. Molecular mechanisms underlying osteoclast formation and activation. Exp Gerontol 2003;38(6):605–14.
[18] Stejskal D, et al. Osteoprotegerin, rank, rankl. Biomed Pap 2001;145(2):61–4.
[19] Han KO, et al. The changes in circulating osteoprotegerin after hormone therapy in postmenopausal women and their relationship with oestrogen responsiveness on bone. Clin Endocrinol 2005;62(3):349–53.
[20] Aubin JE, Bonnelye E. Osteoprotegerin and its ligand: a new paradigm for regulation of osteoclastogenesis and bone resorption. Osteoporosis Int 2000;11(11):905–13.
[21] Zaidi M, et al. Osteoclast function and its control. Exp Physiol 1993;78(6):721–39.
[22] Holtrop ME, et al. The ultrastructure of osteoclasts in microphthalmic mice. Metab Bone Dis Relat Res 1981;3(2):123–9.
[23] Engsig MT, et al. Matrix metalloproteinase 9 and vascular endothelial growth factor are essential for osteoclast recruitment into developing long bones. J Cell Biol 2000;151(4):879–90.
[24] Delaissé J-M, et al. Matrix metalloproteinases (MMP) and cathepsin K contribute differently to osteoclastic activities. Microsc Res Tech 2003; 61(6):504–13.
[25] Bonewald LF, Mundy GR. Role of transforming growth factor-beta in bone remodeling. Clin Orthop Relat Res 1990;(250):261–76.
[26] Locklin R. Effects of TGFβ and BFGF on the differentiation of human bone marrow stromal fibroblasts. Cell Biol Int 1999;23(3):185–94.
[27] Ueland T. GH/IGF-I and bone resorption in vivo and in vitro. Eur J Endocrinol 2005;152(3):327–32.
[28] Fox SW, Lovibond AC. Current insights into the role of transforming growth factor-β in bone resorption. Mol Cell Endocrinol 2005;243(1–2): 19–26.
[29] Huiskes R, et al. Nature 2000;405(6787):704–6.
[30] Chow JWM, et al. Mechanical loading stimulates bone formation by reactivation of bone lining cells in 13-week-old rats. J Bone Miner Res 1998;13(11):1760–7.
[31] Hauge EM, et al. Cancellous bone remodeling occurs in specialized compartments lined by cells expressing osteoblastic markers. J Bone Miner Res 2001;16(9):1575–82.
[32] Garnero P, et al. Increased bone turnover in late postmenopausal women is a major determinant of osteoporosis. J Bone Miner Res 1996;11(3): 337–49.
[33] Reid IR. Menopause. In: Primer on the metabolic bone diseases and disorders of mineral metabolism. Wiley-Blackwell; 2013. p. 165–70.
[34] Bilezikian JP. Osteoporosis in men. J Clin Endocrinol Metab 1999;84(10):3431–4.
[35] Looker AC, et al. Prevalence of low femoral bone density in older U.S. Adults from NHANES III. J Bone Miner Res 1997;12(11):1761–8.
[36] Van Pottelbergh I, et al. Perturbed sex steroid status in men with idiopathic osteoporosis and their sons. J Clin Endocrinol Metab 2004;89(10): 4949–53.
[37] Amin S. Association of hypogonadism and estradiol levels with bone mineral density in elderly men from the Framingham study. Ann Intern Med 2000;133(12):951–63.
[38] Farr JN, et al. Independent roles of estrogen deficiency and cellular senescence in the pathogenesis of osteoporosis: evidence in young adult mice and older humans. J Bone Miner Res 2019;34(8):1407–18.
[39] Garnero P, et al. Markers of bone resorption predict hip fracture in elderly women: the EPIDOS prospective study. J Bone Miner Res 1996; 11(10):1531–8.
[40] Garnero P, et al. Biochemical markers of bone turnover, endogenous hormones and the risk of fractures in postmenopausal women: the OFELY study. J Bone Miner Res 2000;15(8):1526–36.
[41] Ho-Pham LT, et al. Type 2 diabetes is associated with higher trabecular bone density but lower cortical bone density: the Vietnam Osteoporosis Study. Osteoporos Int 2018;29(9):2059–67.
[42] Sheu A, et al. Fractures in type 2 diabetes confer excess mortality: the Dubbo osteoporosis epidemiology study. Bone 2022;159:116373.
[43] Kinsella S, et al. Moderate chronic kidney disease in women is associated with fracture occurrence independently of osteoporosis. Nephron Clin Pract 2010;116(3):c256–62.
[44] Chao CT, et al. Chronic kidney disease-related osteoporosis is associated with incident frailty among patients with diabetic kidney disease: a propensity score-matched cohort study. Osteoporos Int 2020;31(4):699–708.
[45] Chen JS, et al. Effect of age-related chronic immobility on markers of bone turnover. J Bone Miner Res 2005;21(2):324–31.
[46] Weinstein RS. Chapter 58. Glucocorticoid-induced osteoporosis. In: Primer on the metabolic bone diseases and disorders of mineral metabolism. Wiley-Blackwell; 2008. p. 267–72.
[47] Kaji H, et al. The threshold of bone mineral density for vertebral fracture in female patients with glucocorticoid-induced osteoporosis. Endocr J 2006;53(1):27–34.
[48] Kennedy CC, Papaioannou A, Adachi JD. Glucocorticoid-induced osteoporosis. Womens Health (Lond) 2006;2(1):65–74.
[49] Favus MJ, Goltzman D. Chapter 21. Regulation of calcium and magnesium. In: Primer on the metabolic bone diseases and disorders of mineral metabolism. Wiley-Blackwell; 2008. p. 103–8.
[50] Kanis JA. Assessment of fracture risk and its application to screening for postmenopausal osteoporosis: synopsis of a WHO report. WHO Study Group. Osteoporos Int 1994;4(6):368–81.

References

[51] Assessment of fracture risk and its application to screening for postmenopausal osteoporosis. Report of a WHO Study Group. World Health Organ Tech Rep Ser 1994;843:1–129.

[52] van beek E, et al. The role of geranylgeranylation in bone resorption and its suppression by bisphosphonates in fetal bone explants in vitro: a clue to the mechanism of action of nitrogen-containing bisphosphonates. J Bone Miner Res 1999;14(5):722–9.

[53] Kimmel DB. Mechanism of action, pharmacokinetic and pharmacodynamic profile, and clinical applications of nitrogen-containing bisphosphonates. J Dent Res 2007;86(11):1022–33.

[54] Black DM, et al. Atypical femur fracture risk versus fragility fracture prevention with bisphosphonates. N Engl J Med 2020;383(8):743–53.

[55] Cummings SR, et al. Denosumab for prevention of fractures in postmenopausal women with osteoporosis. N Engl J Med 2009;361(8):756–65.

[56] Smith MR, et al. Effects of denosumab on bone mineral density in men receiving androgen deprivation therapy for prostate cancer. J Urol 2009;182(6):2670–5.

[57] Bone HG, et al. Effects of denosumab on bone mineral density and bone turnover in postmenopausal women. J Clin Endocrinol Metab 2008; 93(6):2149–57.

[58] McClung MR, et al. Denosumab in postmenopausal women with low bone mineral density. N Engl J Med 2006;354(8):821–31.

[59] Lipton A, et al. Extended efficacy and safety of denosumab in breast cancer patients with bone metastases not receiving prior bisphosphonate therapy. Clin Cancer Res 2008;14(20):6690–6.

[60] Sato M, et al. Dual-energy x-ray absorptiometry of raloxifene effects on the lumbar vertebrae and femora of ovariectomized rats. J Bone Miner Res 1994;9(5):715–24.

[61] Bowman AR, et al. Raloxifene analog (LY117018 HCL) ameliorates cyclosporin A-induced osteopenia in oophorectomized rats. J Bone Miner Res 1996;11(8):1191–8.

[62] Draper MW, et al. A controlled trial of raloxifene (LY139481) HCl: impact on bone turnover and serum lipid profile in healthy postmenopausal women. J Bone Miner Res 1996;11(6):835–42.

[63] Evans GL, et al. Raloxifene inhibits bone turnover and prevents further cancellous bone loss in adult ovariectomized rats with established osteopenia. Endocrinology 1996;137(10):4139–44.

[64] Taranta A, et al. The selective estrogen receptor modulator raloxifene regulates osteoclast and osteoblast activity in vitro. Bone 2002;30(2): 368–76.

[65] Black LJ, et al. Raloxifene (LY139481 HCI) prevents bone loss and reduces serum cholesterol without causing uterine hypertrophy in ovariectomized rats. J Clin Invest 1994;93(1):63–9.

[66] Sato M, et al. Advantages of raloxifene over alendronate or estrogen on nonreproductive and reproductive tissues in the long-term dosing of ovariectomized rats. J Pharmacol Exp Ther 1996;279(1):298–305.

[67] Lufkin EG, et al. Treatment of established postmenopausal osteoporosis with raloxifene: a randomized trial. J Bone Miner Res 1998;13(11): 1747–54.

[68] Sato M, et al. LY353381.HCl: a novel raloxifene analog with improved SERM potency and efficacy in vivo. J Pharmacol Exp Ther 1998;287(1): 1–7.

[69] Ettinger B, et al. Reduction of vertebral fracture risk in postmenopausal women with osteoporosis treated with raloxifene: results from a 3-year randomized clinical trial. Multiple Outcomes of Raloxifene Evaluation (MORE) Investigators. JAMA 1999;282(7):637–45.

[70] Stump AL, Kelley KW, Wensel TM. Bazedoxifene: a third-generation selective estrogen receptor modulator for treatment of postmenopausal osteoporosis. Ann Pharmacother 2007;41(5):833–9.

[71] Palacios S, et al. Assessment of the safety of long-term bazedoxifene treatment on the reproductive tract in postmenopausal women with osteoporosis: results of a 7-year, randomized, placebo-controlled, phase 3 study. Maturitas 2013;76(1):81–7.

[72] Conjugated estrogens/bazedoxifene (Duavee) for menopausal symptoms and prevention of osteoporosis. Med Lett Drugs Ther 2014; 56(1441):33–4.

[73] Goldberg T, Fidler B. Conjugated estrogens/bazedoxifene (Duavee): a novel agent for the treatment of moderate-to-severe vasomotor symptoms associated with menopause and the prevention of postmenopausal osteoporosis. P T 2015;40(3):178–82.

[74] Dello Russo NM, et al. Osteonecrosis in the jaws of patients who are using oral biphosphonates to treat osteoporosis. Int J Oral Maxillofac Implants 2007;22(1):146–53.

[75] Lee SH, et al. Risk of osteonecrosis in patients taking bisphosphonates for prevention of osteoporosis: a systematic review and meta-analysis. Osteoporos Int 2014;25(3):1131–9.

[76] Drake FH, et al. Cathepsin K, but not cathepsins B, L, or S, is abundantly expressed in human osteoclasts. J Biol Chem 1996;271(21):12511–6.

[77] Forteo approved for osteoporosis treatment. FDA Consum 2003;37(2):4.

[78] Deal C, Gideon J. Recombinant human PTH 1-34 (Forteo): an anabolic drug for osteoporosis. Cleve Clin J Med 2003;70(7):585–6. 589–90, 592–586 passim.

[79] Hutton SF. Forteo (teriparatide): first approved medication to rebuild bone. S D J Med 2003;56(10):423–4.

[80] Matsumoto T, et al. Abaloparatide increases lumbar spine and hip BMD in Japanese patients with osteoporosis: the phase 3 ACTIVE-J study. J Clin Endocrinol Metab 2022;107(10):e4222–31.

[81] Leder BZ, et al. Effects of abaloparatide, a human parathyroid hormone-related peptide analog, on bone mineral density in postmenopausal women with osteoporosis. J Clin Endocrinol Metab 2015;100(2):697–706.

[82] Miller PD, et al. Effect of abaloparatide vs placebo on new vertebral fractures in postmenopausal women with osteoporosis: a randomized clinical trial. JAMA 2016;316(7):722–33.

[83] Vahle JL, et al. Bone neoplasms in F344 rats given teriparatide [rhPTH(1-34)] are dependent on duration of treatment and dose. Toxicol Pathol 2004;32(4):426–38.

[84] Gilsenan A, et al. Assessing the incidence of osteosarcoma among teriparatide users based on Medicare Part D and US State Cancer Registry Data. Pharmacoepidemiol Drug Saf 2020;29(12):1616–26.

[85] Schnoke M, Midura SB, Midura RJ. Parathyroid hormone suppresses osteoblast apoptosis by augmenting DNA repair. Bone 2009;45(3): 590–602.

[86] Chandra A, et al. PTH1-34 blocks radiation-induced osteoblast apoptosis by enhancing DNA repair through canonical Wnt pathway. J Biol Chem 2015;290(1):157–67.

[87] Truswell AS. Osteopetrosis with syndactyly; a morphological variant of Albers-Schonberg's disease. J Bone Joint Surg Br 1958;40-B(2): 209–18.
[88] Van Buchem FS, Hadders HN, Ubbens R. An uncommon familial systemic disease of the skeleton: hyperostosis corticalis generalisata familiaris. Acta Radiol 1955;44(2):109–20.
[89] Balemans W, et al. Increased bone density in sclerosteosis is due to the deficiency of a novel secreted protein (SOST). Hum Mol Genet 2001; 10(5):537–43.
[90] Balemans W, et al. Identification of a 52 kb deletion downstream of the SOST gene in patients with van Buchem disease. J Med Genet 2002; 39(2):91–7.
[91] Li X, et al. Targeted deletion of the sclerostin gene in mice results in increased bone formation and bone strength. J Bone Miner Res 2008;23(6): 860–9.
[92] Li C, et al. Increased callus mass and enhanced strength during fracture healing in mice lacking the sclerostin gene. Bone 2011;49(6):1178–85.
[93] Li X, et al. Sclerostin antibody treatment increases bone formation, bone mass, and bone strength in a rat model of postmenopausal osteoporosis. J Bone Miner Res 2009;24(4):578–88.
[94] Graeff C, et al. Administration of romosozumab improves vertebral trabecular and cortical bone as assessed with quantitative computed tomography and finite element analysis. Bone 2015;81:364–9.
[95] Lewiecki EM, et al. A phase III randomized placebo-controlled trial to evaluate efficacy and safety of romosozumab in men with osteoporosis. J Clin Endocrinol Metab 2018;103(9):3183–93.
[96] Ishibashi H, et al. Romosozumab increases bone mineral density in postmenopausal Japanese women with osteoporosis: a phase 2 study. Bone 2017;103:209–15.
[97] Tominaga A, et al. Early clinical effects, safety, and appropriate selection of bone markers in romosozumab treatment for osteoporosis patients: a 6-month study. Osteoporos Int 2021;32(4):653–61.
[98] Chavassieux P, et al. Bone-forming and antiresorptive effects of romosozumab in postmenopausal women with osteoporosis: bone histomorphometry and microcomputed tomography analysis after 2 and 12 months of treatment. J Bone Miner Res 2019;34(9):1597–608.
[99] Zhang H, et al. A phase III randomized, double-blind, placebo-controlled trial of the denosumab biosimilar QL1206 in postmenopausal Chinese women with osteoporosis and high fracture risk. Acta Pharmacol Sin 2023;44(2):446–53.
[100] Langdahl B, et al. Romosozumab efficacy and safety in European patients enrolled in the FRAME trial. Osteoporos Int 2022;33(12):2527–36.
[101] Tsai JN, et al. Teriparatide and denosumab, alone or combined, in women with postmenopausal osteoporosis: the DATA study randomised trial. Lancet 2013;382(9886):50–6.
[102] Aghamohammadi D, et al. Ginger (Zingiber officinale) and turmeric (Curcuma longa L.) supplementation effects on quality of life, body composition, bone mineral density and osteoporosis related biomarkers and micro-RNAs in women with postmenopausal osteoporosis: a study protocol for a randomized controlled clinical trial. J Complement Integr Med 2020;18(1):131–7.
[103] Ogata T, Noda M. Expression of ID, a negative regulator of helix-loop-helix DNA binding proteins, is down-regulated at confluence and enhanced by dexamethasone in a mouse osteoblastic cell line, MC3T3E1. Biochem Biophys Res Commun 1991;180(3):1194–9.
[104] Lacey DL, et al. Interleukin 4, interferon-gamma, and prostaglandin E impact the osteoclastic cell-forming potential of murine bone marrow macrophages. Endocrinology 1995;136(6):2367–76.
[105] Horwood NJ, et al. IL-12 alone and in synergy with IL-18 inhibits osteoclast formation in vitro. J Immunol 2001;166(8):4915–21.
[106] Mirosavljevic D, et al. T-cells mediate an inhibitory effect of interleukin-4 on osteoclastogenesis. J Bone Miner Res 2003;18(6):984–93.
[107] Owens J, Chambers TJ. Macrophage colony-stimulating factor (M-CSF) induces migration in osteoclasts in vitro. Biochem Biophys Res Commun 1993;195(3):1401–7.
[108] Cao J, et al. Expression of RANKL and OPG correlates with age-related bone loss in male C57BL/6 mice. J Bone Miner Res 2003;18(2):270–7.
[109] Kushlinskii NE, et al. Components of the RANK/RANKL/OPG system, IL-6, IL-8, IL-16, MMP-2, and calcitonin in the sera of patients with bone tumors. Bull Exp Biol Med 2014;157(4):520–3.
[110] Moschen AR, et al. The RANKL/OPG system and bone mineral density in patients with chronic liver disease. J Hepatol 2005;43(6):973–83.
[111] Mountzios G, et al. Abnormal bone remodeling process is due to an imbalance in the receptor activator of nuclear factor-kappaB ligand (RANKL)/osteoprotegerin (OPG) axis in patients with solid tumors metastatic to the skeleton. Acta Oncol 2007;46(2):221–9.
[112] Ozer FF, Dagdelen S, Erbas T. Relation of RANKL and OPG levels with bone resorption in patients with acromegaly and prolactinoma. Horm Metab Res 2018;50(7):562–7.
[113] Gaur T, et al. Canonical WNT signaling promotes osteogenesis by directly stimulating Runx2 gene expression. J Biol Chem 2005;280(39): 33132–40.
[114] Lecka-Czernik B, et al. Inhibition of Osf2/Cbfa1 expression and terminal osteoblast differentiation by PPARγ2. J Cell Biochem 1999;74(3): 357–71.
[115] Vaughan T, et al. Alleles of RUNX2/CBFA1 gene are associated with differences in bone mineral density and risk of fracture. J Bone Miner Res 2002;17(8):1527–34.
[116] Lee B, et al. Missense mutations abolishing DNA binding of the osteoblast-specific transcription factor OSF2/CBFA1 in cleidocranial dysplasia. Nat Genet 1997;16(3):307–10.
[117] Nakashima K, et al. The novel zinc finger-containing transcription factor osterix is required for osteoblast differentiation and bone formation. Cell 2002;108(1):17–29.
[118] Anderson HC. Matrix vesicles and calcification. Curr Rheumatol Rep 2003;5(3):222–6.
[119] Kim SW, et al. Sclerostin antibody administration converts bone lining cells into active osteoblasts. J Bone Miner Res 2017;32(5):892–901.
[120] Hill PA, Tumber A, Meikle MC. Multiple extracellular signals promote osteoblast survival and apoptosis. Endocrinology 1997;138(9): 3849–58.
[121] Jilka RL, et al. Osteoblast programmed cell death (apoptosis): modulation by growth factors and cytokines. J Bone Miner Res 1998;13(5): 793–802.
[122] Almeida M, et al. Wnt proteins prevent apoptosis of both uncommitted osteoblast progenitors and differentiated osteoblasts by beta-catenin-dependent and -independent signaling cascades involving Src/ERK and phosphatidylinositol 3-kinase/AKT. J Biol Chem 2005; 280(50):41342–51.

[123] Coonse KG, et al. Cadmium induces apoptosis in the human osteoblast-like cell line Saos-2. J Toxicol Environ Health 2007;70(7):575–81.
[124] Feng Y, Ding L, Li L. LPS-inducible circAtp9b is highly expressed in osteoporosis and promotes the apoptosis of osteoblasts by reducing the formation of mature miR-17-92a. J Orthop Surg Res 2022;17(1):193.
[125] Chen RM, et al. Nitric oxide induces osteoblast apoptosis through the de novo synthesis of Bax protein. J Orthop Res 2002;20(2):295–302.
[126] Soroceanu MA, et al. Rosiglitazone impacts negatively on bone by promoting osteoblast/osteocyte apoptosis. J Endocrinol 2004;183(1):203–16.
[127] Farr JN, et al. Identification of senescent cells in the bone microenvironment. J Bone Miner Res 2016;31(11):1920–9.
[128] Kim HN, et al. DNA damage and senescence in osteoprogenitors expressing Osx1 may cause their decrease with age. Aging Cell 2017;16(4):693–703.
[129] Chandra A, et al. Targeted reduction of senescent cell burden alleviates focal radiotherapy-related bone loss. J Bone Miner Res 2020;35(6):1119–31.
[130] Jo S, et al. WNT16 elevation induced cell senescence of osteoblasts in ankylosing spondylitis. Arthritis Res Ther 2021;23(1):301.
[131] Dallas SL, Prideaux M, Bonewald LF. The osteocyte: an endocrine cell ... and more. Endocr Rev 2013;34(5):658–90.
[132] Burra S, et al. Dendritic processes of osteocytes are mechanotransducers that induce the opening of hemichannels. Proc Natl Acad Sci U S A 2010;107(31):13648–53.
[133] Buenzli PR, Sims NA. Quantifying the osteocyte network in the human skeleton. Bone 2015;75:144–50.
[134] Mullender MG, Huiskes R. Osteocytes and bone lining cells: which are the best candidates for mechano-sensors in cancellous bone? Bone 1997;20(6):527–32.
[135] Uitterlinden AG, et al. Polymorphisms in the sclerosteosis/van Buchem disease gene (SOST) region are associated with bone-mineral density in elderly whites. Am J Hum Genet 2004;75(6):1032–45.
[136] Kim JM, et al. Osteoblast-osteoclast communication and bone homeostasis. Cells 2020;9(9).
[137] Vashishth D, et al. Decline in osteocyte lacunar density in human cortical bone is associated with accumulation of microcracks with age. Bone 2000;26(4):375–80.
[138] Mullender MG, et al. Osteocyte density changes in aging and osteoporosis. Bone 1996;18(2):109–13.
[139] Tonna EA. A study of osteocyte formation and distribution in aging mice complemented with H3-proline autoradiography. J Gerontol 1966;21(1):124–30.
[140] Tonna EA, Lampen NM. Electron microscopy of aging skeletal cells. I. Centrioles and solitary cilia. J Gerontol 1972;27(3):316–24.
[141] Beyer EC, Paul DL, Goodenough DA. Connexin43: a protein from rat heart homologous to a gap junction protein from liver. J Cell Biol 1987;105(6 Pt 1):2621–9.
[142] Beyer EC, et al. Antisera directed against connexin43 peptides react with a 43-kD protein localized to gap junctions in myocardium and other tissues. J Cell Biol 1989;108(2):595–605.
[143] Cheng B, et al. Expression of functional gap junctions and regulation by fluid flow in osteocyte-like MLO-Y4 cells. J Bone Miner Res 2001;16(2):249–59.
[144] Bivi N, et al. Deletion of Cx43 from osteocytes results in defective bone material properties but does not decrease extrinsic strength in cortical bone. Calcif Tissue Int 2012;91(3):215–24.
[145] Cherian PP, et al. Mechanical strain opens connexin 43 hemichannels in osteocytes: a novel mechanism for the release of prostaglandin. Mol Biol Cell 2005;16(7):3100–6.
[146] Jiang JX, Cheng B. Mechanical stimulation of gap junctions in bone osteocytes is mediated by prostaglandin E2. Cell Commun Adhes 2001;8(4–6):283–8.
[147] Siller-Jackson AJ, et al. Adaptation of connexin 43-hemichannel prostaglandin release to mechanical loading. J Biol Chem 2008;283(39):26374–82.
[148] Davis HM, et al. Cx43 overexpression in osteocytes prevents osteocyte apoptosis and preserves cortical bone quality in aging mice. JBMR Plus 2018;2(4):206–16.
[149] Kar R, et al. Connexin 43 channels protect osteocytes against oxidative stress-induced cell death. J Bone Miner Res 2013;28(7):1611–21.
[150] Moller AMJ, et al. Aging and menopause reprogram osteoclast precursors for aggressive bone resorption. Bone Res 2020;8:27.
[151] Willey JS, et al. Early increase in osteoclast number in mice after whole-body irradiation with 2 Gy X rays. Radiat Res 2008;170(3):388–92.
[152] Perkins SL, et al. Age-related bone loss in mice is associated with an increased osteoclast progenitor pool. Bone 1994;15(1):65–72.
[153] Scheven BA, et al. Comparison of direct and indirect radiation effects on osteoclast formation from progenitor cells derived from different hemopoietic sources. Radiat Res 1987;111(1):107–18.
[154] Chung PL, et al. Effect of age on regulation of human osteoclast differentiation. J Cell Biochem 2014;115(8):1412–9.
[155] Rozman C, et al. Age-related variations of fat tissue fraction in normal human bone marrow depend both on size and number of adipocytes: a stereological study. Exp Hematol 1989;17(1):34–7.
[156] Milisic L, Vegar-Zubovic S, Valjevac A. Bone marrow adiposity is inversely associated with bone mineral density in postmenopausal females. Med Glas 2020;17(1).
[157] Woods GN, et al. Greater bone marrow adiposity predicts bone loss in older women. J Bone Miner Res 2020;35(2):326–32.
[158] Zou W, et al. Ablation of fat cells in adult mice induces massive bone gain. Cell Metab 2020;32(5):801–813 e6.
[159] Cao J, et al. Deletion of PPARgamma in mesenchymal lineage cells protects against aging-induced cortical bone loss in mice. J Gerontol A Biol Sci Med Sci 2020;75(5):826–34.
[160] Chandra A, et al. Bone marrow adiposity in models of radiation- and aging-related bone loss is dependent on cellular senescence. J Bone Miner Res 2022;37(5):997–1011.
[161] Chandra A, et al. Targeted clearance of p21- but not p16-positive senescent cells prevents radiation-induced osteoporosis and increased marrow adiposity. Aging Cell 2022;21(5):e13602.
[162] Farr JN, et al. Targeting cellular senescence prevents age-related bone loss in mice. Nat Med 2017;23(9):1072–9.
[163] Kusumbe AP, Ramasamy SK, Adams RH. Coupling of angiogenesis and osteogenesis by a specific vessel subtype in bone. Nature 2014;507(7492):323–8.
[164] Watson EC, Adams RH. Biology of bone: the vasculature of the skeletal system. Cold Spring Harb Perspect Med 2018;8(7).

[165] Swift ME, Kleinman HK, DiPietro LA. Impaired wound repair and delayed angiogenesis in aged mice. Lab Invest 1999;79(12):1479−87.
[166] Hayflick L, Moorhead PS. The serial cultivation of human diploid cell strains. Exp Cell Res 1961;25:585−621.
[167] Serrano M, et al. Oncogenic ras provokes premature cell senescence associated with accumulation of p53 and p16INK4a. Cell 1997;88(5):593−602.
[168] Rajarajacholan UK, Riabowol K. Aging with ING: a comparative study of different forms of stress induced premature senescence. Oncotarget 2015;6(33):34118−27.
[169] Kural KC, et al. Pathways of aging: comparative analysis of gene signatures in replicative senescence and stress induced premature senescence. BMC Genom 2016;17(Suppl. 14):1030.
[170] Cristofalo VJ, Pignolo RJ. Replicative senescence of human fibroblast-like cells in culture. Physiol Rev 1993;73(3):617−38.
[171] Narita M, et al. Rb-mediated heterochromatin formation and silencing of E2F target genes during cellular senescence. Cell 2003;113(6):703−16.
[172] Sadaie M, et al. Redistribution of the Lamin B1 genomic binding profile affects rearrangement of heterochromatic domains and SAHF formation during senescence. Genes Dev 2013;27(16):1800−8.
[173] Aird KM, Zhang R. Detection of senescence-associated heterochromatin foci (SAHF). Methods Mol Biol 2013;965:185−96.
[174] Baker DJ, et al. Clearance of p16Ink4a-positive senescent cells delays ageing-associated disorders. Nature 2011;479(7372):232−6.
[175] Zhu Y, et al. Identification of a novel senolytic agent, navitoclax, targeting the Bcl-2 family of anti-apoptotic factors. Aging Cell 2016;15(3):428−35.
[176] Sayama K, et al. Possible involvement of p21 but not of p16 or p53 in keratinocyte senescence. J Cell Physiol 1999;179(1):40−4.
[177] Herbig U, et al. Telomere shortening triggers senescence of human cells through a pathway involving ATM, p53, and p21(CIP1), but not p16(INK4a). Mol Cell 2004;14(4):501−13.
[178] Morisaki H, et al. Complex mechanisms underlying impaired activation of Cdk4 and Cdk2 in replicative senescence: roles of p16, p21, and cyclin D1. Exp Cell Res 1999;253(2):503−10.
[179] Serrano M, Hannon GJ, Beach D. A new regulatory motif in cell-cycle control causing specific inhibition of cyclin D/CDK4. Nature 1993;366(6456):704−7.
[180] Alcorta DA, et al. Involvement of the cyclin-dependent kinase inhibitor p16 (INK4a) in replicative senescence of normal human fibroblasts. Proc Natl Acad Sci U S A 1996;93(24):13742−7.
[181] el-Deiry WS, et al. WAF1, a potential mediator of p53 tumor suppression. Cell 1993;75(4):817−25.
[182] Stein GH, et al. Differential roles for cyclin-dependent kinase inhibitors p21 and p16 in the mechanisms of senescence and differentiation in human fibroblasts. Mol Cell Biol 1999;19(3):2109−17.
[183] Haddad MM, et al. Activation of a cAMP pathway and induction of melanogenesis correlate with association of p16(INK4) and p27(KIP1) to CDKs, loss of E2F-binding activity, and premature senescence of human melanocytes. Exp Cell Res 1999;253(2):561−72.
[184] Alexander K, Hinds PW. Requirement for p27(KIP1) in retinoblastoma protein-mediated senescence. Mol Cell Biol 2001;21(11):3616−31.
[185] Garcia-Fernandez RA, et al. Combined loss of p21(waf1/cip1) and p27(kip1) enhances tumorigenesis in mice. Lab Invest 2011;91(11):1634−42.
[186] Matsushita M, et al. Age-related changes in bone mass in the senescence-accelerated mouse (SAM). SAM-R/3 and SAM-P/6 as new murine models for senile osteoporosis. Am J Pathol 1986;125(2):276−83.
[187] Takeda T. Senescence-accelerated mouse (SAM): a biogerontological resource in aging research. Neurobiol Aging 1999;20(2):105−10.
[188] Saul D, et al. Modulation of fracture healing by the transient accumulation of senescent cells. Elife 2021;10.
[189] Xue W, et al. Senescence and tumour clearance is triggered by p53 restoration in murine liver carcinomas. Nature 2007;445(7128):656−60.
[190] Xu M, et al. JAK inhibition alleviates the cellular senescence-associated secretory phenotype and frailty in old age. Proc Natl Acad Sci U S A 2015;112(46):E6301−10.
[191] Hoeijmakers JH. DNA damage, aging, and cancer. N Engl J Med 2009;361(15):1475−85.
[192] Maya-Mendoza A, et al. High speed of fork progression induces DNA replication stress and genomic instability. Nature 2018;559(7713):279−84.
[193] Fragola G, et al. Deletion of Topoisomerase 1 in excitatory neurons causes genomic instability and early onset neurodegeneration. Nat Commun 2020;11(1):1962.
[194] Liu B, et al. Genomic instability in laminopathy-based premature aging. Nat Med 2005;11(7):780−5.
[195] de Paula Rodrigues GH, et al. Severe bone changes in a case of Hutchinson-Gilford syndrome. Ann Genet 2002;45(3):151−5.
[196] de Boer J, et al. Premature aging in mice deficient in DNA repair and transcription. Science 2002;296(5571):1276−9.
[197] Rasheed N, et al. Atm-deficient mice: an osteoporosis model with defective osteoblast differentiation and increased osteoclastogenesis. Hum Mol Genet 2006;15(12):1938−48.
[198] McClintock B. The stability of broken ends of chromosomes in Zea mays. Genetics 1941;26(2):234−82.
[199] Bodnar AG. Extension of life-span by introduction of telomerase into normal human cells. Science 1998;279(5349):349−52.
[200] Bernardes de Jesus B, et al. Telomerase gene therapy in adult and old mice delays aging and increases longevity without increasing cancer. EMBO Mol Med 2012;4(8):691−704.
[201] Jaskelioff M, et al. Telomerase reactivation reverses tissue degeneration in aged telomerase-deficient mice. Nature 2011;469(7328):102−6.
[202] Sfeir A, de Lange T. Removal of shelterin reveals the telomere end-protection problem. Science 2012;336(6081):593−7.
[203] de Lange T. Shelterin-mediated telomere protection. Annu Rev Genet 2018;52:223−47.
[204] Pignolo RJ, et al. Defects in telomere maintenance molecules impair osteoblast differentiation and promote osteoporosis. Aging Cell 2008;7(1):23−31.
[205] Brennan TA, et al. Mouse models of telomere dysfunction phenocopy skeletal changes found in human age-related osteoporosis. Dis Model Mech 2014;7(5):583−92.
[206] Hewitt G, et al. Telomeres are favoured targets of a persistent DNA damage response in ageing and stress-induced senescence. Nat Commun 2012;3:708.
[207] Eckhardt BA, et al. Accelerated osteocyte senescence and skeletal fragility in mice with type 2 diabetes. JCI Insight 2020;5(9).

[208] von Zglinicki T, et al. Mild hyperoxia shortens telomeres and inhibits proliferation of fibroblasts: a model for senescence? Exp Cell Res 1995; 220(1):186−93.
[209] Vizioli MG, et al. Mitochondria-to-nucleus retrograde signaling drives formation of cytoplasmic chromatin and inflammation in senescence. Genes Dev 2020;34(5−6):428−45.
[210] Domazetovic V, et al. Oxidative stress in bone remodeling: role of antioxidants. Clin Cases Miner Bone Metab 2017;14(2):209−16.
[211] Agidigbi TS, Kim C. Reactive oxygen species in osteoclast differentiation and possible pharmaceutical targets of ROS-mediated osteoclast diseases. Int J Mol Sci 2019;20(14).
[212] Altindag O, et al. Total oxidative/anti-oxidative status and relation to bone mineral density in osteoporosis. Rheumatol Int 2008;28(4):317−21.
[213] Zhou Q, et al. Oxidative stress-related biomarkers in postmenopausal osteoporosis: a systematic review and meta-analyses. Dis Markers 2016;2016:7067984.
[214] Dobson PF, et al. Mitochondrial dysfunction impairs osteogenesis, increases osteoclast activity, and accelerates age related bone loss. Sci Rep 2020;10(1):11643.
[215] Geurts J, et al. Prematurely aging mitochondrial DNA mutator mice display subchondral osteopenia and chondrocyte hypertrophy without further osteoarthritis features. Sci Rep 2020;10(1):1296.
[216] Gandhi SS, et al. Risk factors for poor bone health in primary mitochondrial disease. J Inherit Metab Dis 2017;40(5):673−83.
[217] Schriner SE, et al. Extension of murine life span by overexpression of catalase targeted to mitochondria. Science 2005;308(5730):1909−11.
[218] Treuting PM, et al. Reduction of age-associated pathology in old mice by overexpression of catalase in mitochondria. J Gerontol A Biol Sci Med Sci 2008;63(8):813−22.
[219] Schreurs AS, et al. Skeletal tissue regulation by catalase overexpression in mitochondria. Am J Physiol Cell Physiol 2020;319(4):C734−45.
[220] Lee SY, et al. PINK1 deficiency impairs osteoblast differentiation through aberrant mitochondrial homeostasis. Stem Cell Res Ther 2021;12(1):589.
[221] Ling W, et al. Mitochondrial Sirt3 contributes to the bone loss caused by aging or estrogen deficiency. JCI Insight 2021;6(10).
[222] Li Q, et al. Deletion of SIRT3 inhibits osteoclastogenesis and alleviates aging or estrogen deficiency-induced bone loss in female mice. Bone 2021;144:115827.
[223] Zhu S, et al. Sirt3 promotes chondrogenesis, chondrocyte mitochondrial respiration and the development of high-fat diet-induced osteoarthritis in mice. J Bone Miner Res 2022;37(12):2531−47.
[224] Gao J, et al. SIRT3/SOD2 maintains osteoblast differentiation and bone formation by regulating mitochondrial stress. Cell Death Differ 2018;25(2):229−40.
[225] Morris JA, et al. Epigenome-wide association of DNA methylation in whole blood with bone mineral density. J Bone Miner Res 2017;32(8):1644−50.
[226] Reppe S, et al. Distinct DNA methylation profiles in bone and blood of osteoporotic and healthy postmenopausal women. Epigenetics 2017;12(8):674−87.
[227] Cao Y, et al. Expression of sclerostin in osteoporotic fracture patients is associated with DNA methylation in the CpG island of the SOST gene. Int J Genomics 2019;2019:7076513.
[228] Jensen ED, et al. Histone deacetylase 7 associates with Runx2 and represses its activity during osteoblast maturation in a deacetylation-independent manner. J Bone Miner Res 2008;23(3):361−72.
[229] Fu Y, et al. Histone deacetylase 8 suppresses osteogenic differentiation of bone marrow stromal cells by inhibiting histone H3K9 acetylation and RUNX2 activity. Int J Biochem Cell Biol 2014;54:68−77.
[230] Lu J, et al. Osterix acetylation at K307 and K312 enhances its transcriptional activity and is required for osteoblast differentiation. Oncotarget 2016;7(25):37471−86.
[231] McGee-Lawrence ME, et al. Histone deacetylase 3 is required for maintenance of bone mass during aging. Bone 2013;52(1):296−307.
[232] Dudakovic A, et al. Enhancer of zeste homolog 2 (Ezh2) controls bone formation and cell cycle progression during osteogenesis in mice. J Biol Chem 2018;293(33):12894−907.
[233] Dudakovic A, et al. Enhancer of zeste homolog 2 inhibition stimulates bone formation and mitigates bone loss caused by ovariectomy in skeletally mature mice. J Biol Chem 2016;291(47):24594−606.
[234] Dudakovic A, et al. Inhibition of the epigenetic suppressor EZH2 primes osteogenic differentiation mediated by BMP2. J Biol Chem 2020;295(23):7877−93.
[235] Hemming S, et al. Identification of novel EZH2 targets regulating osteogenic differentiation in mesenchymal stem cells. Stem Cells Dev 2016;25(12):909−21.
[236] Delgado-Calle J, et al. Genome-wide profiling of bone reveals differentially methylated regions in osteoporosis and osteoarthritis. Arthritis Rheum 2013;65(1):197−205.
[237] Fernandez-Rebollo E, et al. Primary osteoporosis is not reflected by disease-specific DNA methylation or accelerated epigenetic age in blood. J Bone Miner Res 2018;33(2):356−61.
[238] Cummings BJ, et al. Aggregation of the amyloid precursor protein within degenerating neurons and dystrophic neurites in Alzheimer's disease. Neuroscience 1992;48(4):763−77.
[239] Yoshimoto M, et al. NACP, the precursor protein of the non-amyloid beta/A4 protein (A beta) component of Alzheimer disease amyloid, binds A beta and stimulates A beta aggregation. Proc Natl Acad Sci U S A 1995;92(20):9141−5.
[240] Chauhan A, et al. Aggregation of amyloid beta-protein as function of age and apolipoprotein E in normal and Alzheimer's serum. J Neurol Sci 1998;154(2):159−63.
[241] Trojanowski JQ, et al. Fatal attractions: abnormal protein aggregation and neuron death in Parkinson's disease and Lewy body dementia. Cell Death Differ 1998;5(10):832−7.
[242] Sitte N, et al. Protein oxidation and degradation during cellular senescence of human BJ fibroblasts: part II−aging of nondividing cells. FASEB J 2000;14(15):2503−10.
[243] Sitte N, et al. Protein oxidation and degradation during cellular senescence of human BJ fibroblasts: part I−effects of proliferative senescence. FASEB J 2000;14(15):2495−502.

[244] Grune T, et al. Protein oxidation and degradation during postmitotic senescence. Free Radic Biol Med 2005;39(9):1208–15.
[245] Chondrogianni N, Gonos ES. Proteasome inhibition induces a senescence-like phenotype in primary human fibroblasts cultures. Biogerontology 2004;5(1):55–61.
[246] Garcia-Gomez A, et al. Preclinical activity of the oral proteasome inhibitor MLN9708 in Myeloma bone disease. Clin Cancer Res 2014;20(6):1542–54.
[247] Terpos E, et al. Myeloma bone disease and proteasome inhibition therapies. Blood 2007;110(4):1098–104.
[248] Garrett IR, et al. Selective inhibitors of the osteoblast proteasome stimulate bone formation in vivo and in vitro. J Clin Invest 2003;111(11):1771–82.
[249] Kaiser MF, et al. The proteasome inhibitor bortezomib stimulates osteoblastic differentiation of human osteoblast precursors via upregulation of vitamin D receptor signalling. Eur J Haematol 2013;90(4):263–72.
[250] Giuliani N, et al. The proteasome inhibitor bortezomib affects osteoblast differentiation in vitro and in vivo in multiple myeloma patients. Blood 2007;110(1):334–8.
[251] Chandra A, et al. Proteasome inhibitor bortezomib is a novel therapeutic agent for focal radiation-induced osteoporosis. FASEB J 2018;32(1):52–62.
[252] Medicine MP, Rehabilitation L, University NY. Yoga in the treatment of osteoporosis. January 2005.
[253] University Y, et al. The effect of protein supplementation on bone health in healthy older men and women. February 2007.
[254] Khosla S, et al. Sympathetic beta1-adrenergic signaling contributes to regulation of human bone metabolism. J Clin Invest 2018;128(11):4832–42.
[255] Center UoMM. Effect of GLP-1 receptor agonists on trabecular bone score and visceral adiposity in postmenopausal women with type 2 diabetes mellitus. November 9, 2021.
[256] Fuhrmann-Stroissnigg H, et al. Identification of HSP90 inhibitors as a novel class of senolytics. Nat Commun 2017;8(1):422.
[257] Chang J, et al. Clearance of senescent cells by ABT263 rejuvenates aged hematopoietic stem cells in mice. Nat Med 2016;22(1):78–83.
[258] Zhu Y, et al. New agents that target senescent cells: the flavone, fisetin, and the BCL-XL inhibitors, A1331852 and A1155463. Aging (Albany NY) 2017;9(3):955–63.
[259] Wang Y, et al. Discovery of piperlongumine as a potential novel lead for the development of senolytic agents. Aging (Albany NY) 2016;8(11):2915–26.
[260] Cherif H, et al. Senotherapeutic drugs for human intervertebral disc degeneration and low back pain. Elife 2020;9.
[261] Ritschka B, et al. The senotherapeutic drug ABT-737 disrupts aberrant p21 expression to restore liver regeneration in adult mice. Genes Dev 2020;34(7–8):489–94.
[262] Yoshida S, et al. The CD153 vaccine is a senotherapeutic option for preventing the accumulation of senescent T cells in mice. Nat Commun 2020;11(1):2482.
[263] Bode-Boger SM, et al. Aspirin reduces endothelial cell senescence. Biochem Biophys Res Commun 2005;334(4):1226–32.
[264] Feng M, et al. Aspirin ameliorates the long-term adverse effects of doxorubicin through suppression of cellular senescence. FASEB Bioadv 2019;1(9):579–90.
[265] Robbins PD, et al. Senolytic drugs: reducing senescent cell viability to extend health span. Annu Rev Pharmacol Toxicol 2021;61:779–803.
[266] Houssaini A, et al. mTOR pathway activation drives lung cell senescence and emphysema. JCI Insight 2018;3(3).
[267] Wang R, et al. Rapamycin inhibits the secretory phenotype of senescent cells by a Nrf2-independent mechanism. Aging Cell 2017;16(3):564–74.
[268] Wu J, et al. Rapamycin improves bone mass in high-turnover osteoporosis with iron accumulation through positive effects on osteogenesis and angiogenesis. Bone 2019;121:16–28.
[269] An JY, et al. Rapamycin treatment attenuates age-associated periodontitis in mice. Geroscience 2017;39(4):457–63.

CHAPTER 5

Age-related disease: Joints

Ilyas M. Khan

Swansea University Medical School, Faculty of Medicine, Health & Life Science, Swansea, UK

1. Cartilage structure and function: a primer

Articular cartilage has two main functions, facilitating the smooth and almost frictionless articulation of bones about each other and the transmission of forces through the skeleton [1]. It lines the ends of bones within a fluid-filled sac, the synovial bursa, within which the complementary surfaces from the proximal and distal bearing surfaces are lubricated with a hyaluronic acid-rich synovial fluid containing mucopolysaccharide-containing proteins such as lubricin.

The work of surface lubrication and shock absorption in cartilage is performed by an extensive extracellular matrix (ECM) composed of an insoluble collagen type II fibril network enclosing the proteoglycan aggrecan which has numerous covalently attached glycosaminoglycan carbohydrate chains along its length [2]. The keratan and chondroitin sulphate containing glycosaminoglycans are highly electronegative and draw in water to hydrate the ECM, whereas collagen type II fibrils function to restrain proteoglycan swelling. The water imbibed by the ECM constitutes nearly 80% of the wet weight of the tissue, it effectively pressurises the cartilage and its movement in and out of cartilage upon weight-bearing allows dissipation of applied biomechanical forces and furthermore contributes significantly to smooth articulation at the surface [3].

Histologically, cartilages are defined as containing a high ratio by volume of ECM to resident cells, chondrocytes, whose principal job in adulthood is to maintain the soluble, mainly proteoglycan, component of the ECM. Even though articular cartilage contains a single cell type, the chondrocyte, it is quite apparent early in development that these cells have quite distinct morphologies, mainly as a function of depth in the tissue [4]. Histologically, adult tissue can be subdivided into several layers based on cellular organisation; superficial, mid, and deep zones constitute the soft chondral tissue, and they are separated by a thin interdigitating calcified zone from a cortical subchondral bony plate, below which is trabecular bone of the epiphysis (Fig. 5.1). The shape of cells in the different strata appears to be influenced by the organisation of collagen fibrils which under polarising light microscope align in arches, with the crowns of the arches meeting at the surface [5]. Consequently, superficial zone chondrocytes appear flat and discoid in shape, reflecting the parallel packing of collagen fibrils at the surface, whereas cells in the mid or transitional zone where the fibrils cross over each other are small and round, and in the deep zone as fibrils thicken and are perpendicular in relation to the surface, hypertrophic chondrocytes are arranged in columns [6]. The organisation of adult cartilage is critical for its function and is acquired relatively late in life around the time of puberty up until growth plate closure, which in humans occurs in females between the ages of 13–15 and in males between 15–17.

In newborn animals, the prospective articular cartilage is in a continuum with epiphyseal cartilage, a temporary cartilage which regulates the growth of the underlying epiphysis, and which will eventually be resorbed and replaced by bone [4] (Fig. 5.1). The organisation of cartilage is simpler with the superficial zone being recognisable by its flatten chondrocytes and parallel alignment of collagen but the underlying tissue has no discernible stratified organisation and cells appear randomly organised within it, proabably associated with the seemingly random alignment of the collagenous network [7–9]. Re-organisation of the collagen network occurs during puberty with reconfiguration of isotropically aligned fibres into elegant arches as first described by Benninghoff [5]. Other changes at

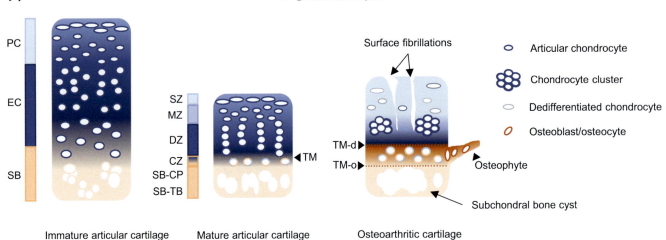

FIGURE 5.1 Growth and development of articular cartilage. Immature articular joints have relatively thick cartilage which is composed of permanent cartilage (PC; expression of the SOX trio 5/6/9 confers permanency), epiphyseal cartilage (EC) which is transient in nature and whose cells are noticeably hypertrophic and express collagen type X. Below is the subchondral bone which slowly subsumes the overlying transient cartilage. Following postnatal maturation, the height of cartilage is reduced by resorption of EC and the remaining PC is reorganised to form four distinct stratified layers: surface (SZ), mid (MZ), deep (DZ), and calcified zones (CZ). The tidemark (TM) is a histological artefact that demarcates non-calcified from calcified cartilage. Connected to the CZ is the subchondral bone cortical plate (SB-CP) under which is found trabecular or spongy bone (SB-TB). In osteoarthritic cartilage the surface is fibrillated and stratified organisation breaks down. Cellular dysfunction is visualised through chondrocyte dedifferentiation and cluster formation and in later stages by osteophyte formation. The creeping advance of the subchondral bone plate is evidenced by duplication of the tidemark (TM-d). Subchondral cysts, bone sclerosis, and changes in bone mineral density are also common radiological features of osteoarthritis.

puberty include the appearance of the stratified organisation of chondrocytes and the chondro-osseous junction, a border which separate the calcified and non-calcified cartilage matrices [10]. This postnatal developmental epoch is important in many respects, not least as it marks the time when no further addition to the collagenous network occurs. Aspartate racemisation studies have shown that the half-life of collagen is approximately 100 years and bomb pulse ^{14}C data analysis of osteoarthritic cartilage shows it does not contain any new collagen over that which was present at postnatal maturation i.e., puberty [11,12]. So, whilst the proteoglycan component is turned over on a monthly basis, the collagen in human joints "when destroyed is never recovered" [13].

From the pathological viewpoint of cartilage breakdown, the fact that no new collagen is detectable in injured or diseased tissues is quite profound and fits well with the paradigm that this tissue has a particularly poor intrinsic repair capacity [14]. The reasons for the lack of regenerative capacity within cartilage is due in part to its avascular nature, which limits transport and ingress of cells required for molecular debridement and repopulation of damaged regions. Equally valid reasons are the sparsity of cells within cartilage and the density of the matrix preventing migration and proliferation to repair injured tissues.

Chronic low-grade inflammation is present in a third of OA patients and contributes to joint remodelling during disease progression [15]. The loss of the soluble proteoglycan component either due to inflammatory insult driven by cytokines such as interleukin-1β (IL1β) and transforming growth factor-α (TNFα) activating catabolic cellular pathways, or lack of maintenance due to chondrocyte cell death, directly impacts ECM functionality, further risking the viability of the remaining cells [16,17]. As important, proteoglycan loss leaves denuded collagen fibrils at greater risk of mechanical or enzymatic degradation (Fig. 5.1). Metalloproteinases-1/8/13 (MMP) degrade collagen fibrils causing an irreversible change in ECM structure, tissue expansion through increased hydration, and the loss of collagen network integrity through surface fibrillation that over time permeates the deep zone causing fissures and fractures [18].

Proteoglycan loss, collagen fibril degradation, cell death, and clustering of the remaining chondrocytes are the microscopic hallmarks of progressive cartilage breakdown clinically classified as osteoarthritic disease [19] (Fig. 5.1). Other important changes involve associated tissues of the joint organ; episodic synovial inflammation, which can lead to joint laxity due to changes in ligament structure, meniscal damage in the knee joint, bone remodelling affecting bone mineral density and/or joint alignment, osteophytosis, and resorption of chondral tissue originating from the calcified zone evidenced by rising and duplicated tidemarks (Fig. 5.1). The extent of disease in joint cartilage can be focal encompassing a few centimetres squared or more diffusely spread with no clear margin [20]. There are currently no clinically validated drug treatments to stabilise or reverse osteoarthritic joint disease, in general physiotherapy and pain relief are prescribed, however when symptoms, especially pain, overwhelm sufferers, they can elect for joint replacement therapy. Arthroplasty is a highly successful surgical procedure allowing patients to regain a degree of pain-free mobility.

Experiments have shown that focal chondral defects with a diameter greater than 2–3 mm do not heal spontaneously [21]. The gold standard repair method is to use autografts cored from a non-weight-bearing region of the joint, or when the defect is large to use topographically matched allografts obtained from cadaveric donors [22]. Cell transplantation has also been used to repair smaller defects, predominantly using a two-step surgical procedure called autologous chondrocyte implantation (ACI); first by isolating autologous chondrocytes from the patient's affected joint and expanding them in vitro, then reimplanting the cells into the defect encased in fibrin glue which is retained in place by suturing a periosteal flap or collagenous membrane over it [23–25].

Further refinements to ACI involve substituting the fibrin glue for more appropriate scaffold matrices, such as hyaluronic acid and collagen-based scaffolds. These monolithic tissue engineering strategies have evolved by incorporating hydroxyapatite or tricalcium phosphate in various guises to produce multi-layered constructs that better mimic native tissue architecture and composition to improve vertical integration [26]. Some cell therapy-based strategies have exploited the biology of chondrocytes, isolating them without inducing dedifferentiation by maintaining their pericellular chondron structure [27]. The pericellular matrix is rich in collagen type VI and heparan sulphate proteoglycans, and with the enclosed chondrocyte constitutes the minimal functional cellular unit of cartilage, its retention stabilises the phenotype of isolated cells. Using intact chondrons combined with autologous mesenchymal stem cells (MSCs) appears to be quite efficacious and has the added benefit of being able to treat larger lesions for several reasons; availability of MSCs is less of an issue, MSCs release immunomodulatory paracrine factors that promote repair, and MSCs stimulate the survival and proliferation of chondrocytes. Other tissue-based strategies use minced autologous cartilage placed under a supporting scaffold matrix and implanted in defects, a technique that appears to promote cellular migration from the edges of diced cartilage pieces causing recellularisation of the remaining matrix [27a].

The overall aim of tissue engineering is to produce an implantable constructs that can substitute and function as well as autograft and allograft osteochondral transplants. The reality of the situation is that we still have more to learn about critical phases of growth and development of articular cartilage, in particular maturation at puberty and chondro-osseous interactions before we can begin generating constructs that will outlive their hosts rather than the reverse [28–31].

In the meantime, the molecular biological dissection of chondrocytes has revealed tantalising glimpses into their lives from birth, death and through disease, such that, druggable targets are emerging that at the very least have the potential to protect tissues from progressive damage and that in the future could be leading candidates for inducing tissue repair and regeneration.

2. Aging and osteoarthritis

Like many age-related diseases, OA is not an inevitable outcome, however the risk of developing disease significantly increases with age. The Centers for Disease Control (CDC) analysed data from a 2-year period and found OA prevalence in 7.6% in people aged 18–44 rising to nearly 30% in those aged between 45%–64% and to 50% in those 65 and older [32]. The latter study also showed 54.4 million people were affected by OA, of which 43.5% had disease-attributable activity limitations. Doctor-diagnosed arthritis in the United States is the leading cause of physical disability. A more stringent measure of OA disease occurrence is the Kellegren-Lawrence (KL) system for radiographic grading scale which is based on the presence and severity of joint space narrowing, subchondral cysts, osteophytosis, and joint line sclerosis [33]. The KL scale ranges from 0 for normal joints to 4 where there is complete loss of joint space. The Framingham Osteoarthritis Study of 1420 subjects aged 60 or higher found an increased prevalence of ipsilateral radiographic knee OA, KL grade 2 and higher, 33% in people aged 60–70 rising to 44% in those aged over 80 [34]. A similar rise in OA prevalence in other joints has been shown; in the hip prevalence rises from 0.7% in the age group aged 40–44 rising to 14% in the people aged 85 or older [35], for the hand the Framingham study found 13% of men and 26% of women over the age of 70 had radiographic evidence of OA [34]. The high prevalence and disabling impact make OA a major social issue, particularly as the increase in socioeconomic status of global populations has had a corresponding positive effect of mean lifespan and a steadily increasing percentage of elders in populations.

The question arises, what biological mechanisms drive this seemingly inexorable breakdown of cartilage structure as we age? Given the association between OA and aging, the transition of aged cells to a well-defined senescent state appears to be a defining feature that characterises the initiation of tissue dysfunction and gradual breakdown [36]. In a tissue where cells contribute <5% of the volume the rest being taken up by ECM, that is continuously maintained by those cells, any loss of differentiated phenotype will eventually have catastrophic consequences for joint function.

3. Replicative senescence in cartilage

The key finding in 1961 by Moorehead and Hayflick [37] that cells have a finite proliferative potential and when they reach an intrinsic limit they undergo cellular stasis or senescence, opened the door to the discovery of the biological mechanisms driving cellular aging. There are two main pathways whereby cells enter senescence, one which limits the proliferative potential of cells called replicative senescence, the other is premature or stress-induced senescence, and whilst not exclusive to aging, both phenomena are highly correlated with aging. Replicative senescence is caused by telomere erosion [38] which activates signalling processes that lead to cell cycle arrest, whereas aberrant signalling and/or damage to cellular components induces premature cellular senescence [39,40]. What type of senescence, replicative or premature, do chondrocytes undergo during aging?

The lack of regenerative capacity of cartilage tissues has already been highlighted and suggests that chondrocytes have a limited proliferative potential in vivo. Lineage tracing studies by Decker et al. in mice showed active chondroblast proliferation during perinatal growth of cartilage, however, limited proliferation postnatally, where volumetric tissue growth occurs as a result of ECM accumulation, topographical cell stacking into columns in the deep zone and cellular hypertrophy [41]. There is however evidence that tissue injury provokes cellular proliferation; studies of cartilage explants removed from bovine joints showed extensive ^{3}H-thymidine incorporation as well as a pronounced anabolic response identified by the incorporation of ^{35}S and ^{3}H-proline in the circumferential wound margin [42,43]. Whilst chondrocytes proliferate in both wounded immature and mature cartilages, in neither case are dividing cells able to regenerate chondral wounds [44]. A histopathological hallmark of osteoarthritic disease is the appearance in fibrillated cartilage of cell clusters, which are groups, often >20, of round chondrocytes nesting within a large lacunae—the inner margin of the cell's pericellular matrix visualised histologically as a round empty space encompassinga shrivelled cell [45]. Whether chondrocyte clusters originate from cellular proliferation has not been directly proved. Early work by Mankin measuring radiolabelled thymidine incorporation in cartilage found proportional increases correlated with histopathological grading of osteoarthritic tissue until tissue loss became significant [46]. Other studies have found the reverse to be true, Rozter and Mohr found ^{3}H-thymidine incorporation occurs rarely in isolated osteoarthritic cartilage explants, one positive doublet per 2000 cells [47]. Chondrocyte clusters can be formed in vitro by treating explanted immature bovine cartilage with fibroblast growth factor-2 (FGF2) to amplify the proliferative response and this dovetails with experimental data showing chondrocytes in injured cartilage release FGF2 to induce tissue remodelling [48,49]. Despite the appearance of cell clusters in diseased cartilage, surprisingly, there is no difference in the chondrocyte cell density between osteoarthritic and normal human articular cartilage [50], suggesting that cellular migration might cause cluster formation [51]. Chondrocytes express the receptor *deleted in colorectal carcinoma* (DCC) which when bound to ligand netrin-1 induces attractive migration of cells; expression of DDC is increased in osteoarthritic cartilage [52]. Indirect evidence for chondrocyte proliferation comes from a study analysing cells isolated from cartilage from a broad range of ages where the mean telomere length correlates with increasing age demonstrating a consistent 20–40 bp decline in telomere length per year [36]. The latter study also found an age-associated decrease in mitotic index and increased senescent-associated β-galactosidase activity in isolated and passaged cells, with no evidence of the expression of human telomerase reverse transcriptase (TERT). Telomerase expression in chondrocytes is undetectable following puberty and this is further evidence for telomere erosion presaging replicative senescence [53]. But without definitive evidence of post-pubertal cellular proliferation in undamaged articular cartilage the causal factor for telomeric attrition being replicative loss is tenuous at best [41,53]. Telomere length decreases as a function of proximity to the central weight-bearing region of human femoral joints, in particular the number of ultra-short telomeres measured by single telomere length analysis (STELA) increases correlating with OA disease severity [54,55]. Ultra-short telomeres occur through additional mutational mechanisms which in the latter circumstance is hypothesised to be biomechanical in origin. At the centre of the weight-bearing region of osteoarthritic femoral heads, cells with longer telomeres are found, positive for Ki67 and mesenchymal stem cell (MSC) progenitor markers such as CD166 and CD34, whereas cells in chondrocyte clusters are negative [54]. Progenitor cells are thought to be recruited to osteoarthritic cartilage to repair damage in the central weight-bearing region, their source may be the synovial lining, articular surface progenitors, or highly migratory MSCs from the underlying bone [56–58]. Whether telomere length attrition itself is a factor in the risk of developing OA disease has been examined using genome-wide association studies (GWAS) of single-nucleotide polymorphisms associated with leukocyte average telomere length, no casual association was found to the risk of developing total OA [59]. This study hypothesised that systemic detection of shorter telomeres which are less resistant to oxidative stress and inflammation could influence development of OA pathology, whilst more direct studies looking at affected tissues show there is a clearer association [54].

Therefore, limited cell turnover in adult cartilage argues for other mechanisms as the causative agent for telomere attrition and/or cellular senescence.

4. Stress-induced senescence in cartilage

In addition to telomeric attrition other hallmarks of aging have been identified: genomic instability, epigenetic alterations, loss of proteostasis, deregulated nutrient sensing, mitochondrial dysfunction, stem cell exhaustion, and altered cellular communication. Less than optimal functionality of any of the latter biological processes leads to degradation of secondary processes, for example, mitochondrial dysfunction can lead to increased levels of reactive oxygen species which impact many cellular functions and cause DNA damage which can induce cellular senescence.

Mitochondrial dysfunction and oxidative stress have been identified as key factors driving osteoarthritic pathology based on the free radical theory published by Harman in 1956 [60]. This theory posits that free radicals such as superoxide, hydrogen peroxide, nitric oxide, hydroxyl radicals, and hydroxyl ions produced by mitochondria as products of normal metabolism can attack cell constituents. The cell has defences against reactive oxygen species (ROS) in the form of antioxidant proteins like superoxide dismutase, catalase, and glutathione peroxidase. ROS affect cellular function in three major ways: by lipid peroxidation, oxidative modification of proteins, and DNA damage, leading to age-related functional decline in tissue activity in, for example, muscle, bone, liver, and brain, changes that are linked to stress-induced cellular senescence [61,62]. In vitro cultured chondrocytes show a decline in mitochondria number and an increase of genomic mutations upon extensive passaging [63]. Studies of OA chondrocytes show they have a reduced capacity for mitochondrial biogenesis [64] leading to the persistence of sub-optimally functioning organelles. A change in balance between ROS generation and the cells' antioxidant capacity can amplify the deleterious effect of ROS. As an example, the expression of superoxide dismutase-2, which is found in the mitochondrial matrix, is down-regulated in osteoarthritic chondrocytes [65], and this is linked to hypermethylation of the gene promoter that occurs before physical manifestation of OA-like lesions in an animal model of spontaneous OA [66].

A vicious cycle of ROS generation will occur if unchecked by cellular antioxidant systems and as mitochondria lack protective histones and have a limited repertoire of available DNA repair mechanisms these organelles are particulary susceptible. Increased ROS generation also has deleterious effects on genomic DNA. Whilst neither superoxide nor hydrogen peroxide efficiently attack DNA, highly reactive hydroxyl radicals (OH*) produced by the Fenton reaction react with either the sugar-phosphate backbone or bases to generate single-strand breaks, DNA–protein cross-links, and DNA–DNA adducts [67]. In the nucleus DNA damage response (DDR) proteins ataxia-telangiectasia mutated kinase (ATM) and ataxia-telangiectasia and Rad3-Related (ATR) respond to double-strand and single-strand DNA breaks, respectively. Through phosphorylation of substrates, ATM/ATR orchestrate a large network of cellular processes to maintain the genomic integrity of cells. Phosphorylation of checkpoint kinase-2 (CHK2) by ATM leads to phosphorylation of histone γH2AX and p53 inducing cell cycle arrest by activating p21^{CIP} and inhibiting cyclin-dependent kinases-4/6 (CDK) [68]. If the DNA damage cannot be removed, persistent damage foci also known as "DNA segments with chromatin alterations reinforcing senescence" (DNA-SCARS) induce permanent cell cycle withdrawal and cellular senescence in a p53/p21^{CIP} and p16^{INK4A}/retinoblastoma protein (pRB)-dependent manner [69]. Acute or repetitive excessive joint loading causes the release of ROS and these events can lead to matrix degradation and OA [70]. Genomic DNA damage (single- and double-strand breaks) measured by comet assay in human OA-derived chondrocytes is greater than is found in chondrocytes from normal cartilage; interestingly there is no significant difference between mean telomere lengths of these cell populations [71,72]. Genotoxic shock of chondrocytes induces a senescent phenotype exemplified by senescent-associated β-galactosidase staining and the suppression of cellular proliferation.

5. The senescence-associated secretory profile in chondrocytes

Increased genomic damage in OA chondrocytes and activation of ATM/ATR together with NF-kappa-B essential modulator (NEMO) activate the NFκB-CEBPβ signalling pathways which lead to the expression of genes involved in inflammation [73]. In cartilage examples of inflammatory factors released by senescent cells include; cytokines interleukin-1β (IL), IL-6, IL-8, IL-17, and tumour necrosis factor-α (TNFα) that act to amplify the inflammatory response, proteases such as MMP-1/3/10/13 and a disintegrin and metalloproteinase with thrombospondin motifs-5 (ADAMTS5) that degrade and remodel the cartilage ECM, and growth factors such as transforming growth factor β1 (TGFβ1) that are involved in fibrosis and osteophyte formation [74–76]. Persistent low-level activation of the inflammatory gene cascade is characteristic of senescent cells and is known as the senescence-associated secretory phenotype (SASP) [77]. The SASP can reinforce cell cycle arrest by autocrine or paracrine signalling, spreading the senescent state to neighbouring healthy, non-damaged cells. The power of the paracrine SASP can be demonstrated through intra-articular injection of senescent MSCs which induce osteoarthritic-like changes in mouse cartilage, changes not evident when non-senescent ones are injected [78]. In a highly informative study, Jeon et al., used a transgenic mouse strain p16^{INK4A}-trimodality reporter (3 MR), which allows detection of p16^{INK4A} expression by luciferase or red fluorescent protein expression, to track the appearance of senescent chondrocytes in the joints of young (2.5-month-old) mice subjected to anterior cruciate ligament transection (ACLT)-induced post-traumatic OA of the knee joint [79]. p16^{INK4A} expression in young mice was localised to the superficial zone but in older 19-month-old mice expression was throughout the cartilage depth, and in the latter case the disease was more severe. Thus, chronologically older or developmentally mature cells are either intrinsically susceptible to senescence or more sensitive to bystander effects, an increased risk that is probably related to the age-related degradation of key biological processes including increased ROS production, reduced antioxidant expression, increased DNA damage, reduced sensitivity to growth factor stimulation, and enforced protective expression of p16^{INK4A} and p53.

6. Autophagy: A complex interrelationship with senescence

Defects in autophagy can also induce cellular senescence in articular chondrocytes. Autophagy is an evolutionary conserved homeostatic cellular process for the removal of damaged or redundant proteins and organelles through a lysosomal-dependent mechanism. In stressful situations autophagy is an adaptive response promoting cell survival however in age-related conditions autophagy is often compromised, and interventions that stimulate autophagy promote longevity [80]. Expression analysis of key proteins that execute the autophagic process, inducer of autophagy unc-51-like kinase-1 (ULK1), regulator complex Beclin-1, and microtubule-associated protein 1A/1B-light chain-3 (LC3) which functions in substrate selection and autophagosome biogenesis, are decreased in mouse models of OA and in aged mice [81,82]. In both situations, proteoglycan loss which is symptomatic of aging or cartilage degradation is associated with reduced protein expression of ULK1, Beclin-1, and LC3, and an increase in apoptosis detected immunohistochemically by cleavage of the p85 subunit of poly(ADP-ribose) polymerase [81,82]. Age-dependent autophagy impairment, evidenced by decreased autophagic degradation, is found in many animals such as worms, mice, and humans, though the reasons for this decline other than reduced expression of components of the system are unclear [83–85]. Chondrocyte-specific deletion of autophagy related-5 (ATG5) in mice induces severe OA especially in males and is again associated with increased apoptosis [86]. Paradoxically, autophagy while capable of reversing senescence in some systems is also required to fully establish and maintain the senescent state [87]. The autophagic caretaker role in senescence is driven by selective autophagy of proteins such as TNFAIP3 Interacting Protein-1 (TNIP1) which inhibits NF-kappa-B (NFκB) activation, thereby activating the SASP of cells. Kelch-like ECH-associated protein-1 (KEAP1) an adaptor protein for the CUL3 ubiquitin ligase complex that promotes proteasomal degradation of nuclear factor erythroid-2 related factor-2 (NRF2), a transcription factor that activates antioxidant-related gene expression, is similarly regulated by selective autophagy to promote redox homeostasis in senescent cells [88]. Lee et al. (2021) found that levels of KEAP1 and TNIP1 are reduced in regions displaying severe OA co-incident with cells expressing the highest levels of p16^{INK4A}, whereas expression is maintained in surrounding intact cartilage. Dysfunctional mitochondria accumulate with age, reducing the efficiency of ATP production and increasing ROS production; the selective engulfment of depolarised and damaged mitochondria by auto-phagosomes and subsequent catabolism by lysosomes is known as mitophagy. In acute settings, such as in monosodium iodoacetate (MIA)-induced OA animal models, there is increased expression of mitophagic regulatory proteins such as PTEN-induced putative protein kinase-1 (PINK1) and Parkin which is associated with apoptotic cell death of chondrocytes [89]. PINK1 accumulates on the outer membrane of damaged mitochondria and recruits Parkin which ubiquitinates outer mitochondrial membrane proteins to trigger selective autophagy. When the joints of PINK1 knockout mice were challenged with MIA there was a marked reduction in the severity of OA and pain-related behaviours in animals. Conversely, overexpression of Parkin can reduce the effects of IL-1β-induced mitochondrial dysfunction and ROS generation in chondrocytes through elevated mitophagy [90].

As we obtain more in-depth knowledge of sub-cellular systems it is clear that their function is highly context-dependent, intricate, and sometimes paradoxical. For example, inhibition of autophagy can promote the senescent state of cells in certain circumstances through the inhibition of p62-mediated selective autophagy of transcription factor GATA4 which is then free to activate transcription factor NFκB to initiate the SASP [91].

7. The role of sirtuins in chondrosenescence

Mitochondrial nicotinamide adenine dinucleotide (NAD)-dependent protein deacetylase sirtuin-3 (SIRT3) is a metabolic sensor which protects against mitochondrial damage by activating mitophagy [92]. SIRT3 expression is reduced in an age-related manner, and this causes increased acetylation of PINK1/Parkin and decreased Parkin expression [93]. Intra-articular SIRT3 overexpression in mice exposed to ACLT-induced OA affords cartilage a measure of protection, conversely, siRNA targeting SIRT3 exacerbates symptoms of disease underlining the importance of redox homeostasis in disease induction [94]. Sirtuin-1 (SIRT1) is involved in mitochondrial biogenesis through the activity of peroxisome proliferator activated receptor-γ co-activator-1α (PGC1α). SIRT1 expression and activity is downregulated in aged chondrocytes, increasing the risk of OA development as shown by conditional deletion of SIRT1 in mouse joints [95]. AMP-activated protein kinase (AMPK) is a master nutrient and energy sensor that maintains energy homeostasis and like SIRT1 its expression is also downregulated upon aging [96]. The importance of AMPK to SIRT1 function lies in its ability to increase SIRT1 expression and directly phosphorylate it promoting its deactylase activity. Overexpression of SIRT1 in chondrocytes increases type II collagen (COL2A1) gene expression through deacetylation-mediated activation of SOX9 [97]. In conditions of pro-inflammatory stress, SIRT1 is cleaved to generate N and C terminal fragments whose serum ratio is an indicator of early OA and chondrosenescence [98]. In the latter study the increased ratio of N-terminal fragments was hypothesised to be attributable to non-senescent chondrocyte apoptosis caused by prolonged inflammatory insult and highlighting the importance of the senescent state in stabilising the disease state. SIRT6 also displays significant protective roles during cartilage aging and disease. SIRT6 depletion by RNAi in human chondrocytes induces chondrosenescence as shown by increased senescence-associated β-gal activity and p16^{INK4A} expression, and presence of γH2AX foci indicative of genomic DNA damage, whereas lentivirus-mediated SIRT6 overexpression in mouse knees alleviates an inflammatory response and chondrosenescence [99,100].

8. The emerging role of circadian clocks in senescence

Intriguingly, cellular senescence can also have cryptic effects such as unbalancing the autonomous circadian clock in chondrocytes [101]. The circadian clock in mammals consists of interlinked transcription/translation feedback loops with genes oscillating in periods of around 24 h. CLOCK:BMAL1 is a transcription factor heterodimer that regulates the expression of period (PER) and cryptochrome (CRY) genes which regulate the function of CLOCK:BMAL1 in a negative feedback loop [102]. NAD$^+$ levels fluctuate in a circadian manner and the rate-limiting enzyme in its production, nicotinamide phosphoribosyltransferase (NAMPT), is regulated by CLOCK:BMAL1 [101]. SIRT1 functions to epigenetically regulate the circadian clock through NAD$^+$-dependent deacetylation of PER and BMAL1 [103]. Therefore, progressive reduction of NAD$^+$ as cells age directly affects SIRT1 function which has a knock-on effect on processes reliant on the circadian clock, with arrhythmic transcriptional profiles out of sync with functional requirements, promoting and stabilising the senescent cell state. Up to 3.9% of the transcriptome of chondrocytes are clock-regulated genes, and it is hypothesised that rhythmic control of anabolic and catabolic gene expression functions to optimise tissue repair and remodelling. For example, in mice peak anabolism is found in the early morning following bouts of nocturnal activity [104]. The amplitude of the autonomous clock in chondrocyte degrades by 40% with aging and is dysregulated in OA cells and BMAL1-deficient mice, which suffer from premature aging and early progressive arthropathy [105]. Interestingly, BMAL1$^{-/-}$ mice also reveal the circadian clock's role in telomere homeostasis, through phasic control of heterochromatin and telomeric repeat-containing RNA (TERRA) expression [106]. Reestablishment of clock functionality should reverse cell senescence or at least forestall its appearance, and this is exactly what happens when aged or BMAL1$^{-/-}$ deficient human MSCs are transfected with lentivirus constructs carrying CLOCK [107]. The latter study found that the function of CLOCK was independent of its transcriptional activity relying solely on interaction with nuclear lamin proteins and stabilisation of repetitive DNA heterochromatin.

9. mTOR, nutrient sensing, and senescence in cartilage

Mechanistic target of rapamycin (mTOR) is a nutrient-sensing signalling cascade that is responsible for the regulation of growth and metabolism of cells as a function of their nutritional background [108]. mTOR, which is a serine/threonine kinase, forms two signalling multiprotein complexes which are structurally similar but differ in function; mTOR complex 1 (mTORC1) is rapamycin-sensitive, whilst mTORC2 is rapamycin-insensitive [109]. mTORC1 is a negative regulator of autography, and promotes cellular growth, protein synthesis, nutrient uptake, and metabolism, thereby controlling the balance between anabolism and catabolism in response to environmental conditions. mTORC1 is also a downstream target of phosphoinositol-3-kinase (PI3K) and protein kinase B (Akt) which are themselves activated by tyrosine kinase growth factors such as insulin growth factor-1 (IGF1) [110]. Inhibition of mTOR by rapamycin induces autophagy which also causes the loss of negative feedback inhibition of Akt phosphorylation promoting cell survival through inhibition of apoptosis. Akt phosphorylates murine double minute-2 (MDM2) upregulating its ubiquitin ligase activity and suppressing p53-mediated apoptosis. Akt also phosphorylates cAMP response element binding protein (CREB) initiating the expression of anti-apoptosis protein B-cell lymphoma protein-2 (BCL2) [111,112]. mTOR signalling is required for chondrocyte proliferation, chondrogenesis, growth, and development [113], whereas rapamycin-mediated inhibition of mTOR signalling causes disruption, especially in the cartilaginous growth plate [114]. mTOR is overexpressed in human OA cartilage and in animal models of OA, and coincides with increased chondrocyte apoptosis and reductions in autophagy [115]. ULK1 phosphorylation by mTORC1 prevents its activation by AMPK, therefore the activities of AMPK and mTORC1 determine the extent of autophagic induction. Zhang et al. showed that cartilage-specific ablation of mTOR causes increased expression of autophagic genes and proteins [115]. When combined with the destabilised medial meniscus (DMM) model of OA in mice, cre-mediated mTOR expression knockdown results in a significant reduction in proteoglycan and cell loss, and protection from cartilage degradation [115]. In the latter study, rapamycin increased autophagy in OA chondrocytes consequent with increased expression of cartilage-specific proteins type II collagen and aggrecan. Senescent cells, though cell cycle arrested, are quite metabolically active, they have an enlarged, hypertrophy morphology and greater cell mass, and mTOR signalling plays a significant role in promoting these characteristics [116]. Rapamycin treatment of p21-IPTG-induced senescent cells can delay the progression of cellular senescence or reverse some aspects of the phenotype such as senescence-associated β-gal activity and even cell cycle arrest [117,118].

10. Stem cell exhaustion?

If cellular senescence pervades stem cell niches it can affect the ability of tissues to replenish connective tissues following injury or as part of normal turnover. Intrinsic stem cell properties can be altered by encroaching senescence; in aging bone marrow mesenchymal stem cells (MSCs), which form a "universal stem cell pool", senescent p16[INK4A]-positive MSCs secrete Dickkopf-related protein-1 (DKK1) triggering premature bone remodelling and initiating osteoarthritic changes in cartilage [119]. Articular cartilage harbours cells which display the hallmarks of MSCs, they are plastic-adherent, colony-forming cells which express the canonical MSC surface markers CD73, CD105, CD166, and STRO-1, and exhibit phenotypic plasticity through tri-lineage differentiation [120]. Articular chondroprogenitors reside in the surface of embryonic and immature postnatal articular cartilage co-localising with high concentrations of fibronectin [65,121,122]. Chondroprogenitors can be enriched through differential adhesion to fibronectin-coated dishes and individual colonies cloned and analysed [123]. Isolated and culture-passaged chondroprogenitors undergo telomere attrition but only after approximately 22 population doublings (PD), resistance to replicative telomere loss is associated with higher telomerase activity in early dividing populations compared to non-selected populations of chondrocytes [124]. High-resolution telomere length analysis using STELA shows human chondrocytes that have undergone >30 PD exhibit a trimodal distribution of telomeres rather than the usual bimodal distribution found in passaged fibroblasts [58]. The upper part of the bimodal telomere distribution represents telomere attrition attributable to replicative loss, whereas the lower part consists of short/ultrashort interallelic variants that may be caused by additional mutational mechanisms [55]. Trimodal distributions include ultralong variants from progenitors enriched by fibronectin adhesion and continual passage. By following the growth kinetics of clonal cell isolates, Fellows et al. (2017), showed OA cartilage-derived chondroprogenitors are composed of two sub-populations, one of which undergoes early replicative senescence [125]. Fifty percent of clonal isolates undergo growth arrest at 30PD, while the remainder maintain doubling rates like clonal cell isolates from normal cartilage [125]. The existence of two types of colony-forming cells in osteoarthritic cartilage may be explained by the inability of daughter cells to undergo differentiation due to bystander effects of SASP originating from surrounding cells, or, that one sub-population is potentially derived from migratory cells recruited to cartilage from surrounding tissues [54]. The fact that a pool of chondroprogenitors with significant proliferative potential (~50PD) is buried within osteoarthritic tissue opens the door to experimental strategies to either reconstitute the stem cell niche or reduce the SASP bystander effects on progenitors in order to induce regenerative processes.

Cell therapeutics, a highly active area of clinical research and translational medicine, is beginning to bear fruit for the repair of focal cartilage lesions in patients [23]. The major limiting factor for treating defects is the availability of cells; tissue engineers generally use 10×10^6 cells per mL to produce cartilage in vitro, clinical work requires 12×10^6 per cm^2 of lesion, which is equivalent to $5-10 \times 10^6$ cells per mL or cm^3—assuming the thickness of cartilage is on average 2 mm [126,127]. There is a widely held belief that autologous chondrocytes, obtained from the non-weight-bearing part of the joint, passaged beyond 3–5 PD, lose their ability to undergo redifferentiation and this imposes a limit on the size of defect that can be treated [128,129]. Given that chondrocytes in allografts enjoy significant protection from immune surveillance, enclosed as they are in a dense ECM, allogenically sourced cells can be a clinically and economically viable proposition for tissue-engineered cartilage repair. From a starting population of 100 colonies isolated by differential adhesion to fibronectin, a cell yield of 1×10^9 cells can be theoretically achieved within 24 PD without significant teleomeric attrition when compared to human bone marrow-derived MSCs [130]. A large defect is generally considered to be >5 cm^2, therefore approximately 200 patients can be treated from a single cell culture run using hollow-fibre bioreactors [131]. Immortalisation of chondrocytes by ectopic expression of telomerase reverse transcriptase (TERT) is an alternative option to generate high numbers of cells, chondrocytes derived in this way increase their proliferative capacity and retain their ability to redifferentiate [132]. Sato et al. (2012) demonstrated that rabbit chondrocytes overexpressing TERT and glucose-regulated protein-78 and tissue engineered to form implants elicit significantly better repair in focal defects than unmodified cells [133]. Redifferentiation characteristics of chondrocytes can be optimised through further genetic modification to express the chondrocyte master transcriptional regulator SOX9 [134,135]. The importance of telomere integrity for mesodermal differentiation is highlighted by the finding that gene ablation of the telomerase RNA component (TERC) in mouse embryonic stem cells produced by somatic cell nuclear transfer severely affects chondrogenic differentiation, TERC$^{-/-}$ mESCs fail to form COL2A1 expressing cells [136]. In a related finding, human MSCs created from BMAL1$^{-/-}$ mESCs also have deficiencies in their ability to undergo chondrogenesis, possibly due to BMAL1's role in telomere homeostasis [106,107] (see Section 8).

11. The role of the extracellular matrix in initiating and stabilising cellular senescence

Thus far we have looked exclusively at cell-intrinsic aspects of cellular senescence, but cells reside in a supportive matrix required for complex multicellular life and it too has an indirect, secondary effect on cellular aging [137]. To put this contribution into some sort of context, the matrisome consists of 1027 genes and counting, producing structural proteins of the extracellular matrix and their modifiers [138]. The ECM of tissues is highly dynamic, with its composition changing dependent on tissue type, developmental stage, and other processes such as biomechanical input. It has a multitude of functions which can be variously summarised to be structural, instructional, protective, informational, and communicative. As mentioned earlier, collagens in cartilage are extremely long-lived molecules having a half-life of over 100 years and therefore are susceptible to age-dependent modifications such as advanced glycation end-products (AGE) produced by the Maillard reaction. Here reducing sugars such as glucose, fructose, and ribose react with lysine or arginine residues leading to the formation of pentosidine cross-links. Increases in AGE cross-links in collagen fibrils lead to matrix stiffening [139], altering the mechanical properties of collagen with the overall effect of making the tissue brittle and increasing the risk of fatigue failure [140]. Changes in the proteoglycan component of the ECM also affect cartilage function. The large aggregating proteoglycan aggrecan experiences age-related changes in composition and structure, there is an increase of lower molecular weight monomers and following enzymatic turnover an accumulation of the hyaluronic acid binding "free region". The intact aggrecan monomer has a half-life of 3.5 years but for the smaller HA-binding "free" region the half-life rises to 25 years, explaining its accumulation during aging [141]. The overall effect of the latter changes is to negatively affect the hydrodynamic properties of cartilage which are reliant on the water-binding capacity of sulphated glycosaminoglycan chains covalently bound to the full-length aggrecan monomer. A change in sulphation pattern of glycosaminoglycan chains in aging and disease can also affect the stem cell niche and the ability of daughter cells to undergo cellular differentiation [142]. Furthermore, cells express receptors for AGEs (RAGE) which can upregulate the expression of catabolic and inflammatory markers, including interleukins IL1β, IL6, IL8, MMP13, COX2, PGE2, and NO, initiating or exacerbating a chronic low-level inflammatory state characteristic of the SASP [143]. Mehta et al. (2021) designed an in vitro system to induce AGE cross-links in cartilage and challenged the model with resveratrol and curcumin, and showed they could effectively inhibit ribose-induced cross-links [144]. Resveratrol and curcumin prevented matrix stiffening, IL1α-induced apoptosis, and cellular senescence. Iijima et al. found that age-related OA in C57/BL6 mice is related to dysregulation of PI3K/Akt signalling and, further upstream, insulin (INS) and insulin receptor (INSR) [145]. They found that reduced α-Klotho expression, a key regulator of INS/INSR/PI3K, significantly accelerated cellular senescence and OA disease in male mice. Age-related declines in α-Klotho are associated with changes in the biomechanical microenvironment, the effects of which are directly coupled by cytoskeletal elements to nuclear envelope shape and structure [146]. By manipulating the stiffness of the matrix Iijima et al. were able to switch the age phenotype of cells from young to old and show that matrix stiffness epigenetically regulates α-Klotho expression [145]. Inhibition of DNA methyltransferase-1 to prevent α-Klotho promoter methylation, actin depolymerisation to prevent mechanotransducive cues to the nucleus, or β-aminopropionitrile to inhibit collagen cross-linking and induce increased matrix elasticity all served to increase α-Klotho expression and attenuate the aging phenotype of cells in vitro. Matrix stiffness also can activate mechanosensitive ion channels such as Piezo1 in nucleus pulposus (NP) in intervertebral disc cartilage, increasing intracellular Ca^{2+} and ROS generation, leading to oxidative stress-induced senescence and apoptosis in NP cells [147]. This finding has been corroborated in articular chondrocytes where Piezo1 also shows an age-dependent increase in expression [148]. By using Yoda1, a chemical agonist of Piezo1, Ren et al. were able to inhibit cellular proliferation, and promote cellular senescence and the expression of the SASP [148].

12. Chondrosenescence: Can the advancing tide be reversed?

Theodosius Dobzhansky famously wrote, *"Nothing in biology makes sense except in the light of evolution"*; can we make sense of cellular senescence in articular cartilage as a function of evolution? First, we must accept that selection pressures act prior to or at the time that reproductive success is in play, then secondly, accept that joint health as it pertains to aging lies well outside these time periods. There is no direct selection pressure to maintain joint function as we age. There are other constraints that must be acknowledged; that cells in cartilage are post-mitotic and that following cartilage maturation at growth plate closure no further incorporation of collagen into the existing framework occurs following puberty, injury or OA [11]. Therefore, natural selection of traits for joint health later in life must be acting through growth and developmental mechanisms earlier in life. One of these mechanisms is thought to be antagonistic pleiotropy, where a gene, allele, or trait that is beneficial in the young becomes detrimental in the aged animal, evolution is compelled to accept the former at a cost to the latter [149]. Examples of this type of trait have been shown for OA genetic risk alleles which are epigenetic modifiers of immature cartilage growth and development and which appear to be counterproductive to joint maintenance later in life [150].

In the theory of "disposable soma", aging results from the accumulation of cell damage as a result of limitations in maintenance and repair [151]. Cellular damage is a ubiquitous aspect of life and many processes have evolved to overcome these problems. Cellular senescence can therefore be viewed as an adaptive protective mechanism to preserve joint health. First, senescent cells are highly resistant to apoptosis through activation of PI3K/Akt signalling inhibiting caspase9 and BAD function [152]. Chondrocytes fill on average 1.6% of the volume of cartilage, with the remainder taken up with ECM, have a mean diameter of 13 μm and are found at a density of 9262 cells per mm^3 [153]. For comparison, the density of neuronal and glial cells in the brain is in the order of 67,000 cells per mm^3 [154]. Preservation of cells that are otherwise susceptible to apoptosis, especially where cell viability is so highly related to tissue viability, is advantageous. Therefore, the use of senolytics, drugs that act to clear tissues of senescent cells through induction of apoptosis, should be used guardedly in this context [155]. Despite Jeon et al. showing that UBX1010, a p53/MDM2 interaction inhibitor drug, attenuated the development of post-traumatic OA following anterior cruciate ligament transection in mice [79], a human phase II clinical trial of UBX0101 in patients with moderate to severe painful OA saw no difference in pain scores after 12 weeks when compared to placebo (NCT04129944).

Senescence of chondrocytes also renders them less likely to undergo dedifferentiation to a fibrocartilaginous phenotype where cells typically express type I collagen and become susceptible to transdifferentiation [156]. In cutaneous wound healing, cellular senescence of fibroblasts restricts fibrosis, the same is true for cardiac myofibroblasts which also enter senescence restricting fibrosis and improving, in the short-term, heart function [157,158]. There is a widely held hypothesis that osteoarthritis occurs due to an imbalance in BMP(ALK1)/TGFβ(ALK5) signalling in articular cartilage that over time due to age-related effects such as reduced growth factor responsiveness becomes biased towards BMP signalling resulting in epiphyseal or osteoblastic transdifferentiation of chondrocytes [159]. In support of this idea, research has shown that BMP5 is overexpressed in human OA and post-traumatic OA in mice and that ablation of BMP5 can inhibit chondrosenescence by inhibiting the expression of matrix-degrading enzymes such as MMP13 and ADAMTS5 [160]. Similarly, ablation of ALK5 in the superficial zone chondrocytes of cartilage using an $Alk5^{Prg4ERT2}$ construct leads to enhanced chondrosenescence of articular cartilage stem cells, decreased proliferation and differentiation, and drastically accelerated cartilage degradation following surgically induced post-traumatic OA [161].

Acute activation of senescence is now recognised as a normal physiological mechanism that regulates embryological development and wound healing through amplification of inflammation and resolution of repair [162]. In the latter circumstances the tissue is cleared of senescent cells by macrophages, natural killer cells, and cytotoxic T-cells of the immune system [163]. Chronic or persistent activation of the SASP state is due to the inability of the body to clear senescent cells and results in progressive degradation of tissue or organ function [164]. As humans age, rather predictably the immune system degrades [165], so the ability to clear senescent cells declines. Furthermore, it is hypothesised that the optimal evolutionary strategy for senescent cell clearance is to leave some cells (false negatives) rather than clear up far too many (false positives) in order to preserve tissue function with the downside being increased "inflammaging", risk of cancer and chronic degenerative diseases [166]. However, whether immune cells such as lymphocytes are able to detect senescent cells in intact ECM of cartilage tissues is questionable due to the shielding effect of ECM where the pore size is on the order of 6 nm, so this form of clearance, along with immunotherapy and immunosensitisation, are probably not relevant here [167]. Modulation of the SASP using senomorphic/senostatic strategies is more apt strategy to control chronic inflammaging in cartilage. For example, metformin activates AMPK and SIRT1, reduces mTORC1 signalling, inhibits NFκB-regulated inflammation pathways, promotes DNA damage repair, and prevents telomeric shortening [168]. In a monosodium iodoacetate-induced model of OA in rats, metformin alleviates pain symptoms in combination with celecoxib whilst reactivating autophagy, and reducing tissue damage and inflammatory-mediated cell death [169]. Again, caution must be exercised, as the use of rapamycin and metformin in a spontaneous guinea pig model of OA exacerbated the condition, a finding associated with hyperglycaemia [170].

From an evolutionary perspective, adaption to a longer lifespan may have required shorter telomeres to enhance tumour suppression, and whilst less problematic for tissues with high cell turnover, tissues with low or undetectable turnover are highly susceptible to stress-induced senescence. Telomeres appear to be sensitive sensors of intrinsic and extrinsic stress and although the DDR resolves relatively quickly (24 h) telomeric damage is persistent, unresolved, and independent of telomere length [171]. Could ectopic expression of telomerase reactivate chondrocyte cell function? Extension of lifespan through ectopic expression of telomerase has been established in in vitro cultured fibroblasts [172]. Beausejour et al. showed lentivirus-directed telomerase expression and Sv40 Large-T Antigen inactivation of p53 reverses the senescent phenotype of fibroblasts with low levels of $p16^{INK4A}$ expression [173]. In vivo, adenovirus-associated virus (AAV)-mediated expression of TERT in 2-year-old mice leads to overall telomere lengthening, an increase of 13% in median lifespan, beneficial effects in insulin sensitivity, neuromuscular coordination, and improvements in metabolism and mitochondrial function [174]. The latter strategy is particularly applicable to articular cartilage owing to the small size, 20–24 nm, of AAV and their ability to infect non-dividing cells. Furthermore, AVV9-TERT overexpression in p53-null K-Ras induction "cancer-prone" mice does not accelerate the onset or progression of tumours, demonstrating that transient TERT therapy has a good safety profile [175]. The rejuvenating effect of TERT has been further confirmed through tamoxifen-inducible expression of TERT in a telomerase-deficient mouse background resulting in telomere extension, reductions in DDR signalling, restoration of neural progenitor proliferation, and elimination of age-associated organ decline [176]. Reprogramming cells can be accomplished using other strategies, Marycz et al. showed that the combination of 5'azacytidine and resveratrol can reverse the senescent phenotype of adipose stem cells with evidence of enhanced chondrogenesis [177]. Senoreversion has been detected in human dermal fibroblasts through inhibiting 3-phosphoinositide-dependent protein kinase 1 (PDK1), when senescence characteristics are lost cells regain their quiescent status [178]. Pharmacological treatment using several phytochemicals which modify telomere length in cells, such as astragaloside IV and its active metabolite cycloastragenol, as a pretreatment prevent replicative aging in nucleus pulposus cells and work by simultaneously activating telomerase and reducing $p16^{INK4a}$ expression [179].

13. Conclusion

From the initial discovery of cellular senescence as a result of replicative failure in vitro, a wealth of discoveries have revealed its complexities and roles in an increasing number of normal and pathological, physiological processes. There is little doubt that the same level of scrutiny applied to cellular senescence in articular cartilage will lead to the same conclusions, that it has a role in regulating the physiological health of this tissue exposed daily to a challenging and occasionally hostile biomechanical environment, and, that subversion of these processes leading to persistent activation can eventually lead to progressive decay of form and function. General anti-aging strategies for controlling cellular senescence may be applicable to joint tissues but how can we know their true benefit without understanding the role of senescent cells in the life cycle of a joint?

There are some tissue-specific developmental aspects that need to be respected when attempting to redress the catabolic events that result from cellular aging, the primary one is the seeming permanence of the collagen network. Spontaneous healing and reconstitution of the collagen fibril can occur in animal models of disease, for example following limb regeneration in urodeles [180]. However, modulating the senescent state of chondrocytes, is not by itself going to help regenerate lost, damaged or diseased cartilage. Regeneration or wound healing require a specialised environment that needs to be recapitulated, therefore, a multi-modal approach will be required which requires an understanding of potential underlying genetic forces at play as well as biomechanical issues. To date, the effects of many targets of chondrosenescence that are druggable, or can be modulated by genetic manipulation, have been tested against anterior cruciate ligament transection and medial meniscus surgical instability models of OA [181]. In general, target modulation shows deceleration of OA disease onset, but this model has no relevance for cartilage regeneration because the initial injury is never repaired. Mouse digit amputation is probably a better representative model of wound healing and regeneration [182], but not of the complex milieu generated by decades of gentle breakdown, repair, remodelling and adaption typified by aging joint cartilage. To model the latter situation, non-invasive post-traumatic models of OA using externally applied mechanical overload are more appropriate, however, even these have confounding factors such as the rapid onset of disease and whether this gives time for chronic rather than acute chondrosenescence to establish [183]. Spontaneous age-associated models of OA in mice or guinea pigs are pathologically similar in their trajectory to the human condition, and whilst no model can be perfect, the Dunkin-Hartley guinea pig develops early-onset disease, and these type of models, particularly in larger species, are probably the most clinically relevant to regeneration and repair [184].

Can reconstitution of optimal chondrocyte function ever regenerate an ulcerated and decrepit human joint? Not yet. If you amputated the limb and could somehow induce limb regeneration, perhaps, though this is not what you want to hear from your local physician when they are staring at an X-ray showing subchondral cysts, sclerotic bone, osteophyte formation, joint space narrowing, malalignment, etc. What we are likely to hear in the next quarter century is the increased availability of tissue-engineered osteochondral allografts that can resurface an entire joint, representing realistic progress in the field. However, in the face of an onslaught of scientific research from highly disparate areas the seemingly intractable disease of osteoarthritis is slowly loosening its grip on inevitability. The emergence of regenerative as opposed to stabilisation strategies for joint health exemplified by the effects of angiopoietin-like-3 in cartilage repair [185] is in the vanguard of techniques that working combinatorically with anti-aging therapies that may soon herald a sea-change in treating or even preventing age-associated cartilage degradation.

References

[1] Archer CW, Caterson B, Benjamin M, Ralphs JR. Biology of the synovial joint. London: Harwood Academics; 1999.
[2] Mow VC, Ratcliffe A, Poole AR. Cartilage and diarthrodial joints as paradigms for hierarchical materials and structures. Biomaterials 1992; 13(2):67–97. https://doi.org/10.1016/0142-9612(92)90001-5.
[3] Cederlund AA, Aspden RM. Walking on water: revisiting the role of water in articular cartilage biomechanics in relation to tissue engineering and regenerative medicine. J R Soc Interface August 2022;19(193):20220364. https://doi.org/10.1098/rsif.2022.0364.
[4] Hunziker EB, Kapfinger E, Geiss J. The structural architecture of adult mammalian articular cartilage evolves by a synchronized process of tissue resorption and neoformation during postnatal development. Osteoarthritis Cartilage April 2007;15(4):403–13. https://doi.org/10.1016/j.joca.2006.09.010.
[5] Benninghoff A. Form und Bau der Gelenkknorpel in ihren Beziehungen zur Funktion. II. Der Aufbau des Gelenkknorpels in seinen Beziehungen zur Funktion. Z für Zellforsch Mikrosk Anat 1925;2:783–862.
[6] Hunziker EB. Articular cartilage repair: basic science and clinical progress. A review of the current status and prospects. Osteoarthritis Cartilage June 2002;10(6):432–63. https://doi.org/10.1053/joca.2002.0801.
[7] Hyttinen MM, Holopainen J, van Weeren PR, Firth EC, Helminen HJ, Brama PA. Changes in collagen fibril network organization and proteoglycan distribution in equine articular cartilage during maturation and growth. J Anat November 2009;215(5):584–91. https://doi.org/10.1111/j.1469-7580.2009.01140.x.
[8] Julkunen P, Iivarinen J, Brama PA, Arokoski J, Jurvelin JS, Helminen HJ. Maturation of collagen fibril network structure in tibial and femoral cartilage of rabbits. Osteoarthritis Cartilage March 2010;18(3):406–15. https://doi.org/10.1016/j.joca.2009.11.007.
[9] Rieppo J, Hyttinen MM, Halmesmaki E, et al. Changes in spatial collagen content and collagen network architecture in porcine articular cartilage during growth and maturation. Osteoarthritis Cartilage April 2009;17(4):448–55. https://doi.org/10.1016/j.joca.2008.09.004.

References

[10] Havelka S, Horn V, Spohrova D, Valouch P. The calcified-noncalcified cartilage interface: the tidemark. Acta Biol Hung 1984;35(2−4):271−9.

[11] Heinemeier KM, Schjerling P, Heinemeier J, et al. Radiocarbon dating reveals minimal collagen turnover in both healthy and osteoarthritic human cartilage. Sci Transl Med July 6, 2016;8(346):346ra90. https://doi.org/10.1126/scitranslmed.aad8335.

[12] Maroudas A, Palla G, Gilav E. Racemization of aspartic acid in human articular cartilage. Connect Tissue Res 1992;28(3):161−9. https://doi.org/10.3109/03008209209015033.

[13] Hunter W. Of the structure and diseases of articular cartilages. Phil Trans Roy Soc Lond 1744;42:514−21.

[14] Khan IM, Gilbert SJ, Singhrao SK, Duance VC, Archer CW. Cartilage integration: evaluation of the reasons for failure of integration during cartilage repair. A review. Eur Cell Mater September 3, 2008;16:26−39. https://doi.org/10.22203/ecm.v016a04.

[15] van Lent PL, Blom AB, Schelbergen RF, et al. Active involvement of alarmins S100A8 and S100A9 in the regulation of synovial activation and joint destruction during mouse and human osteoarthritis. Arthritis Rheum May 2012;64(5):1466−76. https://doi.org/10.1002/art.34315.

[16] Ollivierre F, Gubler U, Towle CA, Laurencin C, Treadwell BV. Expression of IL-1 genes in human and bovine chondrocytes: a mechanism for autocrine control of cartilage matrix degradation. Biochem Biophys Res Commun December 30, 1986;141(3):904−11. https://doi.org/10.1016/s0006-291x(86)80128-0.

[17] Zwerina J, Redlich K, Polzer K, et al. TNF-induced structural joint damage is mediated by IL-1. Proc Natl Acad Sci U S A July 10, 2007;104(28):11742−7. https://doi.org/10.1073/pnas.0610812104.

[18] Caterson B, Flannery CR, Hughes CE, Little CB. Mechanisms involved in cartilage proteoglycan catabolism. Matrix Biol August 2000;19(4):333−44. https://doi.org/10.1016/s0945-053x(00)00078-0.

[19] Grassel S, Zaucke F, Madry H. Osteoarthritis: novel molecular mechanisms increase our understanding of the disease pathology. J Clin Med April 30, 2021;10(9). https://doi.org/10.3390/jcm10091938.

[20] Dell'accio F, Vincent TL. Joint surface defects: clinical course and cellular response in spontaneous and experimental lesions. Eur Cell Mater September 28, 2010;20:210−7. https://doi.org/10.22203/ecm.v020a17.

[21] Jackson DW, Lalor PA, Aberman HM, Simon TM. Spontaneous repair of full-thickness defects of articular cartilage in a goat model. A preliminary study. J Bone Joint Surg Am January 2001;83(1):53−64. https://doi.org/10.2106/00004623-200101000-00008.

[22] Lai WC, Bohlen HL, Fackler NP, Wang D. Osteochondral allografts in knee surgery: narrative review of evidence to date. Orthop Res Rev 2022;14:263−74. https://doi.org/10.2147/ORR.S253761.

[23] Brittberg M, Lindahl A, Nilsson A, Ohlsson C, Isaksson O, Peterson L. Treatment of deep cartilage defects in the knee with autologous chondrocyte transplantation. N Engl J Med October 6, 1994;331(14):889−95. https://doi.org/10.1056/NEJM199410063311401.

[24] Grande DA, Pitman MI, Peterson L, Menche D, Klein M. The repair of experimentally produced defects in rabbit articular cartilage by autologous chondrocyte transplantation. J Orthop Res 1989;7(2):208−18. https://doi.org/10.1002/jor.1100070208.

[25] O'Driscoll SW, Keeley FW, Salter RB. The chondrogenic potential of free autogenous periosteal grafts for biological resurfacing of major full-thickness defects in joint surfaces under the influence of continuous passive motion. An experimental investigation in the rabbit. J Bone Joint Surg Am September 1986;68(7):1017−35.

[26] Dewan AK, Gibson MA, Elisseeff JH, Trice ME. Evolution of autologous chondrocyte repair and comparison to other cartilage repair techniques. BioMed Res Int 2014;2014:272481. https://doi.org/10.1155/2014/272481.

[27] Korpershoek JV, Vonk LA, Kester EC, et al. Efficacy of one-stage cartilage repair using allogeneic mesenchymal stromal cells and autologous chondron transplantation (IMPACT) compared to nonsurgical treatment for focal articular cartilage lesions of the knee: study protocol for a crossover randomized controlled trial. Trials October 9, 2020;21(1):842. https://doi.org/10.1186/s13063-020-04771-8.

[27a] Albrecht FH. Closure of joint cartilage defects using cartilage fragments and fibrin glue. Fortschr Med 1983;101(37):1650−2.

[28] Hunziker EB. The elusive path to cartilage regeneration. Adv Mater September 4, 2009;21(32−33):3419−24. https://doi.org/10.1002/adma.200801957.

[29] Khan IM, Evans SL, Young RD, et al. Fibroblast growth factor 2 and transforming growth factor beta1 induce precocious maturation of articular cartilage. Arthritis Rheum November 2011;63(11):3417−27. https://doi.org/10.1002/art.30543.

[30] Khan IM, Francis L, Theobald PS, et al. In vitro growth factor-induced bio engineering of mature articular cartilage. Biomaterials February 2013;34(5):1478−87. https://doi.org/10.1016/j.biomaterials.2012.09.076.

[31] Morgan BJ, Bauza-Mayol G, Gardner OFW, et al. Bone morphogenetic protein-9 is a potent chondrogenic and morphogenic factor for articular cartilage chondroprogenitors. Stem Cells Dev July 2020;29(14):882−94. https://doi.org/10.1089/scd.2019.0209.

[32] Centers for Disease Control and Prevention. Prevalence of doctor-diagnosed arthritis and arthritis-attributable activity limitation − United States, 2007−2009. MMWR Morb Mortal Wkly Rep October 8, 2010;59(39):1261−5.

[33] Kellgren JH, Lawrence JS. Radiological assessment of osteo-arthrosis. Ann Rheum Dis December 1957;16(4):494−502. https://doi.org/10.1136/ard.16.4.494.

[34] Felson DT, Naimark A, Anderson J, Kazis L, Castelli W, Meenan RF. The prevalence of knee osteoarthritis in the elderly. The Framingham Osteoarthritis Study. Arthritis Rheum August 1987;30(8):914−8. https://doi.org/10.1002/art.1780300811.

[35] Dagenais S, Garbedian S, Wai EK. Systematic review of the prevalence of radiographic primary hip osteoarthritis. Clin Orthop Relat Res March 2009;467(3):623−37. https://doi.org/10.1007/s11999-008-0625-5.

[36] Martin JA, Buckwalter JA. Telomere erosion and senescence in human articular cartilage chondrocytes. J Gerontol A Biol Sci Med Sci April 2001;56(4):B172−9. https://doi.org/10.1093/gerona/56.4.b172.

[37] Hayflick L, Moorhead PS. The serial cultivation of human diploid cell strains. Exp Cell Res December 1961;25:585−621. https://doi.org/10.1016/0014-4827(61)90192-6.

[38] Blackburn EH. Structure and function of telomeres. Nature April 18, 1991;350(6319):569−73. https://doi.org/10.1038/350569a0.

[39] Campisi J. Cancer, aging and cellular senescence. In Vivo January−February 2000;14(1):183−8.

[40] Munoz-Espin D, Serrano M. Cellular senescence: from physiology to pathology. Nat Rev Mol Cell Biol July 2014;15(7):482−96. https://doi.org/10.1038/nrm3823.

[41] Decker RS, Um HB, Dyment NA, et al. Cell origin, volume and arrangement are drivers of articular cartilage formation, morphogenesis and response to injury in mouse limbs. Dev Biol June 1, 2017;426(1):56−68. https://doi.org/10.1016/j.ydbio.2017.04.006.

[42] Redman SN, Dowthwaite GP, Thomson BM, Archer CW. The cellular responses of articular cartilage to sharp and blunt trauma. Osteoarthritis Cartilage February 2004;12(2):106−16. https://doi.org/10.1016/j.joca.2002.12.001.

[43] Tew SR, Kwan AP, Hann A, Thomson BM, Archer CW. The reactions of articular cartilage to experimental wounding: role of apoptosis. Arthritis Rheum January 2000;43(1):215–25. https://doi.org/10.1002/1529-0131(200001)43:1<215::AID-ANR26>3.0.CO;2-X.

[44] Tew S, Redman S, Kwan A, et al. Differences in repair responses between immature and mature cartilage. Clin Orthop Relat Res October 2001;(391 Suppl.):S142–52. https://doi.org/10.1097/00003086-200110001-00014.

[45] Lotz MK, Otsuki S, Grogan SP, Sah R, Terkeltaub R, D'Lima D. Cartilage cell clusters. Arthritis Rheum August 2010;62(8):2206–18. https://doi.org/10.1002/art.27528.

[46] Mankin HJ, Dorfman H, Lippiello L, Zarins A. Biochemical and metabolic abnormalities in articular cartilage from osteo-arthritic human hips. II. Correlation of morphology with biochemical and metabolic data. J Bone Joint Surg Am April 1971;53(3):523–37.

[47] Rotzer A, Mohr W. [3H-thymidine incorporation into chondrocytes of arthritic cartilage]. Z Rheumatol July–August 1992;51(4):172–6. 3H-Thymidin-Inkorporation in Chondrozyten arthrotischen Knorpels.

[48] Khan IM, Palmer EA, Archer CW. Fibroblast growth factor-2 induced chondrocyte cluster formation in experimentally wounded articular cartilage is blocked by soluble Jagged-1. Osteoarthritis Cartilage February 2010;18(2):208–19. https://doi.org/10.1016/j.joca.2009.08.011.

[49] Vincent T, Hermansson M, Bolton M, Wait R, Saklatvala J. Basic FGF mediates an immediate response of articular cartilage to mechanical injury. Proc Natl Acad Sci U S A June 11, 2002;99(12):8259–64. https://doi.org/10.1073/pnas.122033199.

[50] Kouri JB, Jimenez SA, Quintero M, Chico A. Ultrastructural study of chondrocytes from fibrillated and non-fibrillated human osteoarthritic cartilage. Osteoarthritis Cartilage June 1996;4(2):111–25. https://doi.org/10.1016/s1063-4584(05)80320-6.

[51] Chang C, Lauffenburger DA, Morales TI. Motile chondrocytes from newborn calf: migration properties and synthesis of collagen II. Osteoarthritis Cartilage August 2003;11(8):603–12. https://doi.org/10.1016/s1063-4584(03)00087-6.

[52] Bosserhoff AK, Hofmeister S, Ruedel A, Schubert T. DCC is expressed in a CD166-positive subpopulation of chondrocytes in human osteoarthritic cartilage and modulates CRE activity. Int J Clin Exp Pathol 2014;7(5):1947–56.

[53] Wilson B, Novakofski KD, Donocoff RS, Liang YX, Fortier LA. Telomerase activity in articular chondrocytes is lost after puberty. Cartilage October 2014;5(4):215–20. https://doi.org/10.1177/1947603514537518.

[54] Harbo M, Delaisse JM, Kjaersgaard-Andersen P, Soerensen FB, Koelvraa S, Bendix L. The relationship between ultra-short telomeres, aging of articular cartilage and the development of human hip osteoarthritis. Mech Ageing Dev September 2013;134(9):367–72. https://doi.org/10.1016/j.mad.2013.07.002.

[55] Baird DM, Rowson J, Wynford-Thomas D, Kipling D. Extensive allelic variation and ultrashort telomeres in senescent human cells. Nat Genet February 2003;33(2):203–7. https://doi.org/10.1038/ng1084.

[56] Koelling S, Kruegel J, Irmer M, et al. Migratory chondrogenic progenitor cells from repair tissue during the later stages of human osteoarthritis. Cell Stem Cell April 3, 2009;4(4):324–35. https://doi.org/10.1016/j.stem.2009.01.015.

[57] Roelofs AJ, Zupan J, Riemen AHK, et al. Joint morphogenetic cells in the adult mammalian synovium. Nat Commun May 16, 2017;8:15040. https://doi.org/10.1038/ncomms15040.

[58] Williams R, Khan IM, Richardson K, et al. Identification and clonal characterisation of a progenitor cell sub-population in normal human articular cartilage. PLoS One October 14, 2010;5(10):e13246. https://doi.org/10.1371/journal.pone.0013246.

[59] Yang J, Xu H, Cai B, et al. Genetically predicted longer telomere length may reduce risk of hip osteoarthritis. Front Genet 2021;12:718890. https://doi.org/10.3389/fgene.2021.718890.

[60] Harman D. Aging: a theory based on free radical and radiation chemistry. J Gerontol July 1956;11(3):298–300. https://doi.org/10.1093/geronj/11.3.298.

[61] Golden TR, Melov S. Mitochondrial DNA mutations, oxidative stress, and aging. Mech Ageing Dev September 30, 2001;122(14):1577–89. https://doi.org/10.1016/s0047-6374(01)00288-3.

[62] Arnheim N, Cortopassi G. Deleterious mitochondrial DNA mutations accumulate in aging human tissues. Mutat Res September 1992;275(3–6):157–67. https://doi.org/10.1016/0921-8734(92)90020-p.

[63] Martin JA, Buckwalter JA. Aging, articular cartilage chondrocyte senescence and osteoarthritis. Biogerontology 2002;3(5):257–64. https://doi.org/10.1023/a:1020185404126.

[64] Wang Y, Zhao X, Lotz M, Terkeltaub R, Liu-Bryan R. Mitochondrial biogenesis is impaired in osteoarthritis chondrocytes but reversible via peroxisome proliferator-activated receptor gamma coactivator 1alpha. Arthritis Rheumatol May 2015;67(8):2141–53. https://doi.org/10.1002/art.39182.

[65] Aigner T, Fundel K, Saas J, et al. Large-scale gene expression profiling reveals major pathogenetic pathways of cartilage degeneration in osteoarthritis. Arthritis Rheum November 2006;54(11):3533–44. https://doi.org/10.1002/art.22174.

[66] Scott JL, Gabrielides C, Davidson RK, et al. Superoxide dismutase downregulation in osteoarthritis progression and end-stage disease. Ann Rheum Dis August 2010;69(8):1502–10. https://doi.org/10.1136/ard.2009.119966.

[67] Kauppila JH, Stewart JB. Mitochondrial DNA: radically free of free-radical driven mutations. Biochim Biophys Acta November 2015;1847(11):1354–61. https://doi.org/10.1016/j.bbabio.2015.06.001.

[68] Jackson SP, Bartek J. The DNA-damage response in human biology and disease. Nature October 22, 2009;461(7267):1071–8. https://doi.org/10.1038/nature08467.

[69] Rodier F, Munoz DP, Teachenor R, et al. DNA-SCARS: distinct nuclear structures that sustain damage-induced senescence growth arrest and inflammatory cytokine secretion. J Cell Sci January 1, 2011;124(Pt 1):68–81. https://doi.org/10.1242/jcs.071340.

[70] Buckwalter JA, Anderson DD, Brown TD, Tochigi Y, Martin JA. The roles of mechanical stresses in the pathogenesis of osteoarthritis: implications for treatment of joint injuries. Cartilage October 1, 2013;4(4):286–94. https://doi.org/10.1177/1947603513495889.

[71] Rose J, Soder S, Skhirtladze C, et al. DNA damage, discoordinated gene expression and cellular senescence in osteoarthritic chondrocytes. Osteoarthritis Cartilage September 2012;20(9):1020–8. https://doi.org/10.1016/j.joca.2012.05.009.

[72] Davies CM, Guilak F, Weinberg JB, Fermor B. Reactive nitrogen and oxygen species in interleukin-1-mediated DNA damage associated with osteoarthritis. Osteoarthritis Cartilage May 2008;16(5):624–30. https://doi.org/10.1016/j.joca.2007.09.012.

[73] Rodier F, Coppe JP, Patil CK, et al. Persistent DNA damage signalling triggers senescence-associated inflammatory cytokine secretion. Nat Cell Biol August 2009;11(8):973–9. https://doi.org/10.1038/ncb1909.

[74] Clutterbuck AL, Smith JR, Allaway D, Harris P, Liddell S, Mobasheri A. High throughput proteomic analysis of the secretome in an explant model of articular cartilage inflammation. J Proteomics May 1, 2011;74(5):704–15. https://doi.org/10.1016/j.jprot.2011.02.017.

References

[75] Coryell PR, Diekman BO, Loeser RF. Mechanisms and therapeutic implications of cellular senescence in osteoarthritis. Nat Rev Rheumatol January 2021;17(1):47−57. https://doi.org/10.1038/s41584-020-00533-7.

[76] De Ceuninck F, Dassencourt L, Anract P. The inflammatory side of human chondrocytes unveiled by antibody microarrays. Biochem Biophys Res Commun October 22, 2004;323(3):960−9. https://doi.org/10.1016/j.bbrc.2004.08.184.

[77] Krtolica A, Parrinello S, Lockett S, Desprez PY, Campisi J. Senescent fibroblasts promote epithelial cell growth and tumorigenesis: a link between cancer and aging. Proc Natl Acad Sci U S A October 9, 2001;98(21):12072−7. https://doi.org/10.1073/pnas.211053698.

[78] Xu M, Bradley EW, Weivoda MM, et al. Transplanted senescent cells induce an osteoarthritis-like condition in mice. J Gerontol A Biol Sci Med Sci June 1, 2017;72(6):780−5. https://doi.org/10.1093/gerona/glw154.

[79] Jeon OH, Kim C, Laberge RM, et al. Local clearance of senescent cells attenuates the development of post-traumatic osteoarthritis and creates a pro-regenerative environment. Nat Med June 2017;23(6):775−81. https://doi.org/10.1038/nm.4324.

[80] Leidal AM, Levine B, Debnath J. Autophagy and the cell biology of age-related disease. Nat Cell Biol December 2018;20(12):1338−48. https://doi.org/10.1038/s41556-018-0235-8.

[81] Carames B, Olmer M, Kiosses WB, Lotz MK. The relationship of autophagy defects to cartilage damage during joint aging in a mouse model. Arthritis Rheumatol June 2015;67(6):1568−76. https://doi.org/10.1002/art.39073.

[82] Carames B, Taniguchi N, Otsuki S, Blanco FJ, Lotz M. Autophagy is a protective mechanism in normal cartilage, and its aging-related loss is linked with cell death and osteoarthritis. Arthritis Rheum March 2010;62(3):791−801. https://doi.org/10.1002/art.27305.

[83] Chang JT, Kumsta C, Hellman AB, Adams LM, Hansen M. Spatiotemporal regulation of autophagy during *Caenorhabditis elegans* aging. Elife July 4, 2017;6. https://doi.org/10.7554/eLife.18459.

[84] Fernandez AF, Sebti S, Wei Y, et al. Disruption of the beclin 1-BCL2 autophagy regulatory complex promotes longevity in mice. Nature June 2018;558(7708):136−40. https://doi.org/10.1038/s41586-018-0162-7.

[85] Saftig P, Beertsen W, Eskelinen EL. LAMP-2: a control step for phagosome and autophagosome maturation. Autophagy May 2008;4(4):510−2. https://doi.org/10.4161/auto.5724.

[86] Bouderlique T, Vuppalapati KK, Newton PT, Li L, Barenius B, Chagin AS. Targeted deletion of Atg5 in chondrocytes promotes age-related osteoarthritis. Ann Rheum Dis March 2016;75(3):627−31. https://doi.org/10.1136/annrheumdis-2015-207742.

[87] Chan ASL, Narita M. Short-term gain, long-term pain: the senescence life cycle and cancer. Genes Dev February 1, 2019;33(3−4):127−43. https://doi.org/10.1101/gad.320937.118.

[88] Lee Y, Kim J, Kim MS, et al. Coordinate regulation of the senescent state by selective autophagy. Dev Cell May 17, 2021;56(10):1512−1525 e7. https://doi.org/10.1016/j.devcel.2021.04.008.

[89] Shin HJ, Park H, Shin N, et al. Pink1-mediated chondrocytic mitophagy contributes to cartilage degeneration in osteoarthritis. J Clin Med November 2, 2019;8(11). https://doi.org/10.3390/jcm8111849.

[90] Ansari MY, Khan NM, Ahmad I, Haqqi TM. Parkin clearance of dysfunctional mitochondria regulates ROS levels and increases survival of human chondrocytes. Osteoarthritis Cartilage August 2018;26(8):1087−97. https://doi.org/10.1016/j.joca.2017.07.020.

[91] Kang C, Xu Q, Martin TD, et al. The DNA damage response induces inflammation and senescence by inhibiting autophagy of GATA4. Science September 25, 2015;349(6255):aaa5612. https://doi.org/10.1126/science.aaa5612.

[92] Onyango P, Celic I, McCaffery JM, Boeke JD, Feinberg AP. SIRT3, a human SIR2 homologue, is an NAD-dependent deacetylase localized to mitochondria. Proc Natl Acad Sci U S A October 15, 2002;99(21):13653−8. https://doi.org/10.1073/pnas.222538099.

[93] Yu W, Gao B, Li N, et al. Sirt3 deficiency exacerbates diabetic cardiac dysfunction: role of Foxo3A-Parkin-mediated mitophagy. Biochim Biophys Acta, Mol Basis Dis August 2017;1863(8):1973−83. https://doi.org/10.1016/j.bbadis.2016.10.021.

[94] Xu K, He Y, Moqbel SAA, Zhou X, Wu L, Bao J. SIRT3 ameliorates osteoarthritis via regulating chondrocyte autophagy and apoptosis through the PI3K/Akt/mTOR pathway. Int J Biol Macromol April 1, 2021;175:351−60. https://doi.org/10.1016/j.ijbiomac.2021.02.029.

[95] Matsuzaki T, Matsushita T, Takayama K, et al. Disruption of Sirt1 in chondrocytes causes accelerated progression of osteoarthritis under mechanical stress and during ageing in mice. Ann Rheum Dis July 2014;73(7):1397−404. https://doi.org/10.1136/annrheumdis-2012-202620.

[96] Hardie DG, Ross FA, Hawley SA. AMPK: a nutrient and energy sensor that maintains energy homeostasis. Nat Rev Mol Cell Biol March 22, 2012;13(4):251−62. https://doi.org/10.1038/nrm3311.

[97] Dvir-Ginzberg M, Gagarina V, Lee EJ, Hall DJ. Regulation of cartilage-specific gene expression in human chondrocytes by SirT1 and nicotinamide phosphoribosyltransferase. J Biol Chem December 26, 2008;283(52):36300−10. https://doi.org/10.1074/jbc.M803196200.

[98] Batshon G, Elayyan J, Qiq O, et al. Serum NT/CT SIRT1 ratio reflects early osteoarthritis and chondrosenescence. Ann Rheum Dis October 2020;79(10):1370−80. https://doi.org/10.1136/annrheumdis-2020-217072.

[99] Nagai K, Matsushita T, Matsuzaki T, et al. Depletion of SIRT6 causes cellular senescence, DNA damage, and telomere dysfunction in human chondrocytes. Osteoarthritis Cartilage August 2015;23(8):1412−20. https://doi.org/10.1016/j.joca.2015.03.024.

[100] Wu Y, Chen L, Wang Y, et al. Overexpression of Sirtuin 6 suppresses cellular senescence and NF-kappaB mediated inflammatory responses in osteoarthritis development. Sci Rep December 7, 2015;5:17602. https://doi.org/10.1038/srep17602.

[101] Gossan N, Zeef L, Hensman J, et al. The circadian clock in murine chondrocytes regulates genes controlling key aspects of cartilage homeostasis. Arthritis Rheum September 2013;65(9):2334−45. https://doi.org/10.1002/art.38035.

[102] Takahashi JS. Transcriptional architecture of the mammalian circadian clock. Nat Rev Genet March 2017;18(3):164−79. https://doi.org/10.1038/nrg.2016.150.

[103] Nakahata Y, Sahar S, Astarita G, Kaluzova M, Sassone-Corsi P. Circadian control of the NAD+ salvage pathway by CLOCK-SIRT1. Science May 1, 2009;324(5927):654−7. https://doi.org/10.1126/science.1170803.

[104] Gossan N, Boot-Handford R, Meng QJ. Ageing and osteoarthritis: a circadian rhythm connection. Biogerontology April 2015;16(2):209−19. https://doi.org/10.1007/s10522-014-9522-3.

[105] Khapre RV, Kondratova AA, Susova O, Kondratov RV. Circadian clock protein BMAL1 regulates cellular senescence in vivo. Cell Cycle December 1, 2011;10(23):4162−9. https://doi.org/10.4161/cc.10.23.18381.

[106] Park J, Zhu Q, Mirek E, et al. BMAL1 associates with chromosome ends to control rhythms in TERRA and telomeric heterochromatin. PLoS One 2019;14(10):e0223803. https://doi.org/10.1371/journal.pone.0223803.

[107] Liang C, Liu Z, Song M, et al. Stabilization of heterochromatin by CLOCK promotes stem cell rejuvenation and cartilage regeneration. Cell Res February 2021;31(2):187−205. https://doi.org/10.1038/s41422-020-0385-7.
[108] Pal B, Endisha H, Zhang Y, Kapoor M. mTOR: a potential therapeutic target in osteoarthritis? Drugs R March 2015;15(1):27−36. https://doi.org/10.1007/s40268-015-0082-z.
[109] Saxton RA, Sabatini DM. mTOR signaling in growth, metabolism, and disease. Cell April 6, 2017;169(2):361−71. https://doi.org/10.1016/j.cell.2017.03.035.
[110] Sarbassov DD, Guertin DA, Ali SM, Sabatini DM. Phosphorylation and regulation of Akt/PKB by the rictor-mTOR complex. Science February 18, 2005;307(5712):1098−101. https://doi.org/10.1126/science.1106148.
[111] Song G, Ouyang G, Bao S. The activation of Akt/PKB signaling pathway and cell survival. J Cell Mol Med January−March 2005;9(1):59−71. https://doi.org/10.1111/j.1582-4934.2005.tb00337.x.
[112] Du K, Montminy M. CREB is a regulatory target for the protein kinase Akt/PKB. J Biol Chem December 4, 1998;273(49):32377−9. https://doi.org/10.1074/jbc.273.49.32377.
[113] Guan Y, Yang X, Yang W, Charbonneau C, Chen Q. Mechanical activation of mammalian target of rapamycin pathway is required for cartilage development. FASEB J October 2014;28(10):4470−81. https://doi.org/10.1096/fj.14-252783.
[114] Phornphutkul C, Lee M, Voigt C, et al. The effect of rapamycin on bone growth in rabbits. J Orthop Res September 2009;27(9):1157−61. https://doi.org/10.1002/jor.20894.
[115] Zhang Y, Vasheghani F, Li YH, et al. Cartilage-specific deletion of mTOR upregulates autophagy and protects mice from osteoarthritis. Ann Rheum Dis July 2015;74(7):1432−40. https://doi.org/10.1136/annrheumdis-2013-204599.
[116] Blagosklonny MV. Cell senescence: hypertrophic arrest beyond the restriction point. J Cell Physiol December 2006;209(3):592−7. https://doi.org/10.1002/jcp.20750.
[117] Demidenko ZN, Blagosklonny MV. Growth stimulation leads to cellular senescence when the cell cycle is blocked. Cell Cycle November 1, 2008;7(21):3355−61. https://doi.org/10.4161/cc.7.21.6919.
[118] Demidenko ZN, Zubova SG, Bukreeva EI, Pospelov VA, Pospelova TV, Blagosklonny MV. Rapamycin decelerates cellular senescence. Cell Cycle June 15, 2009;8(12):1888−95. https://doi.org/10.4161/cc.8.12.8606.
[119] Malaise O, Tachikart Y, Constantinides M, et al. Mesenchymal stem cell senescence alleviates their intrinsic and seno-suppressive paracrine properties contributing to osteoarthritis development. Aging (Albany NY) October 22, 2019;11(20):9128−46. https://doi.org/10.18632/aging.102379.
[120] Dominici M, Le Blanc K, Mueller I, et al. Minimal criteria for defining multipotent mesenchymal stromal cells. The International Society for Cellular Therapy position statement. Cytotherapy 2006;8(4):315−7. https://doi.org/10.1080/14653240600855905.
[121] Hayes AJ, MacPherson S, Morrison H, Dowthwaite G, Archer CW. The development of articular cartilage: evidence for an appositional growth mechanism. Anat Embryol June 2001;203(6):469−79. https://doi.org/10.1007/s004290100178.
[122] Kozhemyakina E, Zhang M, Ionescu A, et al. Identification of a Prg4-expressing articular cartilage progenitor cell population in mice. Arthritis Rheumatol May 2015;67(5):1261−73. https://doi.org/10.1002/art.39030.
[123] Dowthwaite GP, Bishop JC, Redman SN, et al. The surface of articular cartilage contains a progenitor cell population. J Cell Sci February 29, 2004;117(Pt 6):889−97. https://doi.org/10.1242/jcs.00912.
[124] Khan IM, Bishop JC, Gilbert S, Archer CW. Clonal chondroprogenitors maintain telomerase activity and Sox9 expression during extended monolayer culture and retain chondrogenic potential. Osteoarthritis Cartilage April 2009;17(4):518−28. https://doi.org/10.1016/j.joca.2008.08.002.
[125] Fellows CR, Williams R, Davies IR, et al. Characterisation of a divergent progenitor cell sub-populations in human osteoarthritic cartilage: the role of telomere erosion and replicative senescence. Sci Rep February 2, 2017;7:41421. https://doi.org/10.1038/srep41421.
[126] Mauck RL, Seyhan SL, Ateshian GA, Hung CT. Influence of seeding density and dynamic deformational loading on the developing structure/function relationships of chondrocyte-seeded agarose hydrogels. Ann Biomed Eng September 2002;30(8):1046−56. https://doi.org/10.1114/1.1512676.
[127] Foldager CB, Gomoll AH, Lind M, Spector M. Cell seeding densities in autologous chondrocyte implantation techniques for cartilage repair. Cartilage April 2012;3(2):108−17. https://doi.org/10.1177/1947603511435522.
[128] Benya PD, Padilla SR, Nimni ME. Independent regulation of collagen types by chondrocytes during the loss of differentiated function in culture. Cell December 1978;15(4):1313−21. https://doi.org/10.1016/0092-8674(78)90056-9.
[129] Benya PD, Shaffer JD. Dedifferentiated chondrocytes reexpress the differentiated collagen phenotype when cultured in agarose gels. Cell August 1982;30(1):215−24. https://doi.org/10.1016/0092-8674(82)90027-7.
[130] Wagner W, Horn P, Castoldi M, et al. Replicative senescence of mesenchymal stem cells: a continuous and organized process. PLoS One May 21, 2008;3(5):e2213. https://doi.org/10.1371/journal.pone.0002213.
[131] Frank ND, Jones ME, Vang B, Coeshott C. Evaluation of reagents used to coat the hollow-fiber bioreactor membrane of the Quantum(R) Cell Expansion System for the culture of human mesenchymal stem cells. Mater Sci Eng C Mater Biol Appl March 2019;96:77−85. https://doi.org/10.1016/j.msec.2018.10.081.
[132] Piera-Velazquez S, Jimenez SA, Stokes D. Increased life span of human osteoarthritic chondrocytes by exogenous expression of telomerase. Arthritis Rheum March 2002;46(3):683−93. https://doi.org/10.1002/art.10116.
[133] Sato M, Shin-ya K, Lee JI, et al. Human telomerase reverse transcriptase and glucose-regulated protein 78 increase the life span of articular chondrocytes and their repair potential. BMC Muscoskel Disord April 2, 2012;13:51. https://doi.org/10.1186/1471-2474-13-51.
[134] Gurusinghe S, Hilbert B, Trope G, Wang L, Bandara N, Strappe P. Generation of immortalized equine chondrocytes with inducible Sox9 expression allows control of hypertrophic differentiation. J Cell Biochem May 2017;118(5):1201−15. https://doi.org/10.1002/jcb.25773.
[135] Md Nazir N, Zulkifly AH, Khalid KA, Zainol I, Zamli Z, Sha'ban M. Matrix production in chondrocytes transfected with sex determining region Y-box 9 and telomerase reverse transcriptase genes: an in vitro evaluation from monolayer culture to three-dimensional culture. Tissue Eng Regen Med June 2019;16(3):285−99. https://doi.org/10.1007/s13770-019-00191-1.
[136] Chang WF, Wu YH, Xu J, Sung LY. Compromised chondrocyte differentiation capacity in TERC knockout mouse embryonic stem cells derived by somatic cell nuclear transfer. Int J Mol Sci March 12, 2019;20(5). https://doi.org/10.3390/ijms20051236.

[137] Vidovic T, Ewald CY. Longevity-promoting pathways and transcription factors respond to and control extracellular matrix dynamics during aging and disease. Front Aging 2022;3:935220. https://doi.org/10.3389/fragi.2022.935220.

[138] Hynes RO, Naba A. Overview of the matrisome—an inventory of extracellular matrix constituents and functions. Cold Spring Harb Perspect Biol January 1, 2012;4(1):a004903. https://doi.org/10.1101/cshperspect.a004903.

[139] Verzijl N, DeGroot J, Thorpe SR, et al. Effect of collagen turnover on the accumulation of advanced glycation end products. J Biol Chem December 15, 2000;275(50):39027—31. https://doi.org/10.1074/jbc.M006700200.

[140] Bank RA, Bayliss MT, Lafeber FP, Maroudas A, Tekoppele JM. Ageing and zonal variation in post-translational modification of collagen in normal human articular cartilage. The age-related increase in non-enzymatic glycation affects biomechanical properties of cartilage. Biochem J February 15, 1998;330(Pt 1):345—51. https://doi.org/10.1042/bj3300345.

[141] Maroudas A, Bayliss MT, Uchitel-Kaushansky N, Schneiderman R, Gilav E. Aggrecan turnover in human articular cartilage: use of aspartic acid racemization as a marker of molecular age. Arch Biochem Biophys February 1, 1998;350(1):61—71. https://doi.org/10.1006/abbi.1997.0492.

[142] Hayes AJ, Smith SM, Caterson B, Melrose J. Concise review: stem/progenitor cell proteoglycans decorated with 7-D-4, 4-C-3, and 3-B-3(-) chondroitin sulfate motifs are morphogenetic markers of tissue development. Stem Cell October 2018;36(10):1475—86. https://doi.org/10.1002/stem.2860.

[143] Rasheed Z, Akhtar N, Haqqi TM. Advanced glycation end products induce the expression of interleukin-6 and interleukin-8 by receptor for advanced glycation end product-mediated activation of mitogen-activated protein kinases and nuclear factor-kappaB in human osteoarthritis chondrocytes. Rheumatology May 2011;50(5):838—51. https://doi.org/10.1093/rheumatology/keq380.

[144] Mehta S, Young CC, Warren MR, et al. Resveratrol and curcumin attenuate ex vivo sugar-induced cartilage glycation, stiffening, senescence, and degeneration. Cartilage December 2021;13(2_Suppl. l):1214S—28S. https://doi.org/10.1177/1947603520988768.

[145] Iijima H, Gilmer G, Wang K, et al. Age-related matrix stiffening epigenetically regulates alpha-Klotho expression and compromises chondrocyte integrity. Nat Commun January 10, 2023;14(1):18. https://doi.org/10.1038/s41467-022-35359-2.

[146] Buxboim A, Irianto J, Swift J, et al. Coordinated increase of nuclear tension and lamin-A with matrix stiffness outcompetes lamin-B receptor that favors soft tissue phenotypes. Mol Biol Cell November 7, 2017;28(23):3333—48. https://doi.org/10.1091/mbc.E17-06-0393.

[147] Wang B, Ke W, Wang K, et al. Mechanosensitive ion channel Piezo1 activated by matrix stiffness regulates oxidative stress-induced senescence and apoptosis in human intervertebral disc degeneration. Oxid Med Cell Longev 2021;2021:8884922. https://doi.org/10.1155/2021/8884922.

[148] Ren X, Li B, Xu C, et al. High expression of Piezo1 induces senescence in chondrocytes through calcium ions accumulation. Biochem Biophys Res Commun June 4, 2022;607:138—45. https://doi.org/10.1016/j.bbrc.2022.03.119.

[149] Mitteldorf J. What is antagonistic pleiotropy? Biochemistry (Mosc) December 2019;84(12):1458—68. https://doi.org/10.1134/S0006297919120058.

[150] Aubourg G, Rice SJ, Bruce-Wootton P, Loughlin J. Genetics of osteoarthritis. Osteoarthritis Cartilage May 2022;30(5):636—49. https://doi.org/10.1016/j.joca.2021.03.002.

[151] Kirkwood TB. Evolution of ageing. Nature November 24, 1977;270(5635):301—4. https://doi.org/10.1038/270301a0.

[152] Soto-Gamez A, Quax WJ, Demaria M. Regulation of survival networks in senescent cells: from mechanisms to interventions. J Mol Biol July 12, 2019;431(15):2629—43. https://doi.org/10.1016/j.jmb.2019.05.036.

[153] Hunziker EB, Quinn TM, Hauselmann HJ. Quantitative structural organization of normal adult human articular cartilage. Osteoarthritis Cartilage July 2002;10(7):564—72. https://doi.org/10.1053/joca.2002.0814.

[154] Leuba G, Garey LJ. Comparison of neuronal and glial numerical density in primary and secondary visual cortex of man. Exp Brain Res 1989;77(1):31—8. https://doi.org/10.1007/BF00250564.

[155] Grosse L, Wagner N, Emelyanov A, et al. Defined p16(high) senescent cell types are indispensable for mouse healthspan. Cell Metab July 7, 2020;32(1):87—99 e6. https://doi.org/10.1016/j.cmet.2020.05.002.

[156] Sandell LJ, Aigner T. Articular cartilage and changes in arthritis. An introduction: cell biology of osteoarthritis. Arthritis Res 2001;3(2):107—13. https://doi.org/10.1186/ar148.

[157] Jun JI, Lau LF. The matricellular protein CCN1 induces fibroblast senescence and restricts fibrosis in cutaneous wound healing. Nat Cell Biol July 2010;12(7):676—85. https://doi.org/10.1038/ncb2070.

[158] Meyer K, Hodwin B, Ramanujam D, Engelhardt S, Sarikas A. Essential role for premature senescence of myofibroblasts in myocardial fibrosis. J Am Coll Cardiol May 3, 2016;67(17):2018—28. https://doi.org/10.1016/j.jacc.2016.02.047.

[159] van der Kraan PM, van den Berg WB. Osteoarthritis in the context of ageing and evolution. Loss of chondrocyte differentiation block during ageing. Ageing Res Rev April 2008;7(2):106—13. https://doi.org/10.1016/j.arr.2007.10.001.

[160] Shao Y, Zhao C, Pan J, et al. BMP5 silencing inhibits chondrocyte senescence and apoptosis as well as osteoarthritis progression in mice. Aging (Albany NY) March 19, 2021;13(7):9646—64. https://doi.org/10.18632/aging.202708.

[161] Tan Q, Wang Q, Kuang L, et al. TGF-beta/Alk5 signaling prevents osteoarthritis initiation via regulating the senescence of articular cartilage stem cells. J Cell Physiol July 2021;236(7):5278—92. https://doi.org/10.1002/jcp.30231.

[162] Huang W, Hickson LJ, Eirin A, Kirkland JL, Lerman LO. Cellular senescence: the good, the bad and the unknown. Nat Rev Nephrol October 2022;18(10):611—27. https://doi.org/10.1038/s41581-022-00601-z.

[163] Burton DGA, Stolzing A. Cellular senescence: immunosurveillance and future immunotherapy. Ageing Res Rev May 2018;43:17—25. https://doi.org/10.1016/j.arr.2018.02.001.

[164] Childs BG, Baker DJ, Kirkland JL, Campisi J, van Deursen JM. Senescence and apoptosis: dueling or complementary cell fates? EMBO Rep November 2014;15(11):1139—53. https://doi.org/10.15252/embr.201439245.

[165] Aw D, Silva AB, Palmer DB. Immunosenescence: emerging challenges for an ageing population. Immunology April 2007;120(4):435—46. https://doi.org/10.1111/j.1365-2567.2007.02555.x.

[166] Kowald A, Passos JF, Kirkwood TBL. On the evolution of cellular senescence. Aging Cell December 2020;19(12):e13270. https://doi.org/10.1111/acel.13270.

[167] Linn FC, Sokoloff L. Movement and composition of interstitial fluid of cartilage. Arthritis Rheum August 1965;8:481—94. https://doi.org/10.1002/art.1780080402.

[168] Kulkarni AS, Gubbi S, Barzilai N. Benefits of metformin in attenuating the hallmarks of aging. Cell Metab July 7, 2020;32(1):15–30. https://doi.org/10.1016/j.cmet.2020.04.001.

[169] Na HS, Kwon JY, Lee SY, et al. Metformin attenuates monosodium-iodoacetate-induced osteoarthritis via regulation of pain mediators and the autophagy-lysosomal pathway. Cells March 19, 2021;10(3). https://doi.org/10.3390/cells10030681.

[170] Minton DM, Elliehausen CJ, Javors MA, Santangelo KS, Konopka AR. Rapamycin-induced hyperglycemia is associated with exacerbated age-related osteoarthritis. Arthritis Res Ther October 7, 2021;23(1):253. https://doi.org/10.1186/s13075-021-02637-1.

[171] Hewitt G, Jurk D, Marques FD, et al. Telomeres are favoured targets of a persistent DNA damage response in ageing and stress-induced senescence. Nat Commun February 28, 2012;3:708. https://doi.org/10.1038/ncomms1708.

[172] Bodnar AG, Ouellette M, Frolkis M, et al. Extension of life-span by introduction of telomerase into normal human cells. Science January 16, 1998;279(5349):349–52. https://doi.org/10.1126/science.279.5349.349.

[173] Beausejour CM, Krtolica A, Galimi F, et al. Reversal of human cellular senescence: roles of the p53 and p16 pathways. EMBO J August 15, 2003;22(16):4212–22. https://doi.org/10.1093/emboj/cdg417.

[174] Bernardes de Jesus B, Vera E, Schneeberger K, et al. Telomerase gene therapy in adult and old mice delays aging and increases longevity without increasing cancer. EMBO Mol Med August 2012;4(8):691–704. https://doi.org/10.1002/emmm.201200245.

[175] Munoz-Lorente MA, Martinez P, Tejera A, et al. AAV9-mediated telomerase activation does not accelerate tumorigenesis in the context of oncogenic K-Ras-induced lung cancer. PLoS Genet August 2018;14(8):e1007562. https://doi.org/10.1371/journal.pgen.1007562.

[176] Jaskelioff M, Muller FL, Paik JH, et al. Telomerase reactivation reverses tissue degeneration in aged telomerase-deficient mice. Nature January 6, 2011;469(7328):102–6. https://doi.org/10.1038/nature09603.

[177] Marycz K, Houston JMI, Weiss C, Rocken M, Kornicka K. 5-Azacytidine and resveratrol enhance chondrogenic differentiation of metabolic syndrome-derived mesenchymal stem cells by modulating autophagy. Oxid Med Cell Longev 2019;2019:1523140. https://doi.org/10.1155/2019/1523140.

[178] An S, Cho SY, Kang J, et al. Inhibition of 3-phosphoinositide-dependent protein kinase 1 (PDK1) can revert cellular senescence in human dermal fibroblasts. Proc Natl Acad Sci U S A December 8, 2020;117(49):31535–46. https://doi.org/10.1073/pnas.1920338117.

[179] Hong H, Xiao J, Guo Q, et al. Cycloastragenol and Astragaloside IV activate telomerase and protect nucleus pulposus cells against high glucose-induced senescence and apoptosis. Exp Ther Med November 2021;22(5):1326. https://doi.org/10.3892/etm.2021.10761.

[180] Rux D, Decker RS, Koyama E, Pacifici M. Joints in the appendicular skeleton: developmental mechanisms and evolutionary influences. Curr Top Dev Biol 2019;133:119–51. https://doi.org/10.1016/bs.ctdb.2018.11.002.

[181] Glasson SS, Blanchet TJ, Morris EA. The surgical destabilization of the medial meniscus (DMM) model of osteoarthritis in the 129/SvEv mouse. Osteoarthritis Cartilage September 2007;15(9):1061–9. https://doi.org/10.1016/j.joca.2007.03.006.

[182] Yu L, Dawson LA, Yan M, et al. BMP9 stimulates joint regeneration at digit amputation wounds in mice. Nat Commun February 5, 2019;10(1):424. https://doi.org/10.1038/s41467-018-08278-4.

[183] Christiansen BA, Guilak F, Lockwood KA, et al. Non-invasive mouse models of post-traumatic osteoarthritis. Osteoarthritis Cartilage October 2015;23(10):1627–38. https://doi.org/10.1016/j.joca.2015.05.009.

[184] Veronesi F, Salamanna F, Martini L, Fini M. Naturally occurring osteoarthritis features and treatments: systematic review on the aged Guinea pig model. Int J Mol Sci June 30, 2022;23(13). https://doi.org/10.3390/ijms23137309.

[185] Gerwin N, Scotti C, Halleux C, et al. Angiopoietin-like 3-derivative LNA043 for cartilage regeneration in osteoarthritis: a randomized phase 1 trial. Nat Med December 2022;28(12):2633–45. https://doi.org/10.1038/s41591-022-02059-9.

CHAPTER 6

Age-related disease: Kidneys

Saswat Kumar Mohanty[1], Bhavana Veerabhadrappa[1], Asit Majhi[1], Kitlangki Suchiang[1,2] and Madhu Dyavaiah[1]

[1]Department of Biochemistry and Molecular Biology, School of Life Sciences, Pondicherry University, Kalapet, Pondicherry, India [2]Department of Biochemistry, North Eastern Hill University, Shillong, Meghalaya, India

1. Introduction

There has been a rise in interest in the aging kidney over the past few decades. Like other organ systems, the kidneys experience natural senescence involving morphological and physiological alterations during aging. These unwanted alterations in an aging kidney are distinct from kidney disorders associated with common diseases like diabetic nephropathy of the elderly [1]. However, the two processes of disease-mediated structural and functional alterations, more prevalent in the aged, and inevitable organ-based aging, are difficult to discern. It is also critical to emphasize that age-related illnesses, when added to the effects of natural aging, can drastically modify the pace of functional decline, deplete the reversing renal function, and put these individuals at risk for acute kidney injury [2]. Recently, with the availability of estimated glomerular filtration rate (eGFR), which is widely used to assess kidney function, along with the adoption of an absolute (non-age-calibrated) threshold for defining chronic kidney disease (CKD) based solely on eGFR values (<60 mL/min/1.73 m^2), have led to an unexpectedly higher rate of older adults being diagnosed with CKD. Age-dependent changes in GFR are depicted in Table 6.1.

Additionally, this diagnostic approach led to increased referrals to the nephrology department, notably for those with mild to moderately low eGFR (30–59 mL/min/1.73 m^2) [3]. The recognized criterion of <60 mL/min/1.73 m^2 used to characterize CKD is at or below the mean eGFR among the community-dwelling population aged over 70 [4]. The now "traditional" beliefs that an isolated and sustained (over 3 months) eGFR <60 mL/min/1.73 m^2 identifies CKD are not always true for older people, who make up a particular demographic. Labeling elderly individuals as having CKD also raises psychological issues [4]. Regardless of how a condition is classified, the decline in kidney function with normal aging has clinical implications for drug dosage, the choice of live kidney donors, and the risk of CKD and acute kidney injury with loss of renal reserves.

The likelihood of developing CKD rises with age. CKD has a bidirectional link with aging, where it is believed that CKD can accelerate biological aging through several pathways. Clinically, the diagnosis of early kidney disorders remains difficult, wherein the measurement of sensitive and early markers, such as its ability to produce the reno-protective and anti-aging protein Klotho was not routinely carried out. Usually, global kidney function [i.e., estimated glomerular filtration rate (eGFR)] tends to be normal, even when CKD is already present [e.g., urinary albumin: creatinine ratio (UACR) ≥ 30 mg/g]. If untreated, however, this can lead to the accumulation of uremic toxins and altered homeostasis when the GFR falls below 60 mL/min/1.73 m^2 [5–7].

Thus, to unravel the hidden cause of underlying structural or morphological kidney disorders, different methods of calculating GFR and techniques are currently employed. For example, on computed tomography (CT) imaging, it has been commonly reported that elderly kidneys have more and larger renal cysts [8,9], localized scars [10], increased cortical surface roughness [11], decreased cortical volume, increased medullary volume [12], and more significant renal artery atherosclerosis [10]. Furthermore, the risk of AKI increased with advancing CKD stages, which conferred the highest risk of AKI other than a prior history of AKI [8,9]. Hence, a holistic understanding of the aging process of the kidneys and the ideal care management suitable for older people is required to prevent kidney functions from further deterioration.

TABLE 6.1 Age-associated decline in GFR.

Age (years)	Average estimated GFR
20–29	116
30–39	107
40–49	99
50–59	93
60–69	85
70+	75

GFR declines with age, even in people without kidney disease. Adapted from National Kidney Foundation USA (https://www.kidney.org).

2. Aging kidney

Renal health is assumed to be significantly influenced by nephron endowment. Each kidney has between 700,000 and 1.8 million functioning nephrons; however, as people age, nephrosclerosis increases, and this number gradually declines [13–15]. Uncertainty exists regarding the relationship between nephron endowment at birth and age-related anatomical or functional changes in the kidney. Low-birth-weight individuals most likely have a smaller nephron endowment, which speeds up the age-related loss in kidney function and causes hypertension, CKD, and end-stage renal disease (ESRD) earlier in life [16,17].

Aging results in time-correlated functional decline and degeneration of biological tissues and organ systems. In a healthy aging person, kidney function declines very slowly, with its primary role in expelling excess fluid and metabolic waste from the body also decreasing. On average, the kidneys receive 20%–25% of the cardiac output, filter 200 L of blood daily, and produce 1.5 L of urine with waste. Thus, the kidneys are highly metabolic organs that can sustain significant oxidative stress and are more vulnerable to aging in physiological settings [18]. As we age, structural changes with decreased renal perfusion and parenchyma mass with increased vascular resistance are commonly observed. This predisposes the kidneys to tubular atrophy, glomerulosclerosis, and interstitial fibrosis [19].

The aging kidney has garnered increased attention during the last few decades. The most likely cause of this is the widespread adoption of estimated glomerular filtration rate (eGFR) as a substitute for serum creatinine for assessing kidney function, along with the adoption of an absolute (non-age-calibrated) threshold for defining chronic kidney disease (CKD) based solely on eGFR values (<60 mL/min/1.73 m^2). Unsurprisingly, this has resulted in a rise in the number of older adults diagnosed with CKD. Additionally, this diagnostic approach has increased nephrology referrals, particularly for those with mild to moderately low eGFR (30–59 mL/min/1.73 m^2) [3]. The recognized criterion of <60 mL/min/1.73 m^2 used to characterize CKD is at or below the mean eGFR among the community-dwelling population aged over 70 (Fig. 6.1). The now "traditional" beliefs that an isolated and sustained (over 3 months) eGFR of 60 mL/min/1.73 m^2 identifies CKD are not always true for older people, who make up a special demographic. Labeling elderly individuals as having CKD also raises psychological issues [4]. Regardless of how a condition is classified, the decline in kidney function that occurs with normal aging has clinical implications for drug dosage, the choice of live kidney donors, and the risk of chronic kidney disease (CKD) and acute kidney injury with loss of renal reserves [2].

2.1 Structural changes of the aging kidney

Age-related structural changes observed in the kidney, including those to the glomeruli, tubules, interstitium, and vasculature, are unquestionably linked to aging. A drop in the quantity of non-sclerotic glomeruli (NSG), tubule loss, vascular alterations, and a rise in the frequency of tubular diverticula have been observed in seemingly healthy aging persons by Darmady and colleagues in an early post-mortem investigations 4 decades ago [20]. The structural alterations in the aging kidney may be split into two major categories: micro-anatomical based on renal biopsy findings and macro-anatomical based on imaging investigations such as CT scans.

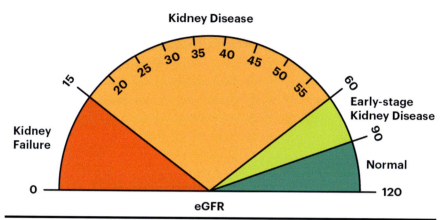

FIGURE 6.1 Representation of the correlation between eGFR and kidney function. *Adapted from National Kidney Foundation USA (https://www.kidney.org).*

2.1.1 Micro-anatomical changes

On microscopic examination, nephrosclerosis and nephron hypertrophy are the principal aging-related alterations.

2.1.1.1 Nephrosclerosis

On a kidney biopsy, the signs of nephrosclerosis can be seen as localized and global but not segmental glomerulosclerosis, tubular atrophy, interstitial fibrosis, and arteriosclerosis (fibro intimal thickening). It is believed that nephron ischemia caused by arteriosclerosis of tiny kidney arteries evolves into worldwide glomerulosclerosis and tubular atrophy. The major characteristics of ischemic-related alterations in the glomerulus are pericapsular fibrosis, capillary tuft wrinkles, and a gradually thicker basement membrane (Fig. 6.2). A broken equilibrium between the creation and degradation of the extracellular matrix in the glomerulus is also responsible for Bowman's space eventually filling with a matrix-like hyaline substance [21]. Finally, glomerular tufts disintegrate, resulting in the growth of globally sclerotic glomeruli (GSG). These GSG could eventually undergo total resorption or atrophy to a size that makes them difficult to distinguish on typical renal biopsy sections [22,23]. Along with GSG, the accompanying tubule atrophies and fibrosis build up in the interstitium around it [24,25].

This aging-related rise in the frequency of GSG has been repeatedly confirmed in a number of investigations, including both autopsy-based studies and studies of living kidney donors [22,24,26–28]. A growing incidence of GSG with advancing age in a group of 1203 healthy live kidney donors was observed [24]. For instance, the frequency of GSG was 19% in the lowest age group of kidney donors (18–29 years) and 82% in the oldest (70–77 years). GSG seems to be universally present in older people [29]. Nephrosclerosis prevalence was just 2.7% among the youngest living donors but rose to 73% among the most aged living donors if it is defined as the presence of two or more of the four major abnormalities. Interestingly, nephrosclerosis's prevalence tends to rise with age independently of the aging-related fall in whole kidney GFR [24].

2.1.1.2 Nephron hypertrophy

Due to glomerular and tubular hypertrophy, nephrons are often bigger in people with diabetes and obesity, as reported [30,31]. It is fair to anticipate that the remaining functioning nephrons will undergo compensatory hypertrophy in response to the aging-related rise in GSG [32,33]. While some studies have found that the average glomerular size decreases with age [20,23], others have shown a gain [13,34]. It is unclear, however, if researchers from these studies used the average glomerular size rather than the number of sclerosed glomeruli in their calculations. While there was no variation in NSG volume with age, profiling of mean tubular area and glomerular density did increase with age, according to research measuring nephron hypertrophy in kidney donors [35]. There also have been reports of inverse relationships between average NSG volume, profile tubular area, and glomerular density [33,35,36]. Nephron hypertrophy causes glomeruli to become more widely spaced, reducing their profile (cross-sectional) density.

Glomerular density is lower in biopsies where less than 10% of all glomeruli are sclerotic, indicating that age-related nephron hypertrophy can be seen in areas with less nephrosclerosis. Glomerular density is greater in areas with more than 10% sclerotic glomeruli, indicating that glomeruli in areas with considerable nephrosclerosis are

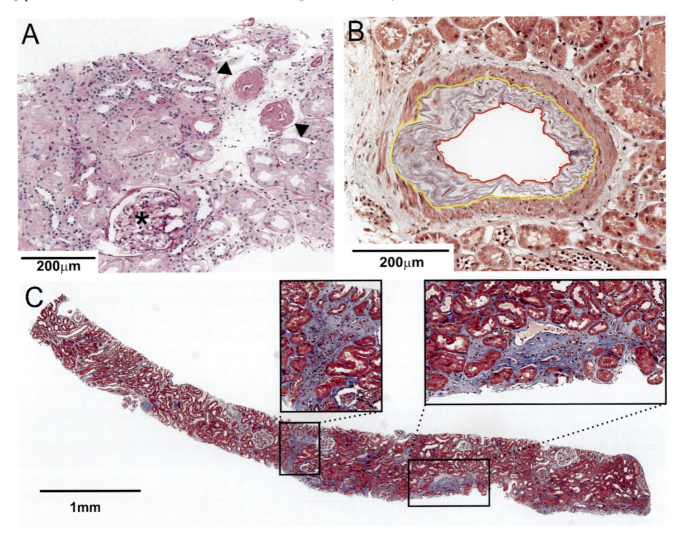

FIGURE 6.2 Representative kidney biopsy images with the key characteristics of nephrosclerosis. (A) Two globally sclerotic glomeruli (GSG) are labeled with black arrowheads. Nonsclerotic glomeruli (NSGs) are labeled with a black star. GSG are surrounded by tubular atrophy. (B) Thickened intima of a small- to medium-sized artery (the area between red and yellow boundaries). (C) Two foci of tubular atrophy and interstitial fibrosis were magnified. *Used with permission from Advances in Chronic Kidney Disease, Elsevier [2].*

brought closer together as a result of nephron atrophy with aging [33]. Although it seems that older age alone only has a minor association with nephron hypertrophy (greater NSG volume, bigger profile tubular area, and lower glomerular density), it does appear to have a stronger association with several comorbidities that are more prevalent as people age, such as obesity and hyperuricemia. Importantly, nephrosclerosis is significantly more strongly related to older age (regardless of comorbidities) than nephron hypertrophy [35].

2.1.2 Macro-anatomical changes

Smaller cortical volume and the occurrence of cysts and tumors, which are frequently benign, fall under the category of macro-anatomical alterations, which are the main aging-related changes detected on computed tomography (CT) scans.

2.1.2.1 Kidney volume

An essential sign of renal impairment is kidney volume. Two early investigations using ultrasound or CT scans on hundreds of adult volunteers and patients without renal disease showed that kidney capacity gradually decreased with aging [37,38]. Regardless of gender, it was projected that the parenchymal thickness of a kidney reduced by 10% with each additional decade of age. Likewise, using magnetic resonance imaging (MRI), Roseman et al. measured the kidney capacity in 1852 individuals and discovered that beyond the age of 60,

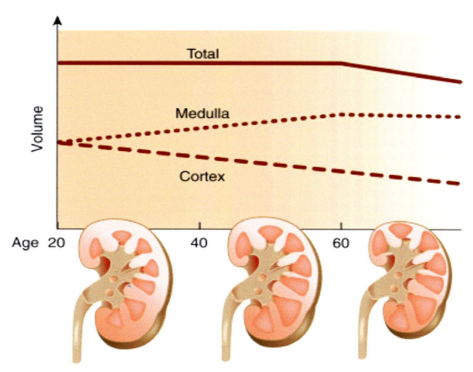

FIGURE 6.3 **Changes in kidney volume and size as a function of age.** Although it is uncertain whether this is because of nephron loss or smaller nephrons, cortical volume does appear to decline during adulthood. Total parenchymal volume is constant, possibly due to a potential compensating increase in medullary volume. After age 60, medullary volume growth stops, resulting in a reduction in overall volume. *Used with permission from Kidney International, Elsevier [41].*

the kidney volume decreases by roughly 16 cm^3 per 10 years [39]. Additionally, Wang et al. used contrast-enhanced CT imaging to assess 1344 potential kidney donors. They found that kidney capacity diminishes in people over 50 years at a rate of 22 cm^3 per decade [40]. Moreover, cortical volume gradually decreases with age while medullary volume increases up to the age of 50 years, resulting in a net drop in total kidney capacity in healthy adults after that age. Between the ages of 30 and 80, there is about a 20%–25% change in kidney mass overall [12] (Fig. 6.3).

2.1.2.2 Kidney cysts and tumors

Most kidney parenchymal cysts are classified as "benign" when they are neither cancerous nor caused by autosomal dominant polycystic kidney disease (ADPKD). With aging, these cysts multiply in number, size, and abundance [8,42]. The abundance of these cortical and medullary cysts is associated with bigger body size, higher blood pressure, male gender, and albuminuria, even in healthy people. Along with cortical and medullary parenchymal cysts, parapelvic, hyperdense cysts, and angiomyolipomas are increasingly prevalent as people become older [8]. According to their lymphatic origin rather than the tubular origin of parenchymal cysts, parapelvic cysts do not correlate with hypertension and albuminuria [43,44].

2.2 Functional changes of the aging kidney

The total glomerular filtration rate (GFR) is the main parameter used to evaluate the functioning state of the kidney. The single-nephron GFR in human individuals cannot yet be measured directly. The average GFR of a single nephron should be calculated by dividing the total GFR of the kidney by the number of functional glomeruli. However, even among individuals with the same total GFR, there can be significant variation in single-nephron GFR. Nephrosclerosis most likely causes the heterogeneity of single-nephron GFR to grow with aging. Since most patients do not frequently undergo renal CT angiograms and biopsies, it is crucial to understand the extent to which total GFR can identify underlying age-related parenchymal alterations in the kidney [2].

2.2.1 Age-related decline in GFR

Dr. Homer Smith, one of the pioneers of kidney physiology, developed the idea of clearance and provided descriptions of how to evaluate GFR (inulin clearance), renal blood flow, and secretion/reabsorption rates. In the same book, Dr. Smith discussed urea clearance's deterioration with age [45]. Findings by Davies and Shock further supported the idea that there is an age-related GFR reduction. GFR showed a linear drop in seemingly healthy persons beyond the age of 30, with an average decrease of 46% from childhood until age 90 [46]. Likewise, longitudinal research revealed that there was an average yearly drop in urinary creatinine clearance (unadjusted for body surface area) of 0.75 mL/min in 254 generally healthy persons (some of whom had diabetes) who were followed up for 14 years [47]. In around a third of these 254 people, aging led to an increase in creatinine clearance rather than a reduction. This discovery, however, may have been caused by temporary hyperfiltration, which is linked to comorbidities of aging, specifically obesity, diabetes, and preclinical cardiovascular disease, as well as the inaccurate measurement of the slopes of creatinine clearance with time [48]. The GFR drop rate of −7.4 mL/min/decade is quite comparable to the GFR decline rate of −6.3 mL/min/decade seen in a cohort of potential kidney donors [24].

According to a series of studies by Fliser et al., as the filtration fraction does not begin to grow until the age of 60 while the GFR starts to drop at 30−40 years of age, the age-related reduction in GFR is primarily driven by a vascular (arterial) mechanism [49,50]. It seems tempting to assume that there is a direct connection between nephrosclerosis and declining GFR [24]. Although several factors can contribute to GFR reduction independently of cortical volume decline, there is evidence that cortical shrinkage with aging is connected to the same mechanism that causes GFR to fall. Lew et al. hypothesized that since decreased protein consumption is typical in older people and is associated with a lower GFR, some age-related GFR reduction may be attributable to diminished protein intake [51].

2.3 Cellular senescence in renal aging

In recent years, numerous experimental pieces of evidence have highlighted how cellular senescence is involved in different physiological and pathological processes and serves as a vital driver of aging and age-related disease in multiple organs, including the kidney [52,53]. Senescence-associated secretory phenotype (SASP), macromolecular damage, and altered metabolism are the three interdependent characteristics of senescence. Cellular senescence is an irreversible cell cycle arrest in response to various types of cellular stresses, resulting in cell phenotypic and functional changes [54,55]. One of the main causes of aging is cellular senescence, which can cause a cell's reparative capacity to be used up [56]. It is now possible to assess cell senescence and age-related alterations in a variety of organ systems, including the kidney, by using a few unifying characteristics [57−59]. The most reliable and widely used biomarker of senescent cells is the acidic senescence-associated β-galactosidase (SA-β-gal) activity detected at pH 6.0, which implies an increase in lysosomal mass [60,61].

Accumulation of senescent cells in tissues and organs in response to a variety of stressors, including metabolic stress, inflammation, mitochondrial dysfunction, etc., triggers cell-cycle inhibition and permanent cell-cycle arrest via pathways either dependent on or independent of the DNA damage response. Thus, the expression of cyclin-dependent kinase (CDK) inhibitors such as $p16^{INK4a}$, $p21^{CIP1}$, $p14^{ARF}$ (in humans), $p19^{ARF}$ (in mice), $p27^{KIP1}$, and $p15^{INK4b}$ significantly increases, whereas proliferation markers like Ki67 are absent in senescent cells, and are often used as parameters to reflect the degree of cellular senescence [62,63]. Moreover, senescent cells can impact nearby cells through secreted soluble factors, sometimes called the senescence-associated secretory phenotype (SASP) [64]. Senescent cells also appear to be more resistant to apoptosis [65]. SA-β-gal and p16INK4a seem detectable in various kidney disorders before the onset of morphologic abnormalities, indicating cellular senescence has a crucial role in kidney fibrogenesis and CKD progression [66]. Increasing evidence suggests a promising new target for therapeutic intervention integration of senotherapy into the care and management of CKD and other age-related diseases [66,67].

2.3.1 Signaling pathways implicated in cellular senescence and renal aging

Cellular senescence and renal aging have been linked to a number of cellular signaling mechanisms, namely the p53/p21 and p16/Rb pathways [59,68−72]. DNA damage response signaling cascades like ataxia-telangiectasia mutated (ATM), ARF (alternative reading frame), or the p53 network are activated in the first response to DNA damages brought on by different stimuli, increasing the expression of $p21^{CIP1}$ and/or inducing the production of $p16^{INK4a}$ [68−70].

Recently, the association of a single-pass transmembrane protein Klotho with cellular senescence and lifespan has been reported [73]. The key regulators of cellular senescence, such as the p53/p21 signaling pathway, are

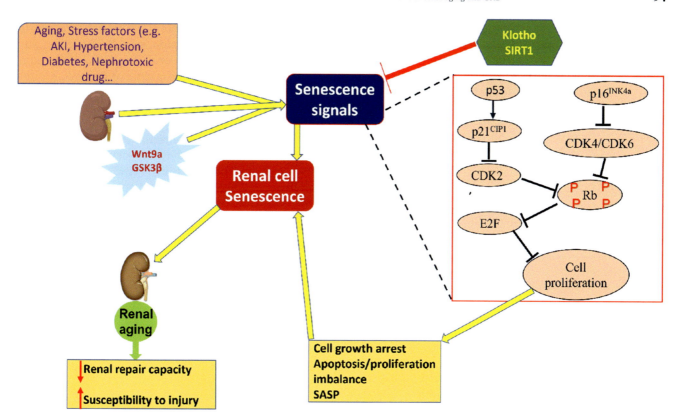

FIGURE 6.4 Cellular signaling mechanisms that contribute to renal aging and cellular senescence. Cell senescence signaling pathways are triggered during renal aging not just by stressors or illnesses such as AKI, hypertension, diabetes, and cytotoxic medications but also by aging itself. Renal cell senescence is thought to be primarily regulated by the p53/p21^{CIP1} and p16^{INK4a}/Rb signaling pathways, which in turn suppress CDK complexes and Rb phosphorylation. These signals ultimately cause cell senescence through Rb inhibition of E2F activity, which is characterized by cell proliferation arrest, an imbalance between apoptosis and proliferation, and the release of SASP factors. As a result, the ability to heal damage decreases and the aging kidney becomes more vulnerable to harm. It has been demonstrated that Klotho and SIRT1 are the primary regulators of cell senescence and suppress cell senescence via modulating the p53/p21^{CIP1} pathway. While Wnt9a promotes cell senescence signaling pathways, which speeds up renal fibrosis, GSK3 accomplishes the same. *AKI*, acute kidney disease; *GSK3β*, glycogen synthase kinase 3β; *CDK*, cyclin-dependent kinase; *SASP*, senescence-associated secretory phenotype; *Rb*, retinoblastoma protein.

regulated by Klotho [74]. The sensitivities of several critical signaling pathways, including fibroblast growth factor 23 (FGF23), transforming growth factor β (TGF- β), cyclic adenosine monophosphate (cAMP), protein kinase C (PKC), Wnt, and insulin/IGF-1 signaling, are also controlled by Klotho [74,75]. Sirtuins 1 (SIRT1) is yet another crucial control of cellular senescence. Age-related glomerulosclerosis and podocyte loss in mouse kidneys were exacerbated by podocyte-specific SIRT1 knockdown [76]. However, several signaling molecules have been shown to accelerate renal aging and senescence as indicated by the elevated expression of p16^{INK4a}, p53, and p21, as well as the enhanced SA-β-gal activity in renal tubules which are induced by Wnt9a/β-catenin signaling mediated renal tubular senescence and renal fibrosis in diseased kidneys [77]. According to recent research, glycogen synthase kinase 3β (GSK3β) accumulation may be important in cell senescence and aging [78]. Many studies in *C. elegans* and *Drosophila* showed that inhibition of GSK3 enhances lifespan [79–81]. These studies suggest that GSK3 is likely a pro-aging factor. The β isoform of GSK3 appears to be primarily expressed in the kidney, especially in glomeruli [82] (Fig. 6.4).

3. Common alterations in cellular and molecular mechanisms of renal aging and CKD

3.1 Renin–angiotensin system (RAS) activity and the kidney function

Studies on humans have revealed that plasma renin and aldosterone levels correlate with a steady drop in systemic RAS activity as people age [83,84]. On the other hand, despite a decline in circulating RAS, in experimental mice, the activity of intrarenal RAS, as shown by a rise in urine angiotensinogen, was also reported [85].

Additionally, angiotensin I intravenous infusion decreased plasma flow rates and glomerular filtration rates, particularly in older animals [86]. In response to RAS stimuli like hypovolemia, hypotension, or salt restriction, such increased sensitivity to RAS activation may accelerate the age-related deterioration in renal function. An altered reaction to RAS blockage may also result from alterations in intrarenal RAS activity with aging. For instance, glomerulosclerosis and albuminuria are less protected against by angiotensin-converting enzyme (ACE) inhibitors in older animals [87]. Additionally, compared to their younger counterparts, the hypotensive effects of these medications are less pronounced in older hypertension patients [88].

Age-related RAS dysregulation has also been linked to alterations in mitochondrial reduction-oxidation (redox) and the production of surplus free radicals, which contribute to renal fibrogenesis and its impact on hemodynamics [89]. The prosurvival genes nicotinamide phosphoribosyltransferase (Nampt) and sirtuin 3 (Sirt3) were reported to be upregulated in the kidney in response to deletion of the *agtr1a* gene, which encodes the angiotensin II type 1a (AT1a) receptor [90]. Additionally, blocking of the RAS pathway using enalapril and losartan shielded old animals against mitochondrial impairment [91] (Fig. 6.5).

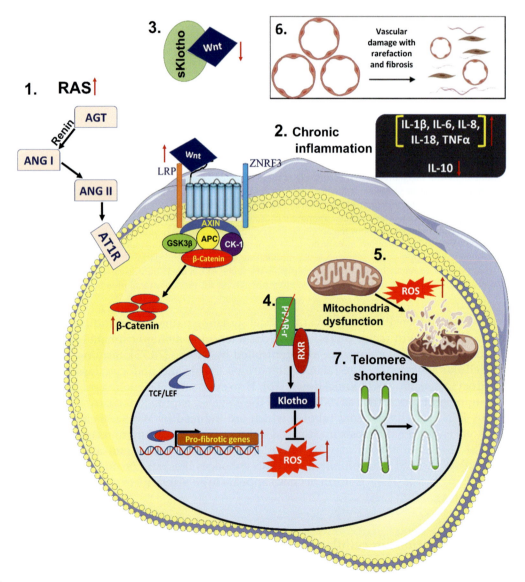

FIGURE 6.5 **Molecular mechanisms of renal aging.** (1) Activation of the RAS system, (2) chronic inflammation, (3) decreased Klotho expression and activation of Wnt/β-catenin signaling, (4) decreased PPAR-γ activity, (5) increased oxidative stress and mitochondrial dysfunction, (6) vascular damage, (7) telomere shortening. *AGT*, angiotensinogen; *ANG I*, angiotensin I; *ANG II*, angiotensin II; *AT1R*, angiotensin II type 1 receptor; *RXR*, retinoid X receptor; *ROS*, reactive oxygen species; *FRZ*, Frizzled family receptor; *GSK*, glycogen synthase kinase; *LRP5/6*, lipoprotein receptor-related protein 5/6; *TCF/LEF*, transcription factors involved in the Wnt signaling pathway.

3.1.1 The connection of RAS with Klotho in the kidney

Renal impairment brought on by aging is also significantly influenced by genetics [92]. The reduction of aging characteristics was discovered to be mediated by the Klotho gene in 1997 [93]. Further understanding of the function of genetics in aging-related renal alterations resulted from the discovery of Klotho. The Klotho gene is mostly expressed in the kidney in a transmembrane form [94], and in the kidney of CKD patients, Klotho expression was found noticeably decreased [95]. Decreased Klotho expression in aging mice increases renal fibrosis and oxidative stress [96]. There is growing evidence that Klotho and the RAS are associated. Long-term Ang II infusion decreased the expression of the renal Klotho gene, and in vivo Klotho gene transfer lessened the effects of Ang II-induced kidney injury [97]. According to a different investigation, Ang-II promotes intrarenal iron deposition and induces oxidative stress, reducing renal Klotho expression [98]. Furthermore, diabetic individuals with CKD who were treated with Ang II type 1 (AT1) receptor antagonists (AT1$_R$A) had higher plasma-soluble Klotho levels than those who were not [99]. In a recent investigation by Yoon et al., chronic cyclosporine-induced nephropathy could result in upregulation of the intrarenal RAS and downregulation of renal Klotho expression. In contrast, treatment with AT1RA upregulates renal Klotho expression and attenuates renal fibrosis and oxidative stress [100].

3.1.2 The role of RAS in AKI-CKD progression

Following AKI, prolonged and excessive RAS activation might lead to CKD through a number of different pathways [101]. First, enhanced Ang II could increase the glomerular arterioles' resistance, especially the efferent one, and hinders afferent arteriole autoregulation, which causes hyperfiltration, glomerular hypertension, and sclerosis [102]. Furthermore, Ang II triggers the Rho kinase pathway, which in turn stimulates the pro-inflammatory transcription factor nuclear factor κ-light-chain-enhancer of activated B cells (NF-κB) [103]. Ang II induces NF-κB activation in mesangial cells by activating Toll-like receptor 4 [104]. Ang II also increases the expression of integrin, vascular cellular adhesion molecule-1, intracellular adhesion molecule-1, and chemokines like monocyte chemoattractant protein-1, which are expressed and secreted by normal T cells and regulated on activation. This results in the recruitment of inflammatory cells into the glomerulus and interstitium [105]. In addition, Ang II directly causes renal fibrosis by increasing the proliferation of renal fibroblasts and the production of fibronectin, connective tissue growth factor, transforming growth factor β1, and type I collagen [106,107]. By upregulating the metalloproteinases plasminogen activator inhibitor-1 and tissue inhibitor of matrix metalloproteinases-1, Ang II also encourages the development of the extracellular matrix [108].

Recent animal investigations have shown that RAS activation plays a crucial role in the AKI-CKD progression during and after injury. An AT1a receptor antagonist, losartan, administered during the reperfusion phase, could prevent glomerular infiltration and hasten renal recovery following ischemia-reperfusion injury (IRI) [109]. AKI brought on by cisplatin is improved by AT1a receptor deletion in renal tubular cells, lowering local and systemic tumor necrosis factor-α levels [110]. Losartan used before IRI, along with the protective effects of RAS inhibitor on the severity of AKI, can also delay the onset of CKD by preserving early renal blood flow, reducing inflammation, and increasing hypoxia-inducible factor-1α activity [111].

3.2 Klotho and Wnt/β-catenin signaling

Klotho loss causes Wnt activation, which is linked to podocyte damage in kidney injuries [112]. The elimination of β-catenin reduces podocyte damage [112]. The transcription factor Snail-1 generated by Wnt through GSK3β reduces the expression of nephrin and hence plays a crucial part in podocyte destruction, even if the precise process is uncertain [113]. The opposite result occurs when the Wilms Tumor 1 (WT1) protein is overexpressed. This protein suppresses the Wnt pathway actively by suppressing the Dishevelled protein in podocytes [114].

Wnt is usually suppressed in the mature kidney, enabling the podocyte to carry out its physiological task. Klotho is a Wnt/β-catenin signaling antagonist whose activation encourages renal fibrogenesis. The fibrogenic effects of TGF-β1 can likewise be countered by overexpressing Klotho [115]. Wnt/β-catenin signaling rises when Klotho levels decrease with age, causing kidney fibrosis and vascular calcification [116,117]. One of the significant impacts of Wnt signaling activation is thought to be the decrease in renal Klotho. Moreover, Klotho inhibits a range of gene transcriptions by binding to several Wnt ligands. The activation of Wnt is stopped by the overexpression of Klotho, which also lowers the production of cytokines and ECM deposition [115]. When Wnt is overexpressed with an increase in TGF-β and collagen type III (Col3), Klotho heterozygous mutant mice exhibit ECM deposition and interstitial fibrosis that are noticeably different from those in wild-type. Klotho could reduce kidney fibrosis via inhibiting Wnt signaling, as demonstrated in vivo [118]. It is interesting to note that high blood sugar levels may affect

Wnt/β-catenin activity. Hence, the involvement of Wnt/β-catenin in the pathophysiology of diabetic nephropathy cannot be ruled out [119]. In this situation, Klotho suppresses glomerular hypertrophy and lowers albuminuria, mitigating podocyte and glomerular injury in a diabetic nephropathy mouse model [120].

3.3 Peroxisome proliferator-activated receptor (PPAR)-γ

The activation of PPAR-γ (peroxisome proliferator-activated receptor-γ) declines with age [121]. According to reports, PPARγ agonists prevent acute inflammation and diffuse tubular necrosis that are brought on by I/R damage and protect the kidney from damage [122]. In addition to raising Klotho expression in old rats, PPAR-γ agonists significantly lower oxidative stress and enhance vascular function [123]. Pioglitazone or baicalin-induced increased PPAR-γ expression reduces age-associated inflammation by preventing pro-inflammatory nuclear factor-κB (NF-κB) activation in aged rat kidneys [123,124].

In human kidney cells (HK-2 cells) exposed to high glucose levels, PPARγ is increased. In these cells, high-glucose-induced TGF-β and MCP-1 levels are suppressed by PPARγ activation, which also reverses G1 phase cell cycle arrest [125]. As a result, it has been hypothesized that PPARγ is implicated in the pathogenesis of diabetic nephropathy. The development of diabetic nephropathy may be prevented by PPARγ stimulation. Patients with diabetic nephropathy have been shown to benefit from PPARγ agonist therapy in terms of improving microalbuminuria, intrarenal nitric oxide bioavailability, and maintaining renal function [126–129].

An increasing body of research demonstrates the beneficial impact of PPARγ activation in renal interstitial fibrosis. Many individuals with chronic kidney disease (CKD) have elevated PPARγ expression in the glomeruli, especially in macrophages, podocytes, and certain parietal epithelial cells [130,131]. By causing Klotho restoration [132] and preventing Wnt signaling-driven fibrogenesis [133], PPARγ activation can slow the course of CKD. When preserving the typical epithelial phenotype and thwarting fibrogenesis, PPARγ plays a critical function in renal tubular epithelial cells [134].

3.4 Chronic inflammation

Numerous studies have shown that immunosenescence, or the deregulation of the immune system caused by aging, is common in healthy older people. For instance, higher levels of interleukin (IL)-1, IL-6, IL-8, IL-18, C-reactive protein, and tumor necrosis factor (TNF) were observed in healthy aged persons, whereas IL-10 was lower [135,136]. The response in older people also shifts from a T helper Th-1 to a Th-2 cytokine [137], with an increase in Th-17 cells and a reduction in regulatory T cells in tandem [138]. In addition, several inflammasomes have been found in aging kidneys, together with elevated concentrations of IL-1 receptor and toll-like receptor-4. The expression of inflammasome components such as the nucleotide-binding domain leucine-rich repeat (NLR), the NLR pyrin domain-containing protein 3 (NLRP3), the NLR family caspase recruitment domain-containing protein 4 (NLRC4), and pro-caspase-1 is markedly elevated in the aging rat kidney [139]. Another investigation by Sato et al. revealed that old mice would generate renal tertiary lymphoid tissues (TLTs) independently, which could encourage inflammatory and fibrotic reactions in the kidney [140]. The age-related deterioration in renal function is known to be aggravated by all of these changes that lead to an imbalance of inflammatory cells and associated cytokines [141].

3.5 Oxidative stress

Increased oxidative stress brought on by aging causes tissue damage and quickens the aging process [142]. A decreased antioxidant capacity in the aging kidney has been associated with increased mitochondrial oxidative phosphorylation, decreased catalase, Cu/Zn-superoxide dismutase, and glutathione reductase levels [143,144]. All of these alterations damage the electron transport chain in mitochondria, which causes the bioenergetics and homeostasis of the organelle to gradually collapse [145]. A nicotinamide adenine dinucleotide$^+$-dependent deacetylase called SIRT1 participates in redox defense; however, as people age, their levels decline. By maintaining renal Sirt1 expression and inhibiting mitochondrial autophagy, Kume et al. showed that caloric restriction protects the aging kidney [146]. The capacity of renal interstitial cells to tolerate the oxidizing medullary environment and provide antifibrotic and antiapoptotic effects may be improved by increased Sirt1 expression [147]. Podocyte-specific Sirt1 knockdown mice were linked to decreased activation of the PPARγ-coactivator-1α (PGC-1α), forkhead box O3, forkhead box O4, and p65 NF-κB through SIRT1-mediated deacetylation, according to a second research work by

Chuang et al. These point to the activation of downstream pathways leading to oxidative stress, cell death, and inflammation in podocytes as a result of SIRT1 dysregulation [76]. Another mitochondrial deacetylase, SIRT3, can boost the expression of prosurvival genes and increase the number of mitochondria, shielding cells from damage brought on by oxidative stress [148]. According to Benigni et al., the AT1 receptor antagonist was able to block angiotensin II's ability to reduce Sirt3 mRNA expression in cultivated tubular epithelial cells [90]. The sirtuin family of proteins may be a therapeutic target for those whose kidney function has decreased with aging.

3.6 Vascular changes

As evidenced by the prevalence of renal artery constriction or atherosclerosis, which rises from 0.4% in the 18—29 age group to 25% in the 60—76 age group, aging kidneys are characterized by a gradual narrowing of the renal vasculature [10]. The circulation of the nephron gradually declines at the microvascular level due to increasing cortical blood shunting to the medullary circulation, atrophy of afferent and efferent arteries, and loss of peritubular capillaries [149]. In all kidney regions, but especially in the cortex, age-related expansion of hypoxia was discovered using the transgenic rats' "hypoxia-responsive" reporter. Vascular endothelial growth factor and glucose transporter-1 were shown to be upregulated in proportion to the degree of hypoxia, emphasizing the role of the hypoxia-inducible factor and its target genes in the aging kidney [150].

Renal vasculature has a propensity to acquire excessive vasoconstriction in the elderly. Increased sympathetic tone, a tendency toward the vasoconstrictor effects of angiotensin II, endothelin, and platelet-activating factor, as well as reduced responsiveness to renal vasodilators such atrial natriuretic peptide, nitric oxide, and amino acids are likely to cause this. Such a vasoconstrictive force-biased imbalance makes it harder for the aging kidney to maintain its normal renal plasma flow, which increases sensitivity to nephrotoxic damage [151,152].

3.7 Telomere shortening

Even though telomere length decreases with age, it is debatable whether it is a sign of aging or a biomarker that might affect biological circumstances, postpone senescence, and increase lifespan [153]. The majority of research points to telomere shortening as a sign of aging and the onset of cellular senescence. Nevertheless, several investigations have shown that telomere shortening increases the age-related vulnerability of the kidney to damage and leads to maladaptive repair. An enzyme complex called telomerase inserts DNA sequence repeats into DNA strands in the telomere regions. The two main parts of telomerase, TerT and TerC, work together to stop telomere shortening. Aged mice treated with telomerase gene therapy have been proved to live longer [154]. Studies have also shown that animals lacking TerC or TerT exhibit delayed recovery from renal ischemia-reperfusion damage, pointing to telomerase's critical function in the repair of injured kidneys [155].

4. Prevalence of CKD in the elderly

Chronic kidney disease is a global burden in the elderly population and represents one of the age-associated diseases with increasing morbidity and mortality trends worldwide. CKD affects over 10% of the population worldwide, accounting for an estimated >800 million individuals [156]. Clinically, CKD can be diagnosed with a marked decline in estimated glomerular filtration rate (eGFR <60 mL/min/1.73 m^2) or the presence of albuminuria (defined as an albumin to creatinine ratio >30 mg/g) persisting for 3 months [157]. Since the kidney performs many functions, including excretory, metabolic, and endocrine, the structural deterioration of the kidney function, as observed in CKD, is parallel with GFR. The prevalence of CKD has also been reported to increase in various patient subgroups and geographic regions. For example, in addition to aging, obesity and diabetes mellitus (DM) represent significant CKD risk factors. The loss of kidney function is also indirectly associated with atherosclerotic disease, systemic hypertension, and possibly obesity. Thus, the prevalence of CKD is partly explained by the increase in CKD risk factors, including the aging population, hypertension, and diagnosed diabetes, reflecting that renal dysfunction can be accelerated both by normal aging and the prevailing health condition of an individual [157].

Senescence is a universal phenomenon [158], and the prevalence of CKD substantially increases in younger adults (aged 20—39 years) to older adults (aged 70 years and more) after numerous renal insults [159]. With the acceleration of age, kidneys exhibit a progressive decline in GFR and renal blood flow. This progressive decline in GFR is due to decreased glomerular capillary plasma flow rate and glomerular capillary ultrafiltration coefficient. The renal mass progressively reduces with advancing age, leading to a decrease in renal weight associated with a significant reduction in renal blood flow [160]. In a recent report from the US population, the prevalence of CKD in the elderly population was 39.4% of persons aged more than 60 years versus 12.6% and 8.5% of persons aged 40—59 years and 20—39 years, respectively [141]. Irrespective of gender, renal impairment increases with advancing age in both men and women [161,162]. In association with adverse complications such as kidney failure, premature death, and cardiovascular diseases (CVD), evidence suggests that death risks increase as the GFR decreases below 60 mL/min/1.73 m^2 of body surface area [163].

4.1 Classification of CKD and staging

Likewise, kidney function can be estimated using parameters of blood creatinine, age, sex, and ethnicity (also known as estimated GFR or eGFR) using the Modification of Diet in Renal Disease (MDRD) study or Cockcroft-Gault estimating equations [164]. A 24-hour urine creatinine and concurrent serum creatinine concentration can also be used to compute the creatinine clearances. Both estimating equations include online tools (MDRD eGFR is accessible at http://www.nkdep.nih.gov/professionals/gfr_calculators/index.htm; Cockcroft-Gault eGFR is available at http://www.mdcalc.com/cockcroftgault). Patients with later stages of CKD are more prone to complications and develop ESRD, necessitating renal replacement therapy. Thus, early detection and management will frequently lessen severe CKD consequences and decrease the disease's course.

The next step after a CKD diagnosis is to identify staging, which is based on GFR, albuminuria, and the etiology of the CKD [165]. Staging of GFR is classified as G1 (GFR $\geq$90 mL/min/1.73 m^2), G2 (GFR 60—89 mL/min/1.73 m^2), G3a (45—59 mL/min/1.73 m^2), G3b (30—44 mL/min/1.73 m^2), G4 (15—29 mL/min/1.73 m^2), and G5 (<15 mL/min/1.73 m^2) [165]. The development of estimating equations (such as the Chronic Kidney Disease Epidemiology Collaboration [CKD-EPI] and MDRD equations) has largely replaced the need for direct measurement in clinical practice [166,167], even though GFR can be directly measured by clearance of agents such as iohexol or iothalamate [168,169]. According to filtration indicators, clinical labs now consistently report estimated GFR (eGFR). Creatinine, a 113 Da byproduct of creatine metabolism, is the most used filtering marker employed [170]. The CKD-EPI 2009 creatinine equation, which is more accurate than the older MDRD equation, is the preferred estimating equation in the United States and much of the rest of the globe, especially for eGFR levels higher than 60 mL/min/1.73 m^2. The CKD-EPI 2012 creatinine-cystatin C equation can be utilized with cystatin C when greater accuracy and precision are required [167]. A high-protein diet, the use of creatinine supplements, the use of medications that impact the tubular secretion of creatinine, and persons with altered creatinine production and/or metabolism may all benefit from adding cystatin C [165,170].

A urine albumin to creatinine ratio (ACR) should preferably be used to measure albuminuria. There are three different levels of albuminuria staging: A1 (urine ACR <30 mg/g), A2 (30—300 mg/g), and A3 (>300 mg/g) [165]. Assays for the former are more likely to be standardized and have superior accuracy at lower values of albuminuria, thus proper guidelines prior to staging of CKD [165] should be followed. Based on The Kidney Disease Improving Global Outcomes (KDIGO), 2022, Clinical Practice Guideline for Diabetes Management in Chronic Kidney Disease, CKD is classified based on cause, GFR category (G1—G5), and albuminuria category (A1—A3) is depicted in Fig. 6.6.

4.2 Etiology

CKD has been recognized for many decades as an underlying cause of death worldwide. CKD patients with reduced kidney functions have markedly declined life expectancy. The etiology of CKD is often not apparent in many patients. A hospital-based study reported that diabetes mellitus and hypertension patients have a greater predisposition to CKD [171]. Furthermore, the prevalence of obese patients, the amplifier with a combination of both hypertension and diabetes mellitus, is reported to be more than 30% in the United States, and it is expected to increase by 50% in 2030 [172]. In addition, patients with CKD have markedly increased cardiovascular risk of overt heart failure, coronary artery diseases, sudden cardiac death, and arrhythmias [173]. CKD and CVD are strongly interrelated. ESRD patients on dialysis are at high risk of mortality due to CVD. Many studies also indicate that cigarette smoking significantly increases the risk factors for the development and progression of CKD. The deleterious effects on kidney function are closely associated with smoking by increasing the risk of microalbuminuria, which facilitates the advancement of microalbuminuria to proteinuria as a result of diabetic nephropathy, which leads

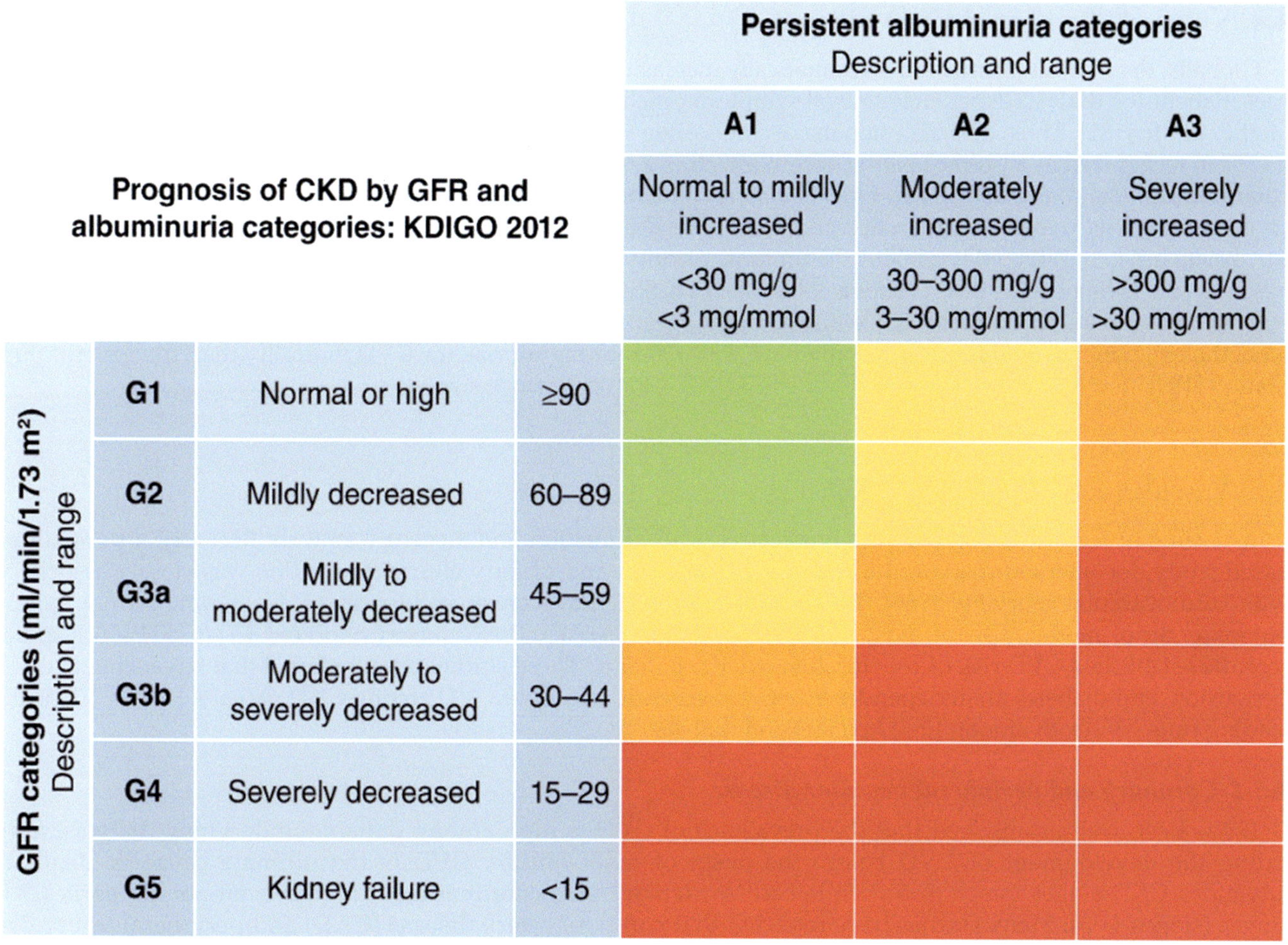

Prognosis of CKD by GFR and albuminuria categories: KDIGO 2012

GFR categories (ml/min/1.73 m²)				Persistent albuminuria categories		
				A1 Normal to mildly increased <30 mg/g <3 mg/mmol	A2 Moderately increased 30–300 mg/g 3–30 mg/mmol	A3 Severely increased >300 mg/g >30 mg/mmol
G1	Normal or high	≥90		Green	Yellow	Orange
G2	Mildly decreased	60–89		Green	Yellow	Orange
G3a	Mildly to moderately decreased	45–59		Yellow	Orange	Red
G3b	Moderately to severely decreased	30–44		Orange	Red	Red
G4	Severely decreased	15–29		Red	Red	Red
G5	Kidney failure	<15		Red	Red	Red

Green: low risk (if no other markers of kidney disease, no CKD); yellow: moderately increased risk; orange: high risk; red: very high risk.

FIGURE 6.6 Classification of CKD and staging. *Adapted from: [165].*

to ESRD [174]. Although genetic variations in certain genes have been linked to an increased risk of developing kidney disease, emerging epidemics of CKD with uncertain etiology (CKDu) worldwide have also been reported, suggesting an important health policy and public health responses [175].

4.3 Progression

The elderly population is at significant risk for further kidney damage and subsequent CKD development, regardless of the underlying etiology of the condition [176,177]. Uncontrolled diabetes and hypertension, which are prevalent in older people, are significant risk factors for advancement. Acute kidney damage (AKI), which is a significant risk factor for advancement, is another issue that puts the elderly at a high risk of developing CKD [178]. First, AKI can be directly caused by a high incidence of co-morbid conditions, such as prostatic hypertrophy or congestive heart failure. Second, drugs and medical procedures frequently employed to address co-morbid illnesses might contribute to or be a cause of AKI. Finally, the structural changes to the kidney that frequently come with aging may make it unable to successfully compensate for sudden drops in GFR. According to data from a large healthcare system, patients who acquire AKI are typically 10 years older than patients who do not [179], and elderly patients who get AKI have a lower chance of recovering kidney functions [180].

4.4 Vulnerability of aged CKD patients

Globally, the dialysis population is dramatically increasing, particularly in low-income and middle-income countries. Patients on dialysis experience several symptoms, including shortened life expectancy and a low health-related quality of life [181]. Thus, dialysis often makes life-saving yet cost-effective treatment among ESRD patients. ESRD has been recognized as a public health concern affecting 2.6 million worldwide with an increased risk of cardiovascular morbidity and mortality [182]. Elderly dialysis patients have a notably higher risk of developing cardiovascular events (such as myocardial infarction and heart failure) and neoplastic events than younger subjects. Thus, across the world, hemodialysis (HD) remains the topmost priority of dialysis. Several studies have demonstrated that HD has better survival benefits than peritoneal dialysis (PD). Still, observational studies showed no significant difference between HD and PD modalities [183], while others suggest one modality associated with a survival advantage [184]. Nonetheless, patients on dialysis generally have a 10−30-fold higher risk for CVD mortality than the general population [185].

4.5 Vascular consequences of CKD

A decline in renal function conversely affects vascular function and structure as well. Recently, a population-based study documented that a reduction in renal function and urinary albumin excretion was closely associated with cardiovascular mortality [186]. The elderly with CKD have fewer incidences of developing kidney failure but rather die of another disease, most commonly vascular diseases. The Chinese Cohort of CKD (C-STRIDE) studies recently established CVD risk factors in Chinese CKD patients. These studies demonstrated that advancing age, hypertension, and diabetes are independently associated with comorbid CVD among CKD patients [187]. CVD is the leading cause of death among ESRD patients on dialysis.

4.5.1 Coronary and peripheral vascular disease

It has been well established that the prevalence of CVD is high among patients undergoing dialysis, indicating the development of CVD before the onset of renal failure. CVD is the ultimate cause of death in advanced CKD stages (stage 4−5), while cardiovascular events are markedly higher even among early CKD stages (stages 1−3). Many studies have reported that coronary artery disease (CAD) and peripheral artery disease (PAD) can cause cardiovascular events, including stroke, heart failure, and arrhythmias. Several studies have revealed that the level of renal function is independently associated with CVD outcomes. Patients with low GFR exhibit high cardiovascular risk factors, driven by both traditional and nontraditional risk factors indicating CAD and sudden cardiac death due to declining renal function. In 2010, Stevens et al. reported that CVD remains the leading cause of mortality among dialysis patients, elucidating nearly 45% of deaths at all ages with two-thirds of cardiovascular deaths [178]. Besides cardiovascular disease, PAD in hemodialysis patients reduces the survival rate.

4.5.2 Cerebrovascular disease and cognitive functioning

In 2018, Tsai et al. reported that CKD was strongly associated with cerebral microbleeds (CMBs), especially in patients with hypertensive intracranial hemorrhage (ICH) [188]. CMB increases the risk of cognitive decline and stroke and is significantly prevalent with declining kidney function. Indeed, CKD contributes to stroke risk factors. Recently, evidence emerged that cognitive decline remains associated with hypertension and diabetes mellitus, reflecting cognitive impairment's close association with CVD [189]. In 2018, Tamura et al. demonstrated that a reduced renal function individual is more likely to develop cognitive impairment than an individual with normal renal function among US adults in a cross-sectional study [190]. In addition, patients undergoing hemodialysis (HD) possibly develop cognitive deficits due to the prevalence of stroke and CVD at an older age.

5. Risk factors for CKD in the elderly

In terms of one's health and the health of the community, determining what variables predispose someone to CKD is crucial since some risk factors are modifiable and can stop or reduce the progression of CKD to ESRD. Some important risk factors for CKD are described below.

5.1 Diabetes

Diseases associated with carbohydrate metabolism are common in elderly individuals. Diabetes mellitus (DM) is among the leading causes of the dramatic rise in the prevalence of impaired renal function with age. In the elderly population, the risk of developing CKD accompanied by DM continues to increase with the rise in life expectancy [191,192]. CKD in type I diabetes is mainly accompanied by diabetes-related microvascular disease, whereas a range of complications contribute to CKD in individuals with type II diabetes. The co-occurrence of DM in CKD elderly patients limits the treatment options for managing CKD [193]. The etiology of the pathogenesis of diabetes in CKD is diverse. It spans the initial mechanism of damage involving adaptive glomerular hyperfiltration leading to long-term damage involving functional impairment of nephrons. Glomerular hyperfiltration is known to occur early in type 1 diabetic patients [192], and persistent glomerular hyperfiltration indicates a decline in kidney function and the risk of developing CKD [194]. It is induced by various mediators such as vascular endothelial growth factor (VEGF), the renin–angiotensin–aldosterone system, nitric oxide, and TGF-β. Other mechanisms include enhanced advanced glycation end product (AGE) formation, activation of the protein kinase C (PKC) pathway, upregulation of the mammalian target of rapamycin (mTOR), and increased inflammation. Besides, the expression of nephrin has been reported to be lowered in CKD patients with diabetes. Further, diabetes and aging-induced vascular aberration are known to result in microalbuminuria and other kidney-related complications in elderly diabetes patients [192].

5.2 Obesity

Obesity contributes significantly to the development of CKD and ESRD. In addition, it increases the risks of developing diabetes and hypertension. Obesity is associated with glomerular hyperfiltration to meet higher metabolic needs due to increased body weight, promoting intraglomerular pressure, and increasing the risk of developing CKD [195]. Many studies have linked obesity to both the development and progression of CKD. A higher body mass index (BMI) indicates proteinuria and is associated with a low GFR. Abdominal obesity with CKD is associated with microalbuminuria, low GFR, or incident ESRD, independent of BMI. Higher visceral tissue is related to the prevalence of albuminuria in men and impaired renal functions; also, visceral adiposity appears to be directly associated with mortality in ESRD patients and those with kidney transplants. The adverse effects of obesity on kidneys also include pathologies such as kidney malignancies and higher chances of developing nephrolithiasis. The association between kidney cancers and obesity is constant among men and women worldwide [195].

While it is true that obesity alone is not sufficient to cause kidney damage, most obese individuals may never develop any complications related to CKD. Some of the adverse effects of obesity may be due to downstream conditions such as diabetes and hypertension. Nevertheless, kidneys are directly affected by endocrine effects such as inflammation, oxidative stress, dysregulated lipid metabolism, activation of the renin–angiotensin–aldosterone system, enhanced insulin production, and increased insulin resistance. These effects are induced by adipose tissue via the production of adipokines such as adiponectin, leptin, resistin, and visfatin, the levels of which are altered due to obesity [195–197]. Downstream of these effects are the kidney's pathological changes that contribute to CKD predisposition. These pathological changes include increased accumulation of ectopic lipid and renal sinus fat, glomerular hyperfiltration, glomerular filtration barrier damage, and the development of glomerulomegaly and glomerulosclerosis. Obesity-related glomerulopathy (ORG) occurs mainly with severe kidney damage conditions in hypertensive or elderly patients [195,198].

5.3 Hypertension and cardiovascular diseases

Hypertension is another risk factor for developing CKD, and progression to ESRD, CVD, and mortality. Early detection and proper treatment of hypertension are recommended to delay its progression to advanced stages and minimize its associated complications. Various studies have indicated varying degrees of risk at which both sexes with hypertension would develop CKD. Nevertheless, a recent meta-analysis reveals that hypertension significantly impacts the risk of developing CKD and ESRD in men more than women [199]. CVD is closely associated with CKD, and both CVD and CKD appear to be risk factors for each other. Interestingly, renal function was more compromised in older men and those with a higher entry diastolic blood pressure. In CKD patients, mortality is frequently linked to the co-occurrence of CVD, and patients with CKD are more likely to die of CVD than due to kidney failure. Further, CVD is highly prevalent in patients who undergo dialysis. The mortality due to CVD is comparatively (about 10–30 times) higher in such patients [141].

Elderly CKD patients undergoing dialysis seem to be at higher risk for mortality, compounded by the common occurrence of congestive heart failure (CHF). The major risk factors involved are systolic dysfunction, diabetes, ischemic heart disease, and older age. Even patients who do not have baseline cardiovascular complications are more likely to develop CHF subsequently, with a significant decrease in the chances of survival over the course of dialysis. Elderly CKD patients were reported to have higher CVD complications than general non-CKD subjects, whereas even CHF incidence in CKD patients is twice the frequency of the same in non-CKD subjects [141].

5.4 Acute kidney injury

There has been an increase in studies that have reported that acute kidney injury (AKI) might act as a driving factor for CKD development and progression, particularly in elderly patients. Studies in animal models of AKI have shown a failure of the potential repair mechanisms following AKI, exemplified by fibrosis, vascular rarefaction, chronic inflammation, and glomerulosclerosis resulting in accelerated kidney aging and a decline in kidney function. Studies suggest that subjects who suffered from AKI showed an about ninefold higher risk for CKD while there was an about threefold increased chance of progression to ESRD. Further, a greater risk of CKD and ESRD relative to the degree of severity of AKI has been observed. Additionally, the increase in the profibrotic and apoptotic factors following AKI-associated inflammation appears to negatively affect the distant organs by leaving an imprint after an AKI episode which also adds to the risk for persistent morbidity and higher mortality [141,161,200].

5.5 Nephrotoxins

Alcohol and several drugs are known to interfere with the function of the kidneys at different levels [161,201]. Drugs such as antianalgesics (e.g., acetaminophen and aspirin), nonsteroid antiinflammatory drugs (NSAIDs), angiotensin II converting enzyme (ACE) inhibitors, and angiotensin receptor blockers (ARBs) reduce the kidneys' ability to regulate glomerular pressure and filtration rate. Calcineurin inhibitors (e.g., cyclosporine and tacrolimus) are known for causing vasoconstriction of afferent arterioles leading to dose-dependent renal functional impairment. Renal tubular cells, such as proximal tubular cells in particular, are susceptible to toxins because of their exposure to circulating toxins. At the same time, they are involved in the concentration and reabsorption of glomerular filtrate.

Drugs such as aminoglycosides, antiretrovirals, and cisplatin cause tubular cell toxicity by hampering mitochondrial function, affecting tubular transport, and elevating oxidative stress by generating ROS. Chronic interstitial nephritis occurs with the consumption of Chinese herbals containing aristocholic acid, high doses of NSAIDs, and antianalgesics, which may progress to ESRD. Certain drugs such as hydralazine, gold therapy, NSAIDs, lithium, and propylthiouracil can cause inflammatory conditions such as glomerulonephritis and fibrosis. Some antibiotics (e.g., ampicillin and ciprofloxacin) and antivirals (e.g., acyclovir), and antiproliferative agents (e.g., methotrexate) that form crystals are insoluble in human urine, causing crystal nephropathy. Individuals with volume depletion and renal insufficiency are at risk for developing this condition. Drugs of abuse such as alcohol, cocaine, ketamine, methamphetamine, and heroin are known to cause rhabdomyolysis, a condition in which the breakdown of skeletal muscle occurs and myoglobin and creatine kinase are released into plasma, which in turn causes renal toxicity and alters GFR. These drugs are the causal agents in most rhabdomyolysis cases, which may eventually progress to CKD. The elderly population over 60 years is particularly vulnerable to drug-induced renal impairment and disease progression [201].

5.6 Smoking

Smoking increases the risk of developing CKD by inducing oxidative stress, generating a proinflammatory state, loss of endothelial function, glomerular sclerosis, and tubular damage [161]. A drastic increase in serum creatinine is associated with excessive daily smoking of cigarettes. The risk of developing CKD is much higher among people who regularly smoke cigarettes. Smoking is associated with microalbuminuria and its progression to proteinuria and impaired renal function. Microalbuminuria is a sensitive marker for glomerular damage, increasing the risk of CKD and ESRD [174]. Excessive smoking contributes to a higher risk of developing CKD in middle-aged healthy individuals with no underlying kidney disease. Smoking cessation could reduce the risk of CKD in former smokers [202]. The risk of CKD further increases with smoking among individuals with nephropathy associated with hypertension and diabetes [174]. Studies in rat models have shown that smoking increases superoxide dismutase activity, promotes renal fibrosis, and the kidneys of rats showed elevated levels of TGF-β, a mediator of renal fibrosis. Further, nicotine has been shown to induce ROS-mediated apoptosis in the podocytes and downstream MAPK signaling. Smoking also causes insulin resistance and AGEs that are harmful to the kidney, which accelerates CKD progression [203].

5.7 Genetic components

CKD is a heritable disease, and genome-wide association studies have identified susceptible loci for glomerular filtration rate of eGFRcrea <60 mL/min per 1.73 min. Mutations in Uromodulin (*UMOD*), which encodes Tamm-Horsfall protein in the urine, were implicated in renal functional variations. Another identified mutation is in *APOL1*, with an autosomal recessive inheritance pattern associated with a greater risk of developing ESRD due to hypertension. The mutation in *APOL1* is more prevalent in those of African origin, making them highly vulnerable to developing CKD. Further, the genes of the renin—angiotensin—aldosterone system are relevant to the development of CKD. A genotype study conducted with Han Chinese in Taiwan by restriction fragment length polymorphism (RFLP) analysis revealed significant relationships with ACE-A2350G (angiotensin I converting enzyme) and AGTR1-C573T (angiotensin II type I receptor) polymorphism in patients with CKD [161]. Another recent study that carried out next-generation sequence-based testing of the broad renal gene panel showed the genetic etiologies related to CKD. Positive pathogenic/likely pathogenic variants were frequently identified in *PKD1*, *COL4A5*, *PKD2*, *COL4A4*, *COL4A3*, and *TTR* genes, among which the highest frequency (34.1%) of variants was identified in *PKD1*. These genetic markers may help determine the underlying inheritable kidney disorders [204].

6. CKD care and management for the elderly

For addressing CKD in the elderly, there is a paucity of evidence-based advice and guidelines. When organizing the care of CKD in the elderly, geriatric concerns such as frailty, quality of life, life expectancy, end-of-life difficulties, pharmacokinetics and pharmacodynamics of medications, and treatment complications must be considered. AKI must be recognized and avoided since it can be triggered by nephrotoxic antibiotics, radiocontrast exposure, angiotensin-converting enzyme inhibitor (ACEI) and angiotensin receptor blockers (ARB) combinations, and nonsteroidal antiinflammatory drugs (NSAIDs) and diuretics [205].

6.1 Blood pressure monitoring

Controlling blood pressure has long been understood to be crucial in efforts to halt the course of CKD. Because of the involvement of angiotensin in the development of CKD and their established favorable effects on proteinuria, the administration of an ACEI or an ARB appears to offer a potential benefit. Lewis et al. published one of the earliest publications that showed type 1 diabetic patients utilizing ACEI to slow the development of CKD [206]. In that research, type 1 diabetics were given either a placebo, Captopril, or other blood pressure medications, resulting in nearly equal blood pressure management in both groups. Lewis discovered those taking Captopril had a significantly lower risk of doubling serum creatinine and concluded that "Captopril protects against deterioration in renal function in insulin-dependent diabetic nephropathy and is significantly more effective than blood-pressure control alone." Later, nondiabetic renal disease confirmed similar results [207,208]. It is unclear what the appropriate blood pressure is for these people. The original targets for blood pressure control were less than 125/75 mmHg for patients with diabetic nephropathy and less than 130/80 mmHg for patients with nondiabetic CKD [209]. However, Upadhyay [210] discovered in a systemic review of more than 2000 patients that there was no evidence that those values were any better than 140/90 mmHg, with the exception of patients with proteinuria.

6.2 Erythropoiesis-stimulating agents

The dosage of erythropoiesis-stimulating agents (ESAs) that should be used to treat anemia caused by renal illness is also in flux. The onset of anemia often occurs in stage 3 CKD, and erythropoietin and its analogs' accessibility have significantly improved several aspects of patients with CKD, including their need for less frequent blood transfusions, their quality of life, and their left ventricular hypertrophy. For dialysis patients, hemoglobin between 11 and 12 g/dL was the ideal range [141]. The correction of hemoglobin and outcomes in renal insufficiency (CHOIR) [211] and the time to reconsider evidence for anemia treatment (TREAT) investigations (median age: 68 years), however, have shown a higher risk of cardiovascular events with hemoglobin levels over 13 g/dL [212,213]. As a result, the 2012 KDIGO recommendations advise against initiating ESA in CKD patients with Hgb >10 g/dL.

6.3 Acidosis

Beginning in stage 3, individuals with CKD have acidosis due to poor ammonia excretion. In recent research by deBrito-Ashurst, 134 patients with metabolic acidosis in stages 3–4 were randomized to receive either conventional therapy or replacement with oral sodium bicarbonate for 2 years. It was shown that those who received the medication had considerably slower CKD development and improved nutritional metrics [214]. Bicarbonate usage was linked to slowing progression, even in early CKD stage 2 patients as compared to controls [215].

6.4 Cardiovascular risk assessment

Numerous conventional CVD risk factors, such as diabetes mellitus and hypertension, are present in CKD patients, and heart disease is more frequent in this group. Patients with CKD are more likely to be affected by metabolic syndrome, high levels of C-reactive protein, and impaired mineral metabolism, particularly calcium. Proteinuria and CKD are regarded as separate risk factors for coronary artery disease (CAD), and cardiovascular mortality is higher in CKD patients. The Rotterdam research examined 4484 seemingly healthy adults with a mean age of 69.6 years to determine if the degree of renal function, as measured by GFR, was related to the risk of acute myocardial infarction. The results of the study revealed that renal function is a graded and independent predictor of the development of myocardial infarction in an aged population and that a reduction in glomerular filtration rate of 10 mL/min/1.73 m^2 was linked to a 32% higher risk of myocardial infarction [216]. The KDIGO guidelines for lowering the risk of CVD in people with CKD recommend interventions to slow GFR decline regardless of age, therapeutic lifestyle changes (such as quitting smoking, losing weight, and increasing physical activity), the specific use of ACEI or ARBs in combination with other medications to control blood pressure, and the management of diabetes and other cardiovascular risk factors [217].

6.5 Insulin and glucose control with oral agents

Numerous studies indicate that maintaining target blood sugar levels slows the development of microvascular problems, including diabetic CKD. According to the United Kingdom Prospective Diabetes Study Group, patients with "tight" control, HgbA1c 7.0%, compared to those with conventional control, HgbA1c 7.9%, had a risk reduction of 11% in all diabetes end goals, including renal failure, during a 10-year period [218]. The Veterans Affairs Diabetes Trial (VADT) also revealed that except for the advancement of albuminuria, intense glucose control in patients with poorly managed type 2 diabetes had no appreciable impact on the incidence of major cardiovascular events, mortality, or microvascular sequelae [219]. The Action to Control Cardiovascular Risk in Diabetes (ACCORD) research in 2010 compared a group of patients with a mean HgbA1c of 7.0%–7.9% to 10,251 individuals who had received extensive treatment and had a HgbA1c of less than 6%. The research was terminated early due to significant mortality in the intensively treated group, even though the tight control delayed the onset of albuminuria [220].

6.6 Prevention of AKI

AKI has been recently recognized as a significant risk factor for the development of CKD and ESRD in the future, even if it is followed by renal recovery [101,221]. According to one report, the hazard ratio for developing ESRD was 41.2 for individuals with AKI and CKD compared to those without renal disease in a sample of 233,803 hospitalized, older patients without prior ESRD or AKI. The fact that 25.2% of patients with advanced ESRD had a history of AKI suggests that instances of AKI while hospitalized may hasten the course of renal disease, especially in individuals with CKD [222]. Implementing preventative steps to safeguard renal function is undoubtedly a preferable plan of action because there are not many effective pharmaceutical therapies for AKI in the elderly [223]. Establishing practical strategies to avoid AKI requires understanding the elderly's heightened sensitivity and possible risk factors. Some of these include looking for appropriate formulas or biomarkers to identify early AKI, controlling dyslipidemia and cardiovascular illnesses, treating diabetes and hypertension, preventing sepsis, and avoiding contrast nephrotoxicity and NSAID usage [224]. Blood pressure levels should be routinely checked to avoid unstable hemodynamics and unintentional AKI, particularly in people with preexisting medical problems [225,226].

6.7 Stabilization of CKD progression

Over the past few decades, there has not been much progress in developing novel drugs for renal illnesses. The gold standard treatment for individuals with proteinuric CKD is blood pressure management with RAS blocking, according to current clinical practice recommendations. However, ACE inhibitors or ARBs block the RAS delay but do not halt the course of renal disease [221,227,228]. The antiquated medication pentoxifylline, a nonselective phosphodiesterase inhibitor, showed additional renoprotective effects when used in conjunction with RAS blockade [228,229]. It could lessen the occurrence of AKI as measured by attenuation of serum creatinine rise [230], associated with the reduction of neonatal sepsis-related metabolic acidosis [229]. The present antidiabetic drug, PPARγ-agonist, may protect against kidney illnesses and renal aging while also improving hyperuricemic nephropathy [231]. A brand-new class of drugs for treating type 2 diabetes patients has been approved: sodium-glucose cotransporter-2 (SGLT2) inhibitors. These medications have been demonstrated to have favorable effects on renal and cardiovascular endpoints, presumably via extra-glycemic mechanisms [232]. The macula densa receives more glucose and salt thanks to SGLT2 inhibition, which also triggers tubuloglomerular feedback. Afferent arteriole vasoconstriction and a decrease in intraglomerular pressure are the outcomes of this [233]. In comparison to starting other glucose-lowering medications, utilizing SGLT2 inhibitor treatment was linked to a slower drop in eGFR and a reduced risk of serious renal events, according to another international observational cohort analysis [234]. It is significant to note that SGLT2 inhibitor studies for nondiabetic progressive renal disease are ongoing [235]. More research is required to validate the clinical efficacy of other new therapeutic medicines, such as Klotho agonists, and medications that target cellular senescence, such as navitoclax, dasatinib, and quercetin [236].

7. Conclusion

The complicated process of renal aging has received enormous interest as the global population grows older. The differences between normal aging and age-related kidney illnesses deserve additional clarification, even though the typical aging kidney changes are widely understood. There is an ongoing debate regarding the precise cellular and molecular signaling processes of renal aging. Cellular senescence plays a significant role in renal aging. The elderly kidney experiences structural and functional changes that are identical to those seen in renal disorders in the early stages. Implementing the available preventative measures to halt the progression of kidney disease requires differentiating between these two. Additionally, CKD is a serious health problem for many older persons. Understanding the risk factors and adopting screening of persons at risk will boost early diagnosis, may lessen the financial burden, and aid in initiating the treatment of modifiable risk factors for ESRD. There are information gaps in the treatment of elderly individuals with CKD and the need for mechanistic study on renal senescence. Because older individuals were often not included in clinical trials for renal progression, there were often conflicting treatment demands, making it difficult to generalize the clinical recommendations to everyone engaged. Currently, the treatment of CKD is based on the etiology, symptoms, and complications. However, when CKD progresses to ESRD, the mortality of CKD patients remains high, and their quality of life is low. Since renal senescence is strongly associated with the development of CKD and its progression, senescence could be a new target for CKD treatment. Targeting senescent cells and their clearance in aging models have shown an improved kidney function, which provided a potential target for the intervention of CKD and then extended to senotherapy that includes senolytics, senomorphics, and rejuvenating agents. Senolytics are a class of compounds that selectively remove senescent cells by promoting the proapoptotic pathways. Preclinical studies have shown that Dasatinib and Quercetin (D+Q) often combine, alleviating senescence-related dysfunction in cell cultures and animal models. In addition, D+Q can also decrease the levels of p16, p21, SA-β-gal, and SASPs in the adipose and skin tissues of patients with diabetic kidney disease. In 2016, the Food and Drug Administration (FDA) approved revised guidelines for metformin treatment in patients with CKD. Although the precise mechanism by which renal-protective abilities of metformin in patients with diabetic nephropathy and nondiabetic kidney disease is unknown, it is suggested that metformin inhibits NF-κB signaling, thus inhibiting SASP. Rejuvenating agents such as resveratrol can improve senescence-related renal injury by activating SIRT1, reducing oxidative stress, and inhibiting the proinflammatory SASPs. Studies have also indicated a correlation between shorter telomere length and CKD onset. Therefore, attention is focused on ways to preserve telomeres' length for treating CKD. Premature senescence was shown to be rescued by transient activation of telomerase reverse transcription expression in mice using telomerase reverse transcription (TERT), and the number of longer telomeres was higher in the kidneys of TERT-treated mice. The human reverse transcriptase telomerase (hTERT), which synthesizes the new telomeric DNA from an RNA template, represents another promising approach in maintaining telomere length and cell function. Consequently, telomerase-based therapies including telomerase activator supplements, and knowledge of maintenance genes involved in the process are being explored for the early detection and therapeutic interventions for the management of age-associated CKD.

Acknowledgment

The author SKM acknowledges the Council of Scientific and Industrial Research (CSIR), Ministry of Science and Technology, Government of India, for providing financial support (File no 09/559(0128)/2019-EMR-I).

References

[1] Glassock RJ, Rule AD. The implications of anatomical and functional changes of the aging kidney: with an emphasis on the glomeruli. Kidney Int 2012;82:270. https://doi.org/10.1038/KI.2012.65.
[2] Denic A, Glassock RJ, Rule AD. Structural and functional changes with the aging kidney. Adv Chron Kidney Dis 2016;23:19−28. https://doi.org/10.1053/j.ackd.2015.08.004.
[3] Hemmelgarn BR, Zhang J, Manns BJ, James MT, Quinn RR, Ravani P, et al. Nephrology visits and health care resource use before and after reporting estimated glomerular filtration rate. JAMA 2010;303:1151−8. https://doi.org/10.1001/JAMA.2010.303.
[4] Schaeffner ES, Ebert N, Delanaye P, Frei U, Gaedeke J, Jakob O, et al. Two novel equations to estimate kidney function in persons aged 70 years or older. Ann Intern Med 2012;157:471−81. https://doi.org/10.7326/0003-4819-157-7-201210020-00003.
[5] Anders HJ, Peired AJ, Romagnani P. SGLT2 inhibition requires reconsideration of fundamental paradigms in chronic kidney disease, 'diabetic nephropathy', IgA nephropathy and podocytopathies with FSGS lesions. Nephrol Dial Transplant 2022;37:1609−15. https://doi.org/10.1093/NDT/GFAA329.
[6] Fernandez-Fernandez B, Izquierdo MC, Valiño-Rivas L, Nastou D, Sanz AB, Ortiz A, et al. Albumin downregulates Klotho in tubular cells. Nephrol Dial Transplant 2018;33:1712−22. https://doi.org/10.1093/NDT/GFX376.
[7] Perez-Gomez MV, Bartsch LA, Castillo-Rodriguez E, Fernandez-Prado R, Fernandez-Fernandez B, Martin-Cleary C, et al. Clarifying the concept of chronic kidney disease for non-nephrologists. Clin Kidney J 2019;12:258−61. https://doi.org/10.1093/CKJ/SFZ007.
[8] Rule AD, Sasiwimonphan K, Lieske JC, Keddis MT, Torres VE, Vrtiska TJ. Characteristics of renal cystic and solid lesions based on contrast-enhanced computed tomography of potential kidney donors. Am J Kidney Dis 2012;59:611−8. https://doi.org/10.1053/J.AJKD.2011.12.022.
[9] Al-Said J, Brumback MA, Moghazi S, Baumgarten DA, O'Neill WC. Reduced renal function in patients with simple renal cysts. Kidney Int 2004;65:2303−8. https://doi.org/10.1111/J.1523-1755.2004.00651.X.
[10] Lorenz EC, Vrtiska TJ, Lieske JC, Dillon JJ, Stegall MD, Li X, et al. Prevalence of renal artery and kidney abnormalities by computed tomography among healthy adults. Clin J Am Soc Nephrol 2010;5:431−8. https://doi.org/10.2215/CJN.07641009.
[11] Duan X, Rule AD, Elsherbiny H, Vrtiska TJ, Avula RT, Alexander MP, et al. Automated assessment of renal cortical surface roughness from computerized tomography images and its association with age. Acad Radiol 2014;21:1441. https://doi.org/10.1016/J.ACRA.2014.05.014.
[12] McLachlan M, Wasserman P. Changes in sizes and distensibility of the aging kidney. Br J Radiol 2014;54:488−91. https://doi.org/10.1259/0007-1285-54-642-488.
[13] Fulladosa X, Moreso F, Narváez JA, Grinyó JM, Serón D. Estimation of total glomerular number in stable renal transplants. J Am Soc Nephrol 2003;14:2662−8. https://doi.org/10.1097/01.ASN.0000088025.33462.B0.
[14] Tan JC, Workeneh B, Busque S, Blouch K, Derby G, Myers BD. Glomerular function, structure, and number in renal allografts from older deceased donors. J Am Soc Nephrol 2009;20(1):181−8. https://doi.org/10.1681/ASN.2008030306.
[15] Tan JC, Busque S, Workeneh B, Ho B, Derby G, Blouch KL, et al. Effects of aging on glomerular function and number in living kidney donors. Kidney Int 2010;78(7):686−92. https://doi.org/10.1038/ki.2010.128.
[16] Vikse BE, Irgens LM, Leivestad T, Hallan S, Iversen BM. Low birth weight increases risk for end-stage renal disease. J Am Soc Nephrol 2008;19(1):151−7. https://doi.org/10.1681/ASN.2007020252.
[17] Luyckx VA, Bertram JF, Brenner BM, Fall C, Hoy WE, Ozanne SE, et al. Effect of fetal and child health on kidney development and long-term risk of hypertension and kidney disease. Lancet 2013;382(9888):273−83. https://doi.org/10.1016/S0140-6736(13)60311-6.
[18] Long DA, Mu W, Price KL, Johnson RJ. Blood vessels and the aging kidney. Nephron Exp Nephrol 2005;101:e95−9. https://doi.org/10.1159/000087146.
[19] Yang HC, Fogo AB. Fibrosis and renal aging. Kidney Int Suppl 2014;4:75. https://doi.org/10.1038/KISUP.2014.14.
[20] Darmady EM, Offer J, Woodhouse MA. The parameters of the ageing kidney. J Pathol 1973;109:195−207. https://doi.org/10.1002/PATH.1711090304.
[21] Martin JE, Sheaff MT. Renal ageing. J Pathol 2007;211:198−205. https://doi.org/10.1002/PATH.2111.
[22] Hoy WE, Douglas-Denton RN, Hughson MD, Cass A, Johnson K, Bertram JF. A stereological study of glomerular number and volume: preliminary findings in a multiracial study of kidneys at autopsy. Kidney Int Suppl 2003;63. https://doi.org/10.1046/J.1523-1755.63.S83.8.X.
[23] Nyengaard JR, Bendtsen TF. Glomerular number and size in relation to age, kidney weight, and body surface in normal man. Anat Rec 1992;232:194−201. https://doi.org/10.1002/AR.1092320205.
[24] Rule AD, Amer H, Cornell LD, Taler SJ, Cosio FG, Kremers WK, et al. The association between age and nephrosclerosis on renal biopsy among healthy adults. Ann Intern Med 2010;152:561−7. https://doi.org/10.7326/0003-4819-152-9-201005040-00006.

References

[25] Mancilla E, Avila-Casado C, Uribe-Uribe N, Morales-Buenrostro LE, Rodríguez F, Vilatoba M, et al. Time-zero renal biopsy in living kidney transplantation: a valuable opportunity to correlate predonation clinical data with histological abnormalities. Transplantation 2008;86: 1684–8. https://doi.org/10.1097/TP.0B013E3181906150.

[26] Vazquez Martul E, Veiga Barreiro A. Importance of kidney biopsy in graft selection. Transplant Proc 2003;35:1658–60. https://doi.org/10.1016/S0041-1345(03)00573-6.

[27] Kaplan C, Pasternack B, Shah H, Gallo G. Age-related incidence of sclerotic glomeruli in human kidneys. Am J Pathol 1975;80:227.

[28] Kubo M, Kiyohara Y, Kato I, Tanizaki Y, Katafuchi R, Hirakata H, et al. Risk factors for renal glomerular and vascular changes in an autopsy-based population survey: the Hisayama study. Kidney Int 2003;63:1508–15. https://doi.org/10.1046/J.1523-1755.2003.00886.X.

[29] Kremers WK, Denic A, Lieske JC, Alexander MP, Kaushik V, Elsherbiny HE, et al. Distinguishing age-related from disease-related glomerulosclerosis on kidney biopsy: the aging kidney anatomy study. Nephrol Dial Transplant 2015;30:2034–9. https://doi.org/10.1093/NDT/GFV072.

[30] Luyckx V, Shukha K, Brenner BM. Low nephron number and its clinical consequences. Rambam Maimonides Med J 2011;2:e0061. https://doi.org/10.5041/RMMJ.10061.

[31] Hayslett JP, Kashgarian M, Epstein FH. Functional correlates of compensatory renal hypertrophy. J Clin Invest 1968;47:774–99. https://doi.org/10.1172/JCI105772.

[32] Grantham JJ. Solitary renal cysts: worth a second look? Am J Kidney Dis 2012;59:593–4. https://doi.org/10.1053/J.AJKD.2012.02.002.

[33] Rule AD, Semret MH, Amer H, Cornell LD, Taler SJ, Lieske JC, et al. Association of kidney function and metabolic risk factors with density of glomeruli on renal biopsy samples from living donors. Mayo Clin Proc 2011;86:282–90. https://doi.org/10.4065/MCP.2010.0821.

[34] Goyal VK. Changes with age in the human kidney. Exp Gerontol 1982;17:321–31. https://doi.org/10.1016/0531-5565(82)90032-8.

[35] Elsherbiny HE, Alexander MP, Kremers WK, Park WD, Poggio ED, Prieto M, et al. Nephron hypertrophy and glomerulosclerosis and their association with kidney function and risk factors among living kidney donors. Clin J Am Soc Nephrol 2014;9:1892–902. https://doi.org/10.2215/CJN.02560314.

[36] Tsuboi N, Kawamura T, Koike K, Okonogi H, Hirano K, Hamaguchi A, et al. Glomerular density in renal biopsy specimens predicts the long-term prognosis of IgA nephropathy. Clin J Am Soc Nephrol 2010;5:39. https://doi.org/10.2215/CJN.04680709.

[37] Emamian SA, Nielsen MB, Pedersen JF, Ytte L. Kidney dimensions at sonography: correlation with age, sex, and habitus in 665 adult volunteers. Am J Roentgenol 2013;160:83–6. https://doi.org/10.2214/AJR.160.1.8416654.

[38] Gourtsoyiannis N, Prassopoulos P, Cavouras D, Pantelidis N. The thickness of the renal parenchyma decreases with age: a CT study of 360 patients. Am J Roentgenol 2013;155:541–4. https://doi.org/10.2214/AJR.155.3.2117353.

[39] Roseman DA, Hwang SJ, Oyama-Manabe N, Chuang ML, O'Donnell CJ, Manning WJ, et al. Clinical associations of total kidney volume: the Framingham heart study. Nephrol Dial Transplant 2017;32:1344–50. https://doi.org/10.1093/NDT/GFW237.

[40] Wang X, Vrtiska TJ, Avula RT, Walters LR, Chakkera HA, Kremers WK, et al. Age, kidney function, and risk factors associate differently with cortical and medullary volumes of the kidney. Kidney Int 2014;85:677–85. https://doi.org/10.1038/KI.2013.359.

[41] O'Neill WC. Structure, not just function. Kidney Int 2014;85:503–5. https://doi.org/10.1038/ki.2013.426.

[42] Eknoyan G. A clinical view of simple and complex renal cysts. J Am Soc Nephrol 2009;20:1874–6. https://doi.org/10.1681/ASN.2008040441.

[43] Baert L, Steg A. Is the diverticulum of the distal and collecting tubules a preliminary stage of the simple cyst in the adult? J Urol 1977;118: 707–10. https://doi.org/10.1016/S0022-5347(17)58167-7.

[44] Schwarz A, Lenz T, Klaen R, Offermann G, Fiedler U, Nussberger J. Hygroma renale: pararenal lymphatic cysts associated with renin-dependent hypertension (Page kidney). Case report on bilateral cysts and successful therapy by marsupialization. J Urol 1993;150:953–7. https://doi.org/10.1016/S0022-5347(17)35660-4.

[45] Smith H. The kidney: structure and function in health and disease. 1951.

[46] Davies DF, Shock NW. Age changes in glomerular filtration rate, effective renal plasma flow, and tubular excretory capacity in adult males. J Clin Invest 1950;29:496–507. https://doi.org/10.1172/JCI102286.

[47] Lindeman RD, Tobin J, Shock NW. Longitudinal studies on the rate of decline in renal function with age. J Am Geriatr Soc 1985;33:278–85. https://doi.org/10.1111/J.1532-5415.1985.TB07117.X.

[48] Eriksen BO, Løchen ML, Arntzen KA, Bertelsen G, Eilertsen BAW, Von Hanno T, et al. Subclinical cardiovascular disease is associated with a high glomerular filtration rate in the nondiabetic general population. Kidney Int 2014;86:146–53. https://doi.org/10.1038/KI.2013.470.

[49] Fliser D, Ritz E. Renal haemodynamics in the elderly. Nephrol Dial Transplant 1996;11(Suppl. 9):2–8. https://doi.org/10.1093/NDT/11.SUPP9.2.

[50] Fliser D, Ritz E. Relationship between hypertension and renal function and its therapeutic implications in the elderly. Gerontology 1998;44: 123–31. https://doi.org/10.1159/000021995.

[51] Lew SQ, Bosch JP. Effect of diet on creatinine clearance and excretion in young and elderly healthy subjects and in patients with renal disease. J Am Soc Nephrol 1991;2:856–65. https://doi.org/10.1681/ASN.V24856.

[52] Docherty MH, O'Sullivan ED, Bonventre JV, Ferenbach DA. Cellular senescence in the kidney. J Am Soc Nephrol 2019;30:726–36. https://doi.org/10.1681/ASN.2018121251.

[53] Tan H, Xu J, Liu Y. Ageing, cellular senescence and chronic kidney disease: experimental evidence. Curr Opin Nephrol Hypertens 2022;31: 235–43. https://doi.org/10.1097/MNH.0000000000000782.

[54] Campisi J, D'Adda Di Fagagna F. Cellular senescence: when bad things happen to good cells. Nat Rev Mol Cell Biol 2007;8(9):729–40. https://doi.org/10.1038/nrm2233.

[55] Gorgoulis V, Adams PD, Alimonti A, Bennett DC, Bischof O, Bishop C, et al. Cellular senescence: defining a path forward. Cell 2019;179: 813–27. https://doi.org/10.1016/J.CELL.2019.10.005.

[56] O'Sullivan ED, Hughes J, Ferenbach DA. Renal aging: causes and consequences. J Am Soc Nephrol 2017;28:407–20. https://doi.org/10.1681/ASN.2015121308.

[57] López-Otín C, Blasco MA, Partridge L, Serrano M, Kroemer G. The hallmarks of aging. Cell 2013;153:1194. https://doi.org/10.1016/j.cell.2013.05.039.

[58] López-Otín C, Blasco MA, Partridge L, Serrano M, Kroemer G. Hallmarks of aging: an expanding universe. Cell 2023;186:243—78. https://doi.org/10.1016/J.CELL.2022.11.001.

[59] Sturmlechner I, Durik M, Sieben CJ, Baker DJ, Van Deursen JM. Cellular senescence in renal ageing and disease. Nat Rev Nephrol 2016;13(2):77—89. https://doi.org/10.1038/nrneph.2016.183.

[60] Dimri GP, Lee X, Basile G, Acosta M, Scott G, Roskelley C, et al. A biomarker that identifies senescent human cells in culture and in aging skin in vivo. Proc Natl Acad Sci USA 1995;92:9363—7. https://doi.org/10.1073/PNAS.92.20.9363.

[61] Kurz DJ, Decary S, Hong Y, Erusalimsky JD. Senescence-associated (beta)-galactosidase reflects an increase in lysosomal mass during replicative ageing of human endothelial cells. J Cell Sci 2000;113:3613—22. https://doi.org/10.1242/JCS.113.20.3613.

[62] Krishnamurthy J, Torrice C, Ramsey MR, Kovalev GI, Al-Regaiey K, Su L, et al. Ink4a/Arf expression is a biomarker of aging. J Clin Invest 2004;114:1299—307. https://doi.org/10.1172/JCI22475.

[63] Sharpless NE, Sherr CJ. Forging a signature of in vivo senescence. Nat Rev Cancer 2015;15(7):397—408. https://doi.org/10.1038/nrc3960.

[64] Hernandez-Segura A, de Jong TV, Melov S, Guryev V, Campisi J, Demaria M. Unmasking transcriptional heterogeneity in senescent cells. Curr Biol 2017;27:2652—2660.e4. https://doi.org/10.1016/J.CUB.2017.07.033.

[65] Matjusaitis M, Chin G, Sarnoski EA, Stolzing A. Biomarkers to identify and isolate senescent cells. Ageing Res Rev 2016;29:1—12. https://doi.org/10.1016/J.ARR.2016.05.003.

[66] Xu J, Zhou L, Liu Y. Cellular senescence in kidney fibrosis: pathologic significance and therapeutic strategies. Front Pharmacol 2020;11. https://doi.org/10.3389/FPHAR.2020.601325.

[67] Raffaele M, Vinciguerra M. The costs and benefits of senotherapeutics for human health. Lancet Healthy Longev 2022;3:e67—77. https://doi.org/10.1016/S2666-7568(21)00300-7.

[68] El-Deiry WS, Tokino T, Velculescu VE, Levy DB, Parsons R, Trent JM, et al. WAF1, a potential mediator of p53 tumor suppression. Cell 1993;75:817—25. https://doi.org/10.1016/0092-8674(93)90500-P.

[69] Rayess H, Wang MB, Srivatsan ES. Cellular senescence and tumor suppressor gene p16. Int J Cancer 2012;130:1715—25. https://doi.org/10.1002/IJC.27316.

[70] Wade Harper J, Adami GR, Wei N, Keyomarsi K, Elledge SJ. The p21 Cdk-interacting protein Cip1 is a potent inhibitor of G1 cyclin-dependent kinases. Cell 1993;75:805—16. https://doi.org/10.1016/0092-8674(93)90499-G.

[71] Zhang H, Xiong Y, Beach D. Proliferating cell nuclear antigen and p21 are components of multiple cell cycle kinase complexes. Mol Biol Cell 1993;4:897—906. https://doi.org/10.1091/MBC.4.9.897.

[72] Young AP, Schisio S, Minamishima YA, Zhang Q, Li L, Grisanzio C, et al. VHL loss actuates a HIF-independent senescence programme mediated by Rb and p400. Nat Cell Biol 2008;10:361—9. https://doi.org/10.1038/NCB1699.

[73] Lee MJ, Feliers D, Mariappan MM, Sataranatarajan K, Mahimainathan L, Musi N, et al. A role for AMP-activated protein kinase in diabetes-induced renal hypertrophy. Am J Physiol Ren Physiol 2007;292:617—27. https://doi.org/10.1152/AJPRENAL.00278.2006/ASSET/IMAGES/LARGE/ZH20020746050010.JPEG.

[74] Sopjani M, Rinnerthaler M, Kruja J, Dermaku-Sopjani M. Intracellular signaling of the aging suppressor protein klotho. Curr Mol Med 2015;15:27—37. https://doi.org/10.2174/1566524015666150114111258.

[75] Yamamoto M, Clark JD, Pastor JV, Gurnani P, Nandi A, Kurosu H, et al. Regulation of oxidative stress by the anti-aging Hormone klotho. J Biol Chem 2005;280:38029. https://doi.org/10.1074/JBC.M509039200.

[76] Chuang PY, Cai W, Li X, Fang L, Xu J, Yacoub R, et al. Reduction in podocyte SIRT1 accelerates kidney injury in aging mice. Am J Physiol Ren Physiol 2017;313:F621—8. https://doi.org/10.1152/AJPRENAL.00255.2017/ASSET/IMAGES/LARGE/ZH20081782940005.JPEG.

[77] Luo C, Zhou S, Zhou Z, Liu Y, Yang L, Liu J, et al. Wnt9a promotes renal fibrosis by accelerating cellular senescence in tubular epithelial cells. J Am Soc Nephrol 2018;29:1238—56. https://doi.org/10.1681/ASN.2017050574/-/DCSUPPLEMENTAL.

[78] Jope RS, Johnson GVW. The glamour and gloom of glycogen synthase kinase-3. Trends Biochem Sci 2004;29:95—102. https://doi.org/10.1016/J.TIBS.2003.12.004.

[79] McColl G, Killilea DW, Hubbard AE, Vantipalli MC, Melov S, Lithgow GJ. Pharmacogenetic analysis of lithium-induced delayed aging in *Caenorhabditis elegans*. J Biol Chem 2008;283:350—7. https://doi.org/10.1074/JBC.M705028200.

[80] Zarse K, Terao T, Tian J, Iwata N, Ishii N, Ristow M. Low-dose lithium uptake promotes longevity in humans and metazoans. Eur J Nutr 2011;50:387—9. https://doi.org/10.1007/S00394-011-0171-X/FIGURES/1.

[81] Castillo-Quan JI, Li L, Kinghorn KJ, Ivanov DK, Tain LS, Slack C, et al. Lithium promotes longevity through GSK3/NRF2-dependent Hormesis. Cell Rep 2016;15:638—50. https://doi.org/10.1016/J.CELREP.2016.03.041.

[82] Zhou S, Wang P, Qiao Y, Ge Y, Wang Y, Quan S, et al. Genetic and pharmacologic targeting of glycogen synthase kinase 3β reinforces the Nrf2 antioxidant defense against podocytopathy. J Am Soc Nephrol 2016;27:2289—308. https://doi.org/10.1681/ASN.2015050565.

[83] Noth RH, Lassman MN, Tan SY, Fernandez Cruz A, Mulrow PJ. Age and the renin-aldosterone system. Arch Intern Med 1977;137:1414—7. https://doi.org/10.1001/ARCHINTE.1977.03630220056014.

[84] Mulkerrin E, Epstein FH, Clark BA. Aldosterone responses to hyperkalemia in healthy elderly humans. J Am Soc Nephrol 1995;6:1459—62. https://doi.org/10.1681/ASN.V651459.

[85] Kobori H, Navar LG. Urinary angiotensinogen as a novel biomarker of intrarenal renin-angiotensin system in chronic kidney disease. Int Rev Thromb 2011;6:108.

[86] Thompson MM, Oyama TT, Kelly FJ, Kennefick TM, Anderson S. Activity and responsiveness of the renin-angiotensin system in the aging rat. Am J Physiol Regul Integr Comp Physiol 2000;279. https://doi.org/10.1152/AJPREGU.2000.279.5.R1787.

[87] Anderson S, Rennke HG, Zatz R. Glomerular adaptations with normal aging and with long-term converting enzyme inhibition in rats. Am J Physiol 1994;267. https://doi.org/10.1152/AJPRENAL.1994.267.1.F35.

[88] Anderson S. Ageing and the renin-angiotensin system. Nephrol Dial Transplant 1997;12:1093—4. https://doi.org/10.1093/NDT/12.6.1093.

[89] Vajapey R, Rini D, Walston J, Abadir P. The impact of age-related dysregulation of the angiotensin system on mitochondrial redox balance. Front Physiol 2014;5. https://doi.org/10.3389/FPHYS.2014.00439.

[90] Benigni A, Corna D, Zoja C, Sonzogni A, Latini R, Salio M, et al. Disruption of the Ang II type 1 receptor promotes longevity in mice. J Clin Invest 2009;119:524—30. https://doi.org/10.1172/JCI36703.

References

[91] de Cavanagh EMV, Piotrkowski B, Basso N, Stella I, Inserra F, Ferder L, et al. Enalapril and losartan attenuate mitochondrial dysfunction in aged rats. FASEB J 2003;17:1096–8. https://doi.org/10.1096/FJ.02-0063FJE.

[92] Ma LJ, Fogo AB. Model of robust induction of glomerulosclerosis in mice: importance of genetic background. Kidney Int 2003;64:350–5. https://doi.org/10.1046/J.1523-1755.2003.00058.X.

[93] Kuro-o M, Matsumura Y, Aizawa H, Kawaguchi H, Suga T, Utsugi T, et al. Mutation of the mouse klotho gene leads to a syndrome resembling ageing. Nature 1997;390:45–51. https://doi.org/10.1038/36285.

[94] Kuro-o M. Klotho and the aging process. Korean J Intern Med (Engl Ed) 2011;26:113–22. https://doi.org/10.3904/KJIM.2011.26.2.113.

[95] Koh N, Fujimori T, Nishiguchi S, Tamori A, Shiomi S, Nakatani T, et al. Severely reduced production of klotho in human chronic renal failure kidney. Biochem Biophys Res Commun 2001;280:1015–20. https://doi.org/10.1006/BBRC.2000.4226.

[96] Lim JH, Kim EN, Kim MY, Chung S, Shin SJ, Kim HW, et al. Age-associated molecular changes in the kidney in aged mice. Oxid Med Cell Longev 2012;2012. https://doi.org/10.1155/2012/171383.

[97] Mitani H, Ishizaka N, Aizawa T, Ohno M, Usui SI, Suzuki T, et al. In vivo klotho gene transfer ameliorates angiotensin II-induced renal damage. Hypertension 2002;39:838–43. https://doi.org/10.1161/01.HYP.0000013734.33441.EA.

[98] Saito K, Ishizaka N, Mitani H, Ohno M, Nagai R. Iron chelation and a free radical scavenger suppress angiotensin II-induced downregulation of klotho, an anti-aging gene, in rat. FEBS Lett 2003;551:58–62. https://doi.org/10.1016/S0014-5793(03)00894-9.

[99] Karalliedde J, Maltese G, Hill B, Viberti G, Gnudi L. Effect of renin-angiotensin system blockade on soluble Klotho in patients with type 2 diabetes, systolic hypertension, and albuminuria. Clin J Am Soc Nephrol 2013;8:1899–905. https://doi.org/10.2215/CJN.02700313.

[100] Yoon HE, Ghee JY, Piao S, Song JH, Han DH, Kim S, et al. Angiotensin II blockade upregulates the expression of Klotho, the anti-ageing gene, in an experimental model of chronic cyclosporine nephropathy. Nephrol Dial Transplant 2011;26:800–13. https://doi.org/10.1093/NDT/GFQ537.

[101] Chou YH, Huang TM, Chu TS. Novel insights into acute kidney injury-chronic kidney disease continuum and the role of renin-angiotensin system. J Formos Med Assoc 2017;116:652–9. https://doi.org/10.1016/J.JFMA.2017.04.026.

[102] Rüster C, Wolf G. Renin-angiotensin-aldosterone system and progression of renal disease. J Am Soc Nephrol 2006;17:2985–91. https://doi.org/10.1681/ASN.2006040356.

[103] Wolf G, Wenzel U, Burns KD, Harris RC, Stahl RAK, Thaiss F. Angiotensin II activates nuclear transcription factor-kappaB through AT1 and AT2 receptors. Kidney Int 2002;61:1986–95. https://doi.org/10.1046/J.1523-1755.2002.00365.X.

[104] Wolf G, Bohlender J, Bondeva T, Roger T, Thaiss F, Wenzel UO. Angiotensin II upregulates toll-like receptor 4 on mesangial cells. J Am Soc Nephrol 2006;17:1585–93. https://doi.org/10.1681/ASN.2005070699.

[105] Rodríguez-Vita J, Sánchez-López E, Esteban V, Rupérez M, Egido J, Ruiz-Ortega M. Angiotensin II activates the Smad pathway in vascular smooth muscle cells by a transforming growth factor-beta-independent mechanism. Circulation 2005;111:2509–17. https://doi.org/10.1161/01.CIR.0000165133.84978.E2.

[106] Wolf G. Link between angiotensin II and TGF-beta in the kidney. Miner Electrolyte Metab 1998;24:174–80. https://doi.org/10.1159/000057367.

[107] Lin SL, Chen RH, Chen YM, Chiang WC, Lai CF, Wu KD, et al. Pentoxifylline attenuates tubulointerstitial fibrosis by blocking Smad3/4-activated transcription and profibrogenic effects of connective tissue growth factor. J Am Soc Nephrol 2005;16:2702–13. https://doi.org/10.1681/ASN.2005040435.

[108] Abrahamsen CT, Pullen MA, Schnackenberg CG, Grygielko ET, Edwards RM, Laping NJ, et al. Effects of angiotensins II and IV on blood pressure, renal function, and PAI-1 expression in the heart and kidney of the rat. Pharmacology 2002;66:26–30. https://doi.org/10.1159/000063252.

[109] Kontogiannis J, Burns KD. Role of AT1 angiotensin II receptors in renal ischemic injury. Am J Physiol 1998;274. https://doi.org/10.1152/AJPRENAL.1998.274.1.F79.

[110] Zhang J, Rudemiller NP, Patel MB, Wei Q, Karlovich NS, Jeffs AD, et al. Competing actions of type 1 angiotensin II receptors expressed on T lymphocytes and kidney epithelium during cisplatin-induced AKI. J Am Soc Nephrol 2016;27:2257–64. https://doi.org/10.1681/ASN.2015060683.

[111] Rodríguez-Romo R, Benítez K, Barrera-Chimal J, Pérez-Villalva R, Gómez A, Aguilar-León D, et al. AT1 receptor antagonism before ischemia prevents the transition of acute kidney injury to chronic kidney disease. Kidney Int 2016;89:363–73. https://doi.org/10.1038/KI.2015.320.

[112] Dai C, Stolz DB, Kiss LP, Monga SP, Holzman LB, Liu Y. Wnt/beta-catenin signaling promotes podocyte dysfunction and albuminuria. J Am Soc Nephrol 2009;20:1997–2008. https://doi.org/10.1681/ASN.2009010019.

[113] Matsui I, Ito T, Kurihara H, Imai E, Ogihara T, Hori M. Snail, a transcriptional regulator, represses nephrin expression in glomerular epithelial cells of nephrotic rats. Lab Invest 2007;87:273–83. https://doi.org/10.1038/labinvest.3700518.

[114] Kim MS, Yoon SK, Bollig F, Kitagaki J, Hur W, Whye NJ, et al. A novel Wilms tumor 1 (WT1) target gene negatively regulates the WNT signaling pathway. J Biol Chem 2010;285:14585–93. https://doi.org/10.1074/jbc.M109.094334.

[115] Zhou L, Li Y, Zhou D, Tan RJ, Liu Y. Loss of Klotho contributes to kidney injury by derepression of Wnt/β-catenin signaling. J Am Soc Nephrol 2013;24:771–85. https://doi.org/10.1681/ASN.2012080865/-/DCSUPPLEMENTAL.

[116] Lim K, Groen A, Molostvov G, Lu T, Lilley KS, Snead D, et al. α-Klotho expression in human tissues. J Clin Endocrinol Metab 2015;100: E1308–18. https://doi.org/10.1210/JC.2015-1800.

[117] Tan RJ, Zhou D, Zhou L, Liu Y. Wnt/β-catenin signaling and kidney fibrosis. Kidney Int Suppl 2014;4:84–90. https://doi.org/10.1038/KISUP.2014.16.

[118] Satoh M, Nagasu H, Morita Y, Yamaguchi TP, Kanwar YS, Kashihara N. Klotho protects against mouse renal fibrosis by inhibiting Wnt signaling. Am J Physiol Ren Physiol 2012;303:F1641. https://doi.org/10.1152/AJPRENAL.00460.2012.

[119] Lin CL, Wang JY, Huang YT, Kuo YH, Surendran K, Wang FS. Wnt/beta-catenin signaling modulates survival of high glucose-stressed mesangial cells. J Am Soc Nephrol 2006;17:2812–20. https://doi.org/10.1681/ASN.2005121355.

[120] Oh HJ, Nam BY, Wu M, Kim S, Park J, Kang S, et al. Klotho plays a protective role against glomerular hypertrophy in a cell cycle-dependent manner in diabetic nephropathy. Am J Physiol Ren Physiol 2018;315:F791–805. https://doi.org/10.1152/AJPRENAL.00462.2017.

[121] Erol A. The functions of PPARs in aging and longevity. PPAR Res 2007;2007. https://doi.org/10.1155/2007/39654.

[122] Reel B, Guzeloglu M, Bagriyanik A, Atmaca S, Aykut K, Albayrak G, et al. The effects of PPAR-γ agonist pioglitazone on renal ischemia/reperfusion injury in rats. J Surg Res 2013;182:176−84. https://doi.org/10.1016/J.JSS.2012.08.020.

[123] Wang P, Li B, Cai G, Huang M, Jiang L, Pu J, et al. Activation of PPAR-γ by pioglitazone attenuates oxidative stress in aging rat cerebral arteries through upregulating UCP2. J Cardiovasc Pharmacol 2014;64:497−506. https://doi.org/10.1097/FJC.0000000000000143.

[124] Lim HA, Lee EK, Kim JM, Park MH, Kim DH, Choi YJ, et al. PPARγ activation by baicalin suppresses NF-κB-mediated inflammation in aged rat kidney. Biogerontology 2012;13:133−45. https://doi.org/10.1007/S10522-011-9361-4.

[125] Panchapakesan U, Pollock CA, Chen XM. The effect of high glucose and PPAR-gamma agonists on PPAR-gamma expression and function in HK-2 cells. Am J Physiol Ren Physiol 2004;287. https://doi.org/10.1152/AJPRENAL.00445.2003.

[126] Bakris G, Viberti G, Weston WM, Heise M, Porter LE, Freed MI. Rosiglitazone reduces urinary albumin excretion in type II diabetes. J Hum Hypertens 2003;17:7−12. https://doi.org/10.1038/SJ.JHH.1001444.

[127] Grossman E. Rosiglitazone reduces blood pressure and urinary albumin excretion in type 2 diabetes: G Bakris et al. J Hum Hypertens 2003;17:5−6. https://doi.org/10.1038/SJ.JHH.1001474.

[128] Pistrosch F, Passauer J, Herbrig K, Schwanebeck U, Gross P, Bornstein SR. Effect of thiazolidinedione treatment on proteinuria and renal hemodynamic in type 2 diabetic patients with overt nephropathy. Horm Metab Res 2012;44:914−8. https://doi.org/10.1055/S-0032-1314836.

[129] Tang SCW, Leung JCK, Chan LYY, Cheng AS, Lan HY, Lai KN. Renoprotection by rosiglitazone in accelerated type 2 diabetic nephropathy: role of STAT1 inhibition and nephrin restoration. Am J Nephrol 2010;32:145−55. https://doi.org/10.1159/000316056.

[130] Revelo MP, Federspiel CC, Helderman H, Fogo AB. Chronic allograft nephropathy: expression and localization of PAI-1 and PPAR-gamma. Nephrol Dial Transplant 2005;20:2812−9. https://doi.org/10.1093/NDT/GFI172.

[131] Paueksakon P, Revelo MP, Ma LJ, Marcantoni C, Fogo AB. Microangiopathic injury and augmented PAI-1 in human diabetic nephropathy. Kidney Int 2002;61:2142−8. https://doi.org/10.1046/J.1523-1755.2002.00384.X.

[132] Lin W, Zhang Q, Liu L, Yin S, Liu Z, Cao W. Klotho restoration via acetylation of peroxisome proliferation-activated receptor γ reduces the progression of chronic kidney disease. Kidney Int 2017;92:669−79. https://doi.org/10.1016/J.KINT.2017.02.023.

[133] Maquigussa E, Paterno JC, Pokorny GH de O, Perez M da S, Varela VA, Novaes A da S, et al. Klotho and PPAR gamma activation mediate the renoprotective effect of losartan in the 5/6 nephrectomy model. Front Physiol 2018;9. https://doi.org/10.3389/FPHYS.2018.01033.

[134] Zhao M, Chen Y, Ding G, Xu Y, Bai M, Zhang Y, et al. Renal tubular epithelium-targeted peroxisome proliferator-activated receptor-γ maintains the epithelial phenotype and antagonizes renal fibrogenesis. Oncotarget 2016;7:64690−701. https://doi.org/10.18632/ONCOTARGET.11811.

[135] Costello-White R, Ryff CD, Coe CL. Aging and low-grade inflammation reduce renal function in middle-aged and older adults in Japan and the USA. Age 2015;37. https://doi.org/10.1007/S11357-015-9808-7.

[136] Castle SC, Uyemura K, Fulop T, Makinodan T. Host resistance and immune responses in advanced age. Clin Geriatr Med 2007;23:463−79. https://doi.org/10.1016/J.CGER.2007.03.005.

[137] Lang PO, Aspinall R. Immunosenescence and herd immunity: with an ever-increasing aging population do we need to rethink vaccine schedules? Expert Rev Vaccines 2012;11:167−76. https://doi.org/10.1586/ERV.11.187.

[138] Schmitt V, Rink L, Uciechowski P. The Th17/Treg balance is disturbed during aging. Exp Gerontol 2013;48:1379−86. https://doi.org/10.1016/J.EXGER.2013.09.003.

[139] Song F, Ma Y, Bai XY, Chen X. The expression changes of inflammasomes in the aging rat kidneys. J Gerontol A Biol Sci Med Sci 2016;71:747−56. https://doi.org/10.1093/GERONA/GLV078.

[140] Sato Y, Mii A, Hamazaki Y, Fujita H, Nakata H, Masuda K, et al. Heterogeneous fibroblasts underlie age-dependent tertiary lymphoid tissues in the kidney. JCI Insight 2016;1. https://doi.org/10.1172/JCI.INSIGHT.87680.

[141] Mallappallil M, Friedman EA, Delano BG, Mcfarlane SI, Salifu MO. Chronic kidney disease in the elderly: evaluation and management. Clin Pract 2014;11:525. https://doi.org/10.2217/CPR.14.46.

[142] Vlassara H, Torreggiani M, Post JB, Zheng F, Uribarri J, Striker GE. Role of oxidants/inflammation in declining renal function in chronic kidney disease and normal aging. Kidney Int Suppl 2009;76. https://doi.org/10.1038/KI.2009.401.

[143] Miyazawa M, Ishii T, Yasuda K, Noda S, Onouchi H, Hartman PS, et al. The role of mitochondrial superoxide anion (O2(-)) on physiological aging in C57BL/6J mice. J Radiat Res 2009;50:73−82. https://doi.org/10.1269/JRR.08097.

[144] Akçetin Z, Erdemli G, Brömme HJ. Experimental study showing a diminished cytosolic antioxidative capacity in kidneys of aged rats. Urol Int 2000;64:70−3. https://doi.org/10.1159/000030494.

[145] Balaban RS, Nemoto S, Finkel T. Mitochondria, oxidants, and aging. Cell 2005;120:483−95. https://doi.org/10.1016/J.CELL.2005.02.001.

[146] Kume S, Uzu T, Horiike K, Chin-Kanasaki M, Isshiki K, Araki SI, et al. Calorie restriction enhances cell adaptation to hypoxia through Sirt1-dependent mitochondrial autophagy in mouse aged kidney. J Clin Invest 2010;120:1043−55. https://doi.org/10.1172/JCI41376.

[147] He W, Wang Y, Zhang MZ, You L, Davis LS, Fan H, et al. Sirt1 activation protects the mouse renal medulla from oxidative injury. J Clin Invest 2010;120:1056−68. https://doi.org/10.1172/JCI41563.

[148] Ansari A, Rahman MS, Saha SK, Saikot FK, Deep A, Kim KH. Function of the SIRT3 mitochondrial deacetylase in cellular physiology, cancer, and neurodegenerative disease. Aging Cell 2017;16:4−16. https://doi.org/10.1111/ACEL.12538.

[149] Takazakura E, Sawabu N, Handa A, Takada A, Shinoda A, Takeuchi J. Intrarenal vascular changes with age and disease. Kidney Int 1972;2:224−30. https://doi.org/10.1038/KI.1972.98.

[150] Tanaka T, Kato H, Kojima I, Ohse T, Son D, Tawakami T, et al. Hypoxia and expression of hypoxia-inducible factor in the aging kidney. J Gerontol A Biol Sci Med Sci 2006;61:795−805. https://doi.org/10.1093/GERONA/61.8.795.

[151] Jerkić M, Vojvodić S, López-Novoa JM. The mechanism of increased renal susceptibility to toxic substances in the elderly. Part I. The role of increased vasoconstriction. Int Urol Nephrol 2001;32:539−47. https://doi.org/10.1023/A:1014484101427.

[152] Hajduczok G, Chapleau MW, Abboud FM. Increase in sympathetic activity with age. II. Role of impairment of cardiopulmonary baroreflexes. Am J Physiol 1991;260. https://doi.org/10.1152/AJPHEART.1991.260.4.H1121.

[153] Shay JW. Telomeres and aging. Curr Opin Cell Biol 2018;52:1−7. https://doi.org/10.1016/J.CEB.2017.12.001.

[154] Bernardes de Jesus B, Vera E, Schneeberger K, Tejera AM, Ayuso E, Bosch F, et al. Telomerase gene therapy in adult and old mice delays aging and increases longevity without increasing cancer. EMBO Mol Med 2012;4:691−704. https://doi.org/10.1002/EMMM.201200245.

[155] Harris RC, Cheng H. Telomerase, autophagy and acute kidney injury. Nephron 2016;134:145–8. https://doi.org/10.1159/000446665.
[156] Kovesdy CP. Epidemiology of chronic kidney disease: an update 2022. Kidney Int Suppl 2011;2022(12):7–11. https://doi.org/10.1016/J.KISU.2021.11.003.
[157] Coresh J, Selvin E, Stevens LA, Manzi J, Kusek JW, Eggers P, et al. Prevalence of chronic kidney disease in the United States. 2007.
[158] Glassock RJ, Winearls C. Ageing and the glomerular filtration rate: truths and consequences. Trans Am Clin Climatol Assoc 2009;120: 419–28.
[159] Chou YH, Chen YM. Aging and renal disease: old questions for new challenges. Aging Dis 2021;12:515–28. https://doi.org/10.14336/AD.2020.0703.
[160] Pradeep A. Chronic kidney disease (CKD) practice essentials. 2023.
[161] Kazancioğlu R. Risk factors for chronic kidney disease: an update. Kidney Int Suppl 2013;3:368–71. https://doi.org/10.1038/kisup.2013.79. Nature Publishing Group.
[162] Levey AS, Eckardt KU, Tsukamoto Y, Levin A, Coresh J, Rossert J, et al. Definition and classification of chronic kidney disease: a position statement from Kidney Disease: Improving Global Outcomes (KDIGO). Kidney Int 2005;67:2089–100.
[163] Go AS, Chertow GM, Fan D, McCulloch CE, Hsu CY. Chronic kidney disease and the risks of death, cardiovascular events, and hospitalization. N Engl J Med 2004;351(13):1296–305. https://doi.org/10.1056/nejmoa041031.
[164] Cockcroft DW, Gault MH. Prediction of creatinine clearance from serum creatinine. Nephron 1976;16:31–41. https://doi.org/10.1159/000180580.
[165] Kidney Disease: Improving Global Outcomes (KDIGO) Diabetes Work Group. KDIGO 2022 clinical practice guideline for diabetes management in chronic kidney disease. Kidney Int 2022;102:S1–127. https://doi.org/10.1016/J.KINT.2022.06.008.
[166] Levey AS, Stevens LA, Schmid CH, Zhang Y, Castro AF, Feldman HI, et al. A new equation to estimate glomerular filtration rate. Ann Intern Med 2009;150:604–12. https://doi.org/10.7326/0003-4819-150-9-200905050-00006.
[167] Inker LA, Schmid CH, Tighiouart H, Eckfeldt JH, Feldman HI, Greene T, et al. Estimating glomerular filtration rate from serum creatinine and cystatin C. N Engl J Med 2012;367:20–9. https://doi.org/10.1056/NEJMOA1114248.
[168] Sigman EM, Elwood CM, Knox F. The measurement of glomerular filtration rate in man with sodium iothalamate 131-I (Conray). J Nucl Med 1966;7:60–8.
[169] Brown SCW, O'Reilly PH. Iohexol clearance for the determination of glomerular filtration rate in clinical practice: evidence for a new gold standard. J Urol 1991;146:675–9. https://doi.org/10.1016/S0022-5347(17)37891-6.
[170] Levey AS, Becker C, Inker LA. Glomerular filtration rate and albuminuria for detection and staging of acute and chronic kidney disease in adults: a systematic review. JAMA 2015;313:837–46. https://doi.org/10.1001/JAMA.2015.0602.
[171] Sharma M, Doley P, Jyoti Das H. Renal data from Asia-Africa etiological profile of chronic kidney disease: a single-center retrospective hospital-based study. Saudi J Kidney Dis Transpl 2018;29.
[172] García-Carro C, Vergara A, Bermejo S, Azancot MA, Sellarés J, Soler MJ. A nephrologist perspective on obesity: from kidney injury to clinical management. Front Med 2021;8. https://doi.org/10.3389/FMED.2021.655871.
[173] Jankowski J, Floege J, Fliser D, Böhm M, Marx N. Cardiovascular disease in chronic kidney disease pathophysiological insights and therapeutic options. Circulation 2021;143:1157–72. https://doi.org/10.1161/CIRCULATIONAHA.120.050686.
[174] Yacoub R, Habib H, Lahdo A, Al Ali R, Varjabedian L, Atalla G, et al. Association between smoking and chronic kidney disease: a case control study. BMC Publ Health 2010;10. https://doi.org/10.1186/1471-2458-10-731.
[175] Lunyera J, Mohottige D, Von Isenburg M, Jeuland M, Patel UD, Stanifer JW. CKD of uncertain etiology: a systematic review. Clin J Am Soc Nephrol 2016;11(3):379–85. https://doi.org/10.2215/cjn.07500715.
[176] Hemmelgarn BR, Zhang J, Manns BJ, Tonelli M, Larsen E, Ghali WA, et al. Progression of kidney dysfunction in the community-dwelling elderly. Kidney Int 2006;69:2155–61. https://doi.org/10.1038/SJ.KI.5000270.
[177] Shlipak MG, Katz R, Kestenbaum B, Fried LF, Newman AB, Siscovick DS, et al. Rate of kidney function decline in older adults: a comparison using creatinine and cystatin C. Am J Nephrol 2009;30:171–8. https://doi.org/10.1159/000212381.
[178] Stevens LA, Viswanathan G, Weiner DE. CKD and ESRD in the elderly: current prevalence, future projections, and clinical significance. Adv Chron Kidney Dis 2010;17:293–301. https://doi.org/10.1053/J.ACKD.2010.03.010.
[179] Hsu CY, Ordõez JD, Chertow GM, Fan D, McCulloch CE, Go AS. The risk of acute renal failure in patients with chronic kidney disease. Kidney Int 2008;74:101–7. https://doi.org/10.1038/KI.2008.107.
[180] Schmitt R, Coca S, Kanbay M, Tinetti ME, Cantley LG, Parikh CR. Recovery of kidney function after acute kidney injury in the elderly: a systematic review and meta-analysis. Am J Kidney Dis 2008;52:262–71. https://doi.org/10.1053/J.AJKD.2008.03.005.
[181] Himmelfarb J, Vanholder R, Mehrotra R, Tonelli M. The current and future landscape of dialysis. Nat Rev Nephrol 2020;16:573–85. https://doi.org/10.1038/s41581-020-0315-4.
[182] Losappio V, Franzin R, Infante B, Godeas G, Gesualdo L, Fersini A, et al. Molecular mechanisms of premature aging in hemodialysis: the complex interplay between innate and adaptive immune dysfunction. Int J Mol Sci 2020;21. https://doi.org/10.3390/ijms21103422.
[183] Bello AK, Okpechi IG, Osman MA, Cho Y, Htay H, Jha V, et al. Epidemiology of haemodialysis outcomes. Nat Rev Nephrol 2022;18:378–95. https://doi.org/10.1038/s41581-022-00542-7.
[184] Van De Luijtgaarden MWM, Noordzij M, Stel VS, Ravani P, Jarraya F, Collart F, et al. Effects of comorbid and demographic factors on dialysis modality choice and related patient survival in Europe. Nephrol Dial Transplant 2011;26:2940–7. https://doi.org/10.1093/ndt/gfq845.
[185] Weiner DE, Tabatabai S, Tighiouart H, Elsayed E, Bansal N, Griffith J, et al. Cardiovascular outcomes and all-cause mortality: exploring the interaction between CKD and cardiovascular disease. Am J Kidney Dis 2006;48:392–401. https://doi.org/10.1053/j.ajkd.2006.05.021.
[186] Hallan S, Astor B, Romundstad S, Aasarød K, Kvenild K, Coresh J. Association of kidney function and albuminuria with cardiovascular mortality in older vs younger individuals the HUNT II study. 2007.
[187] Yuan J, Zou XR, Han SP, Cheng H, Wang L, Wang JW, et al. Prevalence and risk factors for cardiovascular disease among chronic kidney disease patients: results from the Chinese cohort study of chronic kidney disease (C-STRIDE). BMC Nephrol 2017;18. https://doi.org/10.1186/s12882-017-0441-9.
[188] Tsai YH, Lee M, Lin LC, Chang SW, Weng HH, Yang JT, et al. Association of chronic kidney disease with small vessel disease in patients with hypertensive intracerebral hemorrhage. Front Neurol 2018;9. https://doi.org/10.3389/fneur.2018.00284.

[189] Knopman D, Boland LL, Mosley T, Howard G, Liao D, Szklo M, et al. Cardiovascular risk factors and cognitive decline in middle-aged adults. Neurology 2001;56(1):42–8. https://doi.org/10.1212/WNL.56.1.42.
[190] Tamura MK, Desai M, Kapphahn KI, Thomas IC, Asch SM, Chertow GM. Dialysis versus medical management at different ages and levels of kidney function in veterans with advanced CKD. J Am Soc Nephrol 2018;29:2169–77. https://doi.org/10.1681/ASN.2017121273.
[191] Iglesias P, Heras M, Díez JJ. Diabetes mellitus y enfermedad renal en el anciano. Nefrologia 2014;34:285–92. https://doi.org/10.3265/Nefrologia.pre2014.Feb.12319.
[192] Pyram R, Kansara A, Banerji MA, Loney-Hutchinson L. Chronic kidney disease and diabetes. Maturitas 2012;71:94–103. https://doi.org/10.1016/j.maturitas.2011.11.009.
[193] Nordheim E, Jenssen TG. Chronic kidney disease in patients with diabetes mellitus. Endocr Connect 2021;10:R151–9. https://doi.org/10.1530/EC-21-0097.
[194] Moriconi D, Sacchetta L, Chiriacò M, Nesti L, Forotti G, Natali A, et al. Glomerular hyperfiltration predicts kidney function decline and mortality in type 1 and type 2 diabetes: a 21-year longitudinal study. Diabetes Care 2023. https://doi.org/10.2337/dc22-2003.
[195] Kovesdy CP, Furth SL, Zoccali C. Obesity and kidney disease: hidden consequences of the epidemic. Can J Kidney Health Dis 2017;4. https://doi.org/10.1177/2054358117698669.
[196] Stepień M, Stepień A, Wlazeł RN, Paradowski M, Banach M, Rysz M, et al. Obesity indices and adipokines in non-diabetic obese patients with early stages of chronic kidney disease. Med Sci Mon Int Med J Exp Clin Res 2013;19. https://doi.org/10.12659/MSM.889390.
[197] Briffa JF, Mcainch AJ, Poronnik P, Hryciw DH. Adipokines as a link between obesity and chronic kidney disease. Am J Physiol Ren Physiol 2013;305. https://doi.org/10.1152/ajprenal.00263.2013.
[198] Nawaz S, Chinnadurai R, Al-Chalabi S, Evans P, Kalra PA, Syed AA, et al. Obesity and chronic kidney disease: a current review. Obes Sci Pract 2022. https://doi.org/10.1002/osp4.629.
[199] Weldegiorgis M, Woodward M. The impact of hypertension on chronic kidney disease and end-stage renal disease is greater in men than women: a systematic review and meta-analysis. BMC Nephrol 2020;21. https://doi.org/10.1186/s12882-020-02151-7.
[200] Hsu RK, yuan HC. The role of acute kidney injury in chronic kidney disease. Semin Nephrol 2016;36:283–92. https://doi.org/10.1016/j.semnephrol.2016.05.005.
[201] Naughton CA. Drug-induced nephrotoxicity. Am Fam Physician 2008;78. https://doi.org/10.5455/2319-2003.ijbcp20140826.
[202] Jo W, Lee S, Joo YS, Nam KH, Yun HR, Chang TI, et al. Association of smoking with incident CKD risk in the general population: a community-based cohort study. PLoS One 2020;15. https://doi.org/10.1371/journal.pone.0238111.
[203] Choi HS, Han Do K, Oh TR, Kim CS, Bae EH, Ma SK, et al. Smoking and risk of incident end-stage kidney disease in general population: a Nationwide population-based cohort study from Korea. Sci Rep 2019;9. https://doi.org/10.1038/s41598-019-56113-7.
[204] Bleyer AJ, Westemeyer M, Xie J, Bloom MS, Brossart K, Eckel JJ, et al. Genetic etiologies for chronic kidney disease revealed through next-generation renal gene panel. Am J Nephrol 2022. https://doi.org/10.1159/000522226.
[205] Fassett RG. Current and emerging treatment options for the elderly patient with chronic kidney disease. Clin Interv Aging 2014;9:191–9. https://doi.org/10.2147/CIA.S39763.
[206] L EJ, H LG, B RP, R RD. The effect of angiotensin-converting-enzyme inhibition on diabetic nephropathy. Cardiovasc Rev Rep 1994;15:69. https://doi.org/10.1056/nejm199311113292004.
[207] Jafar TH, Stark PC, Schmid CH, Landa M, Maschio G, De Jong PE, et al. Progression of chronic kidney disease: the role of blood pressure control, proteinuria, and angiotensin-converting enzyme inhibition: a patient-level meta-analysis. Ann Intern Med 2003;139. https://doi.org/10.7326/0003-4819-139-4-200308190-00006.
[208] Ruggenenti P, Perna A, Loriga G, Ganeva M, Ene-Iordache B, Turturro M, et al. Blood-pressure control for renoprotection in patients with non-diabetic chronic renal disease (REIN-2): multicentre, randomised controlled trial. Lancet 2005;365:939–46. https://doi.org/10.1016/S0140-6736(05)71082-5.
[209] Kunz R, Friedrich C, Wolbers M, Mann JFE. Meta-analysis: effect of monotherapy and combination therapy with inhibitors of the renin angiotensin system on proteinuria in renal disease. Ann Intern Med 2008;148:30–48. https://doi.org/10.7326/0003-4819-148-1-200801010-00190.
[210] Upadhyay A, Earley A, Haynes SM, Uhlig K. Systematic review: blood pressure target in chronic kidney disease and proteinuria as an effect modifier. Ann Intern Med 2011;154:541–8. https://doi.org/10.7326/0003-4819-154-8-201104190-00335.
[211] Singh AK, Szczech L, Tang KL, Barnhart H, Sapp S, Wolfson M, et al. Correction of anemia with epoetin alfa in chronic kidney disease. N Engl J Med 2006;355:2085–98. https://doi.org/10.1056/NEJMOA065485.
[212] Pfeffer MA, Burdmann EA, Chen CY, Cooper ME, de Zeeuw D, Eckardt KU, et al. A trial of darbepoetin alfa in type 2 diabetes and chronic kidney disease. N Engl J Med 2009;361(21):2019–32. https://doi.org/10.1056/NEJMoa0907845.
[213] Goldsmith D, Covic A. Time to Reconsider Evidence for Anaemia Treatment (TREAT) = Essential Safety Arguments (ESA). Nephrol Dial Transplant 2010;25(6):1734–7. https://doi.org/10.1093/ndt/gfq099.
[214] De Brito-Ashurst I, Varagunam M, Raftery MJ, Yaqoob MM. Bicarbonate supplementation slows progression of CKD and improves nutritional status. J Am Soc Nephrol 2009;20:2075–84. https://doi.org/10.1681/ASN.2008111205.
[215] Mahajan A, Simoni J, Sheather SJ, Broglio KR, Rajab MH, Wesson DE. Daily oral sodium bicarbonate preserves glomerular filtration rate by slowing its decline in early hypertensive nephropathy. Kidney Int 2010;78:303–9. https://doi.org/10.1038/KI.2010.129.
[216] Brugts JJ, Knetsch AM, Mattace-Raso FUS, Hofman A, Witteman JCM. Renal function and risk of myocardial infarction in an elderly population: the Rotterdam study. Arch Intern Med 2005;165:2659–65. https://doi.org/10.1001/ARCHINTE.165.22.2659.
[217] Mann JF, Schmieder RE, McQueen M, Dyal L, Schumacher H, Pogue J, et al. Renal outcomes with telmisartan, ramipril, or both, in people at high vascular risk (the ONTARGET study): a multicentre, randomised, double-blind, controlled trial. Lancet 2008;372:547–53. https://doi.org/10.1016/S0140-6736(08)61236-2.
[218] Turner R. Intensive blood-glucose control with sulphonylureas or insulin compared with conventional treatment and risk of complications in patients with type 2 diabetes (UKPDS 33). Lancet 1998;352:837–53. https://doi.org/10.1016/S0140-6736(98)07019-6.
[219] Duckworth W, Abraira C, Moritz T, Reda D, Emanuele N, Reaven PD, et al. Glucose control and vascular complications in veterans with type 2 diabetes. N Engl J Med 2009;360:129–39. https://doi.org/10.1056/NEJMOA0808431.

[220] Ismail-Beigi F, Craven T, Banerji MA, Basile J, Calles J, Cohen RM, et al. Effect of intensive treatment of hyperglycaemia on microvascular outcomes in type 2 diabetes: an analysis of the ACCORD randomised trial. Lancet 2010;376:419−30. https://doi.org/10.1016/S0140-6736(10)60576-4.

[221] Chou YH, Huang TM, Pan SY, Chang CH, Lai CF, Wu VC, et al. Renin-angiotensin system inhibitor is associated with lower risk of ensuing chronic kidney disease after functional recovery from acute kidney injury. Sci Rep 2017;7. https://doi.org/10.1038/SREP46518.

[222] Ishani A, Xue JL, Himmelfarb J, Eggers PW, Kimmel PL, Molitoris BA, et al. Acute kidney injury increases risk of ESRD among elderly. J Am Soc Nephrol 2009;20:223. https://doi.org/10.1681/ASN.2007080837.

[223] Anderson S, Eldadah B, Halter JB, Hazzard WR, Himmelfarb J, Horne FM, et al. Acute kidney injury in older adults. J Am Soc Nephrol 2011; 22:28−38. https://doi.org/10.1681/ASN.2010090934.

[224] Coca SG. Acute kidney injury in elderly persons. Am J Kidney Dis 2010;56:122−31. https://doi.org/10.1053/J.AJKD.2009.12.034.

[225] Cushman WC, Evans GW, Byington RP, Goff DC, Grimm RH, Cutler JA, et al. Effects of intensive blood-pressure control in type 2 diabetes mellitus. N Engl J Med 2010;362:1575−85. https://doi.org/10.1056/NEJMOA1001286.

[226] Wright JT, Williamson JD, Whelton PK, Snyder JK, Sink KM, Rocco MV, et al. A randomized trial of intensive versus standard blood-pressure control. N Engl J Med 2015;373:2103−16. https://doi.org/10.1056/NEJMOA1511939.

[227] Turgut F, Balogun RA, Abdel-Rahman EM. Renin-angiotensin-aldosterone system blockade effects on the kidney in the elderly: benefits and limitations. Clin J Am Soc Nephrol 2010;5:1330−9. https://doi.org/10.2215/CJN.08611209.

[228] Donate-Correa J, Tagua VG, Ferri C, Martín-Núñez E, Hernández-Carballo C, Ureña-Torres P, et al. Pentoxifylline for renal protection in diabetic kidney disease. A model of old drugs for new Horizons. J Clin Med 2019;8. https://doi.org/10.3390/JCM8030287.

[229] Chen YM, Chiang WC, Lin SL, Tsai TJ. Therapeutic efficacy of pentoxifylline on proteinuria and renal progression: an update. J Biomed Sci 2017;24. https://doi.org/10.1186/S12929-017-0390-4.

[230] Barkhordari K, Karimi A, Shafiee A, Soltaninia H, Khatami MR, Abbasi K, et al. Effect of pentoxifylline on preventing acute kidney injury after cardiac surgery by measuring urinary neutrophil gelatinase − associated lipocalin. J Cardiothorac Surg 2011;6. https://doi.org/10.1186/1749-8090-6-8.

[231] Hong W, Hu S, Zou J, Xiao J, Zhang X, Fu C, et al. Peroxisome proliferator-activated receptor γ prevents the production of NOD-like receptor family, pyrin domain containing 3 inflammasome and interleukin 1β in HK-2 renal tubular epithelial cells stimulated by monosodium urate crystals. Mol Med Rep 2015;12:6221−6. https://doi.org/10.3892/MMR.2015.4145.

[232] McMurray JJV, Solomon SD, Inzucchi SE, Køber L, Kosiborod MN, Martinez FA, et al. Dapagliflozin in patients with heart failure and reduced ejection fraction. N Engl J Med 2019;381:1995−2008. https://doi.org/10.1056/NEJMOA1911303.

[233] DeFronzo RA, Norton L, Abdul-Ghani M. Renal, metabolic and cardiovascular considerations of SGLT2 inhibition. Nat Rev Nephrol 2017; 13:11−26. https://doi.org/10.1038/NRNEPH.2016.170.

[234] Heerspink HJL, Karasik A, Thuresson M, Melzer-Cohen C, Chodick G, Khunti K, et al. Kidney outcomes associated with use of SGLT2 inhibitors in real-world clinical practice (CVD-REAL 3): a multinational observational cohort study. Lancet Diabetes Endocrinol 2020;8:27−35. https://doi.org/10.1016/S2213-8587(19)30384-5.

[235] Wanner C, Brenner S. Credence and delight deliver on renal benefits. Nat Rev Nephrol 2019;15:459−60. https://doi.org/10.1038/S41581-019-0171-2.

[236] Hu MC, Shi M, Gillings N, Flores B, Takahashi M, Kuro-o M, et al. Recombinant α-Klotho may be prophylactic and therapeutic for acute to chronic kidney disease progression and uremic cardiomyopathy. Kidney Int 2017;91:1104−14. https://doi.org/10.1016/J.KINT.2016.10.034.

Further reading

[1] Wei SY, Pan SY, Li B, Chen YM, Lin SL. Rejuvenation: turning back the clock of aging kidney. J Formos Med Assoc 2020;119:898−906. https://doi.org/10.1016/J.JFMA.2019.05.020.

CHAPTER 7

Age-related disease: Immune system

Karin de Punder and Alexander Karabatsiakis
Department of Psychology, Clinical Psychology-II, University of Innsbruck, Innsbruck, Austria

1. Introduction

Since the rise of the first civilizations, modern mankind has been highly interested in the continuous process of better characterizing and understanding the aging process. However, if we look back in history, many of the concepts, expectations, and motivations to unmask the process of aging were already discussed in ancient societies, for example by the Greek physician commonly called *Aelios Galenos*, who lived almost 2000 years ago. In contrast to many others of his time and beyond, he thought about aging as a naturally occurring process rather than a disease that can also be influenced by internal and external factors, for example, caloric restriction. Of course, the methodology and technology of today's research were missing in those days, but it is fascinating to see that these initial, many centuries-old concepts still show relevance for our modern understanding as well as our direction of future-orientated thinking. Through continuously broadening access to a progressively developing field of life science-centered methodology and technology, our knowledge has gained a catalytic benefit for unraveling the underlying mechanisms of aging with utmost importance in terms of life quality, health, and disease. Here, the immune system, which is comprised of a variety of highly specialized, organized, and orchestrated cells, shows special importance for maintaining the health span as long as possible across our lifetime.

It is well established that many aspects of the immune response show a progressive functional decline with the calendar aging of an individual, while others show the opposite effect, manifesting in dysregulated immunity [1]. This age-related immune dysfunction includes the loss of diversity of adaptive immunity, an increase in the number of senescent T cells, and the manifestation of a lingering level of low-grade inflammation, a physiological process referred to as "inflamm-aging" in the literature [2]. Combined, these profound changes exhibited by the aging immune system are termed "immunosenescence", a concept that was first coined by Roy Walford [3]. In his landmark book "The Immunological Theory of Aging" he postulated that the normal aging process in humans and animals is pathogenetically related to faulty immune processes [3,4]. Conceptualized as a progress of immunological imbalance, immunosenescence affects the overall response to immunological challenges, resulting in increased disease susceptibility. Epidemiological data demonstrate a higher incidence of infection, neoplastic, and autoimmune diseases in older individuals [5–7]. More recently, age-driven immunosenescence has been linked to chronic inflammatory systemic diseases such as diabetes mellitus, cardiovascular and cerebrovascular disease, stroke, affective disorders, and neurodegenerative diseases [8], which are anticipated to increase dramatically in the coming decades as the population ages on a global scale. Therefore, a major aim of modern aging research is not only to understand how the composition and functioning of the immune system change over the lifespan, but also to identify targets for future interventions to prevent, slow down, or even reverse its age-associated functional impairments, which often lead elderly people to disease onset with a negative impact on life quality (years lost due to disability), comorbidity risk, and mortality.

2. The aging of the immune system

The hematopoietic stem cell (HSC) system produces all the necessary cells of the immune system. As the HSCs and precursor cells of the immune system share a common cellular nursery in the bone marrow, they are introduced

120   7. Age-related disease: Immune system

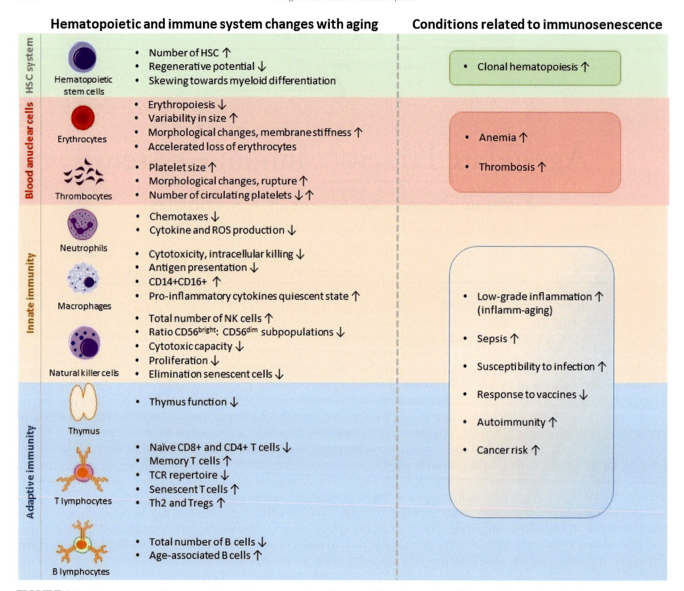

FIGURE 7.1 An overview of hematopoietic and immune system changes with aging and resulting conditions related to immunosenescence. *This figure was created using motifolio image assets.*

and discussed here together. An overview of hematopoietic and immune system changes with aging, including conditions related to immunosenescence, is provided in Fig. 7.1.

2.1 The hematopoietic stem cell system

The HSC system, located in the bone marrow, is a highly active and specialized repertoire of cells responsible for the regulated high-throughput production of mature blood cells into the peripheral blood circulation. These mature cells, although not all can be considered true cells in that they lack a nucleus (e.g., mature erythrocytes, also called red blood cells [RBC], and platelets), have a limited life span, and need to be continuously replaced to ensure optimal oxygen transport (RBCs) and processes of the immune and clotting response (i.e., platelets). The HSCs reside at the top of the hematopoietic hierarchy and possess the potential for both multi-potency and self-renewal, differentiating into all functional blood cells and maintaining the HSC pool of stem cells throughout the life of the individual [9,10], including both the myeloid and lymphoid lineages. Myeloid progenitor cells give rise to RBCs, granulocytes, monocytes, and platelets, whereas lymphoid cells give rise to lymphocytes and natural killer (NK) cells [11].

Rodent studies have shown that during fetal development the liver is the predominant hematopoietic organ for HSC expansion and maturation. Hematopoiesis in the fetal liver depends on the migration of hematopoietic progenitor cells, which may originate from the placenta [12] or the aorta-gonad-mesonephros, a region of the embryonic mesoderm [13]. Following significant expansion in the fetal liver, HSCs migrate to the spleen and later to the bone marrow. After birth, the number of HSCs in the liver sharply declines, and in adults, the major site of hematopoiesis is located in the bone marrow [12].

As we age, the number of most somatic stem cells declines, while HSCs increase by two- to tenfold. However, this increase is not capable of compensating for the loss of function which is associated with diminished regenerative potential and repair [14,15]. Another hallmark of aged HSCs is their skewing toward myeloid cell differentiation, indicated by the general downregulation of lymphoid genes and a general upregulation of myeloid genes with aging [14]. Besides cell-intrinsic factors, extrinsic influences such as the stem cell niche microenvironment and systemic factors also seem to contribute to HSC aging [14].

2.2 Erythrocytes and platelets

Human erythrocytes are anuclear cells containing hemoglobin, the primary carrier protein of molecular oxygen. The iron atom in each heme moiety of the hemoglobin protein can reversibly bind oxygen, allowing its transport from the lungs to all oxygen-dependent cells [16]. Besides their role in oxygen transport, erythrocytes are also involved in maintaining systemic acid/base equilibria in blood and they are well equipped with antioxidant systems, which essentially contribute to their function and integrity [17]. RBCs have a mean full lifespan of around 120 days, after which they are cleared by phagocytosis by reticuloendothelial macrophages [18]. To replace those lost by damage or senescence, about 200 billion erythrocytes are produced daily in the body. Mature circulating erythrocytes are produced from erythroid precursors that differentiate from HSCs residing in the adult bone marrow, a process that is regulated by an oxygen-sensing mechanism to maintain RBC numbers within a physiological range. A key regulator protein in the process of erythropoiesis is erythropoietin (EPO), a cytokine produced in the kidney by promoting the maturation and survival of erythroid progenitors [19]. Physical activity as well as environmental factors like oxygen partial pressure in air also affect the regulation of EPO, therefore contributing to the maintenance of immunological health across the lifespan. The availability of iron and other important nutrients, such as folate and vitamin B12, is essential for the speed and degree of the physiological control of RBC production [17].

Aging and aging-related conditions, including anemia, have been associated with impaired erythropoiesis [20,21], reduced hemoglobin supply, greater variability in RBC size [22], and the accelerated loss of circulating RBCs, mainly due to a form of suicidal cell death called eryptosis [23]. In addition, changes in RBC morphology and membrane stiffness [24] highlight the importance of the structural and functional integrity of the cell membranes that are affected by pericellular stress in terms of loss of unsaturated fatty acids (UFAs) that have to be compensated by nutritional supply. As the outer cell border, the membrane is essential for processes including intercellular attachment, enzymatic activity, transport, cell—cell identification, and signal transduction. Changes in membrane fluidity due to its compositional shifts in fatty acids that also occur as a consequence of biological aging can therefore affect the biological function of a cell and need to be counterbalanced constantly across the lifespan. Whether these changes of membrane stiffness in circulating immune cells can be attributed and traced back to HSCs has not been investigated, but it can be expected to represent a systemic rather than a cell-type specific effect.

The next cell type of the myeloid lineage is the blood platelet, also called thrombocyte. Platelets are multifunctional anuclear cells produced from megakaryocyte precursors that are primarily involved in maintaining hemostasis, thrombosis formation, and wound healing and have an additional role in immunity and inflammation. The lifespan of platelets is limited to between 5 and 7 days following formation and separation from the megakaryocyte [25].

Several effects of aging on platelet morphology and function have been observed. In the course of life, platelet size increases, directly affecting platelet content, including granules and pro-coagulation factors [26]. Also, morphological changes occur, such as a more irregular, less smooth plasma membrane with more frequent rupture. The number of circulating platelets is thought to decrease with advanced age, although inconsistent findings have been reported [27]. Age-driven alterations in platelet (hyper)reactivity can be attributed to a higher incidence of cardiovascular disease and thrombosis, cancer, inflammatory bowel disease, sepsis, rheumatoid arthritis, myeloproliferative disease, Alzheimer's disease, and diabetes [28—32], clearly highlighting platelets as an interesting and important target for a predictive, preventive, and tailored medical intervention approach in aging research.

2.3 Setting the focus to the innate and adaptive immune system

The immune system comprises a complex and finely tuned network of white blood cells, tissues, organs, and the substances they synthesize that help the body fight infections and other diseases. Immunity consists of primarily two interacting components: the innate and the adaptive host defense. The innate immune response serves as the first, nonspecific, and rapid line of defense against environmental pathogens and is mediated mainly by monocytes, NK cells, dendritic cells, and the complement system. As opposed to the innate response, adaptive immunity has a higher latency in reactivity, takes longer to develop, is highly specific, and has the capacity for long-term memory formation. Here, specialized lymphocytes of the T (thymus-derived) and B (maturing in the bone marrow, initially named after the Bursa of fabricius) cell types are essential in the development of an adaptive immune response and mediate the cellular and humoral immune response [33].

The cellular response is orchestrated by T cells that only recognize peptides displayed on antigen-presenting cells via unique and highly specific T cell receptors (TCR). Co-stimulatory signals, such as those delivered by the surface molecule CD28, promote full T cell activation [34]. T cells can further differentiate into various subtypes. CD8+ T cells are mainly cytotoxic cells, which are important for the clearance of intracellular pathogens, whereas CD4+ T-helper cells primarily regulate the cellular and humoral immune responses. Also, T cells can be naive or antigen-experienced and exist in a resting or activated state. Antigens recognized by the B cell receptor activate the humoral response by inducing B cells to proliferate and differentiate into a plasma cell secreting antigen-specific antibodies [33].

The production of antigen-specific receptors in both cell types is the result of random rearrangement and splicing together of multiple DNA segments that encode for the antigen-binding areas of the receptors (complementarity-determining regions). Gene rearrangement occurs early in the development of the cells, before exposure to antigen, which leads to the production of a repertoire of over 10^8 T-cell receptors and 10^{10} antibody specificities, sufficient to cover a large range of pathogens likely to be encountered throughout life [33]. At least as important as antigen receptor generation is clonal selection. In the thymus, the delicate process of positive selection aims to prevent the appearance of self-reactive T cells, which could lead to autoimmunization [33], a health-relevant process, especially in later stages of life.

2.4 Age-related changes in the innate immune system

Aging has an impact on many aspects of the innate immune system, including changes in the architecture and functioning of specific immune cells [1], such as neutrophils, which are the most abundant type of white blood cells in the peripheral circulation. These cells scan the organism for pathogens, and when detected, quickly respond to eliminate the invaders [35]. Neutrophil adhesive capacity and phagocytic activity are similar in older and younger people. However, chemotaxis and the production of free radicals and cytokines have been shown to decline with age [1].

As regulators and effectors of the immune response, monocytes and macrophages also change with age. Although data from humans are scarce, many of the effector functions including cytotoxicity, intracellular killing, and antigen presentation decline during the aging process. The inflammatory monocyte subpopulation CD14+CD16+ becomes more abundant with increasing age and produces more pro-inflammatory cytokines during the quiescent state [36].

Natural killer (NK) cells are considered the primary defense lymphocyte against virally infected and virally transformed cells, but are also involved in the antimicrobial defense, elimination of senescence cells, and resolution of inflammation [37]. They are generally considered to be a component of the innate immune system because of lacking antigen-specific cell surface receptors. However, recent data showed that NK cells can also mount a form of antigen-specific immunologic memory implying its involvement in both innate and adaptive immunity [38]. Age-related decreases are seen in the ratio of $CD56^{bright}:CD56^{dim}$ subpopulations, involved in cytokine production and cytotoxicity, respectively [37]. Additionally, with age, the numbers of NK cells increase though they have reduced cytotoxic activity and proliferation ability. NK cell-mediated elimination of senescent cells declines with aging and results in the accumulation of aged cells in tissue or organs, which impairs tissue homeostasis and their function [37]. This accumulation has attracted very much attention in the aging research field as senolytic approaches try to directly address these senescent cells (see also Section 7.5).

Collectively, when it comes to the innate immune system, aging is characterized by a heightened pro-inflammatory status at the basal level and impaired performance upon immune stimulation when specific functions, like free radical production, are needed. The exact underlying mechanisms of how aging affects innate immunity remain poorly understood, making the identification of effective intervention strategies—based on rather simplified models—still hard to realize.

Age-related changes in the adaptive immune system

The most marked change in the immune system occurring with age is the decline in naïve T cells in the peripheral blood, especially in the CD8+ T cell compartment [1]. This decline might be explained by a reduced thymic output, partially related to thymus involution. While earlier studies reported thymus involution to start at about the third decade of life [39], recent findings state this process starts much earlier [40]. Also, the number of memory T cells increases when antigen-stimulated naïve cells differentiate into effector and memory cells, another hallmark of adaptive immune aging that has been consistently reported in age-progressed individuals [5]. With immunological aging there is a loss of the T cell receptor (TCR) repertoire, resulting in increased susceptibility to, e.g., new infections, decreased vaccine responses, and decreased memory function [5–7]. Additionally, an expansion of senescent CD28− T cells is seen in normal aging, again mainly in the CD8+ compartment [34]. This is likely the result of repeated antigenic stimulation, inducing a progressive reduction in surface CD28 expression [41]. The decline in CD28+ T cells is associated with decreased responses to vaccines in the elderly, reduction in the overall T cell receptor repertoire, and diminished control over infections. Stimulation of CD28 increases the reactivity of cells after antigen exposure, while a lack of CD28 signaling reduces the lifespan of activated T cells [34]. Moreover, with age CD4+ T-cell subpopulations shift with an increase in the number of T-helper type 2 cells (Th2) and regulatory T cells (Tregs) resulting in an inadequate adaptive immune response toward new antigens and altered memory responses [36]. Other functional changes leading to altered activity are related to T-cell membrane composition. As membrane cholesterol content increases with age, intracellular pathway signaling abilities may be altered due to increased membrane viscosity [42].

A recent study by Mittelbrunn and colleagues showed that dysfunctional (TFAM-deficient) T cells act as accelerators of senescence throughout multiple organ systems, suggesting that T cells can regulate organismal fitness and life span [43]. Premature aging in mice with TFAM-deficient T cells was partially rescued by blocking proinflammatory cytokine (tumor necrosis factor [TNF]-α) signaling or by preventing senescence with nicotinamide adenine dinucleotide (NAD) precursors [43,44], which are further discussed in Section 7.4. Mittelbrunn and Kroemer also proposed 10 molecular hallmarks to represent common denominators of T cell aging. These hallmarks are grouped into four primary hallmarks (thymic involution, mitochondrial dysfunction, genetic and epigenetic alterations, and loss of proteostasis) and four secondary hallmarks (reduction of the TCR repertoire, naive–memory imbalance, T cell senescence, and lack of effector plasticity) [44].

Aging has also been associated with an overall decline in B cell production in the bone marrow, which is paralleled with an increase in a subset of B cells, named age-associated B cells. This subset of B cells is likely to be involved in impaired immunity associated with aging and may contribute to the onset and development of autoimmune diseases by the production of autoantibodies [45].

To summarize, a decrease in adaptive immunity is observed during aging, with consequences for the overall fitness and health span of the organism. Over the last few decades, many observations and findings about the aging immune system have revealed more and more puzzle parts necessary for a holistic perspective on the concepts of immunosenescence in healthy and pathological aging. We provide a momentary snapshot that offers glimpses into an ongoing discipline, and as we continue to unravel the changes that occur in the immune system with age, we can expect new insights.

2.6 Inflamm-aging

Changes in the regulation of immune responses with aging are paralleled with a progressive increase in circulating levels of inflammatory cytokines and proteins, a phenomenon also known as "inflamm-aging", a term introduced in 2000 by Franceschi and colleagues [2]. Due to its medical importance in the transition from healthy to pathological aging, the concept of inflamm-aging, characterized by a chronic, sterile low-grade proinflammatory profile, has received significant scientific attention while constituting one of the theories of aging [6]. Inflamm-aging is the long-term result of the chronic stimulation of the innate immune system by antigens originating from pathogens or from transformed tissue and molecular debris, such as advanced glycated end products, that arise as a result of processes driven by inflammation and oxidative stress [1]. Both increases in basal levels as well as ex vivo-stimulated levels of inflammatory cytokine and protein production have been associated with aging, although this has not been shown for all investigated inflammatory measures. It should also be noted that some of the reported associations were strongly influenced by the health status of the individual [8], and therefore, increased inflammation could not always be attributed solely to the consequences of the aging process. Nevertheless, inflammaging as a concept is scientifically followed by research disciplines beyond classic immunology research, underlining the

importance of interdisciplinary approaches to understand the inflammatory
alterations with age on a systemic level.

2.7 Immune risk phenotype

The simple knowledge of basic mechanisms and their functional changes with age is not sa...
challenge, for example during the COVID-19 pandemic. As a result, timely translational research ...
tify efficient ways to implement these findings into clinical practice. Empirical evidence from explo... ...ies
using large databases identified high-risk profiles of inflammation and immunosenescence and showed ... value
in the context of predicting healthy and pathological aging and the probability of higher morbidity risk [46—49].
Going even further, a cluster of immune parameters that predict future mortality has emerged from longitudinal
studies of elderly cohorts and is referred to as the *immune risk phenotype* (IRP). The IRP is characterized by high
CD8+ and low CD4+ numbers, resulting in an inverted CD4+/CD8+ ratio, attenuated mitogen-stimulated lymphoproliferative responses and is associated with seropositivity to cytomegalovirus (CMV) as well as an increased
number of lately differentiated CD8+CD28− effector cells [47,48]. A more recent study longitudinally tracked multiple immune features, including cell-subset phenotyping, cytokine response assays, gene expression profiling, and
CMV seropositivity in a cohort of healthy adult individuals and constructed a high-dimensional trajectory of immune aging (*IMM-AGE*) that described an individual's immune functioning significantly better (and served as a better risk predictor) than did chronological age. The *IMM-AGE* score predicted all-cause mortality, establishing its
potential future use for identifying at-risk patients in clinical settings [49]. Another profile characterized by high
inflammation and low cortisol (stress hormone) levels predicted an increased probability of mortality within the
next 10 consequent years after assessment [46]. The multidisciplinarity of recently available methods and technology
(e.g., multivariate statistics, machine learning, and artificial intelligence combined with new biolaboratory and molecular capabilities) applied to (psychoneuro)immunology and aging research will enable us to gain a better understanding of the physiological and psychological factors contributing to the process of immunological aging and may
provide new insights for prevention, prediction, and innovative treatment interventions.

3. Examples of conditions related to immune system aging

The medical relevance of the immune system in the context of aging will be presented here by selected examples.
A major concern, particularly in intensive care medicine, is sepsis. The World Health Organization (WHO) repeatedly reports sepsis, a pathophysiological state including severe signs of immune reactivity and inflammation, to
cause more than 11 million deaths per year [50]. Sepsis can affect individuals of all ages, but epidemiology also states
that the probability of sepsis significantly increases with compromised immune function, a history of previous inflammatory diseases, and, of course, chronologic age [51]. Besides its direct clinical effects in aging societies, sepsis
further contributes to the significant consumption of healthcare resources [51,52].

3.1 Infectious diseases

It is not very surprising to find evidence in the literature of an increased prevalence and severity of infectious diseases with progressing age [6,52]. However, this process does not follow linear rules but rather is defined by different
developmental stages of an individual associated with differential vulnerabilities and probabilities for disease. According to the relatively health-critical period ranging from infancy to adolescence, epidemiology shows that after
reaching adulthood, a more linear character can be observed for the probability of an individual dying of infections
and other pathogen-driven diseases [53]. This probability doubles around every 8 years of living according to the
Gompertz-Makeham law of human mortality first defined by Benjamin Gompertz [54]. Some of the most common
infectious diseases in the elderly comprise urinary tract infections, influenza, bacterial pneumonia (also due to
methicillin-resistant *Staphylococcus aureus*, MRSA), gastrointestinal infections, chronic viral infection reactivation,
such as CMV and herpes viruses, as well as fungal (e.g., candidiasis, etc.), or parasitic infections [55]. As a consequence of the age-related functional decline of the immune system, aggravated by a generally weaker immune
response to most vaccines, mortality, and morbidity during annual influenza epidemics and the current pandemics
of COVID-19, are increased in elderly individuals [7].

During aging, the increased output of self-reactive T cells by the atrophied thymus and the accumulation of age-associated B cells results in a diminished ability of the immune system to differentiate between pathogens and "self" body tissues [45,56], although the peak age of onset varies among the different autoimmune diseases and can be sex-dependent. For example, giant cell arteritis more often affects elderly individuals and has a peak incidence at 70–80 years of age, while inflammatory bowel disease has two peaks of onset, the first in younger subjects and the second after 60 years of age [57]. Rheumatoid arthritis (RA) is more prevalent among women than men and often coincides with the onset of menopause [58]. Another example of the impact of age on the onset of autoimmune disorders covers the cluster of autoimmune thyroid disease (AITD), including Hashimoto thyroiditis (HT). Besides a family history of AITD, also chronologic age, sex of the individual, and smoking can all influence the onset risk for this autoimmune disease [59]. The chronicity of (auto)immune-related pathophysiology further lowers the probability of clinical remission, increases the risk of clinical relapse, and also affects the risk for secondary autoimmune conditions, making age-driven changes in immune regulation a promising therapeutic target.

3.3 Cancer and other neoplasms

Aging increases the risk for the development of various types of cancer, with special relevance for individuals passing the age of 65 years, showing a peak incidence in the seventh and eighth decades of life [60]. This time window parallels the age-associated decline in immune function, including prolonged low-grade inflammation and a reduced capacity for immunosurveillance, the process by which the immune system detects and eliminates cancerous cells. A characteristic feature in about 90% of all malignancies is the constitutive expression of high levels of the telomere-elongating enzyme telomerase. Targeting telomerase activity pharmacologically has been studied as a promising approach, however, it is also necessary to again remind the reader that the biology of cancer is—in analogy to aging—not sufficiently understood, dynamic, and adaptive and therefore shows many facets that limit the applicability of relatively simplified etiological and treatment approaches [61]. A more detailed introduction and overview of the biomolecular mechanisms involved in age-related cancer development will be provided in Chapter 12.

Other cancer types do not result from somatic cell malfunctioning but are directly linked to failures of the bone marrow. For example, myelodysplastic syndromes (MDS) are a type of rare blood diseases where immature blood cells in the bone marrow do not mature or become normal functional blood cells. Clinically, these syndromes are characterized by dysregulated hematopoiesis, ineffective erythropoiesis, peripheral cytopenia (reduced blood cells), and an increased risk of transformation to acute myeloid leukemia [62]. Clonal hematopoiesis is a consequence of genetic abnormalities in HSCs and results if a mutated clone disproportionately contributes to mature blood cell production over time. It is exacerbated by an aged inflammatory milieu and, in the majority of cases, MDS is due to mutations in genes involved in epigenetic regulation (DNMT3A, TET2, ASXL1). These mutations are rare in the younger but highly prevalent in the elderly population, with about 10%–20% of individuals over 70 years of age harboring a clone of appreciable size [15,63].

For certain cancers, innovative research co-focusing on the immune system has taken out some of the terror. For example, the use of immunotherapy has revolutionized the therapy of certain cancers over the last decade. Immune checkpoint inhibitors have promoted durable remissions across a variety of tumor types by modulation of the patient's existing immune system and by the use of autologous tumor-specific T cells or chimeric antigen-receptor T cells (CAR T cell therapy) has proven effective in treating leukemia and multiple myeloma [64]. Progress in science and technology gives rise to the confidence that a significant improvement also in the treatment of different types of cancers (somatic vs. stem-cell derived) can be achieved using i.a. immunotherapy within the next few decades, although solid tumors are currently somewhat more difficult targets.

3.4 The role of immunosenescence in health and disease: An introduction to telomere biology

Immunosenescence is much more than just the aging of the immune system. Directly or indirectly, it is viewed as playing a role in a multitude of common, age-associated diseases, including cardiovascular disease [65–67], type 2 diabetes mellitus [68], neuroendocrine and neurodegenerative disorders, such as depression [69–71] and Alzheimer's disease [72,73]. Interestingly enough, all of these conditions have been associated with shorter leukocyte telomere length (LTL), a prominently reported, but debated biomarker for diseases of aging and even overall aging processes [74]. The shortening of leukocyte telomeres, the repetitive hexamer repeats with the sequence (TTAGGG)n at the end of linear chromosomes, is thought to reflect systemic influences and the senescent status of immune cells

circulating in blood [75]. As immune cell senescence induces proinfla...
shortening may be a direct contributor to the etiology of these conditions...
implicated, peripheral LTL may also be an indicator of aging-related diseas...
such as those related to lifestyle and pathogen load, impact both telomere biology a...
with aging independently.

The use of peripheral LTL measures raises several issues related to analytic methods. ...
relative LTL measured by quantitative polymerase chain reaction (qPCR), although informatic...
mere length (or lowest decile) might be more relevant to cell function [74,76], but not detectable ...
technology. Alternatively, Southern blot or quantitative fluorescence in situ hybridization (qFISH) pro...
sights into the character of telomeric integrity, but are much more time-consuming and therefore not e... ...pli-
cable for high-throughput analyses. For investigations in larger cohorts, qPCR remains the method of ...noice, although it cannot adequately address the question of whether telomeres are truly shortened or if the shortening of mean telomere length is rather attributable to oxidative stress-induced DNA damage, violating the integrity of telomeres. And, although telomere length is known to vary between different immune cell subtypes [77], studies often measure telomeres in leukocytes or in whole blood, while telomere length data in specific immune cell subtypes could provide a better picture of which immune cell subtypes show the shortest telomere length and therefore might have the strongest contribution on the process of immunosenescence. To further explore the mechanisms underlying the aging of the immune system, in the next sections the role of telomere biology and interrelated biomolecular mechanisms, together with upstream risk factors amplifying the process of immunosenescence, will be highlighted.

4. Biomolecular mechanisms of immunosenescence

4.1 Role of the telomere biology system

To what extent cellular senescence contributes to the functional decline of the immune system remains under debate, but increasing evidence indicates that cellular senescence is, at least in part, a cause of immunosenescence. According to Hayflick's observation of limited cellular replication capacity [78], telomere biology will be presented next as one central mechanism of cellular senescence. The shortening of telomeres is a well-known hallmark of both cellular senescence and organismal aging and is observed in most immune system cells including HSCs. To compensate for cellular turnover and loss of telomeres due to replication, the level of telomerase activity, a reverse transcriptase enzyme that consists of an RNA component (telomerase RNA component, TERC) and a catalytic protein domain (human telomerase reverse transcriptase, TERT), can be upregulated in cells that undergo rapid expansion, such as HSCs and committed hematopoietic progenitor cells (HPCs) [79]. However, although these cells can induce telomerase, their telomeres do shorten with age (if perhaps less rapidly than do cells that lack this ability), as indicated by animal and human studies showing telomere shortening in parallel with the accumulation of markers of cell senescence such as p16INK4a [80–82].

There appears to be a link between the replicative history of HSCs and LTL, but, unlike peripheral leukocytes, which are exposed to factors present in the circulation, HSCs are not required to exhibit telomere shortening at a similar rate [74]. Because, especially in humans, the in vivo behavior of HSC within the bone marrow is difficult to study [83], it remains challenging to determine the exact kinetics of the HSC system and whether interindividual variation in the length of peripheral leukocyte telomeres reflects that of HSC. Evidence indicates that during early life there is a faster attrition rate of leukocyte telomeres as compared to attrition rates seen in adults, which likely corresponds to the fast expansion of the HSC and HPC pools at the beginning of life [84,85]. A model characterizing HSC kinetics from birth to the age of 20 years estimated that HSCs replicate about 17 times by the first year of life and replicate 2.5 times/year between the ages of 3 and 13 years, while in adults replication slows considerably to a rate of 0.6 times/year [85]. Building upon this model of HSC kinetics, a new model of telomere length-dependent T cell clonal expansion capacity with age was developed and its relationship to COVID-19 mortality was examined. The model showed that an individual with an average LTL at the age of 20 years maintains maximal T cell clonal capacity up to the age of 50 years. Thereafter, clonal expansion capacity declines exponentially and coincides with the steep increase in COVID-19 mortality with age [86]. Further evidence supporting this model was provided by a recent study in patients with COVID-19. Here, it was shown that the shortest LTL was associated with low lymphocyte counts, possibly reflecting a failure of T cell proliferation to keep up with the T cell loss from the circulation during the course of the illness [87].

2.5 Age-related changes in the adaptive immune system

The most marked change in the immune system occurring with age is the decline in naïve T cells in the peripheral blood, especially in the CD8+ T cell compartment [1]. This decline might be explained by a reduced thymic output, partially related to thymus involution. While earlier studies reported thymus involution to start at about the third decade of life [39], recent findings state this process starts much earlier [40]. Also, the number of memory T cells increases when antigen-stimulated naïve cells differentiate into effector and memory cells, another hallmark of adaptive immune aging that has been consistently reported in age-progressed individuals [5]. With immunological aging there is a loss of the T cell receptor (TCR) repertoire, resulting in increased susceptibility to, e.g., new infections, decreased vaccine responses, and decreased memory function [5–7]. Additionally, an expansion of senescent CD28− T cells is seen in normal aging, again mainly in the CD8+ compartment [34]. This is likely the result of repeated antigenic stimulation, inducing a progressive reduction in surface CD28 expression [41]. The decline in CD28+ T cells is associated with decreased responses to vaccines in the elderly, reduction in the overall T cell receptor repertoire, and diminished control over infections. Stimulation of CD28 increases the reactivity of cells after antigen exposure, while a lack of CD28 signaling reduces the lifespan of activated T cells [34]. Moreover, with age CD4+ T-cell subpopulations shift with an increase in the number of T-helper type 2 cells (Th2) and regulatory T cells (Tregs) resulting in an inadequate adaptive immune response toward new antigens and altered memory responses [36]. Other functional changes leading to altered activity are related to T-cell membrane composition. As membrane cholesterol content increases with age, intracellular pathway signaling abilities may be altered due to increased membrane viscosity [42].

A recent study by Mittelbrunn and colleagues showed that dysfunctional (TFAM-deficient) T cells act as accelerators of senescence throughout multiple organ systems, suggesting that T cells can regulate organismal fitness and life span [43]. Premature aging in mice with TFAM-deficient T cells was partially rescued by blocking proinflammatory cytokine (tumor necrosis factor [TNF]-α) signaling or by preventing senescence with nicotinamide adenine dinucleotide (NAD) precursors [43,44], which are further discussed in Section 7.4. Mittelbrunn and Kroemer also proposed 10 molecular hallmarks to represent common denominators of T cell aging. These hallmarks are grouped into four primary hallmarks (thymic involution, mitochondrial dysfunction, genetic and epigenetic alterations, and loss of proteostasis) and four secondary hallmarks (reduction of the TCR repertoire, naive–memory imbalance, T cell senescence, and lack of effector plasticity) [44].

Aging has also been associated with an overall decline in B cell production in the bone marrow, which is paralleled with an increase in a subset of B cells, named age-associated B cells. This subset of B cells is likely to be involved in impaired immunity associated with aging and may contribute to the onset and development of autoimmune diseases by the production of autoantibodies [45].

To summarize, a decrease in adaptive immunity is observed during aging, with consequences for the overall fitness and health span of the organism. Over the last few decades, many observations and findings about the aging immune system have revealed more and more puzzle parts necessary for a holistic perspective on the concepts of immunosenescence in healthy and pathological aging. We provide a momentary snapshot that offers glimpses into an ongoing discipline, and as we continue to unravel the changes that occur in the immune system with age, we can expect new insights.

2.6 Inflamm-aging

Changes in the regulation of immune responses with aging are paralleled with a progressive increase in circulating levels of inflammatory cytokines and proteins, a phenomenon also known as "inflamm-aging", a term introduced in 2000 by Franceschi and colleagues [2]. Due to its medical importance in the transition from healthy to pathological aging, the concept of inflamm-aging, characterized by a chronic, sterile low-grade proinflammatory profile, has received significant scientific attention while constituting one of the theories of aging [6]. Inflamm-aging is the long-term result of the chronic stimulation of the innate immune system by antigens originating from pathogens or from transformed tissue and molecular debris, such as advanced glycated end products, that arise as a result of processes driven by inflammation and oxidative stress [1]. Both increases in basal levels as well as ex vivo-stimulated levels of inflammatory cytokine and protein production have been associated with aging, although this has not been shown for all investigated inflammatory measures. It should also be noted that some of the reported associations were strongly influenced by the health status of the individual [8], and therefore, increased inflammation could not always be attributed solely to the consequences of the aging process. Nevertheless, inflammaging as a concept is scientifically followed by research disciplines beyond classic immunology research, underlining the

2.7 Immune risk phenotype

The simple knowledge of basic mechanisms and their functional changes with age is not satisfactory in times of challenge, for example during the COVID-19 pandemic. As a result, timely translational research is needed to identify efficient ways to implement these findings into clinical practice. Empirical evidence from exploratory studies using large databases identified high-risk profiles of inflammation and immunosenescence and showed their value in the context of predicting healthy and pathological aging and the probability of higher morbidity risk [46–49]. Going even further, a cluster of immune parameters that predict future mortality has emerged from longitudinal studies of elderly cohorts and is referred to as the *immune risk phenotype* (IRP). The IRP is characterized by high CD8+ and low CD4+ numbers, resulting in an inverted CD4+/CD8+ ratio, attenuated mitogen-stimulated lymphoproliferative responses and is associated with seropositivity to cytomegalovirus (CMV) as well as an increased number of lately differentiated CD8+CD28− effector cells [47,48]. A more recent study longitudinally tracked multiple immune features, including cell-subset phenotyping, cytokine response assays, gene expression profiling, and CMV seropositivity in a cohort of healthy adult individuals and constructed a high-dimensional trajectory of immune aging (*IMM-AGE*) that described an individual's immune functioning significantly better (and served as a better risk predictor) than did chronological age. The *IMM-AGE* score predicted all-cause mortality, establishing its potential future use for identifying at-risk patients in clinical settings [49]. Another profile characterized by high inflammation and low cortisol (stress hormone) levels predicted an increased probability of mortality within the next 10 consequent years after assessment [46]. The multidisciplinarity of recently available methods and technology (e.g., multivariate statistics, machine learning, and artificial intelligence combined with new biolaboratory and molecular capabilities) applied to (psychoneuro)immunology and aging research will enable us to gain a better understanding of the physiological and psychological factors contributing to the process of immunological aging and may provide new insights for prevention, prediction, and innovative treatment interventions.

3. Examples of conditions related to immune system aging

The medical relevance of the immune system in the context of aging will be presented here by selected examples. A major concern, particularly in intensive care medicine, is sepsis. The World Health Organization (WHO) repeatedly reports sepsis, a pathophysiological state including severe signs of immune reactivity and inflammation, to cause more than 11 million deaths per year [50]. Sepsis can affect individuals of all ages, but epidemiology also states that the probability of sepsis significantly increases with compromised immune function, a history of previous inflammatory diseases, and, of course, chronologic age [51]. Besides its direct clinical effects in aging societies, sepsis further contributes to the significant consumption of healthcare resources [51,52].

3.1 Infectious diseases

It is not very surprising to find evidence in the literature of an increased prevalence and severity of infectious diseases with progressing age [6,52]. However, this process does not follow linear rules but rather is defined by different developmental stages of an individual associated with differential vulnerabilities and probabilities for disease. According to the relatively health-critical period ranging from infancy to adolescence, epidemiology shows that after reaching adulthood, a more linear character can be observed for the probability of an individual dying of infections and other pathogen-driven diseases [53]. This probability doubles around every 8 years of living according to the Gompertz-Makeham law of human mortality first defined by Benjamin Gompertz [54]. Some of the most common infectious diseases in the elderly comprise urinary tract infections, influenza, bacterial pneumonia (also due to methicillin-resistant *Staphylococcus aureus*, MRSA), gastrointestinal infections, chronic viral infection reactivation, such as CMV and herpes viruses, as well as fungal (e.g., candidiasis, etc.), or parasitic infections [55]. As a consequence of the age-related functional decline of the immune system, aggravated by a generally weaker immune response to most vaccines, mortality, and morbidity during annual influenza epidemics and the current pandemics of COVID-19, are increased in elderly individuals [7].

3.2 Autoimmunity

During aging, the increased output of self-reactive T cells by the atrophied thymus and the accumulation of age-associated B cells results in a diminished ability of the immune system to differentiate between pathogens and "self" body tissues [45,56], although the peak age of onset varies among the different autoimmune diseases and can be sex-dependent. For example, giant cell arteritis more often affects elderly individuals and has a peak incidence at 70–80 years of age, while inflammatory bowel disease has two peaks of onset, the first in younger subjects and the second after 60 years of age [57]. Rheumatoid arthritis (RA) is more prevalent among women than men and often coincides with the onset of menopause [58]. Another example of the impact of age on the onset of autoimmune disorders covers the cluster of autoimmune thyroid disease (AITD), including Hashimoto thyroiditis (HT). Besides a family history of AITD, also chronologic age, sex of the individual, and smoking can all influence the onset risk for this autoimmune disease [59]. The chronicity of (auto)immune-related pathophysiology further lowers the probability of clinical remission, increases the risk of clinical relapse, and also affects the risk for secondary autoimmune conditions, making age-driven changes in immune regulation a promising therapeutic target.

3.3 Cancer and other neoplasms

Aging increases the risk for the development of various types of cancer, with special relevance for individuals passing the age of 65 years, showing a peak incidence in the seventh and eighth decades of life [60]. This time window parallels the age-associated decline in immune function, including prolonged low-grade inflammation and a reduced capacity for immunosurveillance, the process by which the immune system detects and eliminates cancerous cells. A characteristic feature in about 90% of all malignancies is the constitutive expression of high levels of the telomere-elongating enzyme telomerase. Targeting telomerase activity pharmacologically has been studied as a promising approach, however, it is also necessary to again remind the reader that the biology of cancer is—in analogy to aging—not sufficiently understood, dynamic, and adaptive and therefore shows many facets that limit the applicability of relatively simplified etiological and treatment approaches [61]. A more detailed introduction and overview of the biomolecular mechanisms involved in age-related cancer development will be provided in Chapter 12.

Other cancer types do not result from somatic cell malfunctioning but are directly linked to failures of the bone marrow. For example, myelodysplastic syndromes (MDS) are a type of rare blood diseases where immature blood cells in the bone marrow do not mature or become normal functional blood cells. Clinically, these syndromes are characterized by dysregulated hematopoiesis, ineffective erythropoiesis, peripheral cytopenia (reduced blood cells), and an increased risk of transformation to acute myeloid leukemia [62]. Clonal hematopoiesis is a consequence of genetic abnormalities in HSCs and results if a mutated clone disproportionately contributes to mature blood cell production over time. It is exacerbated by an aged inflammatory milieu and, in the majority of cases, MDS is due to mutations in genes involved in epigenetic regulation (DNMT3A, TET2, ASXL1). These mutations are rare in the younger but highly prevalent in the elderly population, with about 10%–20% of individuals over 70 years of age harboring a clone of appreciable size [15,63].

For certain cancers, innovative research co-focusing on the immune system has taken out some of the terror. For example, the use of immunotherapy has revolutionized the therapy of certain cancers over the last decade. Immune checkpoint inhibitors have promoted durable remissions across a variety of tumor types by modulation of the patient's existing immune system and by the use of autologous tumor-specific T cells or chimeric antigen-receptor T cells (CAR T cell therapy) has proven effective in treating leukemia and multiple myeloma [64]. Progress in science and technology gives rise to the confidence that a significant improvement also in the treatment of different types of cancers (somatic vs. stem-cell derived) can be achieved using i.a. immunotherapy within the next few decades, although solid tumors are currently somewhat more difficult targets.

3.4 The role of immunosenescence in health and disease: An introduction to telomere biology

Immunosenescence is much more than just the aging of the immune system. Directly or indirectly, it is viewed as playing a role in a multitude of common, age-associated diseases, including cardiovascular disease [65–67], type 2 diabetes mellitus [68], neuroendocrine and neurodegenerative disorders, such as depression [69–71] and Alzheimer's disease [72,73]. Interestingly enough, all of these conditions have been associated with shorter leukocyte telomere length (LTL), a prominently reported, but debated biomarker for diseases of aging and even overall aging processes [74]. The shortening of leukocyte telomeres, the repetitive hexamer repeats with the sequence (TTAGGG)n at the end of linear chromosomes, is thought to reflect systemic influences and the senescent status of immune cells circulating in blood [75]. As immune cell senescence induces proinflammatory processes, immune cell telomere shortening may be a direct contributor to the etiology of these conditions [75]. However, despite being causally implicated, peripheral LTL may also be an indicator of aging-related disease risk, since upstream risk factors, such as those related to lifestyle and pathogen load, impact both telomere biology and the risk of diseases associated with aging independently.

The use of peripheral LTL measures raises several issues related to analytic methods. Most studies report mean relative LTL measured by quantitative polymerase chain reaction (qPCR), although information on the shortest telomere length (or lowest decile) might be more relevant to cell function [74,76], but not detectable with this analytical technology. Alternatively, Southern blot or quantitative fluorescence in situ hybridization (qFISH) provide better insights into the character of telomeric integrity, but are much more time-consuming and therefore not easily applicable for high-throughput analyses. For investigations in larger cohorts, qPCR remains the method of choice, although it cannot adequately address the question of whether telomeres are truly shortened or if the shortening of mean telomere length is rather attributable to oxidative stress-induced DNA damage, violating the integrity of telomeres. And, although telomere length is known to vary between different immune cell subtypes [77], studies often measure telomeres in leukocytes or in whole blood, while telomere length data in specific immune cell subtypes could provide a better picture of which immune cell subtypes show the shortest telomere length and therefore might have the strongest contribution on the process of immunosenescence. To further explore the mechanisms underlying the aging of the immune system, in the next sections the role of telomere biology and interrelated biomolecular mechanisms, together with upstream risk factors amplifying the process of immunosenescence, will be highlighted.

4. Biomolecular mechanisms of immunosenescence

4.1 Role of the telomere biology system

To what extent cellular senescence contributes to the functional decline of the immune system remains under debate, but increasing evidence indicates that cellular senescence is, at least in part, a cause of immunosenescence. According to Hayflick's observation of limited cellular replication capacity [78], telomere biology will be presented next as one central mechanism of cellular senescence. The shortening of telomeres is a well-known hallmark of both cellular senescence and organismal aging and is observed in most immune system cells including HSCs. To compensate for cellular turnover and loss of telomeres due to replication, the level of telomerase activity, a reverse transcriptase enzyme that consists of an RNA component (telomerase RNA component, TERC) and a catalytic protein domain (human telomerase reverse transcriptase, TERT), can be upregulated in cells that undergo rapid expansion, such as HSCs and committed hematopoietic progenitor cells (HPCs) [79]. However, although these cells can induce telomerase, their telomeres do shorten with age (if perhaps less rapidly than do cells that lack this ability), as indicated by animal and human studies showing telomere shortening in parallel with the accumulation of markers of cell senescence such as p16INK4a [80–82].

There appears to be a link between the replicative history of HSCs and LTL, but, unlike peripheral leukocytes, which are exposed to factors present in the circulation, HSCs are not required to exhibit telomere shortening at a similar rate [74]. Because, especially in humans, the in vivo behavior of HSC within the bone marrow is difficult to study [83], it remains challenging to determine the exact kinetics of the HSC system and whether interindividual variation in the length of peripheral leukocyte telomeres reflects that of HSC. Evidence indicates that during early life there is a faster attrition rate of leukocyte telomeres as compared to attrition rates seen in adults, which likely corresponds to the fast expansion of the HSC and HPC pools at the beginning of life [84,85]. A model characterizing HSC kinetics from birth to the age of 20 years estimated that HSCs replicate about 17 times by the first year of life and replicate 2.5 times/year between the ages of 3 and 13 years, while in adults replication slows considerably to a rate of 0.6 times/year [85]. Building upon this model of HSC kinetics, a new model of telomere length-dependent T cell clonal expansion capacity with age was developed and its relationship to COVID-19 mortality was examined. The model showed that an individual with an average LTL at the age of 20 years maintains maximal T cell clonal capacity up to the age of 50 years. Thereafter, clonal expansion capacity declines exponentially and coincides with the steep increase in COVID-19 mortality with age [86]. Further evidence supporting this model was provided by a recent study in patients with COVID-19. Here, it was shown that the shortest LTL was associated with low lymphocyte counts, possibly reflecting a failure of T cell proliferation to keep up with the T cell loss from the circulation during the course of the illness [87].

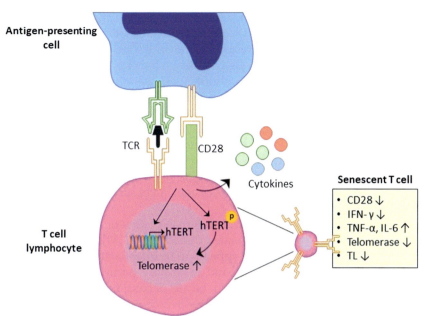

FIGURE 7.2 Telomerase expression and activation upon antigen stimulation in T cells and changes in T cell characteristics with senescence. *This figure was created using motifolio image assets.*

T cells can upregulate telomerase activity upon cross-linking of the TCR-CD3 complex and CD28 upon binding to antigen. This cross-linking initiates a signal transduction cascade that ultimately induces the translocation of transcription factors such as NF-κB to the nucleus, entry into the cell cycle, DNA synthesis, hTERT expression, and telomerase activation [88–90] (see also Fig. 7.2). Telomerase activity then facilitates the immune response and prevents immune cell senescence during fast and profound (clonal) cell expansion. However, the upregulation of telomerase in T cells is not sufficient to indefinitely prevent telomere shortening and senescence, as the shortening of T cell telomeres is observed in human aging and during in vitro long-term culture [89,91].

Senescent T cells bear several trademarks, including shortened telomeres, loss of proliferative capacity, and decreased or lack of expression of the CD28 co-stimulatory receptor. Also, other co-receptors like CD27, CD57, the killer cell-lectin-like receptor G1 and programmed death receptor 1 (PD-1) have been associated with senescence [92–94]. CD28 signaling is required for optimal telomerase up-regulation, as indicated by the paralleled loss of CD28 expression and telomerase activity in T cells after chronic antigen stimulation in vitro [95]. Senescent T cells produce reduced levels of the antiviral cytokine, interferon (IFN)-γ, and increased levels of pro-inflammatory cytokines such as interleukin (IL)-6 and TNF-α [96]. Interestingly, the loss of CD28 in a long-term stimulated CD8+ T cell culture is delayed by inhibition of TNF-α and is paralleled by increased telomerase activity [97]. Of the main leukocyte subsets, senescent CD8+ CD28− T cells have the lowest telomerase activity and shortest telomere length [98]. Moreover, a higher percentage of CD8+CD28− T cells is correlated with shorter mean LTL [98] (see Fig. 7.2). Besides CD28, other co-stimulatory signals, like CD27, seem to be associated with telomerase regulation and telomere maintenance. For example, CD8+CD28− T cells lacking the CD27 co-stimulatory receptor (CD8+CD28−CD27− T cells) showed the shortest telomeres and extremely low telomerase activity upon stimulation [94].

Telomerase upregulation upon antigen presentation might not be the sole mechanism by which T cells delay senescence. Intriguing new evidence indicated that antigen-presenting cells donate telomeres to favor some T cells becoming long-lived memory cells. A model based on these preliminary results proposes that telomere transfer occurs before clonal expansion begins when T cells are still bound to antigen-presenting cells and form immunological synapses and augment certain, probably ultrashort, telomeres by ∼3,000 bp. Telomerase then replenishes telomere loss at all chromosomes during postsynaptic T cell divisions (∼100–200 bp) when T cells undergo massive proliferative clonal expansion [99].

4.2 Regulation of telomerase activity

Telomerase permits the proliferation of resting stem cells and protects cellular proliferation capacity and survival of activated immune cells under conditions of cellular stress [100]. Normally, very low levels of telomerase are detectable in quiescent HSCs or resting immune cells, but the level of telomerase activity can be dramatically increased by antigenic stimulation of lymphocytes or upon proliferation and differentiation of committed HSCs. Hiyama et al. [101] were the first to describe the upregulation of telomerase in human lymphocytes after in vitro stimulation. While in isolated resting peripheral blood mononuclear cells (PBMCs) telomerase was detectable in very low levels, its activity increased up to 300–1000-fold over 1 week in cultured T cells stimulated with phytohemagglutinin (PHA) and interleukin (IL)-2, and increased up to 30-fold over 1 week in cultured B cells stimulated with pokeweed mitogen. In this study, it was additionally shown that unstimulated telomerase activity in PBMCs decreases with increasing age until 60 years, but no longer in older individuals. Evidence from a longitudinal study in humans showed that basal telomerase activity levels in both T and B cells and induced telomerase activity in T cells decline with age [102]. Contrasting findings showed that the capacity for induction of telomerase activity in T or B cells after in vitro stimulation varied significantly across subjects but did not change as a function of the subject's age (the age of study participants ranged between birth and 94 years of age), an observation that has since been replicated by subsequent studies [103–105].

In addition to its relationship with aging, several studies have characterized the association of measures of basal and stimulation-induced immune cell telomerase with stress- and aging-related disease states [89]. In general, these studies have reported that the stimulated leukocyte telomerase response appears to be attenuated in subjects with autoimmune conditions such as systemic lupus erythematosus and RA, and in healthy participants perceiving chronic stress or exhibiting an increased physiological stress response [89,105]. Levels of basal immune cell telomerase activity seem to be less associated with pathological stress and disease states, which may be indicative of the dynamic regulation of telomerase activity by different biological mechanisms [89].

Telomerase activity capacity is known to vary by immune cell subtype, therefore, measuring telomerase activity at the single-cell level can provide useful insights into the role of telomerase in immune-cell senescence. For example, by using a digital droplet PCR system, a robust increase in telomerase activity was only observed in a subset of CD28+ T cells upon immune stimulation and indicated that telomerase is not required for the short-term proliferation of T cells. Yet, solely the CD28+ T cells that induced a robust telomerase response were capable of maintaining their telomere lengths during proliferation [106].

Several kinases can phosphorylate the human telomerase reverse transcriptase (hTERT) protein, thereby controlling its cellular localization and activity. Antigen stimulation causes hTERT to enter the nucleus, which then leads to increased nuclear telomerase activity in lymphocytes. However, only the phosphorylated form of hTERT can be imported into the nucleus, where it can bind to the other components of the telomerase holoenzyme [107]. As part of the PI3K-Akt-mammalian target of rapamycin (mTOR) signal transduction pathway, Akt phosphorylates hTERT and upregulates its activity. In addition, heat shock protein 90 (Hsp90), a protein known to form a complex with hTERT, Akt, and mTOR in NK cells, is essential in the regulation of telomerase activity and is necessary for its stability and activity [108,109]. The PI3K-Akt-mTOR signaling pathway is crucial for immune-cell proliferation and eliciting normal immune responses [109,110]. With aging there are many changes in immune cell signaling, including alterations in mTOR signaling [42], potentially altering hTERT localization and phosphorylation and thereby influencing telomeres and cell function.

In addition, the TERT protein harbors a mitochondrial localization sequence and shuttles out of the nucleus and into mitochondria upon extrinsic (e.g., by applying hydrogen peroxide, hyperoxia or irradiation) or intrinsic (e.g., senescence-associated increase in cellular reactive oxygen species [ROS] levels) oxidative stress to protect mitochondrial DNA and function [111]. Further evidence indicated that the inhibition of mTOR signaling by rapamycin and dietary restriction acts as additional stimuli for mitochondrial localization of TERT in neurons and MCF-7 cells, and that this localization correlates to lower mitochondrial oxidative stress [111]. However, to date, it is not clear if inhibiting the mTOR pathway in human immune cells results in similar observations. Finally, TERC, which serves as an RNA template for the reverse transcriptase TERT, has been shown to have telomerase-independent functions. In CD4+ cells, telomerase-inactive TERC seems to play a role in the regulation of apoptosis [112]. Additionally TERC promoted cellular inflammatory responses in vitro, independent of telomerase activity [113].

4.3 Mitochondria, telomere biology, and cellular aging

Mitochondria, also called bioblasts (in earlier days of science first named by the German pathologist and histologist Richard Altmann [1852–1900]), are central in providing cellular energy, but also play a major role in a multitude of cellular functions including homeostasis, proliferation, biosynthesis, and apoptosis [114]. Although histologically classified as cellular organelles, mitochondria contain their own circular DNA (mtDNA), a biological residual of their bacterial ancestry [115]. As all mitochondria have a provenance approaching 2 billion years in eukaryotes, this particularly raises the question of why we do not inherit dysfunctional mitochondria from our maternal zygote. There is, however, to date, no explanation for the observation that mitochondria did not age retroactively over 2 billion years and then rapidly age within a few decades after fertilization.

Due to their physiological function of ROS-associated energy production (in terms of ATP), mitochondria show structural as well as functional alterations with age [116]. Therefore, "mitochondrial health" has become a new scientific focus of growing interest over the last two decades. Cellular aging correlates with a decline in mitochondrial respiratory function and ATP production, as well as an increase in ROS-induced mutations of mtDNA, thereby making mitochondrial DNA damage a good indicator for the presence of oxidative stress [117]. Moreover, the cellular mtDNA copy number (mtDNAcn) has significant implications for cellular health and function. For example, using an in vitro cell model, it was shown that depletion of mtDNAcn is linked to reduced downstream processes of cellular function, including mitochondrial transcription and lower expression of protein products contributing to the OXPHOS machinery [118]. Interestingly, the administration of acetyl-L-carnitine showed compensatory potential due to its reversibility effects on transcription and translation [119,120]. Also in humans, age-progressed individuals show a decline in the total copy number of peripheral blood cell mtDNA [121], and it is not surprising to find changes in blood cell mtDNAcn also in different age-associated diseases, including cardiovascular disease [122], neurodegenerative diseases, such as dementia and Parkinson's disease [123], cancer [124], as well as in a broad range of mental-health complications [125–128], although opposite findings also were reported for younger cohorts [129]. Therefore, the robustness of these observations with progressing age requires more research. Nevertheless, these findings indicate the potential of immune cell mtDNAcn regulation also as a potential biomarker of healthy and pathological aging and, in analogy to the findings from animal studies, it remains to be investigated whether age-associated changes in mtDNAcn and mitochondrial function in humans can be also compensated or reversed by the intake of substances such as acetyl-L-carnitine [119,120].

Growing evidence supports a direct link between telomeres and mitochondrial biology. To illustrate, telomere dysfunction activates *p53*, a tumor suppressor gene, resulting in the repression of peroxisome proliferator-activated receptor gamma coactivator (PGC)-1α and β, the main promotors of mitochondrial biogenesis, which then leads to the suppression of mitochondrial biosynthesis and function [130]. Also, telomerase has been reported as essential for normal mitochondrial production and function. As previously described, several recent reports have examined the transport of hTERT into mitochondria, which appears to be mediated by a mitochondrial import sequence at the N-terminal of hTERT. Results showed that accumulating endogenous oxidative stress induces the nuclear export of hTERT to the cytosol and decreases nuclear and total telomerase activity, whereas cytosolic hTERT protein increases in cytosolic fractions, a process that can be decelerated by antioxidants [131]. Evidence further indicated that in the mitochondrial matrix TERT binds to mtDNA at ND1 and ND2 regions, coding for proteins of the electron transport chain. TERT also increased respiratory chain activity, reduced ROS production, protected mtDNA, and inhibited apoptosis induction, suggesting a protective function of telomerase in mitochondria under conditions of increased oxidative stress [132]. Subsequent new findings suggest that not only TERT but also TERC, the RNA component of telomerase, can enter the mitochondria. Here, TERC is processed to a shorter form, TERC-53, which is shuttled back to the cytosol and possibly translocates to the nucleus to signal mitochondrial function. In vitro data indicated that TERC-53 regulates gene expression and accelerates cellular senescence without interfering with telomerase activity levels [133]. How these processes are fine-tuned in the process of aging, also in different cells and tissues (contributing to the immune system), remains to be further explored and may lead to the identification of new strategies for slowing down or even reversing the aging progress.

4.4 DNA damage and repair mechanisms

Exceeding levels of the production of ROS and reactive nitrogen species (RNS), occurring during demanding and challenging responses to exogenous as well as endogenous stressors, are associated with the promotion of biological aging and an increased risk for the onset of age-associated diseases [134]. According to Denham Harman's free radical theory of aging postulated in 1956 [135], these mechanisms also contribute to age-associated changes in immune cells. The structural damaging effects of free radicals include single- as well as double-strand breaks that have to be compensated by a complex DNA repair machinery essential for the health and survival of stressed cells, tissues, and organs and activated by the DNA damage response (DDR). There is mounting evidence that age-related DNA damage accumulation in HSCs and changes in the microenvironment contribute to the dysfunction of HSCs with age. The cumulative effects of DNA damage experienced during the lifespan of HSCs are particularly noticeable when the damage is not effectively repaired, as demonstrated by several animal models with DNA repair factor defects [136–138].

How DNA repair mechanisms are fine-tuned and how the structural and functional integrity of the whole DNA is monitored is not fully understood, but its relevance has been highlighted especially for telomeres, which show high vulnerability against free-radical inducible damage [139]. Critical telomere shortening induces the DDR, however, telomeres, which can be considered as DNA ends, have to be "protected" from the action of the DDR machinery to avoid highly detrimental telomere fusions, a process ensured by the shelterin complex [140]. Therefore, telomeres differ from the rest of the genome in their DNA repair capacity and evidence indicates that the inability to repair double-strand breaks at telomeric regions results in a persistent DDR, contributing to the process of cellular senescence [141].

Several lines of evidence demonstrated that all pathways of DNA repair become less efficient with age. Damaged DNA that is not properly repaired can lead to genomic instability, apoptosis, or senescence. The accumulation of damaged or senescent cells contributes to the onset of the senescence-associated secretory phenotype (SASP), associated with the release of pro-inflammatory mediators, affecting immunity and inflammatory regulation. The consequences of SASP can be addressed by intervention programs to rejuvenate the average age of the immune system, for example, by identifying and removing these cells from the body by applying senolytic drugs [142]. For more details and a discussion on the use of senolytics we refer to Section 7.5).

4.5 Crosstalk between telomere biology and epigenetic processes

Epigenetics summarizes a variety of different biological mechanisms and modifications that contribute to the structural and functional adaptation of cells, tissues, and organs across the lifespan. Besides changes in acetylation and phosphorylation of regulatory proteins, remodeling of chromatin sections, or altered expression of noncoding RNA, methylation of so-called CpG units (cytosine followed by guanine) within the genomic sequence can lead to the downregulation or even silencing of gene expression. Epigenetic marks, such as DNA methylation and histone modifications, are established during embryonic development and are stably inherited during mitosis, ensuring cell differentiation and fate. Furthermore, DNA methylation, and also other epigenetic modifications, are highly related to age and correlate with the cell passage number [143]. Since many regions of the genome exhibit predictable hypo- and hypermethylation changes as time progresses, epigenetic research has attracted very strong attention in the fields of healthy and pathological aging and resulted in the development of numerous epigenetic aging clocks based on DNA from e.g., circulating immune cells [143] (see also Section 6.4 on epigenetic aging clocks). Previous research has also linked epigenetic variations with aging in stem cells. For example, aging HSCs show site-specific alterations in DNA methylation and chromatin marks associated with stem cell self-renewal and loss of differentiation capacity [144].

Various studies have indicated that telomeres and telomerase are targets for epigenetic processes that shape the telomere structure and influence its maintenance [145]. However, the shortening of telomeres also directly affects the epigenetic status of telomeres and subtelomeric regions, which influences the telomere position effect (TPE), a phenomenon referring to the ability of mammalian telomeres to silence subtelomeric genes [145]. Moreover, a genome-wide association study indicated a causal TERT-specific role on cell intrinsic DNA methylation age [146]. While to date it is not clear if telomere shortening is the primary driver of cellular aging mechanisms, there appears to be a direct crosstalk between telomere maintenance and epigenetic modifications that accompany aging.

5. Upstream risk factors

Even though a multifold of upstream factors affect immune system aging, here we review several genetic factors, as well as a selection of factors related to our life history amplifying the aging process.

5.1 Genetic factors

It is estimated that about 25% of the variation in the human lifespan is determined by genetics [147]. Contrary to longevity and healthy aging, other genetic variants have been associated with accelerated aging and early immunosenescence. For example, human germline mutations that negatively affect telomerase function or telomere maintenance result in a variety of diseases collectively called telomeropathies. These telomere biology disorders (TBDs) often stem from severely short or dysfunctional telomeres of which most severe disorders eventually lead to HSC failure, the major cause of morbidity and mortality in affected patients [148]. The clinical manifestations of TBDs are thought to arise from severely short or dysfunctional telomeres that limit replicative capacity, leading to the loss of critical stem cell pools. TBD-causative mutations have been mapped to several genes that span numerous macromolecular complexes and pathways important for telomerase maintenance, including the processing and assembly of the telomerase holoenzyme (for detailed reviews see Grill and Nandakumar [148] and Revy et al. [149]).

Another genetic disorder associated with early immunosenescence is trisomy 21, also called Down syndrome. Individuals with trisomy 21 have an increased susceptibility to develop infections, hematologic malignancies, autoimmunity, and autoinflammatory diseases, and show an incomplete efficacy of vaccinations [150]. Several changes in both the architecture and the functioning of the immune system are observed in trisomy 21, including an abnormal (smaller) thymus that is intrinsically deficient, lower absolute T cell counts, and lack of expansion normally seen in the first years of life [150]. Although individuals with trisomy 21 are not born with shorter leukocyte telomeres [151] (one study even suggests they are born with longer telomeres [152]), with advancing age, telomeres shorten more quickly compared with aged-matched controls [152,153].

Strikingly, immunosenescence has not been observed in the Hutchinson—Gilford progeria syndrome (HGPS), a premature aging disease caused by mutations of the LMNA gene leading to increased production of a partially processed form of the nuclear fibrillar protein lamin A [154]. Although most other tissues in HG-progeric children do show shorter telomeres and early cell senescence, the immune systems of children with HGPS appear to function in a normal manner and circulating leukocytes display no subnormal telomerase activity or shorter-than-normal telomeres [155]. Therefore, HGPS might be the result of segmental accelerated senescence of a limited number of cell phenotypes. Affected cells include dermal fibroblasts, arterial endothelial cells, chondrocytes, and perhaps other types. Unaffected cells seem to include glial, neural, and lymphocytic cell lines, corresponding to the lack of clinical findings in these tissues [156].

5.2 Lifestyle and environmental factors

The following upstream risk factors are associated with an individual's life history and offer interesting targets for top-down interventions to attenuate the aging process.

5.2.1 Pathogen load

Accumulating evidence shows that chronic infection by viruses such as human immunodeficiency virus (HIV), CMV, Epstein—Barr virus (EBV), hepatitis B/C/D, human T cell leukemia virus type 1, herpes simplex virus-1/2, and varicella-zoster virus are stimulators of immunosenescence, resulting in the accumulation of senescent specific T cells [157,158]. Many of these viruses can establish lifelong infections and change the architecture of the immune system in a significant way [157].

A virus with a remarkable impact in the immune system and driving the largest antigen-specific immune response of any known microbe within the vascular system is CMV [159]. Chronic infection with human CMV has been associated with the so-called immune risk phenotype that has been predictive of early mortality in several longitudinal studies [47,157]. Human CMV infection is related to the expansion of virus-specific effector memory CD8+ and CD4+ T cells and a concomitant decline in naïve T cells. According to the duration of latent infection, up to 10% of effector memory CD4+ cells and 50% of effector memory CD8+ cells become targeted at CMV antigens, a phenomenon known as "memory inflation." This CMV-driven expansion of CD8+ T cells is accompanied by hallmarks of T cell senescence, including the loss of the costimulatory receptors CD27 and CD28 and increased expression of CD57 [157].

Viral exposure has also been linked to telomerase activity in several studies. For example, hepatitis B and C virus-infected individuals displayed lower levels of leukocyte hTERT mRNA than healthy controls [160] and in individuals suffering a primary infection, the telomere length of EBV-specific T cells was maintained by high telomerase activity, but later in life, these specific T cells showed shorter telomere length and lower telomerase activity [161]. In another study, elevated CMV IgG was associated with lower telomerase activity among CMV seropositive individuals, whereas CMV seropositivity was only associated with decreased telomerase activity among women [162]. The reduction in telomerase activity observed during viral infections might be induced by the release of IFN-α by plasmacytoid dendritic cells, inhibiting telomerase activity by transcriptional and posttranscriptional mechanism and inducing the loss of the co-stimulatory receptors CD27 and CD28 in human T cells both in vitro and in vivo [163].

5.2.2 Mental distress

Mental stressors are part of our daily life, however, when stressors of sufficient intensity, frequency, or chronicity occur, with special relevance for the sensitive period early in life, stress can get "under the skin" or even "inside our cells," affecting all systems of the body, including the immune system [164]. A central mediator of the endocrine stress response system is cortisol, a cholesterol-derived steroid hormone mainly synthetized in the adrenal glands. Cortisol has immunosuppressive properties, making it a substance of the broadest applicability in medical treatment. However, long-term exposure of the immune system to high levels of cortisol has revealed immunological impairments that are reflected by functional, cellular, as well as biomolecular alterations [164].

The exposure of an individual to chronic and traumatic stress, such as forms of social disadvantage and isolation, (childhood) adversity, rape, or caregiving stress has been related to immunosenescence [165] and changes in leukocyte telomere biology [166,167], mitochondrial function [168,169] and accelerated epigenetic aging [170–172]. A link between adult chronic psychosocial stress burden and telomere biology was first proposed and demonstrated by Epel, Blackburn, and co-workers [173]. Since this early work, numerous studies concluded that different types of stress, such as stressful life events, experiencing domestic violence, caregiving stress, and financial worries, but also certain aspects of the stress response, are associated with shorter telomeres [166,167], although there are some exceptions. For example, Rentscher and colleagues [174] found no significant associations between chronic stress exposure and LTL but did observe increased cellular senescence $P16^{INK4a}$ signal expression. Also, early life stress, including prenatal stress, has been connected with shorter leukocyte telomeres, which persist into adulthood [175], indicating that early life stressors have a significant impact on telomere length throughout life. This has been confirmed by a longitudinal study that found that childhood trauma, especially when multiple traumas were reported, predicted adult telomere shortening over a 6 year period [176]. The exact mechanisms of how these early life events are retained into adulthood remain unknown, but one can speculate that when stressful events have an impact on HSCs, e.g., through the release of stress-induced neuroendocrine mediators, like cortisol and adrenaline, inflammatory mediators and oxidative stress, these effects will be amplified when occurring in early life phases at which HSC clonal expansion is accelerated [84,85].

Traditional methods of addressing stress, such as meditation, yoga, and mindfulness training, offer cost-effective and time-efficient interventions. While their beneficial effects on health are "known" and have been accepted for centuries, the biological mechanisms and biochemical pathways involved remain uncharacterized. Nevertheless, the first observations in immune cells report promising results of such interventions on telomere length and telomerase regulation [89,177]. However, the interpretation of such data poses various challenges involving methodological considerations and confounding effects of immune cell distribution [74,89]. Additionally, positive effects have been observed for mitochondrial health [178] and epigenetic aging [179]. How these stress-reducing interventions induce beneficial effects on a cellular and molecular level is not yet completely understood, but may be explained by changes in neuroendocrine and inflammatory signaling. Future studies will provide sufficient evidence to warrant the implementation of such interventions to a broad range of aging patients by demonstrating enduring effects on the immunosenescence process.

5.2.3 Nutrition

It is well established that healthy nutrition and a balanced energy intake are key lifestyle factors that modulate the aging process [180]. Hence, nutritional interventions are promising nonpharmacological therapeutics to attenuate human aging and disease. A reduction in caloric intake while maintaining optimal nutrition, termed calorie restriction (CR), is regarded as the most effective nutritional intervention to prolong health span, slow biological aging, and improve quality of life [181]. Currently, the most widely accepted model is that CR protects from DNA damage through a decrease in metabolism and mitochondrial activity, resulting in a tissue-protective phenotype [182]. The first report on the role of CR in extending the health and lifespan of rodents by McCay dates from 1935 [183]. Later research by Ray Walford and others found that restricting caloric intake in laboratory mice by 30% −60% percent could more than double their normal lifespan, at least when starting CR early in life [182]. In addition to metabolic changes, CR in mice attenuated the age-associated changes in the adaptive immune system, thereby enhancing the immune response to vaccination and reducing the incidence of spontaneous malignancies [184]. On the cellular level, CR upregulated DNA repair, attenuated telomere erosion, increased antioxidant function, and lowered ROS production in the mitochondria [185].

Human research on caloric restriction and intermittent fasting found that CR protocols lower circulating levels of inflammatory markers, such as CRP and TNF-α, reduce white blood cell counts, and lower markers of oxidative stress [182,186]. However, similar frequencies of pre-senescent CD8+ CD28− T cells and CD57- and PD-1-expressing T cells and, surprisingly, shorter LTLs were observed between CR practitioners and a healthy and an overweight control group [187]. To date, one (longitudinal) study observed an inverse significant relationship between baseline energy intake and follow-up LTL in men, but this effect was not seen in women [188]. Taken together, human studies on CR primarily showed beneficial effects on metabolic function, oxidative stress, and inflammation, but there have been very few reports on the association of energy intake with immune cell senescence in humans.

Although caloric restriction can have beneficial effects on healthy aging, malnutrition has severe consequences for health and proper immune functioning. Due to e.g., malabsorption, digestion problems, low appetite and food intake, inadequate or deficient serum levels of micronutrients are frequently observed in the elderly [189,190]. Along with the previously described mechanisms leading to malnutrition, a reduced capacity to produce vitamin D in response to UV light often results in vitamin D deficiency in the elderly and is associated with a dysregulated immune response and lower protection against pathogens. Through the expression of the vitamin D receptor, many immune cells directly respond to vitamin D, resulting in an enhancement of both innate and adaptive immunity [190]. Further, vitamin D serum levels and intake have been associated with telomere biology; data from observational and intervention studies indicate a positive effect of vitamin D on LTL and telomerase activity [191]. In conjunction with vitamin D, the dietary intake of zinc and selenium is essential for normal immune system functioning and has been related to longer LTL [192,193].

Other essential nutrients from our diet, such as essential fatty acids, strongly influence the functioning and aging of our immune system. For example, the intake of n-3 polyunsaturated fatty acids (n-3 PUFA) enhanced antioxidant enzymes and lowered pro-inflammatory mediator production in mice [194] and humans [195,196], including patients with COVID-19 [197]. Observational studies on circulating n-3 PUFA levels, and intervention studies using fish oil supplementation also showed beneficial effects on LTL shortening (although not shown in all studies) [198,199], and a differential effect on telomerase activity levels [200].

Finally, adherence to a diet with antiinflammatory and antioxidative properties, such as the Mediterranean diet, rich in olive oil, fish, vegetables, and fruits, has been associated with longevity [201], lower inflammation [202], improved function of immune cells [203], and longer LTL [204]. Although the effects of nutritional interventions on telomere biology of HSCs remain unknown, it might be expected that a reduction in "inflamm-aging" might also have positive effects on the stem cell niche environment [205], potentially attenuating stem cell aging. Together, these results further highlight that dietary and nutritional strategies to alleviate inflammation and sustain adequate immune balance throughout the lifespan are promising approaches to promote healthy aging of the immune system.

5.2.4 Effects of physical activity and exercise on immune functions

The modern lifestyle in industrialized countries is often characterized by a rather high percentage of time spent sedentarily or inactively in combination with high-caloric and highly processed (fast) food. Especially in this double-hit combination, the lack of physical activity and regular exercise can contribute to the onset and manifestation of a (patho)physiological state associated with low-grade inflammation [206], which increases the risk for impaired immunity. While all these aspects seem to have a severe health-threatening character, there is also good news to share: Both physical activity and exercise are highly economic, making them very attractive for medical interventions to counteract cellular senescence and immunological aging toward the reconstitution of life quality [207]. The effects of physical activity not only help to slow down aging processes in the young, but also support the body to identify aged and damaged cells in the elderly. Exercise leads to changes in the composition of T cell subsets, revealing an elevation of naive subsets of T helper and cytotoxic T cells, accompanied by a drop in senescent and effector memory T cells with exercise [208]. While it is an accepted finding that exercise can change the innate and adaptive immune response by immune cell count and function [208,209], it is, however, also important to note that the findings do show some heterogeneity [207] and the robustness of the effects of physical activity and exercise on the immune composition, immunity and cellular senescence lack evidence from longitudinal studies [210]. Physical activity also shows an association with higher integrity of telomeres as reflected by increased telomere length and telomerase activity [211], at least in circulating PBMCs. The precise biomolecular mechanisms behind these observations need to be further elicited; however, changes in free radical production rates, altered activity patterns of the antioxidative (humoral and enzymatic) potential of cells, as well as the inducible activity of telomerase in immune cells, need to be further investigated.

5.2.5 The importance of sleep for health

The physiology of almost all body cells, tissues, and systems is organized and synchronized in circadian rhythms that regulate the sleep—wake cycle [212]. Regenerative sleep during the night contributes to processes that counteract the driving forces of aging and protects our mental and physical health across the lifespan [213]. Sleep also directly changes several immune parameters involving leukocyte numbers function, proliferation, and cytokine production [214]. As sleep and immunity are considered to communicate bidirectionally, they both contribute to the continuity of health across aging. With progressing age, changes are observed in the regulation of the circadian rhythm (e.g., decrease in circadian amplitude, reduced ability to adjust to phase shifting), with clinically relevant consequences for sleep quality and duration [213,215,216].

The influence of sleep characteristics on telomere length in circulating immune cells has been investigated in different populations and health conditions. Positive associations between sleep duration and immunocellular telomere length were reported in individuals of health and disease, covering both sexes and different decades of lifespan [216]. After dichotomizing sleep at the 7 h cutoff according to the consensus statement of the American Academy of Sleep Medicine and Sleep Research Society [217], shorter sleep duration has been reported to be associated with shorter LTL in a dose-dependent (often linear) manner [218]. Again, these data support peripheral effects only and cannot be directly translated to the stem cell niche. Taken together, sleep is an important modulator of health and, therefore, active manipulation of sleep characteristics may not only promote healthy immune system functioning but may also attenuate immunosenescence across the lifespan.

6. Biomarkers/downstream factors related to aging and immunosenescence

Although biomarkers of aging are not the aging process itself, age-associated downstream biomarkers can be useful tools for setting clinical diagnoses, performing risk assessments, and for research purposes. Therefore, in the next section, we discuss a selection of "state-of-the-art" biomarker candidates for the assessment of immune system aging.

6.1 Indirect and direct measures of free radicals

The technical possibility to assess the production rate of free radicals as well as the neutralization capability by the antioxidative system has received continuously increasing attention over the last few decades. In vivo and in vitro diagnostic approaches have led to a broad range of different readouts of primary and secondary damage due to free radicals and oxidative stress. Prominent candidates used are cumulating members of the isoprostane family (e.g., F2-isoprostane, 8-isoprostane) representing free-radical damage of fatty acids (lipids), 8-hydroxy-deoxy-guanosine (8-OHDG) as markers of genomic damage, as well as changes in the fluidity of the cell membrane. Also, peptides and proteins are affected by free radicals and oxidative stress. Their structure and function can be damaged or even lost, increasing the risk for apoptotic signaling in a dose—response-dependent manner [219]. Damage of protein function can be assessed in various forms, for example by changes in enzyme kinetics using spectrophotometry (ex vivo) or by immunohistochemistry coupled to fluorescence microscopy to detect and estimate ubiquitylation, a process that adds ubiquitin, a small regulatory protein, to a protein with the fate of degradation. While none of these assessments detects free radicals directly, why not look for such possibilities in the chemistry of physics? Electron spin resonance (ESR) spectroscopy is a technology from the field of analytical physics that provides a fascinating way to directly measure the presence and reactivity of various forms of free radicals in biological as well as nonvivid systems. While ESR-based observations of immunological ROS and RNS production are—to the best of our knowledge—missing in the literature, it is only a question of time until this technology will reveal its potential in translational research focusing on biomolecular processes in the context of immunity, inflammation, and aging.

6.2 Protein glycosylation and its changes with age

Glycosylation following the process of translation is one of the hallmarks of protein-specific fine-tuning in their structure and function. The enzymatic addition of glycans, complex oligosaccharide chains, to proteins, can either be observed at the N-terminus of the amino acid asparagine or the O-terminus of the amino acids serine or threonine of the given protein sequence. N-glycosylated proteins detectable in blood play an important role in the regulation of immunological processes as they represent signaling molecules involved in processes like cell—cell recognition, signal transduction within the immune cascade, and host—pathogen interactions [220,221]. Pro-inflammatory cytokines affect the expression of $\alpha 1,3$-fucosyl-transferase, $\alpha 2,3$- sialyl-transferase, and N-acetylglucosaminyl-transferase that belong to glycan-modifying enzymes [222], contributing to the N-terminal glycosylation of immune proteins, such as immunoglobulin G (IgG). Here, even the smallest modifications in the glycosylation pattern of IgG, the most prevalent glycoprotein in peripheral blood [223], are bilaterally connected to changes in cytokine (re)activity [224]. Inflammatory diseases [224], but also age-driven complications of senescence in patients with hepatocellular carcinoma (HCC) [223], dementia, and the progeroid disease Cockayne syndrome [225], have been associated with specific alterations in the blood protein glycosylation profile. Physiological aging has further been linked to characteristic alterations in the N-glycan profile, particularly elevated levels of agalactosylated core-α-1,6-fucosylated biantennary N-glycan (NG0A2F) and lower levels of bigalactosylated, core-α-1,6-fucosylted biantennary N-glycan (NA2F). These findings led to the elaboration of the *GlycoAge* test (=log[NG0A2F/NA2F]) for the potential to estimate the biological age of an individual based on a simple serum sample [225]. Although exploration of the detailed biological and biochemical underpinnings of N-glycan-linked adaptation and signaling has only recently been initiated, the *GlycoAge* test might be a promising biomarker candidate and even a potential target for counteracting immunological aging.

6.3 Mitochondrial health index (MHI) and the Bioenergetic Health Index (BHI)

Two promising candidates for cellular health, function, and integrity have been elaborated and proposed as available biomarkers, with a special focus on the functional integrity of mitochondria. The first indicator of interest introduced is the so-called *Mitochondrial Health Index* (MHI). According to the aspect of changes in mitochondrial content with age and responses to deficiencies in mtDNA integrity, the MHI considers the qualitative ratio of the bioenergetic capacity of cells related to their mitochondrial content [226]. Alterations of immune cell MHI have been reported for conditions related to premature (immunological) aging, including psychological stress [226] and neurodegenerative disorders [227,228].

The second indicator of interest is the *Bioenergetic Health Index (BHI)*. Cellular bioenergetics include three major pathways: glycolysis, the Krebs cycle, and the electron transport chain with ADP/ATP turnover as the central mechanism of OXPHOS. The BHI constitutes an average ratio of OXPHOS to glycolysis. In more detail, it considers characteristic aspects of OXPHOS capacity, including reserve capacity (maximum respiration capacity compared to routine respiration), oxygen consumption of mitochondria related to ATP production, and low leak respiration reflecting low levels of proton leak across the inner mitochondrial membrane due to high integrity of the structure [229]. To date, changes in immune cell BHI have been reported in response to oxidate stress [230] and for nephropathy [231]. Especially when considered together with the MHI, the BHI is discussed as a very powerful tool to better evaluate cellular bioenergetics and mitochondrial health toward the prevention of a bioenergetic crisis of clinical relevance in different patient populations [228].

6.4 Epigenetic aging clocks

In recent years, there has been a growing interest in assessing the process of human aging also at the level of epigenetics, e.g., by examining patterns of DNA methylation (DNAm) to predict an epigenetic or biological age. DNAm-based age (epigenetic age) estimators include the pan-tissue epigenetic clock by Horvath [232], based on a set of 353 CpGs, and an estimator based on 71 CpGs in leukocytes developed by Hannum [233], shown to outperform other blood-based biomarkers regarding the prediction of lifespan, such as LTL [143]. Also relevant to immune system aging is the DNAm *PhenoAge* estimator. This was constructed by first generating a weighted average of clinical characteristics, including chronological age and immunological markers such as CRP levels, lymphocyte percentages, mean cell volume, RBC distribution width, and white blood cell count, which then was regressed on DNA methylation levels in the blood. This approach resulted in the automatic selection of 513 CpGs [234] and stands out in terms of its predictive accuracy for time to death, its association with smoking status, as well as with various markers of immunosenescence [143]. Besides their predicting capacities, epigenetic clocks also allow monitoring of the reversibility of the DNA methylation process across antiaging interventions.

7. Future perspectives

After the previous discussion of upstream intervention targets, in the following text selected examples of upcoming potential interventions that can be attributed to prevent, attenuate, or even reverse immunosenescence are introduced. More detailed discussions on approaches that reverse cell aging in vivo using biotechnological interventions to reset gene expression to that of younger, functional cells will be covered in Chapter 13.

7.1 Genetic therapy and pharmacological interventions to stimulate telomerase activity

Over the years, several attempts to reverse immunosenescence have been made, including telomerase activation as a potential therapeutic strategy. Telomerase gene therapy to reset gene expression was first achieved by delivering mouse TERT with an adeno-associated virus (AAV) into young and old mice. This nonintegrative gene therapy resulted in elongated telomeres, extended lifespans, and had beneficial effects on health and fitness in both age groups [235]. Also, human immune cells, including naïve, melanoma specific, and HIV-specific CD8+ T-cell clones, have been evaluated for the effects of telomerase gene therapy (hTERT transduction) and showed an extension of the ability to proliferate and were more efficient at killing target cells. hTERT transduction did not alter the phenotype or function of the cells, which still required antigenic stimulation to trigger their activation and proliferation [91]. In addition to telomerase gene therapy, a number of studies have investigated the activation of telomerase by specific substances. For example, exposure of CD8+ T lymphocytes from HIV-infected human donors to a small-molecule telomerase activator, TAT2, extracted from *Astragalus membranaceus*, modestly retarded telomere shortening, increased replicative capacity, and enhanced immune function [236]. Another extract from the same plant, TA-65, increased telomerase and proliferative activity of human CD4+ and CD8+ cells in vitro [237], and in humans decreased the number of senescent T cells and NK cells, together with a decline in the percentage of the shortest leukocyte telomeres [238]. Also, other natural compounds, such as *Centella asiatica* extract formulation (08AGTLF), oleanolic acid, and maslinic acid, as well as sex hormones, have been shown to stimulate telomerase activity in vitro [239,240]. However, ongoing research is necessary to further characterize the safety and effectiveness of these compounds in human trials.

7.2 Stem cells from umbilical cord blood as a cellular reservoir for immunological treatments/interventions

Since the first successful transplantation of HSCs from an HLA-identical sibling in a child with severe Fanconi anemia in the late 1980s [241], the collection and cryopreservation of these multipotent cells for later medical applications has attracted high scientific as well as commercial interest. HSCs from umbilical cord blood can be used in patients with malignancies including acute myeloid leukemia [242], bone marrow failure, hemoglobinopathies, X-linked adrenoleukodystrophy (ALD) [243], or in individuals with inborn errors of metabolism or primary immunodeficiencies [244]. In addition to its purposes in hematologic transplantology and other curative or regenerative medical approaches, the transplantation of cord blood and HSCs has been considered for antiaging therapies also to increase life expectancy. Additionally, human cord blood-derived plasma (hUCP) has been demonstrated to show antiaging effects in mice [245]. Whether these findings can be directly transduced into medical applications in humans needs to be further investigated, but the gate of possibilities seems to be wide open for potential therapeutic approaches.

7.3 Rejuvenation effects of blood transfer and parabiosis from the young in the aged

A strong body of literature has provided empirical evidence for the rejuvenating effects of young blood on cells, tissues, and organs of aged individuals, at least in animal studies [246]. It remains to be seen to what extent these findings can be directly translated into humans, but some initial steps have been taken in this direction. Besides the spirit of optimism, initial findings dampen the expectation in a way that the effects observed in the elderly treated with blood from the young do not support (at least) strong effects [247], for example, on survival probability in older individuals that received blood transfusions [248]. An established model to investigate the effects of blood on immune functions is the approach of *parabiosis*, the anatomical joining of two animals sharing one blood circulation system. Especially the effects on the immune system, including the function of the thymus, gave rise to the concern that biology is not as simple as expected. While in the old individuals no evidence was found of any effects on thymic rejuvenation, in the young animal a significant reduction in thymus weight was observed along with no significant effects on thymocyte subpopulations [249]. As stated by Hofmann [250], "eternal youth and endless bliss have always been vital human dreams". Whether blood transfusion is the missing key to opening this gate to rejuvenation will be answered scientifically, but also consequences and concerns have to be considered regarding the applicability of these approaches.

7.4 Supplementation of nicotinamide adenine dinucleotide (NAD)

Essential physiological processes, including the bioenergetic metabolism of cells, depend on the availability and functioning of the coenzyme nicotinamide adenine dinucleotide (NAD+), synthesized from a variety of dietary sources as well as from its major precursors such as tryptophan, nicotinic acid, nicotinamide riboside, nicotinamide mononucleotide, and nicotinamide. NAD+ is a central redox mediator in the transduction of electrons together with protons to contribute to the proton motive force inside mitochondria as well as a contributor to a variety of intracellular signaling processes including cell death, cell signaling, DNA repair, gene expression, aging [251], as well as in central carbon metabolic pathways such as glycolysis, the Krebs cycle, OXPHOS, and fatty acid oxidation [252]. It has been demonstrated that the long-term administration of nicotinamide mononucleotide increases energy metabolism, insulin sensitivity, lipid metabolism, and mitochondrial oxidative metabolism, and protects against age-associated functional decline [253]. In addition, nicotinamide riboside can be converted into NAD+ and has been shown to be beneficial in safeguarding against aging and restoring mitochondrial dysfunction [254] and promoting longevity [254,255]. Moreover, NAD+ metabolism is proposed to be highly relevant for regulating the immune response and is emerging as a key modulator of immune cell exhaustion as well as tumor biology, with implications for immunotherapy [256].

7.5 The use of senolytics

One relatively new conceptual approach addressing the consequences of cellular aging includes senolytics. Senolytics comprise both senolytic compounds (synthetic or plant-derived pharmaceuticals and drugs, for example, theaflavins found in black tea extracts, quercetin, and fisetin present in various plants/fruits, resorcinol, cardiac glycosides, or dasatinib), as well as cell-clearing therapies with the core aim of removing senescent cells [257]. At best, senolytic treatment allows the specific and targeted removal of senescent cells without affecting the integrity of surrounding healthy cells [258]. The number of clinical trials using new senolytic compounds has increased exponentially within the last decade, making the discovery of significant interventions—if clinically effective—more likely. In the same vein, the removal of senescent immune cells should also be considered. The ability to address immunosenescence by directly targeting circulating senescent cells, is already available, for example using cell cytometry-centered approaches in combination with appropriate surface markers to actively filter and remove these cells from the body [259,260]. Besides potential benefits, both theoretical considerations and published data suggest long-term adverse effects of senolytic treatments. The removal of senescent cells could accelerate stem cell division and consequent cell senescence, inducing premature cell and tissue aging, with subsequent acceleration of age-related clinical disease [261]. Therefore, the long-term evaluation of the safety and efficacy of senolytics has to be tested and confirmed in humans.

8. Conclusion

Over the last few decades, aging research has provided important insights into the alterations in the anatomy and function of the immune system over the lifespan. These changes do not only affect the immune system, but have systemic consequences for the overall fitness and health span of the organism, making the immune system an attractive target for future interventions to prevent, slow down, or even reverse age-associated functional impairments. To what extent the decline of the immune system with biological aging can be manipulated and whether such approaches are efficient and safe remains to be further explored in humans.

References

[1] Fulop T, Witkowski JM, Le Page A, Fortin C, Pawelec G, Larbi A. Intracellular signalling pathways: targets to reverse immunosenescence. Clin Exp Immunol 2017;187(1):35–43. https://doi.org/10.1111/cei.12836.

[2] Franceschi C, Bonafè M, Valensin S, et al. Inflamm-aging. An evolutionary perspective on immunosenescence. Ann N Y Acad Sci 2000;908: 244–54. https://doi.org/10.1111/j.1749-6632.2000.tb06651.x.

[3] Walford RL. The immunologic theory of aging. Immunol Rev 1969;2(1). https://doi.org/10.1111/j.1600-065X.1969.tb00210.x. 171–171.

[4] Effros RB. Roy Walford and the immunologic theory of aging. Immun Ageing 2005;2(1):7. https://doi.org/10.1186/1742-4933-2-7.

[5] Pawelec G, Bronikowski A, Cunnane SC, et al. The conundrum of human immune system "senescence". Mech Ageing Dev 2020;192:111357. https://doi.org/10.1016/j.mad.2020.111357.

[6] Fulop T, Larbi A, Pawelec G, et al. Immunology of aging: the birth of inflammaging. Clin Rev Allergy Immunol 2023;64(2):109–22. https://doi.org/10.1007/s12016-021-08899-6.

[7] Witkowski JM. Immune system aging and the aging-related diseases in the COVID-19 era. Immunol Lett 2022;243:19–27. https://doi.org/10.1016/j.imlet.2022.01.005.

[8] Furman D, Campisi J, Verdin E, et al. Chronic inflammation in the etiology of disease across the life span. Nat Med 2019;25(12):1822–32. https://doi.org/10.1038/s41591-019-0675-0.

[9] Gao X, Xu C, Asada N, Frenette PS. The hematopoietic stem cell niche: from embryo to adult. Development 2018;145(2). https://doi.org/10.1242/dev.139691. dev139691.

[10] Seita J, Weissman IL. Hematopoietic stem cell: self-renewal versus differentiation. WIREs Mech Dis 2010;2(6):640–53. https://doi.org/10.1002/wsbm.86.

[11] Kondo M. Lymphoid and myeloid lineage commitment in multipotent hematopoietic progenitors. Immunol Rev 2010;238(1):37–46. https://doi.org/10.1111/j.1600-065X.2010.00963.x.

[12] Jackson JD. Hematopoietic stem cell properties, markers, and therapeutics. In: Principles of regenerative medicine. Elsevier; 2019. p. 191–204. https://doi.org/10.1016/B978-0-12-809880-6.00013-8.

[13] Mikkola HKA, Orkin SH. The journey of developing hematopoietic stem cells. Development 2006;133(19):3733–44. https://doi.org/10.1242/dev.02568.

[14] Geiger H, de Haan G, Florian MC. The ageing haematopoietic stem cell compartment. Nat Rev Immunol 2013;13(5):376–89. https://doi.org/10.1038/nri3433.

[15] Brunet A, Goodell MA, Rando TA. Ageing and rejuvenation of tissue stem cells and their niches. Nat Rev Mol Cell Biol 2023;24(1):45–62. https://doi.org/10.1038/s41580-022-00510-w.

[16] Farid Y, Bowman NS, Lecat P. Biochemistry, hemoglobin synthesis. In: StatPearls. StatPearls Publishing; 2022. http://www.ncbi.nlm.nih.gov/books/NBK536912/. [Accessed 30 September 2022].

[17] Kuhn V, Diederich L, Keller TCS, et al. Red blood cell function and dysfunction: redox regulation, nitric oxide metabolism, anemia. Antioxid Redox Signal 2017;26(13):718–42. https://doi.org/10.1089/ars.2016.6954.

[18] Thiagarajan P, Parker CJ, Prchal JT. How do red blood cells die? Front Physiol 2021;12:655393. https://doi.org/10.3389/fphys.2021.655393.

[19] Bhoopalan SV, Huang LJS, Weiss MJ. Erythropoietin regulation of red blood cell production: from bench to bedside and back. F1000Res 2020;9:F1000. https://doi.org/10.12688/f1000research.26648.1. Faculty Rev-1153.

[20] Price EA. Aging and erythropoiesis: current state of knowledge. Blood Cell Mol Dis 2008;41(2):158–65. https://doi.org/10.1016/j.bcmd.2008.04.005.

[21] Chen S, Liu Y, Cai L, et al. Erythropoiesis changes with increasing age in the elderly Chinese. Int J Lab Hematol 2021;43(5):1168–73. https://doi.org/10.1111/ijlh.13615.

[22] Kim KM, Lui LY, Browner WS, et al. Association between variation in red cell size and multiple aging-related outcomes. J Gerontol A Biol Sci Med Sci 2021;76(7):1288–94. https://doi.org/10.1093/gerona/glaa217.

[23] Lupescu A, Bissinger R, Goebel T, et al. Enhanced suicidal erythrocyte death contributing to anemia in the elderly. Cell Physiol Biochem 2015;36(2):773–83. https://doi.org/10.1159/000430137.

[24] Strijkova-Kenderova V, Todinova S, Andreeva T, et al. Morphometry and stiffness of red blood cells-signatures of neurodegenerative diseases and aging. Int J Mol Sci 2021;23(1):227. https://doi.org/10.3390/ijms23010227.

[25] Holinstat M. Normal platelet function. Cancer Metastasis Rev 2017;36(2):195–8. https://doi.org/10.1007/s10555-017-9677-x.

[26] Le Blanc J, Lordkipanidzé M. Platelet function in aging. Front Cardiovasc Med 2019;6:109. https://doi.org/10.3389/fcvm.2019.00109.

[27] Faria AVS, Andrade SS, Peppelenbosch MP, Ferreira-Halder CV, Fuhler GM. Platelets in aging and cancer-"double-edged sword". Cancer Metastasis Rev 2020;39(4):1205–21. https://doi.org/10.1007/s10555-020-09926-2.

[28] Cameron-Vendrig A, Reheman A, Siraj MA, et al. Glucagon-like peptide 1 receptor activation attenuates platelet aggregation and thrombosis. Diabetes 2016;65(6):1714–23. https://doi.org/10.2337/db15-1141.

[29] Murphy AJ, Bijl N, Yvan-Charvet L, et al. Cholesterol efflux in megakaryocyte progenitors suppresses platelet production and thrombocytosis. Nat Med 2013;19(5):586–94. https://doi.org/10.1038/nm.3150.

[30] Casoli T, Di Stefano G, Balietti M, Solazzi M, Giorgetti B, Fattoretti P. Peripheral inflammatory biomarkers of Alzheimer's disease: the role of platelets. Biogerontology 2010;11(5):627–33. https://doi.org/10.1007/s10522-010-9281-8.

[31] Xu XR, Yousef GM, Ni H. Cancer and platelet crosstalk: opportunities and challenges for aspirin and other antiplatelet agents. Blood 2018;131(16):1777–89. https://doi.org/10.1182/blood-2017-05-743187.

[32] Xu XR, Zhang D, Oswald BE, et al. Platelets are versatile cells: new discoveries in hemostasis, thrombosis, immune responses, tumor metastasis and beyond. Crit Rev Clin Lab Sci 2016;53(6):409–30. https://doi.org/10.1080/10408363.2016.1200008.

[33] Parkin J, Cohen B. An overview of the immune system. Lancet 2001;357(9270):1777–89. https://doi.org/10.1016/S0140-6736(00)04904-7.

[34] Chou JP, Effros RB. T cell replicative senescence in human aging. Curr Pharm Des 2013;19(9):1680–98. https://doi.org/10.2174/138161213805219711.

[35] Rosales C. Neutrophil: a cell with many roles in inflammation or several cell types? Front Physiol 2018;9:113. https://doi.org/10.3389/fphys.2018.00113.

[36] Ray D, Yung R. Immune senescence, epigenetics and autoimmunity. Clin Immunol 2018;196:59–63. https://doi.org/10.1016/j.clim.2018.04.002.

[37] Gounder SS, Abdullah BJJ, Radzuanb NEIBM, et al. Effect of aging on NK cell population and their proliferation at ex vivo culture condition. Anal Cell Pathol 2018;2018:7871814. https://doi.org/10.1155/2018/7871814.

[38] Vivier E, Raulet DH, Moretta A, et al. Innate or adaptive immunity? the example of natural killer cells. Science 2011;331(6013):44–9. https://doi.org/10.1126/science.1198687.

[39] Moore AV, Korobkin M, Olanow W, et al. Age-related changes in the thymus gland: CT-pathologic correlation. AJR Am J Roentgenol 1983;141(2):241–6. https://doi.org/10.2214/ajr.141.2.241.

[40] Palmer S, Albergante L, Blackburn CC, Newman TJ. Thymic involution and rising disease incidence with age. Proc Natl Acad Sci USA 2018;115(8):1883–8. https://doi.org/10.1073/pnas.1714478115.

[41] Mou D, Espinosa J, Lo DJ, Kirk AD. CD28 negative T cells: is their loss our gain? Am J Transplant 2014;14(11):2460–6. https://doi.org/10.1111/ajt.12937.

[42] Fülöp T, Dupuis G, Witkowski JM, Larbi A. The role of immunosenescence in the development of age-related diseases. Rev Invest Clin 2016; 68(2):84—91.

[43] Desdín-Micó G, Soto-Heredero G, Aranda JF, et al. T cells with dysfunctional mitochondria induce multimorbidity and premature senescence. Science 2020;368(6497):1371—6. https://doi.org/10.1126/science.aax0860.

[44] Mittelbrunn M, Kroemer G. Hallmarks of T cell aging. Nat Immunol 2021;22(6):687—98. https://doi.org/10.1038/s41590-021-00927-z.

[45] Ma S, Wang C, Mao X, Hao Y. B cell dysfunction associated with aging and autoimmune diseases. Front Immunol 2019;10:318. https://doi.org/10.3389/fimmu.2019.00318.

[46] Bertele N, Karabatsiakis A, Talmon A, Buss C. Biochemical clusters predict mortality and reported inability to work 10 years later. Brain Behav Immun Health 2022;21:100432. https://doi.org/10.1016/j.bbih.2022.100432.

[47] Pawelec G, Akbar A, Caruso C, Solana R, Grubeck-Loebenstein B, Wikby A. Human immunosenescence: is it infectious? Immunol Rev 2005; 205:257—68. https://doi.org/10.1111/j.0105-2896.2005.00271.x.

[48] Wikby A, Ferguson F, Forsey R, et al. An immune risk phenotype, cognitive impairment, and survival in very late life: impact of allostatic load in Swedish octogenarian and nonagenarian humans. J Gerontol A Biol Sci Med Sci 2005;60(5):556—65. https://doi.org/10.1093/gerona/60.5.556.

[49] Alpert A, Pickman Y, Leipold M, et al. A clinically meaningful metric of immune age derived from high-dimensional longitudinal monitoring. Nat Med 2019;25(3):487—95. https://doi.org/10.1038/s41591-019-0381-y.

[50] Global report on the epidemiology and burden of sepsis. https://www.who.int/publications-detail-redirect/9789240010789. [Accessed 13 December 2022].

[51] Martín S, Pérez A, Aldecoa C. Sepsis and immunosenescence in the elderly patient: a review. Front Med 2017;4. https://doi.org/10.3389/fmed.2017.00020.

[52] Monneret G, Gossez M, Venet F. Sepsis and immunosenescence: closely associated in a vicious circle. Aging Clin Exp Res 2021;33(3):729—32. https://doi.org/10.1007/s40520-019-01350-z.

[53] Glynn JR, Moss PAH. Systematic analysis of infectious disease outcomes by age shows lowest severity in school-age children. Sci Data 2020; 7(1):329. https://doi.org/10.1038/s41597-020-00668-y.

[54] Gompertz B. XXIV. On the nature of the function expressive of the law of human mortality, and on a new mode of determining the value of life contingencies. In a letter to Francis Baily, Esq. F. R. S. &c. Phil Trans R Soc. 1825;115:513—83. https://doi.org/10.1098/rstl.1825.0026.

[55] Kline KA, Bowdish DM. Infection in an aging population. Curr Opin Microbiol 2016;29:63—7. https://doi.org/10.1016/j.mib.2015.11.003.

[56] Thomas R, Wang W, Su DM. Contributions of age-related thymic involution to immunosenescence and inflammaging. Immun Ageing 2020; 17(1):2. https://doi.org/10.1186/s12979-020-0173-8.

[57] Watad A, Bragazzi NL, Adawi M, et al. Autoimmunity in the elderly: insights from basic science and clinics — a mini-review. Gerontology 2017;63(6):515—23. https://doi.org/10.1159/000478012.

[58] Eun Y, Jeon KH, Han K, et al. Menopausal factors and risk of seropositive rheumatoid arthritis in postmenopausal women: a nationwide cohort study of 1.36 million women. Sci Rep 2020;10(1):20793. https://doi.org/10.1038/s41598-020-77841-1.

[59] Manji N, Carr-Smith JD, Boelaert K, et al. Influences of age, gender, smoking, and family history on autoimmune thyroid disease phenotype. J Clin Endocrinol Metab 2006;91(12):4873—80. https://doi.org/10.1210/jc.2006-1402.

[60] White MC, Holman DM, Boehm JE, Peipins LA, Grossman M, Jane Henley S. Age and cancer risk. Am J Prev Med 2014;46(3):S7—15. https://doi.org/10.1016/j.amepre.2013.10.029.

[61] Fossel M, Whittemore K. Telomerase and cancer: a complex relationship. OBM Geriatrics 2021;5(1). https://doi.org/10.21926/obm.geriatr.2101156.

[62] Cazzola M. Myelodysplastic syndromes. N Engl J Med 2020;383(14):1358—74. https://doi.org/10.1056/NEJMra1904794.

[63] Jaiswal S, Ebert BL. Clonal hematopoiesis in human aging and disease. Science 2019;366(6465):eaan4673. https://doi.org/10.1126/science.aan4673.

[64] Whiteside TL, Demaria S, Rodriguez-Ruiz ME, Zarour HM, Melero I. Emerging opportunities and challenges in cancer immunotherapy. Clin Cancer Res 2016;22(8):1845—55. https://doi.org/10.1158/1078-0432.CCR-16-0049.

[65] Haycock PC, Heydon EE, Kaptoge S, Butterworth AS, Thompson A, Willeit P. Leucocyte telomere length and risk of cardiovascular disease: systematic review and meta-analysis. BMJ 2014;349:g4227. https://doi.org/10.1136/bmj.g4227.

[66] D'Mello MJJ, Ross SA, Briel M, Anand SS, Gerstein H, Paré G. Association between shortened leukocyte telomere length and cardiometabolic outcomes: systematic review and meta-analysis. Circ Cardiovasc Genet 2015;8(1):82—90. https://doi.org/10.1161/CIRCGENETICS.113.000485.

[67] Fossel M, Bean J, Khera N, Kolonin MG. A unified model of age-related cardiovascular disease. Biology 2022;11(12):1768. https://doi.org/10.3390/biology11121768.

[68] Tamura Y, Takubo K, Aida J, Araki A, Ito H. Telomere attrition and diabetes mellitus. Geriatr Gerontol Int 2016;16(1):66—74. https://doi.org/10.1111/ggi.12738.

[69] Ridout KK, Ridout SJ, Price LH, Sen S, Tyrka AR. Depression and telomere length: a meta-analysis. J Affect Disord 2016;191:237—47. https://doi.org/10.1016/j.jad.2015.11.052.

[70] Schutte NS, Malouff JM. The association between depression and leukocyte telomere length: a meta-analysis. Depress Anxiety 2015;32(4): 229—38. https://doi.org/10.1002/da.22351.

[71] Lin PY, Huang YC, Hung CF. Shortened telomere length in patients with depression: a meta-analytic study. J Psychiatr Res 2016;76:84—93. https://doi.org/10.1016/j.jpsychires.2016.01.015.

[72] Hackenhaar FS, Josefsson M, Adolfsson AN, et al. Short leukocyte telomeres predict 25-year Alzheimer's disease incidence in non-APOE ε4-carriers. Alzheimers Res Ther 2021;13(1):130. https://doi.org/10.1186/s13195-021-00871-y.

[73] Fossel M. A unified model of dementias and age-related neurodegeneration. Alzheimers Dement. 2020;16(2):365—83. https://doi.org/10.1002/alz.12012.

[74] Fossel M. Use of telomere length as a biomarker for aging and age-related disease. Curr Tran Geriatr Exp Gerontol Rep 2012;1(2):121—7. https://doi.org/10.1007/s13670-012-0013-6.

[75] Blackburn EH, Epel ES, Lin J. Human telomere biology: a contributory and interactive factor in aging, disease risks, and protection. Science 2015;350(6265):1193–8. https://doi.org/10.1126/science.aab3389.

[76] Hemann MT, Strong MA, Hao LY, Greider CW. The shortest telomere, not average telomere length, is critical for cell viability and chromosome stability. Cell 2001;107(1):67–77. https://doi.org/10.1016/S0092-8674(01)00504-9.

[77] Weng N ping. Telomere and adaptive immunity. Mech Ageing Dev 2008;129(1–2):60–6. https://doi.org/10.1016/j.mad.2007.11.005.

[78] Hayflick L. The limited in vitro lifetime of human diploid cell strains. Exp Cell Res 1965;37(3):614–36. https://doi.org/10.1016/0014-4827(65)90211-9.

[79] Hiyama E, Hiyama K. Telomere and telomerase in stem cells. Br J Cancer 2007;96(7):1020–4. https://doi.org/10.1038/sj.bjc.6603671.

[80] Schultz MB, Sinclair DA. When stem cells grow old: phenotypes and mechanisms of stem cell aging. Development 2016;143(1):3–14. https://doi.org/10.1242/dev.130633.

[81] Brümmendorf TH, Balabanov S. Telomere length dynamics in normal hematopoiesis and in disease states characterized by increased stem cell turnover. Leukemia 2006;20(10):1706–16. https://doi.org/10.1038/sj.leu.2404339.

[82] Drummond MW, Balabanov S, Holyoake TL, Brummendorf TH. Concise review: telomere biology in normal and leukemic hematopoietic stem cells. Stem Cell 2007;25(8):1853–61. https://doi.org/10.1634/stemcells.2007-0057.

[83] Shepherd BE, Guttorp P, Lansdorp PM, Abkowitz JL. Estimating human hematopoietic stem cell kinetics using granulocyte telomere lengths. Exp Hematol 2004;32(11):1040–50. https://doi.org/10.1016/j.exphem.2004.07.023.

[84] Kimura M, Gazitt Y, Cao X, Zhao X, Lansdorp PM, Aviv A. Synchrony of telomere length among hematopoietic cells. Exp Hematol 2010;38(10):854–9. https://doi.org/10.1016/j.exphem.2010.06.010.

[85] Sidorov I, Kimura M, Yashin A, Aviv A. Leukocyte telomere dynamics and human hematopoietic stem cell kinetics during somatic growth. Exp Hematol 2009;37(4):514–24. https://doi.org/10.1016/j.exphem.2008.11.009.

[86] Anderson JJ, Susser E, Arbeev KG, et al. Telomere-length dependent T-cell clonal expansion: a model linking ageing to COVID-19 T-cell lymphopenia and mortality. EBioMedicine 2022;78:103978. https://doi.org/10.1016/j.ebiom.2022.103978.

[87] Benetos A, Lai TP, Toupance S, et al. The nexus between telomere length and lymphocyte count in seniors hospitalized with COVID-19. J Gerontol Series A 2021;76(8):e97–101. https://doi.org/10.1093/gerona/glab026.

[88] Buchkovich KJ, Greider CW. Telomerase regulation during entry into the cell cycle in normal human T cells. Mol Biol Cell 1996;7(9):1443–54.

[89] de Punder K, Heim C, Wadhwa PD, Entringer S. Stress and immunosenescence: the role of telomerase. Psychoneuroendocrinology 2019;101: 87–100. https://doi.org/10.1016/j.psyneuen.2018.10.019.

[90] Patrick M, Weng NP. Expression and regulation of telomerase in human T cell differentiation, activation, aging and diseases. Cell Immunol 2019;345:103989. https://doi.org/10.1016/j.cellimm.2019.103989.

[91] Effros RB, Dagarag M, Valenzuela HF. In vitro senescence of immune cells. Exp Gerontol 2003;38(11–12):1243–9. https://doi.org/10.1016/j.exger.2003.09.004.

[92] Henson SM, Franzese O, Macaulay R, et al. KLRG1 signaling induces defective Akt (ser473) phosphorylation and proliferative dysfunction of highly differentiated CD8+ T cells. Blood 2009;113(26):6619–28. https://doi.org/10.1182/blood-2009-01-199588.

[93] Dedeoglu B, de Weerd AE, Huang L, et al. Lymph node and circulating T cell characteristics are strongly correlated in end-stage renal disease patients, but highly differentiated T cells reside within the circulation. Clin Exp Immunol 2017;188(2):299–310. https://doi.org/10.1111/cei.12934.

[94] Plunkett FJ, Franzese O, Finney HM, et al. The loss of telomerase activity in highly differentiated CD8+CD28-CD27- T cells is associated with decreased Akt (Ser473) phosphorylation. J Immunol 2007;178(12):7710–9. https://doi.org/10.4049/jimmunol.178.12.7710.

[95] Valenzuela HF, Effros RB. Divergent telomerase and CD28 expression patterns in human CD4 and CD8 T cells following repeated encounters with the same antigenic stimulus. Clin Immunol 2002;105(2):117–25.

[96] Effros RB. Kleemeier Award Lecture 2008—the canary in the coal mine: telomeres and human healthspan. J Gerontol A Biol Sci Med Sci 2009; 64(5):511–5. https://doi.org/10.1093/gerona/glp001.

[97] Parish ST, Wu JE, Effros RB. Modulation of T lymphocyte replicative senescence via TNF-{alpha} inhibition: role of caspase-3. J Immunol 2009;182(7):4237–43. https://doi.org/10.4049/jimmunol.0803449.

[98] Lin J, Epel E, Cheon J, et al. Analyses and comparisons of telomerase activity and telomere length in human T and B cells: insights for epidemiology of telomere maintenance. J Immunol Methods 2010;352(1–2):71–80. https://doi.org/10.1016/j.jim.2009.09.012.

[99] Lanna A, Vaz B, D'Ambra C, et al. An intercellular transfer of telomeres rescues T cells from senescence and promotes long-term immunological memory. Nat Cell Biol 2022;24(10):1461–74. https://doi.org/10.1038/s41556-022-00991-z.

[100] Sahin E, Depinho RA. Linking functional decline of telomeres, mitochondria and stem cells during ageing. Nature 2010;464(7288):520–8. https://doi.org/10.1038/nature08982.

[101] Hiyama K, Hirai Y, Kyoizumi S, et al. Activation of telomerase in human lymphocytes and hematopoietic progenitor cells. J Immunology 1995;155(8):3711–5.

[102] Lin Y, Damjanovic A, Metter EJ, et al. Age-associated telomere attrition of lymphocytes in vivo is co-ordinated with changes in telomerase activity, composition of lymphocyte subsets and health conditions. Clinical science 2015;128(6):367–77. https://doi.org/10.1042/CS20140481.

[103] Benko AL, Olsen NJ, Kovacs WJ. Estrogen and telomerase in human peripheral blood mononuclear cells. Mol Cell Endocrinol 2012; 364(1–2):83–8. https://doi.org/10.1016/j.mce.2012.08.012.

[104] Thewissen M, Linsen L, Geusens P, Raus J, Stinissen P. Impaired activation-induced telomerase activity in PBMC of early but not chronic rheumatoid arthritis patients. Immunol Lett 2005;100(2):205–10. https://doi.org/10.1016/j.imlet.2005.03.007.

[105] de Punder K, Heim C, Przesdzing I, Wadhwa PD, Entringer S. Characterization in humans of in vitro leucocyte maximal telomerase activity capacity and association with stress. Philos Trans R Soc Lond B Biol Sci 2018;373(1741):20160441. https://doi.org/10.1098/rstb.2016.0441.

[106] Huang EE, Tedone E, O'Hara R, et al. The maintenance of telomere length in CD28+ T cells during T lymphocyte stimulation. Sci Rep 2017; 7(1):6785. https://doi.org/10.1038/s41598-017-05174-7.

[107] Liu K, Hodes RJ, Weng N. Cutting edge: telomerase activation in human T lymphocytes does not require increase in telomerase reverse transcriptase (hTERT) protein but is associated with hTERT phosphorylation and nuclear translocation. J Immunol 2001;166(8):4826–30.

[108] Haendeler J, Hoffmann J, Rahman S, Zeiher AM, Dimmeler S. Regulation of telomerase activity and anti-apoptotic function by protein-protein interaction and phosphorylation. FEBS Lett 2003;536(1−3):180−6. https://doi.org/10.1016/s0014-5793(03)00058-9.

[109] Kawauchi K, Ihjima K, Yamada O. IL-2 increases human telomerase reverse transcriptase activity transcriptionally and posttranslationally through phosphatidylinositol 3′-kinase/Akt, heat shock protein 90, and mammalian target of rapamycin in transformed NK cells. J Immunol 2005;174(9):5261−9. https://doi.org/10.4049/jimmunol.174.9.5261.

[110] Sundin T, Hentosh P. InTERTesting association between telomerase, mTOR and phytochemicals. Expert Rev Mol Med 2012;14:e8. https://doi.org/10.1017/erm.2012.1.

[111] Miwa S, Saretzki G. Telomerase and mTOR in the brain: the mitochondria connection. Neural Regen Res 2017;12(3):358−61. https://doi.org/10.4103/1673-5374.202922.

[112] Gazzaniga FS, Blackburn EH. An antiapoptotic role for telomerase RNA in human immune cells independent of telomere integrity or telomerase enzymatic activity. Blood 2014;124(25):3674−84. https://doi.org/10.1182/blood-2014-06-582254.

[113] Liu H, Yang Y, Ge Y, Liu J, Zhao Y. TERC promotes cellular inflammatory response independent of telomerase. Nucleic Acids Res 2019;47(15):8084−95. https://doi.org/10.1093/nar/gkz584.

[114] Rath E, Moschetta A, Haller D. Mitochondrial function — gatekeeper of intestinal epithelial cell homeostasis. Nat Rev Gastroenterol Hepatol 2018;15(8):497−516. https://doi.org/10.1038/s41575-018-0021-x.

[115] Yan D, Zeng L, Song Z. Mitochondrial DNA: distribution, mutations, and elimination. Cells 2019;8(4):379. https://doi.org/10.3390/cells8040379.

[116] Byrne E, Dennett X, Trounce I. Oxidative energy failure in post-mitotic cells: a major factor in senescence. Rev Neurol (Paris) 1991;147(6−7):532−5.

[117] Saretzki G. Telomerase, mitochondria and oxidative stress. Exp Gerontol 2009;44(8):485−92. https://doi.org/10.1016/j.exger.2009.05.004.

[118] Jeng JY, Yeh TS, Lee JW, Lin SH, Fong TH, Hsieh RH. Maintenance of mitochondrial DNA copy number and expression are essential for preservation of mitochondrial function and cell growth. J Cell Biochem 2008;103(2):347−57. https://doi.org/10.1002/jcb.21625.

[119] Gadaleta MN, Petruzzella V, Daddabbo L, et al. Mitochondrial DNA transcription and translation in aged rat: effect of acetyl-L-carnitinea. Ann N Y Acad Sci 1994;717(1):150−60. https://doi.org/10.1111/j.1749-6632.1994.tb12082.x.

[120] Gadaleta MN, Petruzzella V, Renis M, Fracasso F, Cantatore P. Reduced transcription of mitochondrial DNA in the senescent rat. Tissue dependence and effect of l-carnitine. Eur J Biochem 1990;187(3):501−6. https://doi.org/10.1111/j.1432-1033.1990.tb15331.x.

[121] Mengel-From J, Thinggaard M, Dalgård C, Kyvik KO, Christensen K, Christiansen L. Mitochondrial DNA copy number in peripheral blood cells declines with age and is associated with general health among elderly. Hum Genet 2014;133(9):1149−59. https://doi.org/10.1007/s00439-014-1458-9.

[122] Luo J, Noordam R, Jukema JW, et al. Low leukocyte mitochondrial DNA abundance drives atherosclerotic cardiovascular diseases: cohort and Mendelian randomization study. Cardiovasc Res 2023;119(4):98−1007. https://doi.org/10.1093/cvr/cvac182.

[123] Pyle A, Anugrha H, Kurzawa-Akanbi M, Yarnall A, Burn D, Hudson G. Reduced mitochondrial DNA copy number is a biomarker of Parkinson's disease. Neurobiol Aging 2016;38:216−216.e10. https://doi.org/10.1016/j.neurobiolaging.2015.10.033.

[124] O'Hara R, Tedone E, Ludlow A, et al. Quantitative mitochondrial DNA copy number determination using droplet digital PCR with single-cell resolution. Genome Res 2019;29(11):1878−88. https://doi.org/10.1101/gr.250480.119.

[125] Kageyama Y, Kasahara T, Kato M, et al. The relationship between circulating mitochondrial DNA and inflammatory cytokines in patients with major depression. J Affect Disord 2018;233:15−20. https://doi.org/10.1016/j.jad.2017.06.001.

[126] Hummel EM, Piovesan K, Berg F, et al. Mitochondrial DNA as a marker for treatment-response in post-traumatic stress disorder. Psychoneuroendocrinology 2023;148:105993. https://doi.org/10.1016/j.psyneuen.2022.105993.

[127] Angrand L, Boukouaci W, Lajnef M, et al. Low peripheral mitochondrial DNA copy number during manic episodes of bipolar disorders is associated with disease severity and inflammation. Brain Behav Immun 2021;98:349−56. https://doi.org/10.1016/j.bbi.2021.09.003.

[128] Wang D, Li Z, Liu W, et al. Differential mitochondrial DNA copy number in three mood states of bipolar disorder. BMC Psychiatr 2018;18(1):149. https://doi.org/10.1186/s12888-018-1717-8.

[129] He Y, Tang J, Li Z, et al. Leukocyte mitochondrial DNA copy number in blood is not associated with major depressive disorder in young adults. PLoS ONE 2014;9(5):e96869. https://doi.org/10.1371/journal.pone.0096869.

[130] Sahin E, Colla S, Liesa M, et al. Telomere dysfunction induces metabolic and mitochondrial compromise. Nature 2011;470(7334):359−65. https://doi.org/10.1038/nature09787.

[131] Haendeler J, Hoffmann J, Diehl JF, et al. Antioxidants inhibit nuclear export of telomerase reverse transcriptase and delay replicative senescence of endothelial cells. Circ Res 2004;94(6):768−75. https://doi.org/10.1161/01.RES.0000121104.05977.F3.

[132] Haendeler J, Drose S, Buchner N, et al. Mitochondrial telomerase reverse transcriptase binds to and protects mitochondrial DNA and function from damage. Arterioscler Thromb Vasc Biol 2009;29(6):929−35. https://doi.org/10.1161/ATVBAHA.109.185546.

[133] Zheng Q, Liu P, Gao G, et al. Mitochondrion-processed TERC regulates senescence without affecting telomerase activities. Protein Cell 2019;10(9):631−48. https://doi.org/10.1007/s13238-019-0612-5.

[134] Salmon AB, Richardson A, Pérez VI. Update on the oxidative stress theory of aging: does oxidative stress play a role in aging or healthy aging? Free Radic Biol Med 2010;48(5):642−55. https://doi.org/10.1016/j.freeradbiomed.2009.12.015.

[135] Harman D. Aging: a theory based on free radical and radiation chemistry. J Gerontol 1956;11(3):298−300. https://doi.org/10.1093/geronj/11.3.298.

[136] Rivera-Munoz P, Malivert L, Derdouch S, et al. DNA repair and the immune system: from V(D)J recombination to aging lymphocytes. Eur J Immunol 2007;37(S1):S71−82. https://doi.org/10.1002/eji.200737396.

[137] Rossi DJ, Bryder D, Seita J, Nussenzweig A, Hoeijmakers J, Weissman IL. Deficiencies in DNA damage repair limit the function of haematopoietic stem cells with age. Nature 2007;447(7145):725−9. https://doi.org/10.1038/nature05862.

[138] Nijnik A, Woodbine L, Marchetti C, et al. DNA repair is limiting for haematopoietic stem cells during ageing. Nature 2007;447(7145):686−90. https://doi.org/10.1038/nature05875.

[139] Armstrong E, Boonekamp J. Does oxidative stress shorten telomeres in vivo? A meta-analysis. Ageing Res Rev 2023;85:101854. https://doi.org/10.1016/j.arr.2023.101854.

[140] de Lange T. Shelterin: the protein complex that shapes and safeguards human telomeres. Genes Dev 2005;19(18):2100—10. https://doi.org/10.1101/gad.1346005.

[141] Hewitt G, Jurk D, Marques FDM, et al. Telomeres are favoured targets of a persistent DNA damage response in ageing and stress-induced senescence. Nat Commun 2012;3(1):708. https://doi.org/10.1038/ncomms1708.

[142] Birch J, Gil J. Senescence and the SASP: many therapeutic avenues. Genes Dev 2020;34(23—24):1565—76. https://doi.org/10.1101/gad.343129.120.

[143] Horvath S, Raj K. DNA methylation-based biomarkers and the epigenetic clock theory of ageing. Nat Rev Genet 2018;19(6):371—84. https://doi.org/10.1038/s41576-018-0004-3.

[144] Sun D, Luo M, Jeong M, et al. Epigenomic profiling of young and aged HSCs reveals concerted changes during aging that reinforce self-renewal. Cell Stem Cell 2014;14(5):673—88. https://doi.org/10.1016/j.stem.2014.03.002.

[145] Adwan-Shekhidem H, Atzmon G. The epigenetic regulation of telomere maintenance in aging. In: Epigenetics of aging and longevity. Elsevier; 2018. p. 119—36. https://doi.org/10.1016/B978-0-12-811060-7.00005-X.

[146] Lu AT, Xue L, Salfati EL, et al. GWAS of epigenetic aging rates in blood reveals a critical role for TERT. Nat Commun 2018;9(1):387. https://doi.org/10.1038/s41467-017-02697-5.

[147] Is longevity determined by genetics?. MedlinePlus Genetics. https://medlineplus.gov/genetics/understanding/traits/longevity/. [Accessed 12 December, 2022].

[148] Grill S, Nandakumar J. Molecular mechanisms of telomere biology disorders. J Biol Chem 2021;296:100064. https://doi.org/10.1074/jbc.REV120.014017.

[149] Revy P, Kannengiesser C, Bertuch AA. Genetics of human telomere biology disorders. Nat Rev Genet 2023;24(2):86—108. https://doi.org/10.1038/s41576-022-00527-z.

[150] Gensous N, Bacalini MG, Franceschi C, Garagnani P. Down syndrome, accelerated aging and immunosenescence. Semin Immunopathol 2020;42(5):635—45. https://doi.org/10.1007/s00281-020-00804-1.

[151] Nakamura KI, Ishikawa N, Izumiyama N, et al. Telomere lengths at birth in trisomies 18 and 21 measured by Q-FISH. Gene 2014;533(1):199—207. https://doi.org/10.1016/j.gene.2013.09.086.

[152] Bhattacharya M, Bhaumik P, Ghosh P, Majumder P, Kumar Dey S. Telomere length inheritance and shortening in trisomy 21. Fetal Pediatr Pathol 2020;39(5):390—400. https://doi.org/10.1080/15513815.2019.1661049.

[153] Vaziri H, Schächter F, Uchida I, et al. Loss of telomeric DNA during aging of normal and trisomy 21 human lymphocytes. Am J Hum Genet 1993;52(4):661—7.

[154] Ahmed MS, Ikram S, Bibi N, Mir A. Hutchinson-gilford progeria syndrome: a premature aging disease. Mol Neurobiol 2018;55(5):4417—27. https://doi.org/10.1007/s12035-017-0610-7.

[155] Fossel M. Human aging and progeria. J Pediatr Endocrinol Metab 2000;13(s2):1477—82. https://doi.org/10.1515/jpem-2000-s622.

[156] Fossel M. Cells, aging, and human disease. Oxford University Press; 2004.

[157] Ford BN, Savitz J. Effect of cytomegalovirus on the immune system: implications for aging and mental health. In: Current topics in behavioral neurosciences. Springer Berlin Heidelberg; 2022. https://doi.org/10.1007/7854_2022_376.

[158] Bellon M, Nicot C. Telomere dynamics in immune senescence and exhaustion triggered by chronic viral infection. Viruses 2017;9(10):E289. https://doi.org/10.3390/v9100289.

[159] Moss P. "From immunosenescence to immune modulation": a re-appraisal of the role of cytomegalovirus as major regulator of human immune function. Med Microbiol Immunol 2019;208(3—4):271—80. https://doi.org/10.1007/s00430-019-00612-x.

[160] Satra M, Dalekos GN, Kollia P, Vamvakopoulos N, Tsezou A. Telomerase reverse transcriptase mRNA expression in peripheral lymphocytes of patients with chronic HBV and HCV infections. J Viral Hepat 2005;12(5):488—93. https://doi.org/10.1111/j.1365-2893.2005.00550.x.

[161] Akbar AN, Vukmanovic-Stejic M. Telomerase in T lymphocytes: use it and lose it? J Immunol 2007;178(11):6689—94. https://doi.org/10.4049/jimmunol.178.11.6689.

[162] Dowd JB, Bosch JA, Steptoe A, et al. Cytomegalovirus is associated with reduced telomerase activity in the Whitehall II cohort. Exp Gerontol 2013;48(4):385—90. https://doi.org/10.1016/j.exger.2013.01.016.

[163] Lanna A, Coutavas E, Levati L, et al. IFN-α inhibits telomerase in human CD8$^+$ T cells by both hTERT downregulation and induction of p38 MAPK signaling. J Immunol 2013;191(7):3744—52. https://doi.org/10.4049/jimmunol.1301409.

[164] Danese A, Baldwin JR. Hidden wounds? Inflammatory links between childhood trauma and psychopathology. Annu Rev Psychol 2017;68:517—44. https://doi.org/10.1146/annurev-psych-010416-044208.

[165] Merz MP, Turner JD. Is early life adversity a trigger towards inflammageing? Exp Gerontol 2021;150:111377. https://doi.org/10.1016/j.exger.2021.111377.

[166] Oliveira BS, Zunzunegui MV, Quinlan J, Fahmi H, Tu MT, Guerra RO. Systematic review of the association between chronic social stress and telomere length: a life course perspective. Ageing Res Rev 2016;26:37—52. https://doi.org/10.1016/j.arr.2015.12.006.

[167] Rentscher KE, Carroll JE, Mitchell C. Psychosocial stressors and telomere length: a current review of the science. Annu Rev Public Health 2020;41:223—45. https://doi.org/10.1146/annurev-publhealth-040119-094239.

[168] Boeck C, Gumpp AM, Calzia E, et al. The association between cortisol, oxytocin, and immune cell mitochondrial oxygen consumption in postpartum women with childhood maltreatment. Psychoneuroendocrinology 2018;96:69—77. https://doi.org/10.1016/j.psyneuen.2018.05.040.

[169] Boeck C, Koenig AM, Schury K, et al. Inflammation in adult women with a history of child maltreatment: the involvement of mitochondrial alterations and oxidative stress. Mitochondrion 2016;30:197—207. https://doi.org/10.1016/j.mito.2016.08.006.

[170] Hamlat EJ, Prather AA, Horvath S, Belsky J, Epel ES. Early life adversity, pubertal timing, and epigenetic age acceleration in adulthood. Dev Psychobiol 2021;63(5):890—902. https://doi.org/10.1002/dev.22085.

[171] Tang R, Howe LD, Suderman M, Relton CL, Crawford AA, Houtepen LC. Adverse childhood experiences, DNA methylation age acceleration, and cortisol in UK children: a prospective population-based cohort study. Clin Epigenetics 2020;12(1):55. https://doi.org/10.1186/s13148-020-00844-2.

[172] Rampersaud R, Protsenko E, Yang R, et al. Dimensions of childhood adversity differentially affect biological aging in major depression. Transl Psychiatry 2022;12(1):431. https://doi.org/10.1038/s41398-022-02198-0.

[173] Epel ES, Blackburn EH, Lin J, et al. Accelerated telomere shortening in response to life stress. Proc Natl Acad Sci U S A 2004;101(49):17312—5. https://doi.org/10.1073/pnas.0407162101.

[174] Rentscher KE, Carroll JE, Repetti RL, Cole SW, Reynolds BM, Robles TF. Chronic stress exposure and daily stress appraisals relate to biological aging marker p16INK4a. Psychoneuroendocrinology 2019;102:139—48. https://doi.org/10.1016/j.psyneuen.2018.12.006.

[175] Entringer S, Epel ES, Kumsta R, et al. Stress exposure in intrauterine life is associated with shorter telomere length in young adulthood. Proc Natl Acad Sci U S A 2011;108(33):E513—8. https://doi.org/10.1073/pnas.1107759108.

[176] Révész D, Milaneschi Y, Terpstra EM, Penninx BWJH. Baseline biopsychosocial determinants of telomere length and 6-year attrition rate. Psychoneuroendocrinology 2016;67:153—62. https://doi.org/10.1016/j.psyneuen.2016.02.007.

[177] Schutte NS, Malouff JM, Keng SL. Meditation and telomere length: a meta-analysis. Psychol Health 2020;35(8):901—15. https://doi.org/10.1080/08870446.2019.1707827.

[178] Gautam S, Kumar U, Kumar M, Rana D, Dada R. Yoga improves mitochondrial health and reduces severity of autoimmune inflammatory arthritis: a randomized controlled trial. Mitochondrion 2021;58:147—59. https://doi.org/10.1016/j.mito.2021.03.004.

[179] Kaliman P. Epigenetics and meditation. Curr Opin Psychol 2019;28:76—80. https://doi.org/10.1016/j.copsyc.2018.11.010.

[180] Roberts SB, Rosenberg I. Nutrition and aging: changes in the regulation of energy metabolism with aging. Physiol Rev 2006;86(2):651—67. https://doi.org/10.1152/physrev.00019.2005.

[181] Dorling JL, Martin CK, Redman LM. Calorie restriction for enhanced longevity: the role of novel dietary strategies in the present obesogenic environment. Ageing Res Rev 2020;64:101038. https://doi.org/10.1016/j.arr.2020.101038.

[182] Fontana L. Aging, adiposity, and calorie restriction. JAMA 2007;297(9):986. https://doi.org/10.1001/jama.297.9.986.

[183] McCay CM, Crowell MF, Maynard LA. The effect of retarded growth upon the length of life span and upon the ultimate body size. J Nutr 1935;10(1):63—79. https://doi.org/10.1093/jn/10.1.63.

[184] White MJ, Beaver CM, Goodier MR, et al. Calorie restriction attenuates terminal differentiation of immune cells. Front Immunol 2016;7:667. https://doi.org/10.3389/fimmu.2016.00667.

[185] Zheng Q, Huang J, Wang G. Mitochondria, telomeres and telomerase subunits. Front Cell Dev Biol 2019;7:274. https://doi.org/10.3389/fcell.2019.00274.

[186] Dorling JL, van Vliet S, Huffman KM, et al. Effects of caloric restriction on human physiological, psychological, and behavioral outcomes: highlights from CALERIE phase 2. Nutr Rev 2021;79(1):98—113. https://doi.org/10.1093/nutrit/nuaa085.

[187] Tomiyama AJ, Milush JM, Lin J, et al. Long-term calorie restriction in humans is not associated with indices of delayed immunologic aging: a descriptive study. Nutr Healthy Aging 2017;4(2):147—56. https://doi.org/10.3233/NHA-160017.

[188] Kark JD, Goldberger N, Kimura M, Sinnreich R, Aviv A. Energy intake and leukocyte telomere length in young adults. Am J Clin Nutr 2012;95(2):479—87. https://doi.org/10.3945/ajcn.111.024521.

[189] Rémond D, Shahar DR, Gille D, et al. Understanding the gastrointestinal tract of the elderly to develop dietary solutions that prevent malnutrition. Oncotarget 2015;6(16):13858—98. https://doi.org/10.18632/oncotarget.4030.

[190] Calder PC, Ortega EF, Meydani SN, et al. Nutrition, immunosenescence, and infectious disease: an overview of the scientific evidence on micronutrients and on modulation of the gut microbiota. Adv Nutr 2022;13(5):S1—26. https://doi.org/10.1093/advances/nmac052.

[191] Zarei M, Zarezadeh M, Hamedi Kalajahi F, Javanbakht MH. The relationship between vitamin D and telomere/telomerase: a comprehensive review. J Frailty Aging 2021;10(1):2—9. https://doi.org/10.14283/jfa.2020.33.

[192] Shu Y, Wu M, Yang S, Wang Y, Li H. Association of dietary selenium intake with telomere length in middle-aged and older adults. Clin Nutr 2020;39(10):3086—91. https://doi.org/10.1016/j.clnu.2020.01.014.

[193] Shi H, Li X, Yu H, Shi W, Lin Y, Zhou Y. Potential effect of dietary zinc intake on telomere length: a cross-sectional study of US adults. Front Nutr 2022;9:993425. https://doi.org/10.3389/fnut.2022.993425.

[194] Fernandes G. Progress in nutritional immunology. Immunol Res 2008;40(3):244—61. https://doi.org/10.1007/s12026-007-0021-3.

[195] Heshmati J, Morvaridzadeh M, Maroufizadeh S, et al. Omega-3 fatty acids supplementation and oxidative stress parameters: a systematic review and meta-analysis of clinical trials. Pharmacol Res 2019;149:104462. https://doi.org/10.1016/j.phrs.2019.104462.

[196] Kavyani Z, Musazadeh V, Fathi S, Hossein Faghfouri A, Dehghan P, Sarmadi B. Efficacy of the omega-3 fatty acids supplementation on inflammatory biomarkers: an umbrella meta-analysis. Int Immunopharmacol 2022;111:109104. https://doi.org/10.1016/j.intimp.2022.109104.

[197] Taha AM, Shaarawy AS, Omar MM, et al. Effect of Omega-3 fatty acids supplementation on serum level of C-reactive protein in patients with COVID-19: a systematic review and meta-analysis of randomized controlled trials. J Transl Med 2022;20(1):401. https://doi.org/10.1186/s12967-022-03604-3.

[198] Farzaneh-Far R, Lin J, Epel ES, Harris WS, Blackburn EH, Whooley MA. Association of marine omega-3 fatty acid levels with telomeric aging in patients with coronary heart disease. JAMA 2010;303(3):250—7. https://doi.org/10.1001/jama.2009.2008.

[199] Ali S, Scapagnini G, Davinelli S. Effect of omega-3 fatty acids on the telomere length: a mini meta-analysis of clinical trials. Biomol Concepts 2022;13(1):25—33. https://doi.org/10.1515/bmc-2021-0024.

[200] da Silva A, Silveira BKS, Hermsdorff HHM, da Silva W, Bressan J. Effect of omega-3 fatty acid supplementation on telomere length and telomerase activity: a systematic review of clinical trials. Prostaglandins Leukot Essent Fatty Acids 2022;181:102451. https://doi.org/10.1016/j.plefa.2022.102451.

[201] Vasto S, Scapagnini G, Rizzo C, Monastero R, Marchese A, Caruso C. Mediterranean diet and longevity in Sicily: survey in a Sicani Mountains population. Rejuvenation Res 2012;15(2):184—8. https://doi.org/10.1089/rej.2011.1280.

[202] Koelman L, Egea Rodrigues C, Aleksandrova K. Effects of dietary patterns on biomarkers of inflammation and immune responses: a systematic review and meta-analysis of randomized controlled trials. Adv Nutr 2022;13(1):101—15. https://doi.org/10.1093/advances/nmab086.

[203] Clements SJ, Maijo M, Ivory K, Nicoletti C, Carding SR. Age-associated decline in dendritic cell function and the impact of mediterranean diet intervention in elderly subjects. Front Nutr 2017;4:65. https://doi.org/10.3389/fnut.2017.00065.

[204] Canudas S, Becerra-Tomás N, Hernández-Alonso P, et al. Mediterranean diet and telomere length: a systematic review and meta-analysis. Adv Nutr 2020;11(6):1544—54. https://doi.org/10.1093/advances/nmaa079.

[205] Matteini F, Mulaw MA, Florian MC. Aging of the hematopoietic stem cell niche: new tools to answer an old question. Front Immunol 2021;12:738204. https://doi.org/10.3389/fimmu.2021.738204.

[206] Simpson RJ, Guy K. Coupling aging immunity with a sedentary lifestyle: has the damage already been done?—a mini-review. Gerontology 2010;56(5):449—58. https://doi.org/10.1159/000270905.

[207] de Araújo AL, Silva LCR, Fernandes JR, Benard G. Preventing or reversing immunosenescence: can exercise be an immunotherapy? Immunotherapy 2013;5(8):879—93. https://doi.org/10.2217/imt.13.77.

[208] Walsh NP, Gleeson M, Shephard RJ, et al. Position statement. Part one: immune function and exercise. Exerc Immunol Rev 2011;17:6—63.

[209] Campbell JP, Turner JE. Debunking the myth of exercise-induced immune suppression: redefining the impact of exercise on immunological health across the lifespan. Front Immunol 2018;9:648. https://doi.org/10.3389/fimmu.2018.00648.

[210] Simpson RJ, Lowder TW, Spielmann G, Bigley AB, LaVoy EC, Kunz H. Exercise and the aging immune system. Ageing Res Rev 2012;11(3):404—20. https://doi.org/10.1016/j.arr.2012.03.003.

[211] Puterman E, Weiss J, Lin J, et al. Aerobic exercise lengthens telomeres and reduces stress in family caregivers: a randomized controlled trial — Curt Richter Award Paper 2018. Psychoneuroendocrinology 2018;98:245—52. https://doi.org/10.1016/j.psyneuen.2018.08.002.

[212] Avidan AY, Alessi C. Geriatric sleep medicine. Informa Healthcare USA; 2008.

[213] Zisapel N. Sleep and sleep disturbances: biological basis and clinical implications. Cell Mol Life Sci 2007;64(10):1174—86. https://doi.org/10.1007/s00018-007-6529-9.

[214] Besedovsky L, Lange T, Haack M. The sleep-immune crosstalk in health and disease. Physiol Rev 2019;99(3):1325—80. https://doi.org/10.1152/physrev.00010.2018.

[215] Landry GJ, Best JR, Liu-Ambrose T. Measuring sleep quality in older adults: a comparison using subjective and objective methods. Front Aging Neurosci 2015;7. https://doi.org/10.3389/fnagi.2015.00166.

[216] Tempaku PF, Mazzotti DR, Tufik S. Telomere length as a marker of sleep loss and sleep disturbances: a potential link between sleep and cellular senescence. Sleep Med 2015;16(5):559—63. https://doi.org/10.1016/j.sleep.2015.02.519.

[217] Watson NF, Badr MS, Belenky G, et al. Recommended amount of sleep for a healthy adult: a joint consensus statement of the American Academy of sleep medicine and sleep research society. Sleep 2015;38(6):843—4. https://doi.org/10.5665/sleep.4716.

[218] Sabot D, Lovegrove R, Stapleton P. The association between sleep quality and telomere length: a systematic literature review. Brain Behav Immun Health 2023;28:100577. https://doi.org/10.1016/j.bbih.2022.100577.

[219] Kannan K, Jain SK. Oxidative stress and apoptosis. Pathophysiology 2000;7(3):153—63. https://doi.org/10.1016/S0928-4680(00)00053-5.

[220] Comelli EM, Head SR, Gilmartin T, et al. A focused microarray approach to functional glycomics: transcriptional regulation of the glycome. Glycobiology 2006;16(2):117—31. https://doi.org/10.1093/glycob/cwj048.

[221] Ding N, Nie H, Sun X, et al. Human serum N-glycan profiles are age and sex dependent. Age Ageing 2011;40(5):568—75. https://doi.org/10.1093/ageing/afr084.

[222] Higai K, Aoki Y, Azuma Y, Matsumoto K. Glycosylation of site-specific glycans of α1-acid glycoprotein and alterations in acute and chronic inflammation. Biochim Biophys Acta Gen Subj 2005;1725(1):128—35. https://doi.org/10.1016/j.bbagen.2005.03.012.

[223] Debruyne EN, Vanderschaeghe D, Van Vlierberghe H, Vanhecke A, Callewaert N, Delanghe JR. Diagnostic value of the hemopexin N-glycan profile in hepatocellular carcinoma patients. Clin Chem 2010;56(5):823—31. https://doi.org/10.1373/clinchem.2009.139295.

[224] Gornik O, Lauc G. Glycosylation of serum proteins in inflammatory diseases. Dis Markers 2008;25(4—5):267—78. https://doi.org/10.1155/2008/493289.

[225] Vanhooren V, Dewaele S, Libert C, et al. Serum N-glycan profile shift during human ageing. Exp Gerontol 2010;45(10):738—43. https://doi.org/10.1016/j.exger.2010.08.009.

[226] Picard M, Prather AA, Puterman E, et al. A mitochondrial health Index sensitive to mood and caregiving stress. Biol Psychiatr 2018;84(1):9—17. https://doi.org/10.1016/j.biopsych.2018.01.012.

[227] Liskova A, Samec M, Koklesova L, Kudela E, Kubatka P, Golubnitschaja O. Mitochondriopathies as a clue to systemic disorders—analytical tools and mitigating measures in context of predictive, preventive, and personalized (3P) medicine. IJMS 2021;22(4):2007. https://doi.org/10.3390/ijms22042007.

[228] Koklesova L, Samec M, Liskova A, et al. Mitochondrial impairments in aetiopathology of multifactorial diseases: common origin but individual outcomes in context of 3P medicine. EPMA J 2021;12(1):27—40. https://doi.org/10.1007/s13167-021-00237-2.

[229] Chacko BK, Kramer PA, Ravi S, et al. The Bioenergetic Health Index: a new concept in mitochondrial translational research. Clinical Science 2014;127(6):367—73. https://doi.org/10.1042/CS20140101.

[230] Chacko BK, Zhi D, Darley-Usmar VM, Mitchell T. The Bioenergetic Health Index is a sensitive measure of oxidative stress in human monocytes. Redox Biol 2016;8:43—50. https://doi.org/10.1016/j.redox.2015.12.008.

[231] Czajka A, Ajaz S, Gnudi L, et al. Altered mitochondrial function, mitochondrial DNA and reduced metabolic flexibility in patients with diabetic nephropathy. EBioMedicine 2015;2(6):499—512. https://doi.org/10.1016/j.ebiom.2015.04.002.

[232] Horvath S. DNA methylation age of human tissues and cell types. Genome Biol 2013;14(10):R115. https://doi.org/10.1186/gb-2013-14-10-r115.

[233] Hannum G, Guinney J, Zhao L, et al. Genome-wide methylation profiles reveal quantitative views of human aging rates. Mol Cell 2013;49(2):359—67. https://doi.org/10.1016/j.molcel.2012.10.016.

[234] Levine ME, Lu AT, Quach A, et al. An epigenetic biomarker of aging for lifespan and healthspan. Aging 2018;10(4):573—91. https://doi.org/10.18632/aging.101414.

[235] Bernardes de Jesus B, Vera E, Schneeberger K, et al. Telomerase gene therapy in adult and old mice delays aging and increases longevity without increasing cancer. EMBO Mol Med 2012;4(8):691—704. https://doi.org/10.1002/emmm.201200245.

[236] Fauce SR, Jamieson BD, Chin AC, et al. Telomerase-based pharmacologic enhancement of antiviral function of human CD8+ T lymphocytes. J Immunol 2008;181(10):7400—6. https://doi.org/10.4049/jimmunol.181.10.7400.

[237] Molgora B, Bateman R, Sweeney G, et al. Functional assessment of pharmacological telomerase activators in human T cells. Cells 2013;2(1):57—66. https://doi.org/10.3390/cells2010057.

[238] Harley CB, Liu W, Blasco M, et al. A natural product telomerase activator as part of a health maintenance program. Rejuvenation Res 2011;14(1):45—56. https://doi.org/10.1089/rej.2010.1085.

[239] Tsoukalas D, Fragkiadaki P, Docea AO, et al. Discovery of potent telomerase activators: unfolding new therapeutic and anti-aging perspectives. Mol Med Rep 2019;20(4):3701—8. https://doi.org/10.3892/mmr.2019.10614.

[240] Calado RT, Yewdell WT, Wilkerson KL, et al. Sex hormones, acting on the TERT gene, increase telomerase activity in human primary hematopoietic cells. Blood 2009;114(11):2236–43. https://doi.org/10.1182/blood-2008-09-178871.
[241] Gluckman E, Broxmeyer HE, Auerbach AD, et al. Hematopoietic reconstitution in a patient with Fanconi's anemia by means of umbilical-cord blood from an HLA-identical sibling. N Engl J Med 1989;321(17):1174–8. https://doi.org/10.1056/NEJM198910263211707.
[242] Ruggeri A, Ciceri F, Gluckman E, Labopin M, Rocha V. Alternative donors hematopoietic stem cells transplantation for adults with acute myeloid leukemia: umbilical cord blood or haploidentical donors? Best Pract Res Clin Haematol 2010;23(2):207–16. https://doi.org/10.1016/j.beha.2010.06.002.
[243] Beam D, Poe MD, Provenzale JM, et al. Outcomes of unrelated umbilical cord blood transplantation for X-linked adrenoleukodystrophy. Biol Blood Marrow Transplant 2007;13(6):665–74. https://doi.org/10.1016/j.bbmt.2007.01.082.
[244] Rocha V, Gluckman E. Clinical use of umbilical cord blood hematopoietic stem cells. Biol Blood Marrow Transplant 2006;12(1):34–41. https://doi.org/10.1016/j.bbmt.2005.09.006.
[245] Lee BC, Kang I, Lee SE, et al. Human umbilical cord blood plasma alleviates age-related olfactory dysfunction by attenuating peripheral TNF-α expression. BMB Rep 2019;52(4):259–64. https://doi.org/10.5483/BMBRep.2019.52.4.124.
[246] Scudellari M. Ageing research: blood to blood. Nature 2015;517(7535):426–9. https://doi.org/10.1038/517426a.
[247] Smith SC, Zhang X, Zhang X, et al. GDF11 does not rescue aging-related pathological hypertrophy. Circ Res 2015;117(11):926–32. https://doi.org/10.1161/CIRCRESAHA.115.307527.
[248] Edgren G, Ullum H, Rostgaard K, et al. Association of donor age and sex with survival of patients receiving transfusions. JAMA Intern Med 2017;177(6):854. https://doi.org/10.1001/jamainternmed.2017.0890.
[249] Pishel I, Shytikov D, Orlova T, Peregudov A, Artyuhov I, Butenko G. Accelerated aging versus rejuvenation of the immune system in heterochronic parabiosis. Rejuvenation Res 2012;15(2):239–48. https://doi.org/10.1089/rej.2012.1331.
[250] Hofmann B. Young blood rejuvenates old bodies: a call for reflection when moving from mice to men. Transfus Med Hemother 2018;45(1):67–71. https://doi.org/10.1159/000481828.
[251] Fang EF, Lautrup S, Hou Y, et al. NAD + in aging: molecular mechanisms and translational implications. Trends Mol Med 2017;23(10):899–916. https://doi.org/10.1016/j.molmed.2017.08.001.
[252] Yaku K, Okabe K, Nakagawa T. NAD metabolism: implications in aging and longevity. Ageing Res Rev 2018;47:1–17. https://doi.org/10.1016/j.arr.2018.05.006.
[253] Mills KF, Yoshida S, Stein LR, et al. Long-term administration of nicotinamide mononucleotide mitigates age-associated physiological decline in mice. Cell Metabol 2016;24(6):795–806. https://doi.org/10.1016/j.cmet.2016.09.013.
[254] Hou Y, Lautrup S, Cordonnier S, et al. NAD$^+$ supplementation normalizes key Alzheimer's features and DNA damage responses in a new AD mouse model with introduced DNA repair deficiency. Proc Natl Acad Sci USA 2018;115(8). https://doi.org/10.1073/pnas.1718819115.
[255] Belenky P, Racette FG, Bogan KL, McClure JM, Smith JS, Brenner C. Nicotinamide riboside promotes Sir2 silencing and extends lifespan via Nrk and Urh1/Pnp1/Meu1 pathways to NAD+. Cell 2007;129(3):473–84. https://doi.org/10.1016/j.cell.2007.03.024.
[256] Navarro MN, Gómez de Las Heras MM, Mittelbrunn M. Nicotinamide adenine dinucleotide metabolism in the immune response, autoimmunity and inflammageing. Br J Pharmacol 2022;179(9):1839–56. https://doi.org/10.1111/bph.15477.
[257] Lagoumtzi SM, Chondrogianni N. Senolytics and senomorphics: natural and synthetic therapeutics in the treatment of aging and chronic diseases. Free Radic Biol Med 2021;171:169–90. https://doi.org/10.1016/j.freeradbiomed.2021.05.003.
[258] Baker DJ, Wijshake T, Tchkonia T, et al. Clearance of p16Ink4a-positive senescent cells delays ageing-associated disorders. Nature 2011;479(7372):232–6. https://doi.org/10.1038/nature10600.
[259] Mogilenko DA, Shpynov O, Andhey PS, et al. Comprehensive profiling of an aging immune system reveals clonal GZMK+ CD8+ T cells as conserved hallmark of inflammaging. Immunity 2021;54(1):99–115.e12. https://doi.org/10.1016/j.immuni.2020.11.005.
[260] Takaya K, Asou T, Kishi K. Selective elimination of senescent fibroblasts by targeting the cell surface protein ACKR3. IJMS 2022;23(12):6531. https://doi.org/10.3390/ijms23126531.
[261] Fossel M. Cell senescence, telomerase, and senolytic therapy. Obm Geriatr 2018;3(1). https://doi.org/10.21926/obm.geriatr.1901034. 1–1.

CHAPTER 8

Age-related disease: Skin

Saranya P. Wyles[1], Krishna Vyas[2], J. Roscoe Wasserburg[3], Ryeim Ansaf[4] and James L. Kirkland[5]

[1]Department of Dermatology, Mayo Clinic, Rochester, MN, United States [2]Division of Plastic Surgery, Department of Surgery, Mayo Clinic, Rochester, MN, United States [3]State University of New York, Downstate College of Medicine, Brooklyn, NY, United States [4]Colorado State University Pueblo, Pueblo, CO, United States [5]Division of General Internal Medicine, Department of Medicine, Mayo Clinic, Rochester, MN, United States

1. Geriatric dermatology

By the year 2050, adults older than 65 years will comprise one-fifth of the global population [1]. The incidence of elderly skin conditions is rising in parallel with this increase, with more than 27 million visits to dermatologists and more than 5 million new skin cancers each year, most in older adults [2]. As the population ages, the likelihood of developing skin-related disorders increase, giving rise to a medical sub-specialty called geriatric dermatology that is receiving particular attention. There are two types of skin aging: *intrinsic* aging, which includes those changes that are due to normal maturity and occur in all individuals, and *extrinsic* aging, produced by extrinsic factors such as ultraviolet (UV) light exposure, smoking, and environmental pollutants. Indeed, dermatological focus has shifted to decreasing the morbidity associated with problems of the aging skin including pruritus, or itchy skin, and xerosis, or dry skin. The central considerations for care of old persons with skin disease include life expectancy, multimorbidity, polypharmacy, function, cognition, mobility, social support, and patient preferences [3]. For example, a patient in their last year of life may not benefit from a total body skin examination. Age is a crude measure of life expectancy and should not be used as the main predictor for dermatological screening or treatment decisions. In this chapter, we explore the principles of skin aging.

2. Fountain of youth and skin health

The fabled *fountain of youth*, a spring that allegedly restores the youth of those who drink or bathe in its waters, is a recurrent theme across the test of time, appearing and reappearing in folklore and mythological stories across many cultures. Ancient Greeks were fascinated with the theme of youth. Greek historian Herodotus in the 5th century BCE wrote of a fountain of youth in the land of Macrobians, giving people of the region exceptional longevity. It is said that the 16th-century Spanish explorer and conquistador, Juan Ponce de León, was intrigued by a legend told by the Taino Indians of the Caribbean, which piqued his interest to stage multiple expeditions across the Caribbean, eventually leading to his discovery of Florida. While the *fountain of youth* was never discovered, the interest in understanding the molecular and cellular basis of aging has been a burgeoning area of intense research that continues to fascinate and inspire many, especially as related to their outward skin appearance and overall skin health.

3. Chronological versus biological skin aging

Skin is a readily accessible portal of knowledge on aging. In general, aging is a complex and multi-factorial process that leads to a progressive decline in physiological tissue function (Fig. 8.1). Fundamentally, it is an

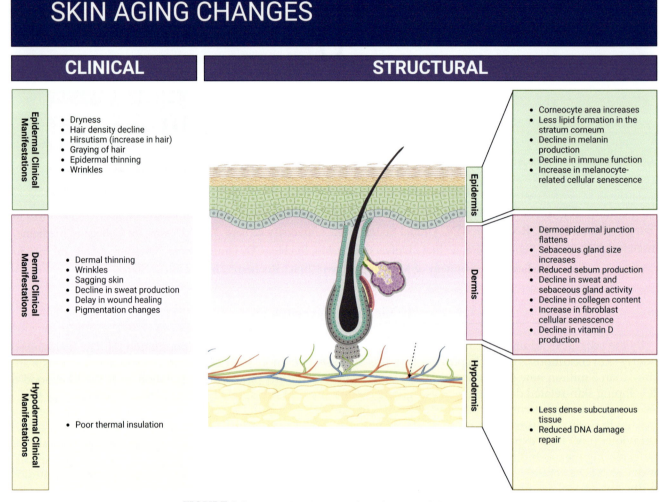

FIGURE 8.1 Clinical and structural implications of skin aging.

organism-wide loss of homeostasis and innate repair mechanisms, with an accumulation of cellular damage and a decline in regenerative capacity. The hallmarks of skin aging include genomic instability, telomere attrition, epigenetic alterations, loss of proteostasis, deregulated nutrient-sensing, mitochondrial dysfunction, stem cell exhaustion, altered intercellular communication, and cellular senescence (Fig. 8.2) [4]. The unitary theory of fundamental aging processes suggests that targeting one fundamental aging process (e.g., cellular senescence) could impact several or all other aspects of aging [5]. Given its well-documented interactions with other aging hallmarks, cellular senescence is an ideal target for pharmacological intervention [6].

Life expectancy has increased by 3 decades since the mid-20th century [7]. A matched expansion in healthspan has, however, not occurred, largely due to the pandemic of chronic diseases afflicting the elderly population. This lag in quality of life is a known challenge that calls for prioritization of disease-free longevity [8]. Age-related diseases range from cardiovascular disease to neurodegenerative disorders, osteoarthritis, diabetes, renal dysfunction, and most cancers, thereby contributing to morbidity and mortality. Understanding the hallmarks of aging has prompted further research to better understand the implications of targeting fundamental processes that contribute to skin aging.

In the field of translational gerontology, there is often a distinction between *biological* and *chronological* age. Biological age refers to epigenetic alteration and DNA methylation which quantifies cellular function and tissue ability when an individual accrues age-related diseases [9]. Chronological age refers to the actual amount of time an individual has existed. Chronological age increases at the same rate for all individuals, in contrast to biological age, which can be altered based on various factors such as genes and lifestyle choices [10]. Age-related epigenetic

FIGURE 8.2　Hallmarks of skin aging.

alterations seen in DNA methylation have been utilized to develop the molecular versus chronological age of human skin [11]. Such models can help support further investigation on how skin ages, as well as impacted by experimental treatments.

4. Clinical manifestations of skin aging

Skin is one of the most studied tissue types in aging research. Like other organs, the human skin ages and undergoes clearly distinguishable changes due to age. Specifically, skin aging is a process influenced by a combination of intrinsic (i.e., genetic) and extrinsic (i.e., UV radiation) factors. The factors contributing to skin aging are multifactorial and include genetics, gravity, photoaging, medical co-morbidities, diet, smoking or alcohol use, obesity, menopause, stress, environmental factors, and changes to tissue elasticity or underlying structural support, which cause tissue redistribution, atrophy, and blunted anatomic contours.

Skin aging reflects the cumulative, dynamic, and interdependent structural changes of the skin, soft tissues (subcutaneous fat, muscle, fascia, and ligamentous structures), and structural support (bone and teeth) of the face, which impact facial topography and skin appearance [12]. Humans consider an aged face to be associated with the descent of soft tissues, resulting in an inverted triangle, which contrasts with the "Triangle of Youth" defined by high cheekbones, full cheeks, and a well-defined jawline. The visible signs of skin aging can include deepened horizontal and vertical rhytides, temporal atrophy, enlarged bony orbital volume and loss of malar volume, volume loss and descent of the midface soft tissues, deepened nasolabial folds, and blunted cervicomental angle.

Aging skin manifests related to impaired skin barriers as seen in histology. These include change of skin constituents or change in the physiological function of collagen and matrix proteins, thinning of the epidermis, and altered homeostasis. Clinically, aged skin may present with rhytids, xerosis, elastosis, dyschromia, atrophy, loss of elasticity, sagging, fragility, telangiectasias, and dry rough skin and loss of collagen and supporting architecture, resulting in a compromised skin barrier. Accumulated DNA damage and genomic instability can trigger cellular senescence. For example, senescent fibroblasts can lose their ability to produce collagen, elastin, and fibronectin, resulting in a

150 8. Age-related disease: Skin

dysfunctional extracellular matrix; these senescent cells also secrete matrix metalloproteinase-1 and other pro-inflammatory factors which impact surrounding tissue structure and function.

Traditional approaches to address skin aging as evidenced by facial aging have emphasized the restoration of youthful facial anatomy with prevention (e.g., neuromodulators or sun protective factor [SPF]), and with skin rejuvenation and volume restoration (e.g., hyaluronic acid fillers). Pharmacologic and minimally invasive anti-aging treatments target different components of facial aging. Rejuvenation techniques range from noninvasive energy-based devices; to chemical peels and lasers to improve skin quality; to volumizing procedures with minimally invasive fillers; to neuromodulators such as botulinum neurotoxin that prevent rhytids; to surgical interventions involving fat grafting, or skin/composite resections, lifting, or tightening.

5. Regenerative aesthetics

Regenerative aesthetics is a burgeoning field within regenerative medicine aimed at recapturing and restoring youthful tissue structure and function [13]. The aging process results in a decline in tissue regenerative potential that may reflect a decrease in the number of epidermal stem cells or their functionality. Potential applications include improvement of skin texture and tone, reducing or improving the appearance of scars, and treating hair loss. Technologies include both cell-based (e.g., stem cells) and acellular therapies (e.g., extracellular vesicle or exosome therapies), or other biologically active substances to stimulate natural healing and regenerative properties (Fig. 8.3). These are emerging strategies that still warrant FDA regulation and currently require further clinical validation through rigorous clinical trials.

Stem cells are cellular regenerative therapies characterized by their self-renewal and differentiation, allowing for the ability to functionally restore tissues. For example, adipose-derived stem cells (ADSCs) and stromal vascular fraction (SVF), harvested through the process of liposuction, represent the most commonly used stem cells in aesthetics given the ease of accessibility, abundance, and sustained quality with age [14]. ADSCs represent a multipotential stem cell population that can differentiate into adipogenic, osteogenic, chondrogenic, and other mesenchymal lineages. Adipose tissue, SVF, and ADSCs have been utilized in regenerative bioaesthetics as isolated treatments or in combination with other modalities. ADSCs can stimulate fibroblasts and keratinocytes to proliferate and migrate, contributing to epidermal homeostasis and wound healing [15,16]. The anti-aging effects of ADSCs are multi-varied and may result from glycation suppression, antioxidation, and trophic effect [17]. However, potential side effects related to adult stem cells include undesired immune response or rejection, unexpected cell contamination, or

CELLULAR

Allogeneic & Autologous Cell Therapy

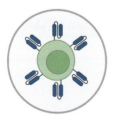
Immune-mediated Effector Cell Therapy (CAR-T)

Mesenchymal Stem Cells

Stem Cells and Matrix

CELL FREE

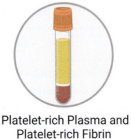
Platelet-rich Plasma and Platelet-rich Fibrin

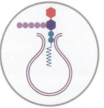
Lipid-bound Vesicles

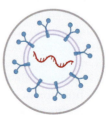
Exosomes

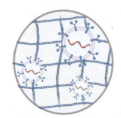
Exosomes and Matrix

FIGURE 8.3 Regenerative aesthetics toolkit.

undesired clinical effects [18]. Other sources of stem cells include umbilical cord-derived or bone marrow-derived mesenchymal stem cells, which are multipotent. Pluripotent stem cells (e.g., embryonic stem cells or bioengineered induced pluripotent stem cells) can give rise to all three germ layers (endoderm, ectoderm, or mesoderm), and are limited in their clinical use given the high-risk side-effect profile of teratoma formation.

Exosomes, or extracellular vesicles, are "cell-free" or acellular nano-sized vesicles that contain bioactive cargo with crucial roles in exerting biological activity on cells and fostering intercellular communication [19,20]. The utility of exosomes varies greatly based on source (e.g., blood products such as platelets, mesenchymal stem cells, or others). Preclinical exosome studies have targeted root causes of skin aging to improve tissue homeostasis and enhance skin quality, scar healing, fat grafting, hair loss, and as a drug-delivery platform. In some instances, exosome technology has certain advantages over adult stem cells in terms of production, stability, storage, and delivery. Yet, exosome therapy is currently challenged by heterogeneity in exosome isolation, purification, dose effects, reproducibility, scalability, and lack of high-quality clinical trials to confirm its safety, effectiveness, and long-term effects. Currently, there are no FDA-approved exosome therapies for skin rejuvenation or hair restoration.

Induced pluripotent stem cells (iPSCs) are pluripotent stem cells derived from mature, somatic cells upon ectopic expression of transcription reprogramming factors. In a study investigating UVB-photoaged dermal fibroblasts treated with iPSC-derived exosomes, there was increased viability, increased collagen 1 gene expression, and decreased expression level of senescence-associated β-galactosidase (SA-β-gal) and MMP1 and MMP3 [21]. In senescent subcultured cell lines, iPSC-derived exosomes reduced SA-beta-Gal activity and had similar effects on gene expression. Similarly, Deng et al. found that exosome therapy protected fibroblasts from UVB-associated senescence, likely through antioxidant activity [22]. This group compared exosomes derived from human dermal fibroblasts (HDF-EVs) to those derived from umbilical cord-MSCs (UC-MSC-EVs) in vitro. Exosomes were found to protect against UVB photoaging through antioxidant activity and protect cells against UVB-induced cell death and cell cycle arrest. Specifically, exosomes increased expression of glutathione peroxidase 1 and collagen 1, decreased reactive oxygen species (ROS) production, decreased pMMP-1, and promoted cell proliferation.

Photoaged skin can clinically appear as uneven and thinning epidermis with irregular pigmentation, cellular atypia, elastic fiber degeneration, loss of glycosaminoglycans, and decreased collagen matrix. Dermal fibroblasts are responsible for the synthesis of structural components such as procollagen and elastic fibers. Aging causes decreased number and proliferation of fibroblasts, resulting in reduced collagen synthesis and accelerated degradation of the skin matrix. Ultraviolet-induced DNA damage to epidermal cells called keratinocytes results in aberrant cellular proliferation. Regenerative therapies have been utilized to ameliorate the effects of photoaging in a preclinical model. For example, Xu et al. investigated the effects of ADSC-derived extracellular vesicles on mice in a UVB-induced photoaging model [23]. Subcutaneous injection of these vesicles resulted in epidermal cell proliferation, attenuated macrophage infiltration and reactive oxygen species production, resulting in reduced skin wrinkling appearance. In vitro, the extracellular vesicles increased human dermal fibroblast activity and enhanced protection from UVB-induced senescence. Furthermore, there was reduced collagen type I, increased MMP-3, decreased M0 to M1 macrophage differentiation, reduced intracellular ROS production, and improved antioxidant activity, contributing to restoration of skin dermal function.

Other factors that impair skin collagen homeostasis with aging include increases in matrix metalloproteinases (MMPs) and decreases in tissue inhibitors of metalloproteinases (TIMPs). Inflammation and ROS species work synergistically to increase MMPs that degrade the extracellular matrix. TIMPs have been shown to suppress apoptosis and enhance the proliferation of fibroblasts. Photoaging also results in interference of the TGF-β signaling pathway, which is important for cellular proliferation, differentiation, and matrix integrity by MMPs. Dermal fibroblasts are principally responsible for the synthesis of structural components such as procollagen and elastic fibers. Collectively, these changes lead to collagen and elastic fiber disorganization, causing skin tissue atrophy. Understanding the mechanisms of aging helps to guide potential targets for anti-aging therapies.

6. Cellular senescence

Cellular senescence is defined as a cell cycle arrest characterized by the exit of the cell from its replicative cycle in response to various stressors such as DNA damage, telomere shortening, and oncogenic signaling. Senescent cells stop cell division and cease their normal function yet they remain metabolically active and secrete pro-inflammatory/pro-fibrotic cytokines, growth factors, and extracellular matrix proteins, collectively known as the senescence-associated secretory phenotype (SASP). Cellular senescence is a natural phenomenon that has been evolutionarily conserved given its ability to prevent the proliferation of damaged or abnormal cells, which could lead to cancer. However, with age, the accumulation of senescent cells in various tissues and organs can have negative effects on tissue function and repair. Research has shown that senescent cells contribute to age-related diseases such as osteoarthritis. Studies in animal models have shown that elimination of senescent cells, a process called senolysis, can improve tissue function and delay the onset of age-related diseases. There are several ways to eliminate senescent cells, such as the use of senolytic drugs that selectively target senescent cells, genetic manipulations to promote cellular apoptosis, or clearance of senescent cells by the immune system. These are discussed later in the chapter.

Senescent cells are also defined by their increased resistance to apoptosis [24–26]. Cell fate may occur as continuity of the replicative cycle, apoptosis, or senescence. Specifically, in response to stress, cells may undergo apoptosis as a programmed cell death mechanism to prevent the passage of damaged DNA or organelles. Routine cell damage may result in apoptosis, in contrast, senescence is a similar response to external or internal stressors, which acts as an antitumorigenic response, but does not result in apoptosis.

The concept of cellular senescence dates to the 1960s when Hayflick and Moorhead observed the cessation of cellular replication in cells grown in vitro that stopped proliferation after several replication cycles [27]. This number of replications (the "Hayflick limit") produces the senescent phenotype in cells, which was postulated to correlate with chromosomal anomalies. Since this discovery, cellular senescence has been further characterized as a complex cell state, driven by both internal and external signals.

Internal signals of cellular senescence are multifactorial and involve overlapping pathways that produce oxidative or genotoxic stress [28,29], telomere shortening [30], mitochondrial dysfunction [29], oncogene activation, DNA damage [31–33], and resultant genomic instability [34,35]. Oxidation of DNA occurs via reactive oxygen species (ROS) (e.g., superoxide radicals, hydroxyl radicals, etc.), introducing adduct radicals which, depending on the nucleotide base, form new haphazard interactions between bases [36]. These new chemical structures interfere with DNA replication and ultimately introduce genotoxic mutations into subsequent generations of daughter cells or simply interrupt routine physiological functions in the impacted cell. Telomere shortening occurs as a natural process in aging and the lack of telomere shortening has been well described in oncogenic transformation. Telomeres shorten by a few base pairs with each round of cellular replication due to the intrinsic activity of telomerase, an enzyme that repairs and extends telomeres [37]. This telomeric shortening has been correlated to increased aging, age-related diseases, and decreased lifespan. Furthermore, mitochondrial dysfunction is mediated similarly by ROS due to normal aerobic respiration. While cells do produce antioxidants, oxidation of mitochondrial membrane and mitochondrial DNA still occurs. Mitochondrial dysfunction has been demonstrated to introduce a senescence phenotype alone [38]. In addition, mitochondrial dysfunction is itself implicated in several diseases [39–42]. As such, oxidative stress is a major driver of various hallmarks of aging.

DNA damage, a prominent contributor to the cellular senescence state, can be mediated by chemical reactions from ionizing radiation to oxidative stress; this cumulative damage and decreased ability for cell repair through nucleotide excision repair, base excision repair, or double-strand repair, creates an environment that halts cell replication. In this state, the cell is genomically unstable, leading to an increased potential for replicative error or progression to malignancy.

The senescent phenotype has been implicated in several age-related disorders, including but not limited to congestive heart failure, cerebrovascular accidents, dementia, various malignancies, diabetes, chronic lung disease, and arthritis [43–45]. This chapter will focus on the involvement of senescent cells in skin aging and age-related skin pathologies.

Skin aging occurs phenotypically as epidermal thinning, decreased elasticity, hyper-/hypopigmentation (also known as solar lentigines or sun spots), and wrinkles. Skin thinning is caused by decreased regenerative capacity due to reduced proliferation of epidermal stem cells or keratinocytes found in the stratum basalis [46,47]. Loss of skin elasticity occurs secondary to changes in the extracellular membrane, mediated by aging or senescent fibroblasts [48]. These aged skin fibroblasts contribute to an overall decrease in collagen and change from type I to type III collagen, leading to decreased elasticity and skin wrinkling [49]. This change is not limited to the skin, with several tissues throughout the body displaying similar type I to type III collagen proportion changes [50–52].

7. Function of senescent cells

The contribution of cellular senescence to life starts as early as in utero. Embryonic development occurs through an array of processes including proliferation, differentiation, migration, apoptosis, and cellular senescence. Cellular senescence occurs during normal embryogenesis to allow for proper morphogenesis. It is necessary to promote tissue remodeling, common in mammalian limb patterning [53,54]. Similarly, senescent cells are implicated in the acute wound-healing response [55]. In fact, removal of senescent cells in early damaged tissue can reduce regenerative ability [56–60]. This response varies for chronic wound healing as senescent cells are found to be inhibitory in later stages of wound healing [58,61].

Cellular senescence also occurs as a response to mitigate potential oncogenic (limitless cellular replication) processes after cellular insult. While the initial senescence response may be successful to suppress tumorigenesis and cell proliferation, the resultant phenotype presents a new challenge for the body and itself may be deleterious. The deleterious nature of cellular senescence is credited to its senescence-associated secretory phenotype (SASP). SASP is a complex microenvironment or complication of secreted factors whereby the expression of proinflammatory interleukins, chemokines, cytokines, matrix metalloproteinases, and growth factors produce localized damage to surrounding cells and contribute to the aging process [62,63]. The SASP itself has been implicated in chronic disease states, which also correlate with aging [64–66] such as idiopathic pulmonary fibrosis [67], osteoarthritis [68], posttransplant state, inflammatory bowel disease, hepatocellular carcinoma, obesity [69], and hepatitis C [70], among others. However, the correlation of senescent cells to disease state has not been proved to necessarily be causal.

Several theories exist regarding the mechanism by which senescent cells act in disparate ways, causing and alternatively preventing disease. The "threshold" theory proposed by Kirkland et al. postulates that there may be an accrual point at which the human body may be unable to resist the clinical effect of senescent cells [5,71,72]. Similarly, Tripathi et al. propose the idea of various subtypes of senescent cells including "helper" versus "deleterious" senescent cells that vary in function [73]. Support for this idea is found in the actions of macrophages expressing particular senescent cell markers following an immune stimulus [74]. There is also support for the idea that the temporality of senescent cells has a direct impact on the human body [75,76]. Senescent cells are often transiently upregulated in acute wounds [77,78]. In fact, their transient expression has been postulated as necessary for acute wound healing [77]. Additionally, the immune response to tissue damage can be seen by transient upregulation of biomarkers of cellular senescence [74,79]. Delayed acute wound healing can also be introduced through the transfer of senescent fibroblasts to young mice [75]. This finding reinforces that there is upregulation of pro-senescence genes after an immediate acute cutaneous wound in young compared to elderly humans [78]. In mice, transient suppression of p21 (a pro-senescence marker) removed the delay in acute wound healing for aged mice [80]. This is in contrast to chronic or delayed wounds where persistent expression of cellular senescence may be detrimental to the wound-healing process.

The idea of cellular senescence temporality (early vs. late senescence) is evoked by understanding the role of senescent cells in chronic wounds. Chronic and nonhealing wounds have persistently elevated levels of senescent cells [81,82]. Higher concentrations of senescent cells in wounds were proven to increase healing time [83]. Nonhealing wounds are best understood as continually stalled at a pro-inflammatory stage, which ties to the existence of the SASP role of senescent cells. By exhibiting a pro-inflammatory phenotype, senescent cells contribute to the microenvironment that prolongs this healing stage without resolution [84]. This concept was demonstrated by SASP exposure to dermal fibroblasts, resulting in a matrix-degrading phenotype and poor wound healing [85]. Chronic wounds also have higher levels of senescent cells and lower levels of collagen, an important component of healthy skin [86]. Senescent cells are also found in high quantities in diabetic ulcers, a common pathology of poor wound healing [61,87,88].

In the context of skin, senescent cells further play an important role in scarring and fibrosis. They are implicated in the formation of keloids and are investigated as potential therapeutic targets [89–91]. Senescent cells are also part of the pathological process of capsular contracture [92]. However, impaired induction of cellular senescence is linked to increased fibrosis [93]. There is also evidence that the removal of senescent cells via knockout p16 in mice can result in increased liver fibrosis compared to normal mice [94]. As such, the functionality of cellular senescence spans from cancer evasion, wound healing, and fibrosis, all depending on the temporality and totality of senescent cell accumulation.

8. Features of senescent cells

Senescent cells are morphologically unique compared to their nonsenescent counterparts. They demonstrate consistently distinct features that allow for their identification and classification [95]. Hallmarks include cellular morphology, mitochondrial changes, lysosomal changes, among others which have been credited to the underlying dysregulation of cellular metabolism. Morphologically, the cell body significantly increases in size and granularity, with certain cells demonstrating a 33% growth [96,97]. Organelles also appear to change in the senescent state. Mitochondria similarly increase in number, size, reactive oxygen species (ROS) production, and decreased membrane integrity [98,99]. Lysosomes in senescent cells also increase in number and contain elevated SA-beta-gal activity which has been described thoroughly as a hallmark of this phenotype [100–102]. Further, SA-beta-gal is used as a biomarker for the identification of senescent cells. Lipofuscin accumulation has also been described, however use of this as an identifying marker of senescent cells may not be ideal as these cells can have temporarily increased lipofuscin from the degradation of lipids [103].

9. Biochemical markers of cellular senescence

Biochemical markers of senescence have been discovered and described thoroughly in the literature [46,104]. Chiefly, senescent cells upregulate markers of cell cycle blockade allowing for unlimited replication, namely p21/CIP1, p53, p16/INK4A, and RB [99,105]. These cell cycle blockade signals are shared features between tumorigenic cells and senescent cells. There is also decreased DNA replication due to genotoxic damage and the cumulative effect of telomere shortening [105,106]. This theory is supported by evidence of DNA damage foci in telomeres, structural collapse of replication forks, and elevated levels of γ-H2AX, a sensitive marker for DNA damage [107]. Lamin B1, an intermediate filament used as a structural protein in the nucleus, is decreased independently of other structural proteins and signaling pathways in cells that underwent senescence [108]. This decrease is mediated at the level of transcription, by promoting mRNA instability. Downregulation of transcription is a common theme in senescent cells, with entire regions becoming systematically repressed through the induction of heterochromatin. This is characterized as senescence-associated heterochromatin foci (SAHF), which permanently prevent replication even when induced by mitogenic stimulation [109].

10. Senescence in skin aging

In relation to plastic surgery and the skin aging process, these characteristics and behaviors have been well described and further explored [46]. Skin aging is characterized by cellular atrophy, reduced proliferative capacity, and degradation of the cellular matrix microenvironment including collagen, elastin, and glycosaminoglycans [110,111]. This insult to the cellular matrix is noted in senescent dermal fibroblasts which display a phenotype that increases matrix metalloproteinase (MMP) expression [112]. These aged fibroblasts, compared to young fibroblasts, are implicated in the formation and continuation of the SASP by expressing MMP and pro-inflammatory markers including IL6 and IL8 (Fig. 8.4) [46,113–115]. Further, they decrease IGF1 which decreases epidermal cell proliferation and collagen expression [116]. The increased p16/INK4a expression common in senescent cells is correlated with a loss of elastic fibers, increased wrinkles, and higher perceived age [117].

11. Targeting senescent cells in the skin

For skin wound healing, targeting cellular senescence has been considered, yet there is active debate on the precise role of senescent cells during the dynamic wound-healing cascade. There seems to be a dichotomy that has developed to conceptualize their exact role and implications of their existence. Early senescent cells are protective in preventing cancer cell proliferation and their transient expression seems to play an integral role in acute wound healing [43,77]. Yet, late senescent cells, that exist in chronic nonhealing wounds, are detrimental, and found in diabetic ulcers [101,118–120].

For skin health, the role of a senotherapeutic peptide on skin aging has been explored, especially in the context of senescent fibroblasts that retard collagen production [121].

Potential strategies for targeting senescent cells are rooted in the physiological mechanisms as senescent cells do not undergo apoptosis. In tissue remodeling, embryogenesis, and development, senescent cells are routinely cleared by macrophages [53]. Indeed, clearance of senescent cells has been shown to reduce inflammation, and enhance stem cell and progenitor cell activity [122,123]. There has been vast development in the tools utilized to target senescent cells with the advent of senolytics and senomorphics [63,71,124–127]. Senolytics are substances that selectively

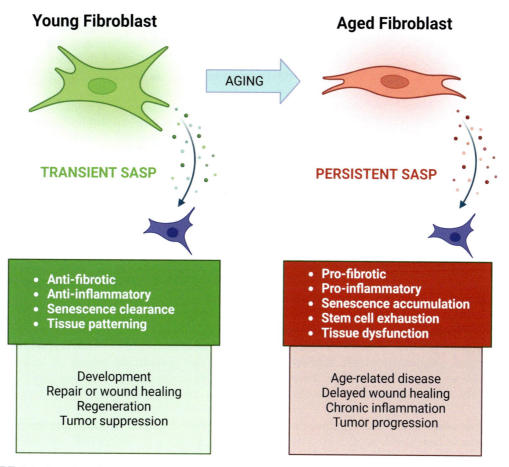

FIGURE 8.4 Sequalae of young versus aged skin fibroblasts and their senescence-associated secretory phenotype (SASP).

eliminate or reduce senescent cells, while senomorphics are substances that suppress the senescence-associated secretory phenotype (SASP) [128]. Both modalities can mitigate the negative physiological effect from cellular senescence on the skin microenvironment.

12. Senotherapeutic agents for skin

Biomarkers of cellular senescence correlate with a decline in longevity, and an accelerated aging process [129]. Elevated INK4a levels can causally induce aging. Further, their expression correlates with SA-β-galactosidase activity, conferring their holistic senescence profile. In mice, induced INK4a deficiency reduces age-related immune decline and tumorigenesis [130].

Senescent cells compensate for cellular damage and microenvironment damage, and evade apoptosis through the development of the senescent cell anti-apoptotic pathways (SCAPs). These pathways protect the cell from apoptosis and allow for survival despite a heavy burden of DNA mutations or organelle damage. This feature of senescent cells allows for greater resistance to apoptosis than cells participating in the cell cycle. It is through this SCAPs paradigm that the first senolytics were developed and tested [131,132].

13. First- and second-generation senolytics

The first generation of small-molecule drugs were designed through a mechanism-based solution, targeting the features of senescent cells and eliminating their survival. Dasatinib, a small-molecule inhibitor of the Src family kinases (SFKs), regulates cell proliferation, differentiation, and survival. SFKs are activated in various types of cancer and therefore considered therapeutic targets. Dasatinib was originally designed to treat Philadelphia chromosome-positive chronic myeloid leukemia (CML) [133]. It functions to selectively bind and thus reduce the activity of tyrosine kinase receptors related to the BCR-ABL kinase. In mice, dasatinib demonstrated the proof-of-concept in reducing the activity of tyrosine kinases and eliminating senescent cells [131]. Further, dasatinib has senolytic effects in mouse models of accelerated aging and age-related diseases, reducing the number of senescent cells in multiple tissue types, and improving overall health and lifespan [134].

Flavonoids, such as quercetin and fisetin, have also been identified as therapeutic options to reduce senescent cell burden. Quercetin is a flavonoid found naturally in certain plants, which acts as an antioxidant in the cell, reducing free radicals and thus serving to protect cells and organelles from oxidative damage. For chronic diabetic foot ulcers, combination treatment with quercetin and oleic acid reduced healing time and had no reported adverse reactions [135]. Fisetin, an analog of quercetin, can reduce senescent cell viability and senescent cell number in vitro without harming nonsenescent cells [136–141].

There has been an increasing effort to combine senolytic drugs in order to potentiate their individual effects and to reduce the chances of cellular resistance, similar to therapeutic strategies currently employed by combined antiretrovirals in HIV/AIDS and combination chemotherapy in certain cancers. For instance, dasatinib and quercetin (D + Q) senolytic combination has shown promising results including increased lifespan, treatment of diabetes, osteoporosis, neurodegeneration, pulmonary fibrosis, and cardiovascular disease in mice [72,131,142,143]. D + Q combination is efficient in downregulating the antiapoptotic pathways that senescent cells employ [144]. Other flavonoids including resveratrol, luteolin, curcumin, and procyanidin C1 (PCC1) are similarly being explored [145–153]. New senolytic agents have been discovered through library screenings of antitumorigenic agents such as piperlongumine. Piperlongumine induces apoptosis by upregulating the caspase pathway of apoptosis [154].

Targeting the natural apoptotic response of senescent cells is a common theme in senolytics, with some aiming to target BCL2-mediated cellular apoptosis. Navitoclax (ABT263) can inhibit the action of BCL-2 and BCL-XL [136,155–157]. However, navitoclax has been shown to cause a potentially devastating side effect of thrombocytopenia [158]. Similarly, BCL-X inhibitors (A1331852, A1155463) are being explored as well to target the apoptotic pathway [136,159].

There are multiple other small senolytic molecules that have been identified for potential therapeutic use. The FOXO4-related peptide, heat shock protein 90 (HSP90) inhibitors, metformin, rapamycin, and ruxolitinib have all been shown to have senotherapeutic effects in vitro. For instance, FOXO4-related peptide acts by preventing binding with p53, which prevents p53 from entering the nucleus and therefore eliminates the antiapoptotic effect [160–162]. HSP90 inhibitors, including geldanamycin, tanespimycin, and alvespimycin, act by inducing apoptosis [163]. Rapamycin inhibits mTOR signaling, suppressing pro-inflammatory cytokine mRNA expression, reducing NF-kappaB activity, and thus reducing the SASP [164]. Ruxolitinib targets the JAK pathway, a key regulator of pro-survival and pro-inflammatory phenotypes. It has been shown to reduce inflammation and frailty in mice by limiting the SASP [165].

Metformin is a biguanide that is a first-line therapy for type 2 diabetes mellitus. Research shows that metformin also has a long history of antitumorigenic and senotherapeutic effects. Metformin has been shown to target the energy-sensing enzyme AMP-activated protein kinase (AMPK), which is involved in the regulation of cell growth and metabolism. Activation of AMPK has been shown to promote the elimination of senescent cells by a process called autophagy. This has been demonstrated in a study that treated senescent human fibroblasts with metformin and found that it effectively eliminated senescent cells as well as reduced the secretion of pro-inflammatory cytokines. Other studies have found that metformin treatment in mice increased the activity of a protein called sirtuin 1 (SIRT1), which is known to play a role in aging and longevity. Furthermore, metformin treatment has improved the health and lifespan of mice. It may reduce the number of senescent cells in various tissues and improve the healthspan of mice as well as the incidence of cancer. In addition to its senolytic effects, metformin has also been shown to have antiinflammatory and antioxidant properties, which may contribute to its ability to improve skin health. In the context of the skin, metformin improved skin elasticity and reduced the appearance of wrinkles in patients with type 2 diabetes. Overall, the evidence suggests that metformin may have potential as a senotherapeutic agent, not only for skin health but also for overall healthspan. Large-scale clinical trials are ongoing to validate these findings and to determine the optimal dosage and duration of treatment as a senotherapeutic agent.

Second-generation senotherapeutics represent a new direction of therapy. While the first generation of senotherapeutics, or "senolytics," targeted and eliminated senescent cells, the second-generation senotherapeutics invoke a comprehensive approach to target aging by modulating multiple cellular pathways. These drugs aim to improve the function of cells, tissues, and organs, rather than just eliminating senescent cells. Some examples of second-generation senotherapeutics include drugs that target epigenetic mechanisms, inflammation, and metabolic pathways. These second-generation drugs are typically identified in high-throughput screening techniques [163,166]. These drugs are currently in preclinical and early-stage clinical development. Their safety and efficacy profile will be defined in the upcoming years.

There are also novel techniques to target cellular senescence, borrowed from successful treatments of other diseases, including chimeric antigen receptor (CAR) T-cell therapy, nanoparticles, vaccines, and immunomodulators.

CAR T-cell therapy is a type of immunotherapy that involves genetically engineering a patient's T cells to target and destroy cancer cells. This is done by introducing a new gene, called a chimeric antigen receptor (CAR), into the patient's T cells. The CAR is designed to recognize and bind to a specific protein found on the surface of cancer cells. Once T cells are engineered with CAR, they are infused back into the patient's body, where they can locate and destroy cancer cells that express the targeted protein. CART therapy has been studied as a treatment for a variety of solid tumors as well as leukemia and lymphoma. CAR T-cell therapy utilizes a cell surface protein (urokinase plasminogen activator receptor) expressed during cellular senescence. uPAR CAR T cells can eliminate senescent cells both in vitro and in vivo in mice [167]. While CAR T-cell therapy is currently utilized in cancer treatments, targeting senescent cells is a novel application. The major limitations of this strategy to target cells are the financial burden, potential adverse reactions including off-target effects, and immunosuppression.

Nanoparticles are another avenue of exploration into specific targeting of senescent cells by delivering chemically bound small-molecule cargos. Taking advantage of the increased presence of SA-β-galactosidase in senescent cell lysosomes, small-molecule cargos attached to galacto-oligosaccharides are preferentially released within these cells. Cells without this upregulation of SA-β-galactosidase are not able to digest the galacto-oligosaccharides and thus receive lower levels of the senolytic cargo [166]. Topical nanoparticle delivery will be particularly effective for the skin.

Vaccines have also been proposed as a potential therapeutic avenue to target cellular senescence. The vaccine itself takes a similar approach as CAR T-cell therapy, utilizing the endogenous immune system to trigger apoptosis [168]. Scientists have successfully engineered cells to upregulate programmed death receptor 1 (PD-1), an immune checkpoint inhibitor, which was targeted by T cells at higher levels. Other immunomodulators have been proposed, which allow for an array of intracellular pathways to be halted within senescent cells to halt or bypass the antiapoptotic survival mechanism [169]. Future research could focus on the combination and dosing of these senotherapeutic agents to garner their ideal effect, especially in the context of a highly absorbent and actively secretory organ such as the skin.

14. Translation to clinical practice

In the United States, a consortium has been developed under the National Institutes of Health (NIH), which acts to monitor and advocate for the advancement of translational approaches from bench to bedside [170]. The NIH Translational Geroscience Network is a national resource of aging research centers working toward conducting early-phase translational clinical trials of therapeutic agents that target fundamental aging processes (Fig. 8.5). The first clinical trial for senolytics explored the effect of dasatinib and quercetin (D + Q) in idiopathic pulmonary fibrosis (NCT02874989). This open-label pilot trial provided initial evidence that senolytics may safely alleviate physical dysfunction in idiopathic pulmonary fibrosis, warranting the evaluation of DQ in larger randomized controlled trials [171]. This finding was supported by an associated decrease in SASP-related MMPs, micro-RNA expression, and pro-inflammatory cytokines.

Success was also found in a clinical trial exploring the use of D + Q for targeting senescent cells in diabetic kidney disease (NCT02848131). The study demonstrated a similar safety profile with no serious adverse events. While the study did not measure the effect on outcomes of D + Q on diabetic kidney disease, the findings were substantial. D + Q reduced senescent cell burden, measured by decreased p16- and p21-expressing cells and decreased SA-beta-gal activity. Further, adipocyte progenitor cells showed lower replicative potential and decreased activation of surrounding macrophages to fuel the SASP. In skin, these findings were replicated, and in serum, IL-1α, IL-6, MMP-9, and MMP-12 were decreased, indicating that there were significantly lower levels of senescent cells in treated patients [172,173].

Other clinical trials are ongoing with D + Q which similarly explore its use in ablating senescent cells to relieve the burden of senescence on the disease process. Diabetic nephropathy (NCT02848131), Alzheimer's disease (NCT04785300, NCT04585590), accelerated age-like state post BMT (NCT02652052), accelerated age-like state in childhood cancer survivors (NCT04733534), and age-related osteoporosis (NCT04313634) are diseases under

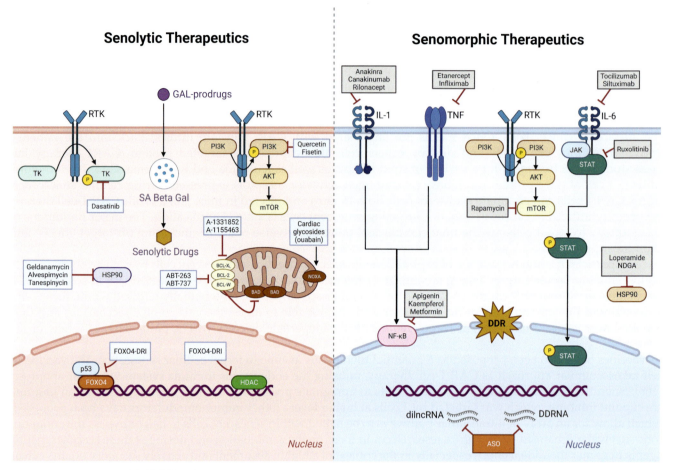

FIGURE 8.5 Targeting the root cause of skin aging with senolytics and senomorphics.

exploration. Oral fisetin is similarly being explored to decrease the senescent cell burden for frailty in older women (NCT03430037), diabetic and chronic kidney disease (NCT03325322), age-related osteoporosis (NCT04313634), and COVID-19 in hospitalized patients (NCT04476953).

15. Targeting cellular senescence with senotherapies

Senolytics are drugs or compounds applied to the skin that selectively target and remove senescent cells such as fibroblasts or melanocytes. These cells accumulate with age and are thought to contribute to age-related disease and decline in function. Several compounds have been identified as senolytics, such as navitoclax and dasatinib, as well as the natural compound fisetin. These compounds have been shown to selectively target and eliminate senescent cells in preclinical studies and are being tested in clinical trials for a variety of indications, including cardiovascular disease, frailty, and cancer.

There is an emerging field of research related to the use of topical senolytics, including several clinical trials investigating their potential use in various indications. A phase I clinical trial evaluated the safety and effectiveness of a topical senolytic gel in reducing skin wrinkles and improving skin elasticity in healthy volunteers. A phase I–II clinical trial examined the safety and efficacy of a topical senolytic cream in improving skin texture and reducing the appearance of cellulite in post-menopausal women. A phase II clinical trial investigated the use of a topical senolytic agent in treating androgenetic alopecia (male pattern baldness) in men.

Overall, the results of these clinical trials suggest that topical senolytics may have potential as a treatment for a variety of indications; however, more research is needed to fully understand their mechanisms of action and to determine their long-term safety, efficacy, and utility. It is important to note that senotherapies are still in the early stages of development and have not yet been approved for use in humans. Further research is needed to fully understand their potential benefits and risks prior to clinical deployment. Each patient should be thoroughly analyzed to provide a tailored healthy aging plan with the correct combination of senotherapy treatments to maximize outcomes.

16. Conclusion

Human skin, like other organs, ages and undergoes clinical, histological, and molecular changes due to aging. The skin aging process is influenced by the hallmarks of aging which include cellular senescence, genomic instability, telomere attrition, epigenetic alterations, loss of proteostasis, deregulated nutrient-sensing, mitochondrial dysfunction, stem cell exhaustion, and altered intercellular communication. Cellular senescence, an ideal pharmacological target, is a complex and dynamic process with significant roles in skin aging and age-related diseases. Skin senescent cells, such as epidermal melanocytes or dermal fibroblasts, can secrete a range of pro-inflammatory and matrix-degrading factors, which may contribute to age-related skin damage and inflammation. The clearance of senescent cells has been shown to extend the lifespan in animal models, suggesting that targeting senescent cells may be a viable topical delivery strategy for preventing or treating age-related skin diseases. Some research has proposed that senescent cells may serve as a protective mechanism against skin cancer, as they can inhibit the proliferation of damaged or mutant cells. However, the role of senescent cells in skin cancer is still not fully understood and is an area of ongoing research. Further research is needed to understand the mechanisms underlying the hallmarks of skin aging and to develop topical strategies for targeting senescent cells in the skin.

References

[1] Jackson S. The epidemiology of aging. In: Hazzard WR, Blass JP, Ettinger Jr WH, Halter JB, Ouslander JG, editors. Principles of geriatric medicine and gerontology. 4th ed. New York, NY: McGraw Hill; 1999. p. 203—25.

[2] Rui P, Hing E, Okeyode T. National ambulatory medical care survey: 2014 state and national summary tables. 2014. https://www.cdc.gov/nchs/data/ahcd/namcs_summary/2014_namcs_web_tables.pdf.

[3] Linos E, Chren MM, Covinsky K. Geriatric dermatology-a framework for caring for older patients with skin disease. JAMA Dermatol 2018;154(7):757—8. https://doi.org/10.1001/jamadermatol.2018.0286.

[4] López-Otín C, Blasco MA, Partridge L, Serrano M, Kroemer G. The hallmarks of aging. Cell 2013;153(6):1194. https://doi.org/10.1016/j.cell.2013.05.039.

[5] Prata LGPL, Ovsyannikova IG, Tchkonia T, Kirkland JL. Senescent cell clearance by the immune system: emerging therapeutic opportunities. Semin Immunol 2018;40. https://doi.org/10.1016/j.smim.2019.04.003.

[6] Wissler Gerdes EO, Misra A, Netto JME, Tchkonia T, Kirkland JL. Strategies for late phase preclinical and early clinical trials of senolytics. Mech Ageing Dev 2021;200:111591. https://doi.org/10.1016/j.mad.2021.111591.

[7] United Nations, Department of Economic and Social Affairs PD. World population prospects 2019: highlights. 2019. ST/ESA/SER.A/423.

[8] Garmany A, Yamada S, Terzic A. Longevity leap: mind the healthspan gap. Npj Regen Med 2021;6(1). https://doi.org/10.1038/s41536-021-00169-5.

[9] Ahadi S, Zhou W, Schüssler-Fiorenza Rose SM, et al. Personal aging markers and ageotypes revealed by deep longitudinal profiling. Nat Med 2020;26(1):83—90. https://doi.org/10.1038/s41591-019-0719-5.

[10] Sillanpää E, Ollikainen M, Kaprio J, et al. Leisure-time physical activity and DNA methylation age—a twin study. Clin Epigenet 2019;11(1). https://doi.org/10.1186/s13148-019-0613-5.

[11] Boroni M, Zonari A, Reis De Oliveira C, et al. Highly accurate skin-specific methylome analysis algorithm as a platform to screen and validate therapeutics for healthy aging. Clin Epigenet 2020;12(1). https://doi.org/10.1186/s13148-020-00899-1.

[12] Vyas K, Vasconez H. Principles of facial aesthetics. Plastic surgery key. https://plasticsurgerykey.com/6-principles-of-facial-aesthetics/. Published February 21, 2021. Accessed February 2, 2023.

[13] Zarbafian M, Fabi SG, Dayan S, Goldie K. The emerging field of regenerative aesthetics-where we are now. Dermatol Surg 2022;48(1):101—8. https://doi.org/10.1097/DSS.0000000000003239.

[14] Vyas KS, Vasconez HC, Morrison S, et al. Fat graft enrichment strategies: a systematic review. Plast Reconstr Surg 2020;145(3):827—41. https://doi.org/10.1097/PRS.0000000000006557.

[15] Choi JS, Cho WL, Choi YJ, et al. Functional recovery in photo-damaged human dermal fibroblasts by human adipose-derived stem cell extracellular vesicles. J Extracell Vesicles 2019;8(1):1565885. https://doi.org/10.1080/20013078.2019.1565885.

[16] Li X, Xie X, Lian W, et al. Exosomes from adipose-derived stem cells overexpressing Nrf2 accelerate cutaneous wound healing by promoting vascularization in a diabetic foot ulcer rat model. Exp Mol Med 2018;50(4):1—14. https://doi.org/10.1038/s12276-018-0058-5.

[17] Liang J-X, Liao X, Li S-H, et al. Antiaging properties of exosomes from adipose-derived mesenchymal stem cells in photoaged rat skin. BioMed Res Int 2020;2020:6406395. https://doi.org/10.1155/2020/6406395.

[18] Marks PW, Witten CM, Califf RM. Clarifying stem-cell therapy's benefits and risks. N Engl J Med 2017;376(11):1007—9. https://doi.org/10.1056/nejmp1613723.

[19] Harding C, Heuser J, Stahl P. Receptor-mediated endocytosis of transferrin and recycling of the transferrin receptor in rat reticulocytes. J Cell Biol 1983;97(2):329—39. https://doi.org/10.1083/jcb.97.2.329.

[20] Vyas KS, Kaufman J, Munavalli GS, Robertson K, Behfar A, Wyles SP. Exosomes: the latest in regenerative aesthetics. Regen Med 2023;18(2): 181–94. https://doi.org/10.2217/rme-2022-0134.

[21] Oh M, Lee J, Kim YJ, Rhee WJ, Park JH. Exosomes derived from human induced pluripotent stem cells ameliorate the aging of skin fibroblasts. Int J Mol Sci 2018;19(6). https://doi.org/10.3390/ijms19061715.

[22] Deng M, Yu Z, Li D, et al. Human umbilical cord mesenchymal stem cell-derived and dermal fibroblast-derived extracellular vesicles protect dermal fibroblasts from ultraviolet radiation-induced photoaging: in vitro. Photochem Photobiol Sci 2020;19(3):406–14. https://doi.org/10.1039/c9pp00421a.

[23] Xu P, Xin Y, Zhang Z, et al. Extracellular vesicles from adipose-derived stem cells ameliorate ultraviolet B-induced skin photoaging by attenuating reactive oxygen species production and inflammation. Stem Cell Res Ther 2020;11(1):264. https://doi.org/10.1186/s13287-020-01777-6.

[24] Tchkonia T, Zhu Y, Van Deursen J, Campisi J, Kirkland JL. Cellular senescence and the senescent secretory phenotype: therapeutic opportunities. J Clin Invest 2013;123(3):966–72. https://doi.org/10.1172/JCI64098.

[25] Lebrasseur NK, Tchkonia T, Kirkland JL. Cellular senescence and the biology of aging, disease, and frailty. Nestle Nutr Inst Workshop Ser 2015;83:11–8. https://doi.org/10.1159/000382054.

[26] González-Gualda E, Baker AG, Fruk L, Muñoz-Espín D. A guide to assessing cellular senescence in vitro and in vivo. FEBS J 2021;288(1): 56–80. https://doi.org/10.1111/febs.15570.

[27] Hayflick L. The limited in vitro lifetime of human diploid cell strains. Exp Cell Res 1965;37(3):614–36. https://doi.org/10.1016/0014-4827(65)90211-9.

[28] Mitchell JR, Hoeijmakers JHJ, Niedernhofer LJ. Divide and conquer: nucleotide excision repair battles cancer and ageing. Curr Opin Cell Biol 2003;15(2):232–40. https://doi.org/10.1016/S0955-0674(03)00018-8.

[29] Robinson AR, Yousefzadeh MJ, Rozgaja TA, et al. Spontaneous DNA damage to the nuclear genome promotes senescence, redox imbalance and aging. Redox Biol 2018;17:259–73. https://doi.org/10.1016/j.redox.2018.04.007.

[30] Goyns MH. Genes, telomeres and mammalian ageing. Mech Ageing Dev 2002;123(7):791–9. https://doi.org/10.1016/S0047-6374(01)00424-9.

[31] Yu Y, Cui Y, Niedernhofer LJ, Wang Y. Occurrence, biological consequences, and human health relevance of oxidative stress-induced DNA damage. Chem Res Toxicol 2016;29(12):2008–39. https://doi.org/10.1021/acs.chemrestox.6b00265.

[32] Kruk PA, Rampino NJ, Bohr VA. DNA damage and repair in telomeres: relation to aging. Proc Natl Acad Sci U S A 1995;92(1):258–62. https://doi.org/10.1073/pnas.92.1.258.

[33] Gorbunova V, Seluanov A, Mao Z, Hine C. Changes in DNA repair during aging. Nucleic Acids Res 2007;35(22):7466–74. https://doi.org/10.1093/nar/gkm756.

[34] Niedernhofer LJ, Gurkar AU, Wang Y, Vijg J, Hoeijmakers JHJ, Robbins PD. Nuclear genomic instability and aging. Annu Rev Biochem 2018; 87:295–322. https://doi.org/10.1146/annurev-biochem-062917-012239.

[35] Yousefzadeh M, Henpita C, Vyas R, Soto-Palma C, Robbins P, Niedernhofer L. DNA damage—how and why we age? Elife 2021;10:1–17. https://doi.org/10.7554/eLife.62852.

[36] Cooke MS, Evans MD, Dizdaroglu M, Lunec J. Oxidative DNA damage: mechanisms, mutation, and disease. FASEB J 2003;17(10):1195–214. https://doi.org/10.1096/fj.02-0752rev.

[37] Shammas MA. Telomeres, lifestyle, cancer, and aging. Curr Opin Clin Nutr Metab Care 2011;14(1):28–34. https://doi.org/10.1097/MCO.0b013e32834121b1.

[38] Wiley CD, Velarde MC, Lecot P, et al. Mitochondrial dysfunction induces senescence with a distinct secretory phenotype. Cell Metab 2016; 23(2):303–14. https://doi.org/10.1016/j.cmet.2015.11.011.

[39] Herbst A, Pak JW, McKenzie D, Bua E, Bassiouni M, Aiken JM. Accumulation of mitochondrial DNA deletion mutations in aged muscle fibers: evidence for a causal role in muscle fiber loss. J Gerontol A Biol Sci Med Sci. 2007;62(3):235–45. https://doi.org/10.1093/gerona/62.3.235.

[40] Lee HY, Choi CS, Birkenfeld AL, et al. Targeted expression of catalase to mitochondria prevents age-associated reductions in mitochondrial function and insulin resistance. Cell Metab 2010;12(6):668–74. https://doi.org/10.1016/j.cmet.2010.11.004.

[41] Safdar A, Hamadeh MJ, Kaczor JJ, Raha S, deBeer J, Tarnopolsky MA. Aberrant mitochondrial homeostasis in the skeletal muscle of sedentary older adults. PLoS One 2010;5(5). https://doi.org/10.1371/journal.pone.0010778.

[42] Wallace DC. Mitochondrial DNA mutations in disease and aging. Environ Mol Mutagen 2010;51(5):440–50. https://doi.org/10.1002/em.20586.

[43] Baker DJ, Childs BG, Durik M, et al. Naturally occurring p16 Ink4a-positive cells shorten healthy lifespan. Nature 2016;530(7589):184–9. https://doi.org/10.1038/nature16932.

[44] Baker DJ, Wijshake T, Tchkonia T, et al. Clearance of p16 Ink4a-positive senescent cells delays ageing-associated disorders. Nature 2011; 479(7372):232–6. https://doi.org/10.1038/nature10600.

[45] Wyles SP, Tchkonia T, Kirkland JL. Targeting cellular senescence for age-related diseases: path to clinical translation. Plast Reconstr Surg 2022;150:20S–6S. https://doi.org/10.1097/PRS.0000000000009669.

[46] Wang AS, Dreesen O. Biomarkers of cellular senescence and skin aging. Front Genet 2018;9:247. https://doi.org/10.3389/fgene.2018.00247.

[47] Poonawalla T. Anatomy of the Skin. https://www.utmb.edu/pedi_ed/CoreV2/Dermatology/page_03.htm. Accessed January 31, 2023.

[48] Shuster S, Black MM, McVITIE E. The influence of age and sex on skin thickness, skin collagen and density. Br J Dermatol 1975;93(6):639–43. https://doi.org/10.1111/j.1365-2133.1975.tb05113.x.

[49] Lovell CR, Smolenski KA, Duance VC, Light ND, Young S, Dyson M. Type I and III collagen content and fibre distribution in normal human skin during ageing. Br J Dermatol 1987;117(4):419–28. https://doi.org/10.1111/j.1365-2133.1987.tb04921.x.

[50] Buckley MR, Evans EB, Matuszewski PE, et al. Distributions of types I, II and III collagen by region in the human supraspinatus tendon. Connect Tissue Res 2013;54(6):374–9. https://doi.org/10.3109/03008207.2013.847096.

[51] Gonçalves-Neto J, Witzel SS, Teodoro WR, Carvalho-Junior AE, Fernandes TD, Yoshinari HH. Changes in collagen matrix composition in human posterior tibial tendon dysfunction. Jt Bone Spine 2002;69(2):189–94. https://doi.org/10.1016/S1297-319X(02)00369-X.

[52] Riley GP, Harrall H, Constant CR, Chard MD, Cawston TE, Hazleman BL. Tendon degeneration and chronic shoulder pain: changes in the collagen composition of the human rotator cuff tendons in rotator cuff tendinitis. Ann Rheum Dis 1994;53(6):359—66. https://doi.org/10.1136/ard.53.6.359.

[53] Muñoz-Espín D, Cañamero M, Maraver A, et al. Programmed cell senescence during mammalian embryonic development. Cell 2013;155(5):1104. https://doi.org/10.1016/j.cell.2013.10.019.

[54] Storer M, Mas A, Robert-Moreno A, et al. Senescence is a developmental mechanism that contributes to embryonic growth and patterning. Cell 2013;155(5):1119. https://doi.org/10.1016/j.cell.2013.10.041.

[55] Pratsinis H, Mavrogonatou E, Kletsas D. Scarless wound healing: from development to senescence. Adv Drug Deliv Rev 2019;146:325—43. https://doi.org/10.1016/j.addr.2018.04.011.

[56] Da Silva-Álvarez S, Guerra-Varela J, Sobrido-Cameán D, et al. Cell senescence contributes to tissue regeneration in zebrafish. Aging Cell 2020;19(1):e13052. https://doi.org/10.1111/acel.13052.

[57] Antelo-Iglesias L, Picallos-Rabina P, Estévez-Souto V, Da Silva-Álvarez S, Collado M. The role of cellular senescence in tissue repair and regeneration. Mech Ageing Dev 2021;198:111528. https://doi.org/10.1016/j.mad.2021.111528.

[58] Majewska J, Krizhanovsky V. Breathe it in—spotlight on senescence and regeneration in the lung. Mech Ageing Dev 2021;199:111550. https://doi.org/10.1016/j.mad.2021.111550.

[59] Saxena S, Vekaria H, Sullivan PG, Seifert AW. Connective tissue fibroblasts from highly regenerative mammals are refractory to ROS-induced cellular senescence. Nat Commun 2019;10(1):1—16. https://doi.org/10.1038/s41467-019-12398-w.

[60] Durant F, Whited JL. Finding solutions for fibrosis: understanding the innate mechanisms used by super-regenerator vertebrates to combat scarring. Adv Sci 2021;8(15):2100407. https://doi.org/10.1002/advs.202100407.

[61] Wilkinson HN, Hardman MJ. Senescence in wound repair: emerging strategies to target chronic healing wounds. Front Cell Dev Biol 2020;8:773. https://doi.org/10.3389/fcell.2020.00773.

[62] Coppé JP, Desprez PY, Krtolica A, Campisi J. The senescence-associated secretory phenotype: the dark side of tumor suppression. Annu Rev Pathol Mech Dis 2010;5:99—118. https://doi.org/10.1146/annurev-pathol-121808-102144.

[63] Kirkland JL, Tchkonia T, Zhu Y, Niederhofer LJ, Robbins PD. The clinical potential of senolytic drugs. J Am Geriatr Soc 2017;65(10):2297—301. https://doi.org/10.1111/jgs.14969.

[64] Kirkland JL, Tchkonia T. Cellular senescence: a translational perspective. EBioMedicine 2017;21:21—8. https://doi.org/10.1016/j.ebiom.2017.04.013.

[65] Tchkonia T, Kirkland JL. Aging, cell senescence, and chronic disease: emerging therapeutic strategies. JAMA J Am Med Assoc 2018;320(13):1319—20. https://doi.org/10.1001/jama.2018.12440.

[66] Kirkland JL. Translating the science of aging into therapeutic interventions. Cold Spring Harb Perspect Med 2016;6(3):a025908. https://doi.org/10.1101/cshperspect.a025908.

[67] Meiners S, Lehmann M. Senescent cells in IPF: locked in repair? Front Med 2020;7:1002. https://doi.org/10.3389/fmed.2020.606330.

[68] Jeon OH, David N, Campisi J, Elisseeff JH. Senescent cells and osteoarthritis: a painful connection. J Clin Invest 2018;128(4):1229—37. https://doi.org/10.1172/JCI95147.

[69] Paez-Ribes M, González-Gualda E, Doherty GJ, Muñoz-Espín D. Targeting senescent cells in translational medicine. EMBO Mol Med 2019;11(12):e10234. https://doi.org/10.15252/emmm.201810234.

[70] Barathan M, Mohamed R, Saeidi A, et al. Increased frequency of late-senescent T cells lacking CD127 in chronic hepatitis C disease. Eur J Clin Invest 2015;45(5):466—74. https://doi.org/10.1111/eci.12429.

[71] Kirkland JL, Tchkonia T. Senolytic drugs: from discovery to translation. J Intern Med 2020;288(5):518—36. https://doi.org/10.1111/joim.13141.

[72] Xu M, Pirtskhalava T, Farr JN, et al. Senolytics improve physical function and increase lifespan in old age. Nat Med 2018;24(8):1246—56. https://doi.org/10.1038/s41591-018-0092-9.

[73] Tripathi U, Misra A, Tchkonia T, Kirkland JL. Impact of senescent cell subtypes on tissue dysfunction and repair: importance and research questions. Mech Ageing Dev 2021;198:111548. https://doi.org/10.1016/j.mad.2021.111548.

[74] Hall BM, Balan V, Gleiberman AS, et al. p16(Ink4a) and senescence-associated β-galactosidase can be induced in macrophages as part of a reversible response to physiological stimuli. Aging (Albany NY) 2017;9(8):1867—84. https://doi.org/10.18632/aging.101268.

[75] Samdavid Thanapaul RJR, Shvedova M, Shin GH, Crouch J, Roh DS. Elevated skin senescence in young mice causes delayed wound healing. GeroScience 2022;44(3):1871—8. https://doi.org/10.1007/s11357-022-00551-1.

[76] Resnik SR, Egger A, Abdo Abujamra B, Jozic I. Clinical implications of cellular senescence on wound healing. Curr Dermatol Rep 2020;9(4):286—97. https://doi.org/10.1007/s13671-020-00320-3.

[77] Demaria M, Ohtani N, Youssef SA, et al. An essential role for senescent cells in optimal wound healing through secretion of PDGF-AA. Dev Cell 2014;31(6):722—33. https://doi.org/10.1016/j.devcel.2014.11.012.

[78] Chia CW, Sherman-Baust CA, Larson SA, et al. Age-associated expression of p21and p53 during human wound healing. Aging Cell 2021;20(5):e13354. https://doi.org/10.1111/acel.13354.

[79] Hall BM, Balan V, Gleiberman AS, et al. Aging of mice is associated with p16(Ink4a)- and β-galactosidase-positive macrophage accumulation that can be induced in young mice by senescent cells. Aging (Albany NY) 2016;8(7):1294—315. https://doi.org/10.18632/aging.100991.

[80] Jiang D, de Vries JC, Muschhammer J, et al. Local and transient inhibition of p21 expression ameliorates age-related delayed wound healing. Wound Repair Regen 2020;28(1):49—60. https://doi.org/10.1111/wrr.12763.

[81] Vande Berg JS, Rose MA, Haywood-Reid PL, Rudolph R, Payne WG, Robson MC. Cultured pressure ulcer fibroblasts show replicative senescence with elevated production of plasmin, plasminogen activator inhibitor-1, and transforming growth factor-β1. Wound Repair Regen 2005;13(1):76—83. https://doi.org/10.1111/j.1067-1927.2005.130110.x.

[82] Wilkinson HN, Clowes C, Banyard KL, Matteuci P, Mace KA, Hardman MJ. Elevated local senescence in diabetic wound healing is linked to pathological repair via CXCR2. J Invest Dermatol 2019;139(5):1171—81. https://doi.org/10.1016/j.jid.2019.01.005.

[83] Stanley A, Osler T. Senescence and the healing rates of venous ulcers. J Vasc Surg 2001;33(6):1206—11. https://doi.org/10.1067/mva.2001.115379.

[84] Pulido T, Velarde MC, Alimirah F. The senescence-associated secretory phenotype: fueling a wound that never heals. Mech Ageing Dev 2021; 199:111561. https://doi.org/10.1016/j.mad.2021.111561.
[85] Ho CY, Dreesen O. Faces of cellular senescence in skin aging. Mech Ageing Dev 2021;198:111525. https://doi.org/10.1016/j.mad.2021.111525.
[86] Lim DXE, Richards T, Kanapathy M, et al. Extracellular matrix and cellular senescence in venous leg ulcers. Sci Rep 2021;11(1):1—12. https://doi.org/10.1038/s41598-021-99643-9.
[87] Berlanga-Acosta JA, Guillén-Nieto GE, Rodríguez-Rodríguez N, et al. Cellular senescence as the pathogenic hub of diabetes-related wound chronicity. Front Endocrinol 2020;11:661. https://doi.org/10.3389/fendo.2020.573032.
[88] Tomic-Canic M, DiPietro LA. Cellular senescence in diabetic wounds: when too many retirees stress the system. J Invest Dermatol 2019; 139(5):997—9. https://doi.org/10.1016/j.jid.2019.02.019.
[89] Blažić TM, Brajac I. Defective induction of senescence during wound healing is a possible mechanism of keloid formation. Med Hypotheses 2006;66(3):649—52. https://doi.org/10.1016/j.mehy.2005.09.033.
[90] Varmeh S, Egia A, McGrouther D, Tahan SR, Bayat A, Pandolfi PP. Cellular senescence as a possible mechanism for halting progression of keloid lesions. Genes and Cancer 2011;2(11):1061—6. https://doi.org/10.1177/1947601912440877.
[91] Ji J, Tian Y, Zhu YQ, et al. Ionizing irradiation inhibits keloid fibroblast cell proliferation and induces premature cellular senescence. J Dermatol 2015;42(1):56—63. https://doi.org/10.1111/1346-8138.12702.
[92] Chung L, Maestas DR, Lebid A, et al. Interleukin 17 and senescent cells regulate the foreign body response to synthetic material implants in mice and humans. Sci Transl Med 2020;12(539). https://doi.org/10.1126/scitranslmed.aax3799.
[93] Jun JI, Lau LF. The matricellular protein CCN1 induces fibroblast senescence and restricts fibrosis in cutaneous wound healing. Nat Cell Biol 2010;12(7):676—85. https://doi.org/10.1038/ncb2070.
[94] Grosse L, Wagner N, Emelyanov A, et al. Defined p16High senescent cell types are indispensable for mouse healthspan. Cell Metab 2020; 32(1):87—99. https://doi.org/10.1016/j.cmet.2020.05.002.
[95] Gasek NS, Kuchel GA, Kirkland JL, Xu M. Strategies for targeting senescent cells in human disease. Nat Aging 2021;1(10):870—9. https://doi.org/10.1038/s43587-021-00121-8.
[96] Biran A, Zada L, Abou Karam P, et al. Quantitative identification of senescent cells in aging and disease. Aging Cell 2017;16(4):661—71. https://doi.org/10.1111/acel.12592.
[97] Gosselin K, Deruy E, Martien S, et al. Senescent keratinocytes die by autophagic programmed cell death. Am J Pathol 2009;174(2):423—35. https://doi.org/10.2353/ajpath.2009.080332.
[98] Passos JF, Saretzki G, Ahmed S, et al. Mitochondrial dysfunction accounts for the stochastic heterogeneity in telomere-dependent senescence. PLoS Biol 2007;5(5):1138—51. https://doi.org/10.1371/journal.pbio.0050110.
[99] Summer R, Shaghaghi H, Schriner DL, et al. Activation of the mTORC1/PGC-1 axis promotes mitochondrial biogenesis and induces cellular senescence in the lung epithelium. Am J Physiol Lung Cell Mol Physiol 2019;316(6):L1049—60. https://doi.org/10.1152/ajplung.00244.2018.
[100] Robbins E, Levine EM, Eagle H. Morphologic changes accompanying senescence of cultured human diploid cells. J Exp Med 1970;131(6): 1211—22. https://doi.org/10.1084/jem.131.6.1211.
[101] Dimri GP, Lee X, Basile G, et al. A biomarker that identifies senescent human cells in culture and in aging skin in vivo. Proc Natl Acad Sci U S A 1995;92(20):9363—7. https://doi.org/10.1073/pnas.92.20.9363.
[102] Debacq-Chainiaux F, Erusalimsky JD, Campisi J, Toussaint O. Protocols to detect senescence-associated beta-galactosidase (SA-βgal) activity, a biomarker of senescent cells in culture and in vivo. Nat Protoc 2009;4(12):1798—806. https://doi.org/10.1038/nprot.2009.191.
[103] Salmonowicz H, Passos JF. Detecting senescence: a new method for an old pigment. Aging Cell 2017;16(3):432—4. https://doi.org/10.1111/acel.12580.
[104] Kohli J, Wang B, Brandenburg SM, et al. Algorithmic assessment of cellular senescence in experimental and clinical specimens. Nat Protoc 2021;16(5):2471—98. https://doi.org/10.1038/s41596-021-00505-5.
[105] Martínez-Zamudio RI, Robinson L, Roux PF, Bischof O. SnapShot: cellular senescence pathways. Cell 2017;170(4):816. https://doi.org/10.1016/j.cell.2017.07.049.
[106] Gorgoulis V, Adams PD, Alimonti A, et al. Cellular senescence: defining a path forward. Cell 2019;179(4):813—27. https://doi.org/10.1016/j.cell.2019.10.005.
[107] Bonner WM, Redon CE, Dickey JS, et al. γH2AX and cancer. Nat Rev Cancer 2008;8(12):957—67. https://doi.org/10.1038/nrc2523.
[108] Freund A, Laberge RM, Demaria M, Campisi J. Lamin B1 loss is a senescence-associated biomarker. Mol Biol Cell 2012;23(11):2066—75. https://doi.org/10.1091/mbc.E11-10-0884.
[109] Aird KM, Zhang R. Detection of senescence-associated heterochromatin foci (SAHF). Methods Mol Biol 2013;965:185—96. https://doi.org/10.1007/978-1-62703-239-1_12.
[110] Haydont V, Bernard BA, Fortunel NO. Age-related evolutions of the dermis: clinical signs, fibroblast and extracellular matrix dynamics. Mech Ageing Dev 2019;177:150—6. https://doi.org/10.1016/j.mad.2018.03.006.
[111] Sgonc R, Gruber J. Age-related aspects of cutaneous wound healing: a mini-review. Gerontology 2013;59(2):159—64. https://doi.org/10.1159/000342344.
[112] Campisi J. The role of cellular senescence in skin aging. J Investig Dermatol Symp Proc 1998;3(1):1—5. https://doi.org/10.1038/jidsymp.1998.2.
[113] Wlaschek M, Maity P, Makrantonaki E, Scharffetter-Kochanek K. Connective tissue and fibroblast senescence in skin aging. J Invest Dermatol 2021;141(4):985—92. https://doi.org/10.1016/j.jid.2020.11.010.
[114] Kuilman T, Michaloglou C, Vredeveld LCW, et al. Oncogene-induced senescence relayed by an interleukin-dependent inflammatory network. Cell 2008;133(6):1019—31. https://doi.org/10.1016/j.cell.2008.03.039.
[115] Coppé JP, Rodier F, Patil CK, Freund A, Desprez PY, Campisi J. Tumor suppressor and aging biomarker p16 INK4a induces cellular senescence without the associated inflammatory secretory phenotype. J Biol Chem 2011;286(42):36396—403. https://doi.org/10.1074/jbc.M111.257071.
[116] Singh K, Maity P, Krug L, et al. Superoxide anion radicals induce IGF-1 resistance through concomitant activation of PTP 1 B and PTEN. EMBO Mol Med 2015;7(1):59—77. https://doi.org/10.15252/emmm.201404082.

[117] Waaijer MEC, Gunn DA, Adams PD, et al. P16INK4a positive cells in human skin are indicative of local elastic fiber morphology, facial wrinkling, and perceived age. J Gerontol A Biol Sci Med Sci. 2016;71(8):1022–8. https://doi.org/10.1093/gerona/glv114.
[118] Herranz N, Gil J. Mechanisms and functions of cellular senescence. J Clin Invest 2018;128(4):1238–46. https://doi.org/10.1172/JCI95148.
[119] Lewis DA, Travers JB, Machado C, Somani AK, Spandau DF. Reversing the aging stromal phenotype prevents carcinoma initiation. Aging (Albany NY) 2011;3(4):407–16. https://doi.org/10.18632/aging.100318.
[120] Weinmüllner R, Zbiral B, Becirovic A, et al. Organotypic human skin culture models constructed with senescent fibroblasts show hallmarks of skin aging. Npj Aging Mech Dis 2020;6(1):1–7. https://doi.org/10.1038/s41514-020-0042-x.
[121] Zonari A, Brace LE, Al-Katib KZ, et al. Senotherapeutic peptide reduces skin biological age and improves skin health markers. bioRxiv November 2020. https://doi.org/10.1101/2020.10.30.362822.
[122] Xu M, Palmer AK, Ding H, et al. Targeting senescent cells enhances adipogenesis and metabolic function in old age. Elife 2015;4. https://doi.org/10.7554/eLife.12997.
[123] Zhu Y, Armstrong JL, Tchkonia T, Kirkland JL. Cellular senescence and the senescent secretory phenotype in age-related chronic diseases. Curr Opin Clin Nutr Metab Care 2014;17(4):324–8. https://doi.org/10.1097/MCO.0000000000000065.
[124] Boccardi V, Mecocci P. Senotherapeutics: targeting senescent cells for the main age-related diseases. Mech Ageing Dev 2021;197:111526. https://doi.org/10.1016/j.mad.2021.111526.
[125] Niedernhofer LJ, Robbins PD. Senotherapeutics for healthy ageing. Nat Rev Drug Discov 2018;17(5):377. https://doi.org/10.1038/nrd.2018.44.
[126] Robbins PD, Jurk D, Khosla S, et al. Senolytic drugs: reducing senescent cell viability to extend health span. Annu Rev Pharmacol Toxicol 2021;61:779–803. https://doi.org/10.1146/annurev-pharmtox-050120-105018.
[127] Dolgin E. Send in the senolytics. Nat Biotechnol 2020;38(12):1371–7. https://doi.org/10.1038/s41587-020-00750-1.
[128] Birch J, Gil J. Senescence and the SASP: many therapeutic avenues. Genes Dev 2020;34(23–24):1565–76. https://doi.org/10.1101/gad.343129.120.
[129] Krishnamurthy J, Torrice C, Ramsey MR, et al. Ink4a/Arf expression is a biomarker of aging. J Clin Invest 2004;114(9):1299–307. https://doi.org/10.1172/JCI22475.
[130] Sharpless NE, Bardeesy N, Lee KH, et al. Loss of p16Ink4a with retention of p19 predisposes mice to tumorigenesis. Nature 2001;413(6851):86–91. https://doi.org/10.1038/35092592.
[131] Zhu Y, Tchkonia T, Pirtskhalava T, et al. The achilles' heel of senescent cells: from transcriptome to senolytic drugs. Aging Cell 2015;14(4):644–58. https://doi.org/10.1111/acel.12344.
[132] Subramanian A, Kuehn H, Gould J, Tamayo P, Mesirov JP. GSEA-P: a desktop application for gene set enrichment analysis. Bioinformatics 2007;23(23):3251–3. https://doi.org/10.1093/bioinformatics/btm369.
[133] Miura M. Therapeutic drug monitoring of imatinib, nilotinib, and dasatinib for patients with chronic myeloid leukemia. Biol Pharm Bull 2015;38(5):645–54. https://doi.org/10.1248/bpb.b15-00103.
[134] Baker JR, Vuppusetty C, Colley T, et al. MicroRNA-570 is a novel regulator of cellular senescence and inflammaging. FASEB J 2019;33(2):1605–16. https://doi.org/10.1096/fj.201800965R.
[135] Gallelli G, Cione E, Serra R, et al. Nano-hydrogel embedded with quercetin and oleic acid as a new formulation in the treatment of diabetic foot ulcer: a pilot study. Int Wound J 2020;17(2):485–90. https://doi.org/10.1111/iwj.13299.
[136] Zhu Y, Doornebal EJ, Pirtskhalava T, et al. New agents that target senescent cells: the flavone, fisetin, and the BCL-XL inhibitors, A1331852 and A1155463. Aging (Albany NY) 2017;9(3):955–63. https://doi.org/10.18632/aging.101202.
[137] Yousefzadeh MJ, Zhu Y, McGowan SJ, et al. Fisetin is a senotherapeutic that extends health and lifespan. EBioMedicine 2018;36:18–28. https://doi.org/10.1016/j.ebiom.2018.09.015.
[138] Chamcheu JC, Esnault S, Adhami VM, et al. Fisetin, a 3,7,3′,4′-tetrahydroxyflavone inhibits the PI3K/Akt/mTOR and MAPK pathways and ameliorates psoriasis pathology in 2D and 3D organotypic human inflammatory skin models. Cells 2019;8(9):1089. https://doi.org/10.3390/cells8091089.
[139] Ren Q, Guo F, Tao S, Huang R, Ma L, Fu P. Flavonoid fisetin alleviates kidney inflammation and apoptosis via inhibiting Src-mediated NF-κB p65 and MAPK signaling pathways in septic AKI mice. Biomed Pharmacother 2020;122:109772. https://doi.org/10.1016/j.biopha.2019.109772.
[140] Kim N, Kang MJ, Lee SH, et al. Fisetin enhances the cytotoxicity of gemcitabine by down-regulating ERK-MYC in MiaPaca-2 human pancreatic cancer cells. Anticancer Res 2018;38(6):3527–33. https://doi.org/10.21873/anticanres.12624.
[141] Ghani MFA, Othman R, Nordin N. Molecular docking study of naturally derived flavonoids with antiapoptotic BCL-2 and BCL-XL proteins toward ovarian cancer treatment. J Pharm BioAllied Sci 2020;12(6):676. https://doi.org/10.4103/JPBS.JPBS_272_19.
[142] Schafer MJ, White TA, Iijima K, et al. Cellular senescence mediates fibrotic pulmonary disease. Nat Commun 2017;8(1):1–11. https://doi.org/10.1038/ncomms14532.
[143] Novais EJ, Tran VA, Johnston SN, et al. Long-term treatment with senolytic drugs Dasatinib and Quercetin ameliorates age-dependent intervertebral disc degeneration in mice. Nat Commun 2021;12(1):1–17. https://doi.org/10.1038/s41467-021-25453-2.
[144] Zhu Y, Tchkonia T, Fuhrmann-Stroissnigg H, et al. Identification of a novel senolytic agent, navitoclax, targeting the Bcl-2 family of anti-apoptotic factors. Aging Cell 2016;15(3):428–35. https://doi.org/10.1111/acel.12445.
[145] Strong R, Miller RA, Astle CM, et al. Evaluation of resveratrol, green tea extract, curcumin, oxaloacetic acid, and medium-chain triglyceride oil on life span of genetically heterogeneous mice. J Gerontol A Biol Sci Med Sci. 2013;68(1):6–16. https://doi.org/10.1093/gerona/gls070.
[146] Rascón B, Hubbard BP, Sinclair DA, Amdam GV. The lifespan extension effects of resveratrol are conserved in the honey bee and may be driven by a mechanism related to caloric restriction. Aging (Albany NY) 2012;4(7):499–508. https://doi.org/10.18632/aging.100474.
[147] Price NL, Gomes AP, Ling AJY, et al. SIRT1 is required for AMPK activation and the beneficial effects of resveratrol on mitochondrial function. Cell Metab 2012;15(5):675–90. https://doi.org/10.1016/j.cmet.2012.04.003.
[148] Armour SM, Baur JA, Hsieh SN, Land-Bracha A, Thomas SM, Sinclair DA. Inhibition of mammalian S6 kinase by resveratrol suppresses autophagy. Aging (Albany NY) 2009;1(6):515–28. https://doi.org/10.18632/aging.100056.
[149] Pearson KJ, Baur JA, Lewis KN, et al. Resveratrol delays age-related deterioration and mimics transcriptional aspects of dietary restriction without extending life span. Cell Metab 2008;8(2):157–68. https://doi.org/10.1016/j.cmet.2008.06.011.

[150] Pollack RM, Barzilai N, Anghel V, et al. Resveratrol improves vascular function and mitochondrial number but not glucose metabolism in older adults. J Gerontol A Biol Sci Med Sci. 2017;72(12):1703–9. https://doi.org/10.1093/gerona/glx041.

[151] Baur JA, Sinclair DA. Therapeutic potential of resveratrol: the in vivo evidence. Nat Rev Drug Discov 2006;5(6):493–506. https://doi.org/10.1038/nrd2060.

[152] Baur JA, Pearson KJ, Price NL, et al. Resveratrol improves health and survival of mice on a high-calorie diet. Nature 2006;444(7117):337–42. https://doi.org/10.1038/nature05354.

[153] Xu Q, Fu Q, Li Z, et al. The flavonoid procyanidin C1 has senotherapeutic activity and increases lifespan in mice. Nat Metab 2021;3(12):1706–26. https://doi.org/10.1038/s42255-021-00491-8.

[154] Wang Y, Chang J, Liu X, et al. Discovery of piperlongumine as a potential novel lead for the development of senolytic agents. Aging (Albany NY) 2016;8(11):2915–26. https://doi.org/10.18632/aging.101100.

[155] He Y, Zhang X, Chang J, et al. Using proteolysis-targeting chimera technology to reduce navitoclax platelet toxicity and improve its senolytic activity. Nat Commun 2020;11(1):1–14. https://doi.org/10.1038/s41467-020-15838-0.

[156] Chang J, Wang Y, Shao L, et al. Clearance of senescent cells by ABT263 rejuvenates aged hematopoietic stem cells in mice. Nat Med 2016;22(1):78–83. https://doi.org/10.1038/nm.4010.

[157] Mérino D, Khaw SL, Glaser SP, et al. Bcl-2, Bcl-xL, and Bcl-w are not equivalent targets of ABT-737 and navitoclax (ABT-263) in lymphoid and leukemic cells. Blood 2012;119(24):5807–16. https://doi.org/10.1182/blood-2011-12-400929.

[158] Afreen S, Bohler S, Müller A, et al. BCL-XL expression is essential for human erythropoiesis and engraftment of hematopoietic stem cells. Cell Death Dis 2020;11(1):1–15. https://doi.org/10.1038/s41419-019-2203-z.

[159] Leverson JD, Phillips DC, Mitten MJ, et al. Exploiting selective BCL-2 family inhibitors to dissect cell survival dependencies and define improved strategies for cancer therapy. Sci Transl Med 2015;7(279). https://doi.org/10.1126/scitranslmed.aaa4642.

[160] Baar MP, Brandt RMC, Putavet DA, et al. Targeted apoptosis of senescent cells restores tissue homeostasis in response to chemotoxicity and aging. Cell 2017;169(1):132–47. https://doi.org/10.1016/j.cell.2017.02.031.

[161] Huang Y, He Y, Makarcyzk MJ, Lin H. Senolytic peptide FOXO4-DRI selectively removes senescent cells from in vitro expanded human chondrocytes. Front Bioeng Biotechnol 2021;9:351. https://doi.org/10.3389/fbioe.2021.677576.

[162] Zhang C, Xie Y, Chen H, et al. FOXO4-DRI alleviates age-related testosterone secretion insufficiency by targeting senescent Leydig cells in aged mice. Aging (Albany NY) 2020;12(2):1272–84. https://doi.org/10.18632/aging.102682.

[163] Fuhrmann-Stroissnigg H, Ling YY, Zhao J, et al. Identification of HSP90 inhibitors as a novel class of senolytics. Nat Commun 2017;8(1):1–14. https://doi.org/10.1038/s41467-017-00314-z.

[164] Laberge RM, Sun Y, Orjalo AV, et al. MTOR regulates the pro-tumorigenic senescence-associated secretory phenotype by promoting IL1A translation. Nat Cell Biol 2015;17(8):1049–61. https://doi.org/10.1038/ncb3195.

[165] Xu M, Tchkonia T, Ding H, et al. JAK inhibition alleviates the cellular senescence-associated secretory phenotype and frailty in old age. Proc Natl Acad Sci U S A 2015;112(46):E6301–10. https://doi.org/10.1073/pnas.1515386112.

[166] Muñoz-Espín D, Rovira M, Galiana I, et al. A versatile drug delivery system targeting senescent cells. EMBO Mol Med 2018;10(9):e9355. https://doi.org/10.15252/emmm.201809355.

[167] Amor C, Feucht J, Leibold J, et al. Senolytic CAR T cells reverse senescence-associated pathologies. Nature 2020;583(7814):127–32. https://doi.org/10.1038/s41586-020-2403-9.

[168] Chen Z, Hu K, Feng L, et al. Senescent cells re-engineered to express soluble programmed death receptor-1 for inhibiting programmed death receptor-1/programmed death ligand-1 as a vaccination approach against breast cancer. Cancer Sci 2018;109(6):1753–63. https://doi.org/10.1111/cas.13618.

[169] Nakagami H. Cellular senescence and senescence-associated T cells as a potential therapeutic target. Geriatr Gerontol Int 2020;20(2):97–100. https://doi.org/10.1111/ggi.13851.

[170] Burd CE, Gill MS, Niedernhofer LJ, et al. Barriers to the preclinical development of therapeutics that target aging mechanisms. J Gerontol A Biol Sci Med Sci. 2016;71(11):1388–94. https://doi.org/10.1093/gerona/glw112.

[171] Justice JN, Nambiar AM, Tchkonia T, et al. Senolytics in idiopathic pulmonary fibrosis: results from a first-in-human, open-label, pilot study. EBioMedicine 2019;40:554–63. https://doi.org/10.1016/j.ebiom.2018.12.052.

[172] Hickson LTJ, Langhi Prata LGP, Bobart SA, et al. Senolytics decrease senescent cells in humans: preliminary report from a clinical trial of Dasatinib plus Quercetin in individuals with diabetic kidney disease. EBioMedicine 2019;47:446–56. https://doi.org/10.1016/j.ebiom.2019.08.069.

[173] Hickson LTJ, Langhi Prata LGP, Bobart SA, et al. Corrigendum to 'Senolytics decrease senescent cells in humans: preliminary report from a clinical trial of Dasatinib plus Quercetin in individuals with diabetic kidney disease' (EBioMedicine (2019) 47 (446–456). EBioMedicine 2020;52. https://doi.org/10.1016/j.ebiom.2019.12.004.

CHAPTER 9

Age-related disease: Lungs

Joshua O. Owuor and Ayman O. Soubani

Department of Pulmonary & Critical Care Medicine, Wayne State University/Detroit Medical Center, Detroit, MI, United States

1. Introduction and relevant anatomy

Advanced age has been correlated with the development of a wide range of diseases as described in earlier chapters. The respiratory system, comprised of the lungs and accompanying pulmonary vascular structures, thoracic cage, and diaphragm, is unfortunately not exempt. The lung is the third largest organ by mass in humans, after the skin and liver. It is integral in facilitating vital gas exchange in the form of oxygen, nitrogen, and carbon dioxide between the body and the outside world, serves as a reservoir for blood, and maintains the body's acid—base balance in concert with the kidneys, a process that is vital to the function of the fundamental unit of each organ in the body, the cell. In essence, it supports life as we know it.

Breathing begins at the nose and mouth (oropharynx) and extends into the lungs in a tree branch-and-stem-like pattern. Humans have a right and left lung, each encased within the chest and separated from the ribs by a thin layer of tissue known as the parietal pleura. Each lung is attached to the trachea (windpipe) at an intersection known as the main carina and from here, it divides further into several smaller stems referred to as the bronchi (bronchus—singular). Both lungs are individually covered by a thin layer of tissue known as the visceral pleura. Grossly, the right lung has three lobes or segments—an upper, middle, and lower, of which each is then further subdivided by the bronchus into subsegments—with the upper having three, the middle containing two, and the lower having five for a total of 10 total subsegments. The left lung is slightly different as it shares the left thoracic cavity with the heart. Unlike the right lung, it only has two segments, an upper and lower lobe. Unlike the right, the upper and lower lobes are each subdivided into four segments, for a total of eight main segments. There is no "middle lobe" on the left as it integrates into the upper lobe during development to make space for the heart (and is part of the upper lobe as a subsegment termed the lingula). As the airway subdivides the volume of cartilage, a tough material that offers structural support beginning at the trachea, sequentially decreases while the percentage of smooth muscle, Type 1 (ATI) and Type II (ATII) epithelial cells, and specialized cells (goblet cells, Clara cells, immune cells [neutrophils, macrophages, and dendritic cells]) increase. Functionally, this upperpart of the respiratory system is thought of as the conducting zone, working to humidify (by serous and mucous secretions), warm (by underlying blood vessels), and filter air (by trapping particles in mucous secretions and transporting back toward the throat by action of cilia for expectoration or swallowing into the gut) (Fig. 9.1).

Each division and subdivision of the lung is accompanied by blood vessels that begin in the right ventricle of the heart as the pulmonary artery. It then becomes the pulmonary trunk which then divides into the right and left main arteries that feed into each lung respectively. From there, the subdivision of these vessels follows the airway segments and together are termed the bronchopulmonary tree.

As the branches of the bronchus are further divided into smaller and smaller branches deeper into each lung, they become known as bronchioles (start of the respiratory section), which then terminate into a sack of alveoli, the fundamental unit where gas exchange (respiration) between the body and outside world occurs. Oxygenated blood is returned to the heart through sequentially enlarging veins that also course along the bronchopulmonary tree back to the heart, ready to be pumped to the rest of the body to support the body's every function. Neural

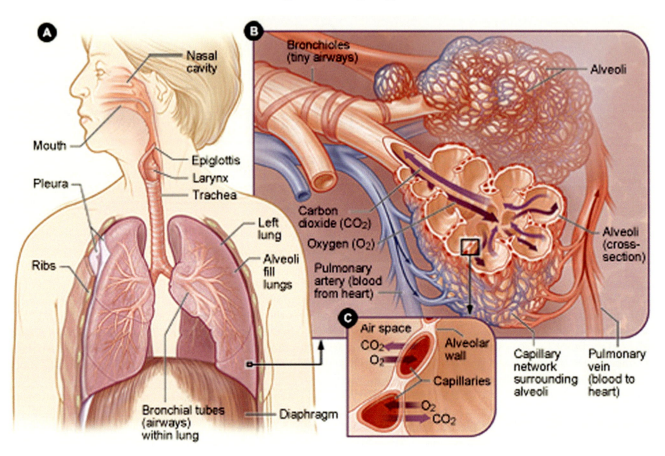

FIGURE 9.1 Functional anatomy and physiology of the respiratory system. (A) Large structures involved in conducting air. Note how the trachea branches into the right and left lungs, then further subdivides into smaller and smaller segments. The lung is encased with a cavity protected externally by the rib cage. The pleura encloses each lung within the chest, providing lubrication and protection from the bony rib cage. (B) The bronchopulmonary tree is depicted. Note further subdivision of the airway and accompanying blood vessels carrying carbon dioxide-rich deoxygenated blood (blue) and oxygenated blood (red), leading to the alveoli. (C) Interface between blood vessels and the alveoli where ventilation occurs, allowing inspired oxygen to move from the alveoli into blood cells and carbon dioxide to travel vice versa. *From Ref. [1].*

innervation of the diaphragm via the phrenic nerve maintains the harmonious rhythm of breathing while sympathetic and parasympathetic systems maintain the delicate balance of ventilation and perfusion to keep up with the body's demand.

Pathology at any point of this system can easily cause alterations in the body's physiology and, subsequently, an individual's ability to function. As we age, the likelihood of this dysfunction increases. In this chapter we explore what is known up to date about the mechanisms of this dysfunction through discussion of conditions in pulmonary medicine with a higher prevalence in the older population.

2. The aging lung

Age is emerging as a risk factor for chronic non-communicable lung diseases such as chronic obstructive pulmonary disease (COPD), some forms of lung cancer, and idiopathic pulmonary fibrosis (IPF). In the United States alone, the proportion of adults aged 65 years or older increased from 12% to 15% over the last decade and is projected to increase to 20% by 2030 [1a]. This increase unfortunately also correlates with increased susceptibility to chronic lung disease and associated morbidity and mortality [2]. As the lung ages, functional and structural changes become evident. As examples, peak airflow (how quickly someone can exhale), vital capacity (the amount of air an individual can exhale after maximum inhalation), and diffusion capacity (ease of exchange of oxygen and carbon dioxide) tend to progressively decline which is thought to be a function of loss of elastic fibers, scarring, or inflammation around the alveolar ducts [3]. Additionally, there's a trend towards weakening of respiratory muscles and decline in the effectiveness of airway clearance and lung defense mechanisms. Although our understanding of the biology of aging has advanced remarkably in recent years, few genetic or molecular mechanisms linking aging to pulmonary diseases have been identified. Clarifying these processes is an important step toward improving our management strategies of chronic lung diseases and setting the stage for the future of targeted individualized therapies.

3. Role of cell aging in age-related lung disease

The lung is at all times exposed to the outside world and, as such, constantly challenged by it. This is in the form of allergens, infectious organisms (such as viruses and bacteria), and occupational and indoor irritants, just to name a few. The lung in turn has evolved defenses like those of the skin to protect and counteract these stressors and to maintain homeostasis. However, sustained, or overwhelming exposure to these challenges which occurs as we age, together with epigenetic changes and genetic predisposition, are what is postulated to tip the balance of physiological repair and renewal toward the development of disease [4].

The aging process is a complicated cascade of pathology. Carlos Lopez-Otin and his colleagues adopted the concept of "nine hallmarks of aging" building on the work of Douglas Hanahan and Robert Weinberg's description of the biology of cancer. They defined these as genomic instability, telomere attrition, epigenetic alterations, loss of protein regulation (proteostasis), deregulated nutrient sensing, mitochondrial dysfunction, cellular senescence, stem cell exhaustion, and altered intercellular communication [5,6]. Silke Miners and her colleagues adopted this conceptual framework for age-related pulmonary disease and expanded on it by subdividing the ideas into cell-intrinsic factors, cell-extrinsic factors (altered intercellular communication, stem cell exhaustion), and introduced extracellular matrix dysregulation as an additional area of investigation. Conceptually, the intrinsic mechanisms are what predetermine a cell's replicative capacity and endurance, while the extrinsic factors determine more globally the environment and stressors that the cell must navigate. We will use this conceptualization in this chapter to demonstrate the role of some of these processes as supported by the latest research for our discussion on how their contribution to lung dysfunction and the development of disease as we age, with a focus on emphysema, COPD, and IPF.

4. Specific lung diseases correlated with advanced age

4.1 Emphysema and COPD

The lung fully matures by the age of 20–25 years and thereafter physiologically begins to decline incrementally. Normal lung aging (senile lung) is associated with alveolar enlargement from loss of supporting wall integrity without destruction and is seen starting around 50 years of age. Imagine a balloon that loses its tensile wall strength and requires less and less force to inflate with each subsequent attempt. Some of these changes can result in premature closure of small airways during normal breathing and potentially cause air trapping and hyperinflation (enlargement), leading to "senile emphysema" [7]. The exact mechanisms leading to these changes are still under investigation. Teramoto and his colleagues demonstrated this concept using age-accelerated mice termed senescence prone mice (SAM-P). They found that unlike their senescence resistance (SAM R) counterparts, the SAM-P mice tended toward the development of airspace enlargement significantly earlier in life, without a significant difference in airway destruction [8–10]. Kuro and his colleagues described the "klotho" gene, encoding a membrane protein thought to be involved in cell signaling and regulation of senesce as a potential culprit. Mice deficient in this gene product developed airspace enlargement, in addition to having a shorter lifespan, infertility, arteriosclerosis, skin atrophy, osteoporosis—a syndrome like that demonstrated in aging humans [11]. These studies, although pivotal to our understanding of the aging lung, focused on homogeneous alterations of lung structures and processes, which unlike age-related changes in human aging and effect on the lungs are very heterogeneous.

COPD, like emphysema, is also a heterogenous lung condition and more often than not, presents with clinical symptoms. This is often in the form of shortness of breath, chronic cough, chest congestion, and above average mucus production. It is due to abnormalities within the bronchus (bronchitis), bronchioles (bronchiolitis), and/or within the alveoli (emphysema), known as small airway remodeling (SAR). It causes thickening of the airways, accumulation of fibrous tissue, and accumulation of damaged alveolar epithelial cells and mesenchymal cells that leads to persistent, often progressive obstruction of airflow [12,13]. The rate of newly diagnosed adults with COPD increases from around 200 cases in 10, 000 in people aged ≤45 years to 1200 cases in 10,000 in those ≥65 years, with the steepest increase between patients aged 65−74 years. In the past decade (2011−20), the age-adjusted prevalence of COPD in the United States has remained unchanged (6% of adults >18 years old) with higher estimates among women. The overall age-standardized death rates in adults ≥45 years in the past 2 decades (1999−2019) has slightly decreased from 120 to 115/100,000, with the largest decrease among men (from 160 to 120 per 100,000) (CDC COPD website). The structural alterations in COPD, unlike those in senile emphysema, are pathologic [14,15] (Fig. 9.2).

Aside from advancing age, smoking is the other major risk factor associated with COPD and emphysema. Other well-known but less studied risk factors include exposure to burning biomass fuels, which is more prevalent in the developing world, household indoor smoke, and outdoor pollution [16]. Interestingly however, it is estimated that only 10%−15% of those who smoke eventually develop COPD, suggesting there could be an interplay, among other processes, that increase susceptibility in some people and not others [17]. Studies in mouse models mentioned earlier showed that exposure to cigarette smoke hastened the development of emphysema and increased susceptibility to alveolar wall damage, like that seen in COPD compared to matched cohorts. This, as a matter of concept, supports a likely "two (or more)-hit" hypothesis. Cell factors predisposing to the development of COPD supported by available research are: (1) loss of protective telomere sequences at the end of chromosomes (telomere attrition), (2) mitochondrial dysfunction, (3) epigenetic alterations in the form of DNA methylation and histone modification (epigenetic alterations), (4) loss of protein integrity and thus homeostasis (proteostasis), cellular senescence, and (5) dysregulated extracellular membrane (ECM).

These factors are in themselves heterogeneous and as we explore them, we start to appreciate how complex the process of disease development with advancing age becomes. Telomere attrition is well known in the field of aging to limit the proliferative capacity of cells after a critical value of base pair loss and has been suggested as a more precise biomarker for age-related disease and progression and severity [18]. This attrition triggers DNA damage responses that can cause cell cycle arrest, senescence, or apoptosis. In the development of COPD, the causal link between smoking and induction of telomere dysfunction or accelerated attrition and senescence has only been demonstrated in mouse models. Human studies have, on the other hand, demonstrated that telomeres of alveolar, endothelial, smooth muscle cells and lymphocyte cells from patients who smoke or have smoked, and those with COPD are appreciably shorter than those of their healthier counterparts [19−24]. This shortening, especially in circulating lymphocytes, although controversial as a marker for organ-specific changes, correlates with the severity of

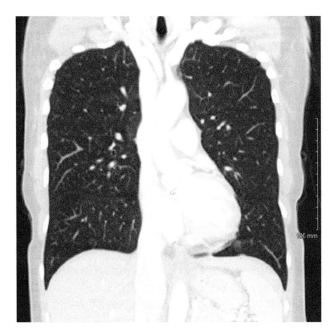
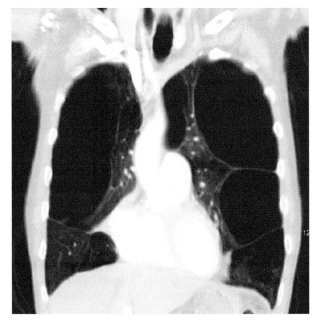

FIGURE 9.2 Left: Hyperinflation and "barrel chest" seen in patients with COPD. Right: Patient with extensive emphysematous changes—noted as areas of lung parenchyma surrounded by large areas of thin-walled air-filled sacs.

COPD as measured by lung function, clinical symptoms, and the number of years of smoking [25,26]. Further support of an association is seen in studies of familial combined emphysema and pulmonary fibrosis syndrome showing that mutations encoding telomerase reverse transcriptase (TERT) and telomerase RNA (TR), which provides the template for telomerase repeat addition, might cause a haploinsufficiency that causes shortening of telomeres that can lead to bone marrow failure, liver disease, and osteoporosis. Interestingly, these manifestations were noted only in members who smoked, further emphasizing that maybe exposure reaches far deeper into cellular DNA mechanics than we have been able to explain with certainty just yet [27]. A similar trend has been reported in families with alpha-1 trypsin deficiency who develop COPD and emphysema, but a correlation with telomere attrition is yet to be established [17].

Our understanding of the exact mechanisms by which smoking or other pollutants predispose to COPD is still evolving. On our journey so far however, we have discovered that these irritants induce a pro-inflammatory and pro-fibrotic state through activation of transcription factor nuclear factor-κB (NF-κB), p38 mitogen-activated protein kinase (MAPK), generation of autoantibodies, and production of reactive oxygen species (ROS). This leads to impairment of the epithelial barrier, sustained immune cell recruitment, mucus hypersecretion, and small airway remodeling and fibrosis that are classically seen in affected patients. Club cells (former known as Clara cells), which promote regeneration of bronchiolar epithelium, have been shown to be reduced or eliminated because of inciting airway injury, leading to further lung tissue decline. The exogenous and endogenously produced ROS from the inflammatory immune response by activated immune cells and lung epithelial cell mitochondrial respiration (oxidative phosphorylation) is significant and is measurably higher in patients with chronic lung disease such as COPD and IPF than in matched cohorts [28].

Cellular mitochondrion function is essential in homeostasis within the lung environment as the endothelial cells that produce mucin, surfactant, and power cilia that beat in waves to rid the airway of pathogen and inhaled particles are bioenergetically active and energy intensive. In addition to being a powerhouse for the cell, the mitochondria of these cells serve a regulatory role in mitophagy and cell growth [17,29]. Its dysfunction is known to play a role in aging and age-related disease but the exact mechanism contributing to chronic lung disease has only recently been realized. In chronic disease research, we now know that exposure to tobacco increases mitochondrial ROS (mtROS), damages mitochondrial DNA (mtDNA), and triggers the release of abnormal intracellular calcium build up and release, altering mitochondrial homeostasis [30]. Furthermore, Prohibins (PHB1 and PHB2), proteins found within the inner mitochondrial membrane associated with NADPH dehydrogenase complex (complex 1), which is vital in mitigating antioxidant stress, are reported to have lower expressions in patients with COPD. It has been suggested as a marker for diagnosis of early COPD as its levels correlate with the degree of airway obstruction, impaired mitophagy, increased risk of mitochondrial damage from ROS, and cell death [31]. Other mitochondrial-related molecules of interest to COPD and under investigation are OPA1 and Drp1, which function in mitochondrial fusion and fission and are observed to be reduced in COPD, NLRP3 inflammasome which is known to cause alveolar inflammation and is elevated in COPD, and endogenous antioxidants such as nuclear factor erythroid 3-related factor 2 (Nrf2) are also increased [32]. It remains unclear thus far if these agents are triggers or lead to the progression of lung aging and disease.

Given the proximity of mitochondria to the cell's protein synthesis and processing complex, the endoplasmic reticulum (ER), we can postulate that any process that affects the mitochondria will eventually affect protein homeostasis (proteostasis). In addition to its role in protein synthesis, it is vital for calcium storage homeostasis and lipid synthesis, and maintains an optimal environment for proper cell and tissue function. Proteostasis is maintained by the unfolded protein response (UPR) which is activated when the cell recognizes unfolded or damaged proteins [33]. Aging and cigarette smoke have been implicated in inciting ER damage and with repeated insults overwhelm the UPRs, causing the accumulation of aberrant proteins and loss of function of a myriad of cellular functions. Smoke exposure has been implicated independently for causing an ER stress response by interfering with the redox environment and promoting S-glutathionylation, the covalent attachment of glutathione to protein cysteines. This dysregulation is of considerable interest as it is speculated to contribute to the pathogenesis of chronic lung diseases by affecting cellular pathways that govern proliferation, migration, plasticity, inflammation, production of pro-fibrotic factors, and cell death [34,35].

Genomic changes known to correlate with advancing age have not been well established in current COPD and emphysema research. Epigenetic alterations via DNA methylation, histone modification, and microRNAs which regulate gene expression, however, have been reported in the blood, sputum, and tissues of patients with COPD, increasing their potential as targets for therapy. It is unclear however whether they arise because of the disease process or if they are the drivers of the pathology. DNA methylation changes have been reported in relation to exposure to cigarette and wood smoke with reports that the induced changes increase the risk of developing COPD [36].

Genome and epigenetic studies have identified variation in 187 cytosine-phosphate-guanine (CpG) sites between smokers and non-smokers with overall hypomethylation in smokers [37]. These changes have been observed to be reversible depending on the dose and time of smoke exposure among patients with COPD. The role of methyltransferase is also under investigation in relation to chronic lung disease.

Histone modifications through acetylation, deacetylation, methylation, and demethylation in the regulation of gene transcription and protein expression, especially in their role in the inflammatory pathways, have also been implicated in COPD pathogenesis. Histone deacetylase (HDAC) levels, particularly HDAC-2, have lower activity and expression in patients with COPD compared to healthy cohorts. This reduction is triggered by cigarette smoke, which causes phosphorylation, ubiquitination, and it degradation. Its significance is further highlighted by finding that lower levels are associated with inflammation and corticosteroid resistance [38–40].

The cellular and tissue environment as we have established in patients with COPD is one in which inflammation is a major driver of the disease process. COPD, in fact, is a disease of systemic inflammation that often associates with other diseases that have been linked to inflammation such as cardiovascular disease, metabolic syndrome, and protein calorie malnutrition [41]. Environmental triggers such as cigarette smoke, as previously discussed, induce alveolar cell apoptosis and stimulate the recruitment of leukocytes (polymorphonuclear neutrophils, macrophages, CD4+ and CD+8 T lymphocytes, and B lymphocytes), which in turn further the inflammatory response. The effect of this inflammatory deluge on the extracellular matrix is a point of interest in the pathogenesis of COPD as activation of these cells either through injury or smoking-induced inflammation or other age-related pathways, leads to the release of proteinase that degrades alveolar septae, as has been demonstrated in familial cases of emphysema.

4.2 Interstitial lung diseases/interstitial pulmonary fibrosis

Interstitial lung disease (ILD) is a group of lung diseases that disrupt the anatomic space between the alveolar epithelial and capillary endothelial cells. Idiopathic interstitial pneumonias (IIPs) are a subset of rare ILDs of unknown etiology characterized by inflammation and fibrosis that involve both lungs, are chronic and progressive, and cause restrictive lung impairment, which then leads to significant disruption of gas transfer and clinically significant hypoxia in affected individuals. Idiopathic pulmonary fibrosis (IPF) is the most common of the IIPs and is characterized by chronic and progressive fibrosis of unknown etiology. Pulmonary hypertension (PH) is a common complication in IPF that presents a special challenge in managing these patients [42]. This cohort tends to have greater severity of dyspnea, functional impairment, and gas exchange abnormalities, leading to extensive morbidity and mortality compared to other types of IIPs (Fig. 9.3).

Aging is a significant risk factor in the development of this condition, with an average age of onset of 60 years and a higher incidence in men than women. The average life expectancy is 3–5 years with the great burden of morbidity and mortality from progressive hypoxic respiratory failure due to destruction of the lung parenchyma from progressive fibrosis and scarring. Co-morbidities such as gastroesophageal reflux disease, obstructive sleep apnea, diabetes, and herpes virus infections have also been suggested in addition to environmental exposure such as smoke and

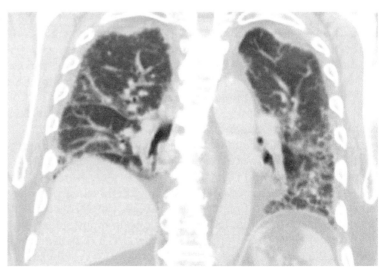

FIGURE 9.3 Patient with bilateral fibrotic changes and honeycombing typical of patients with IPF.

other pollutants [43]. It remains unclear however if each or any of these factors are individually or collectively culpable, or how the interplay leads to the development of IPF with advancing age. We again employ the framework suggested by Carlos-Otin et al. and adopted by Silke Miners and her colleagues to explore the interplay between forces that drive aging and its correlates as supported by available data in the development of interstitial lung disease and potential targets for therapy. Suspected cell intrinsic factors include telomere attrition and senescence and mitochondrial dysfunction. Stem cell exhaustion and extracellular matrix dysfunction are the extrinsic factors we discuss in the pathogenesis of IPF.

The implication of telomere attrition in IPF derives from studies of patients with dyskeratosis congenita, a disease of accelerated aging, and in cases of familial IPF, which presents in patients younger than 50 years of age. It has been observed that between 5% and 15% of familial cases and 1%—3% of sporadic cases of IPF are associated with mutations in telomerase genes [TERT, TERC, and dyskerine (DKC1)] or in telomere-protective proteins known as the shelterin complex made up of TRF1, TRF2, TIN2, POT1, TPP1, and RAP1 proteins that function to activate DNA damage response pathways and induce cellular senescence through activation of the p53 pathway [44—47]. The cell-specific effect of this telomere attrition is variable but is known to result in cellular/replicative senescence—also known as the Hayflick effect. Alveolar type II cells (ATII) and fibroblasts are the two lung resident cell types most studied in the pathogenesis of IPF. As described earlier in this chapter, ATII cells are localized within the alveolar and function in the secretion of surfactant that helps to maintain surface tension and prevent alveolar collapse. It also plays a vital role as a stem cell reservoir for damaged cells. Likewise, fibroblasts are resident cells in between the alveolar and blood vessels. During normal cell development, they are a vital support cell and function to direct formation and extension of alveolar septa, and alveolar epithelial proliferation and differentiation [48]. Telomere shortening has been shown to be significant in ATII cells and leukocytes, but not in fibroblasts in the pathogenesis of IPF. The link between this attrition and IPF has been shown in mouse studies to be a function of the replicative arrested but metabolically active ATII cells. This senescence phenotype secretes factors (SASP) that spread cellular senesce signals to other ATII cells via a paracrine pathway. In addition, they release cytokines and profibrotic factors such as TGFB1 that promote differentiation of fibroblasts to scar-forming myofibroblasts which release an exaggerated amount of matrix components such as collagen and fibronectin that leads to and promotes further fibrosis [49,50]. These senescent cells also have been shown to have reduced proliferative capacity, which further decreases the lung's capacity for repair. Maria Blasco and colleagues recently demonstrated in a mouse model the significance of ATII cells by demonstrating that TRF1 deletions in lung fibroblasts, club cells, and basal cells although they could be a contributor to the fibrotic process, do not independently lead to interstitial fibrosis, further underscoring ATII cells as the likely cell type driving the fibrotic process [51].

Mitochondrial dysfunction as described in our discussion of COPD is also of interest in IPF pathogenesis. Alveolar macrophages and ATII cells have been shown to have dysmorphic and dysfunctional mitochondria in both human and mouse models [52,53]. Mitochondrial stress particularly in the form of mitochondrial reactive oxygen species (mtROS) induces mtDNA damage, which in turn can induce mitophagy and subsequent senesce or activation of the NLRP3 inflammasome which has been shown to promote the release of pro-inflammatory factors, pyroptosis, and subsequent pulmonary fibrosis [32]. Exposure to cigarette smoke, silica, and carbon nanotubes are known to trigger inflammasome activity, although their correlation with IPF remains unclear. Interestingly, mitochondrial homeostasis regulators such as PINK1, PARK2, and NRF1/2 have all been shown to be downregulated in patients with IPF.

Massive ECM deposition by fibroblast and tissue architectural damage is the hallmark of this disease. This fibroblastic focus has an immature hyaluronic acid-rich matrix underneath the epithelial layer, loss of AT1 cell differentiation, and increased alpha SMA (+) myofibroblasts. A new subpopulation of senescent basal epithelial cell type localized in the honeycombing structure of the lungs has been described using single-cell RNA sequencing, supporting the notion of other intermediary cells at play in promoting fibrosis [54—58]. The fibrotic deposition is postulated to be intimately tied to exhaustion of ATII cells, which are also the lung's resident stem cells, sustained myofibroblast activation, and failure of these cells to undergo apoptosis or revert to a quiescent phenotype. So far, the available evidence suggests prolonged stress during the transitional state of ATII to ATI cells, and abnormal paracrine signaling between ATII and fibroblasts could be the mechanisms that trigger and subsequently promote fibrosis [59].

5. Current research, future direction, and potential interventions

Elucidating the molecular mechanisms of aging is essential to understanding the drastic decline in lung function with age, and the development and progression of acute and chronic lung diseases discussed so far. Our understanding of the biology of aging has advanced remarkably in the last 2 decades and the molecular mechanisms linking aging to age-related pulmonary disease are opening new avenues for therapeutics, especially in the case of IPF. The novel antifibrotic agents pirfenidone and nintedanib, which work by reducing the profibrotic activity of fibroblasts and limiting ECM deposition, have been shown to slow lung function decline and ameliorate the severity of disease. They are, however, not curative.

Senotherapeutics is an emerging field targeting senescent cells, now suspected to play a role in the development of lung fibrosis. Senolytics are agents that induce apoptosis of senescent cells by inhibiting senescent cell antiapoptotic pathways (SCAPS). Combination of Desantib, a tyrosine kinase inhibitor with a long history of use in treating myeloproliferative syndromes, and Quercetin, a flavanol and nonspecific kinase inhibitor targeting PI3K/AKT, BCL-2, insulin/IGF-1, and HIF-1 alpha SCAP network components, recently completed a phase I trial in humans after showing improved pulmonary outcomes in mice [60,61]. A phase II clinical trial evaluating Setaxinib (formerly GKT137831), a NAPDH oxidase 1/4 inhibitor, is currently recruiting patients. Targeting altered redox balance and inducing senescent myofibroblasts into apoptosis has been shown to reverse established lung fibrosis in mouse models [62]. Targeting of activated myofibroblasts and inducing apoptosis with BH3 mimetics such as navitoclax (ABT-263) have also been shown in mouse models to not only prevent progression of fibrosis but also reverse established disease. Activated myofibroblasts are primed for death with pro-apoptotic BH3 molecules, but by the same token, express anti-apoptotic BCL-2 proteins that can sustain them in a senescent state. The BH3 mimetic tips the balance into an apoptotic state [63–65]. No clinical trials are currently being undertaken in IPF, but studies are underway in cancer therapeutics. Senomorphics are agents capable of modulating the activity and function of senescent cells without inducing apoptosis. As such, they can prevent cells from adopting a senescent state, revert to such a state, or modulate SASP components that promote fibrosis [66].

Gene therapy is another avenue of therapy under investigation for IPF. Maria Blasco and her colleagues recently demonstrated in a mouse model that treating *TERT* improves lung function, decreases inflammation, and accelerates fiber disappearance in fibrotic lungs [67]. The implication of this breakthrough cannot be overstated as it has the potential to revolutionize medical practice. No human trials testing the concept were underway at the time of writing.

Stem cell transplantation is another avenue of investigation for both IPF and COPD. However, clinical studies so far have been small and underpowered to show efficacy or feasibility [68]. Several clinical trials are currently registered on clinicaltrials.gov, which will hopefully shed more light on this therapy.

Smoking cessation, of course, remains the cornerstone of management as it has been implicated in almost all lung conditions.

References

[1] National heart, lung, and blood Institute. National Institutes of Health. U.S. Department of Health and Human Services.
[1a] Vincent GK, Velkoff VA. The next four decades: the older population in the United States: 2010–2050. Population estimates and projections. Washington, DC: U.S. Department of Commerce Census Bureau; 2010.
[2] Thannickal VJ, Murthy M, Balch WE, Chandel NS, Meiners S, Eickelberg O, et al. Blue journal conference. Aging and susceptibility to lung disease. Am J Respir Crit Care Med February 1, 2015;191(3):261–9. https://doi.org/10.1164/rccm.201410-1876PP.
[3] Lamb D, Gillooly M, Farrow AS. Microscopic emphysema and its variations with age, smoking, and site within the lungs. Ann N Y Acad Sci 1991;624:339–40. https://doi.org/10.1111/j.1749-6632.1991.tb17040.x.
[4] Meiners S, Eickelberg O, Königshoff M. Hallmarks of the ageing lung. Eur Respir J 2015;45:807–27. https://doi.org/10.1183/09031936.00186914.
[5] Lopez-Otin C, Blasco MA, Partidge L, Serrano M, Kroemer G. The hallmarks of aging. Cell Rev 2013;153(6):1194–217. https://doi.org/10.1016/j.cell.2013.05.039.
[6] Hannan D, Weinberg RA. Hallmarks of Cancer: the next generation. Cell Rev 2011;144(5):646–74. https://doi.org/10.1016/j.cell.2011.02.013.
[7] Sharma G, Goodwin J. Effect of aging on respiratory system physiology and immunology. Clin Interv Aging 2006;1(3):253–60. https://doi.org/10.2147/ciia.2006.1.3.253.
[8] Teramoto S, Fukuchi Y, Uejima Y, Teramoto K, Oka T, Orimo H. A novel model of senile lung: senescence-accelerated mouse (SAM). Am J Respir Crit Care Med July 1994;150(1):238–44. https://doi.org/10.1164/ajrccm.150.1.8025756.
[9] Teramoto S, Uejima Y, Oka T, Teramoto K, Fukuchi Y. Effects of chronic cigarette smoke inhalation on the development of senile lung in senescence-accelerated mouse. Res Exp Med 1997;197:1–11. https://doi.org/10.1007/s004330050050.
[10] Teramoto S. Age-related changes in lung structure and function in the senescence-accelerated mouse (SAM): SAM-P/1 as a new model of senile hyperinflation of lung. Am J Respir Crit Care Med October 1997;156(4 Pt 1):1361.
[11] Kuro -OM, Matsumara Y, Aizawa H, Kawaguchi H, Suga T, Utsugi T, et al. Mutation of the mouse Klotho gene leads to a syndrome resembling ageing. Nature 1997;390(6655):45–51. https://doi.org/10.1038/36285.
[12] Alder JK, Guo N, Kembou F, Parry EM, Anderson CJ, Gorgy AI, et al. Telomere length is a determinant of emphysema susceptibility. Am J Respir Crit Care Med 2011;184:904–12.

References

[13] Tiendrébéogo AJF, Soumagne T, Pellegrin F, Dagouassat M, Nhieu JTV, Caramelle P, et al. The telomerase activator TA-65 protects from cigarette smoke-induced small airway remodeling in mice through extra-telomeric effects. Sci Rep 2023;13:25. https://doi.org/10.1038/s41598-022-25993-7.

[14] Rycroft CE, Heyes A, Lanza L, Becker K. Epidemiology of chronic obstructive pulmonary disease: a literature review. Int J Chronic Obstr Pulm Dis 2012;7:457—94. https://doi.org/10.2147/COPD.S32330.

[15] Mannino DM. Fifty years of progress in the epidemiology of chronic obstructive pulmonary disease: a review of national heart, lung, and blood institute-sponsored studies. Chronic Obstr Pulm Dis 2019;6:350—8.

[16] Patel AR, Patel AR, Singh S, Singh S, Khawaja I. Global initiative for chronic obstructive lung disease: the changes made. Cureus June 24, 2019; 11(6):e4985. https://doi.org/10.7759/cureus.4985.

[17] Wan ES, Silverman EK. Genetics of COPD and emphysema. Chest 2009;136(3):859—66. https://doi.org/10.1378/chest.09-0555.

[18] Fossel M. Use of telomere length as a biomarker for aging and age-related disease. Curr Transl Geriatr Exp Gerontol Rep 2012;1:121—7.

[19] Sorrentino JA, Krishnamurthy J, Tilley S, Alb Jr JG, Burd CE, Sharpless NE. p16INK4a reporter mice reveal age-promoting effects of environmental toxicants. J Clin Investig 2014;124:169—73.

[20] Amsellem V, Gary-Bobo G, Marcos E, Maitre B, Chaar V, Validire P, et al. Telomere dysfunction causes sustained inflammation in chronic obstructive pulmonary disease. Am J Respir Crit Care Med 2011;184:1358—66.

[21] Chan SR, Blackburn EH. Telomeres and telomerase. Philos Trans R Soc Lond B Biol Sci 2004;359:109—21.

[22] Morla M, Busquets X, Pons J, Sauleda J, MacNee W, Agusti AG. Telomere shortening in smokers with and without COPD. Eur Respir J 2006; 27:525—8.

[23] Astuti Y, Wardhana A, Watkins J, Wulaningsih W, PILAR Research Network. Cigarette smoking and telomere length: a systematic review of 84 studies and meta-analysis. Environ Res October 2017;158:480—9. https://doi.org/10.1016/j.envres.2017.06.038.

[24] Tsuji T, Aoshiba K, Nagai A. Cigarette smoke induces senescence in alveolar epithelial cells. Am J Respir Cell Mol Biol 2004;31:643—9.

[25] Nunes H, Monnet I, Kannengiesser C, Uzunhan Y, Valeyre D, Kambouchner M, et al. Is telomeropathy the explanation for combined pulmonary fibrosis and emphysema syndrome? report of a family with TERT mutation. Am J Respir Crit Care Med 2014;189:753—4.

[26] Valdes AM, Andrew T, Gardner JP, Kimura M, Oelsner E, Cherkas LF, et al. Obesity, cigarette smoking, and telomere length in women. Lancet 2005;366:662—4.

[27] Stanley SE, Chen JJ, Podlevsky JD, Alder JK, Hansel NN, Mathias RA, et al. Telomerase mutations in smokers with severe emphysema. J Clin Invest February 2015;125(2):563—70. https://doi.org/10.1172/JCI78554.

[28] Barnes PJ. Oxidative stress-based therapeutics in COPD. Redox Biol 2020;33:101544.

[29] Fang T, Wang M, Xiao H, Wei X. Mitochondrial dysfunction, and chronic lung disease. Cell Biol Toxicol 2019;35:493—502.

[30] Chellappan DK, Paudel KR, Tan NW, Cheong KS, Khoo SSQ, Seow SM, et al. Targeting the mitochondria in chronic respiratory diseases. Mitochondrion 2022;67:15—37.

[31] Yue L, Yao H. Mitochondrial dysfunction in inflammatory responses and cellular senescence: pathogenesis and pharmacological targets for chronic lung diseases. Br J Pharmacol August 2016;173(15):2305—18. https://doi.org/10.1111/bph.13518.

[32] Chen Y, Zhang Y, Li N, Jiang Z, Li X. Role of mitochondrial stress and the NLRP3 inflammasome in lung diseases. Inflamm Res 2023;72: 829—46. https://doi.org/10.1007/s00011-023-01712-4.

[33] Meiners S, Greene CM. Protein quality control in lung disease: it's all about cloud networking. Eur Respir J 2014;44:846—9.

[34] Janssen-Heininger YM, Nolin JD, Hoffman SM, van der Velden JL, Tully JE, Lahue KG, et al. Emerging mechanisms of glutathione-dependent chemistry in biology and disease. J Cell Biochem 2013;114:1962—8.

[35] Janssen-Heininger Y, Reynaert NL, Viliet A, Anathy V. Endoplasmic reticulum stress and glutathione therapeutics in chronic lung disease. Redox Biol 2020;33:1010516.

[36] Wan ES, Qiu W, Baccarelli A, Carey VJ, Bacherman H, Rennard SI, et al. Cigarette smoking behaviors and time since quitting are associated with differential DNA methylation across the human genome. Hum Mol Genet 2012;21:3073—82.

[37] Zeilinger S, Kühnel B, Klopp N, Baurecht H, Kleinschmidt A, Gieger C, et al. Tobacco smoking leads to extensive genome-wide changes in DNA methylation. PLoS One 2013;8.

[38] Chen Y, Huang P, Ai W, Li X, Guo W, Zhang J, et al. Histone deacetylase activity is decreased in peripheral blood monocytes in patients with COPD. J Inflamm 2012;9:10.

[39] Adenuga D, Yao H, March TH, Seagrave J, Rahman I. Histone deacetylase 2 is phosphorylated, ubiquitinated, and degraded by cigarette smoke. Am J Respir Cell Mol Biol 2009;40:464—73.

[40] Barnes PJ. Targeting histone deacetylase 2 in chronic obstructive pulmonary disease treatment. Expert Opin Ther Targets 2005;9:1111—21.

[41] Wang A, Li Z, Sun Z, Liu Y, Zhang D, Ma X. Potential mechanisms between HF and COPD: new insights from bioinformatics. Curr Probl Cardiol 2023;48(3):01539.

[42] Raghu G, Remy-Jardin M, Myers JL, Richeldi L, Ryerson CJ, Lederer DJ, et al. Diagnosis of idiopathic pulmonary fibrosis an official ATS/ ERS/JRS/ALAT clinical practice guideline. Am J Respir Crit Care Med 2018;198:e44—68.

[43] Zaman T, Lee JS. Risk factors for the development of idiopathic pulmonary fibrosis: a review. Curr Pulmonol Rep December 2018;7(4):118—25. https://doi.org/10.1007/s13665-018-0210-7.

[44] Chibbar R, Gjevre JA, Shih F, Neufeld H, Lemere EG, Fladeland DA, et al. Familial interstitial pulmonary fibrosis: a large family with atypical clinical features. Cancer Res J 2010;17:269—74.

[45] Cronkhite JT, Xing C, Raghu G, Chin KM, Torres F, Rosenblatt RL, et al. Telomere shortening in familial and sporadic pulmonary fibrosis. Am J Respir Crit Care Med 2008;178:729—37.

[46] Alder JK, Chen JJ-L, Lancaster L, Danoff S, Su S, Cogan JD, et al. Short telomeres are a risk factor for idiopathic pulmonary fibrosis. Proc Natl Acad Sci USA 2008;105:13051—6.

[47] Armanios MY, Chen JJ-L, Cogan JD, Alder JK, Ingersoll RG, Markin C, et al. Telomerase mutations in families with idiopathic pulmonary fibrosis. N Engl J Med 2007;356:1317—26.

[48] George Ushakumary M, Riccetti M, Anne-Karina TP. Resident interstitial lung fibroblasts and their role in alveolar stem cell niche development, homeostasis, injury, and regeneration. Stem Cells Transl Med July 2021;10(7):1021—32. https://doi.org/10.1002/sctm.20-0526.

[49] Álvarez D, Cárdenes N, Sellarés J, Bueno M, Corey C, Hanumanthu VS, Rojas M. IPF lung fibroblasts have a senescent phenotype. Am J Physiol Lung Cell Mol Physiol 2017;313(6):L1164−73.
[50] Wiley CD, Brumwell AN, Davis SS, Jackson JR, Valdovinos A, Calhoun C, Le Saux CJ. Secretion of leukotrienes by senescent lung fibroblasts promotes pulmonary fibrosis. JCI Insight 2019;4(24).
[51] Piñeiro-Hermida S, Martínez P, Bosso G, Flores JM, Saraswati S, Connor J, et al. Consequences of telomere dysfunction in fibroblasts, club and basal cells for lung fibrosis development. Nat Commun 2022;13:5656. https://doi.org/10.1038/s41467-022-32771-6.
[52] Tsitoura E, Vasarmidi E, Bibaki E, Trachalaki A, Koutoulaki C, Papastratigakis P, et al. Accumulation of damaged mitochondria in alveolar macrophages with reduced OXPHOS related gene expression in IPF. Respir Res 2019;20:264. https://doi.org/10.1186/s12931-019-1196-6.
[53] Bueno M, Lai YC, Romero Y, Brands J, St Croix CM, Kamga C, et al. PINK1 deficiency impairs mitochondrial homeostasis and promotes lung fibrosis. J Clin Invest 2015;125:521−38.
[54] Habermann AC, Gutierrez AJ, Bui LT, Yahn SL, Winters NI, Calvi CL, et al. Single-cell RNA sequencing reveals profibrotic roles of distinct epithelial and mesenchymal lineages in pulmonary fibrosis. Sci Adv July 8, 2020;6(28):eaba1972. https://doi.org/10.1126/sciadv.aba1972.
[55] Adams TS, Schupp JC, Poli S, Ayaub EA, Neumark N, Ahangari F, et al. Single-cell RNA-seq reveals ectopic and aberrant lung-resident cell populations in idiopathic pulmonary fibrosis. Sci Adv 2020;6(28):eaba1983. https://doi.org/10.1126/sciadv.aba1983.
[56] Kathiriya JJ, Wang C, Zhou M, Brumwell A, Cassandras M, Le Saux CJ, et al. Human alveolar type 2 epithelium transdifferentiates into metaplastic KRT5(+) basal cells. Nat Cell Biol 2022;24(1):10−23. https://doi.org/10.1038/s41556-021-00809-4.
[57] Strunz M, Simon LM, Ansari M, Kathiriya JJ, Angelidis I, Mayr CH, et al. Alveolar regeneration through a Krt8+ transitional stem cell state that persists in human lung fibrosis. Nat Commun 2020;11(1):3559. https://doi.org/10.1038/s41467-020-17358-3.
[58] Fang Y, Tian J, Fan Y, Cao P. Latest progress on the molecular mechanisms of idiopathic pulmonary fibrosis. Mol Biol Rep 2020;47:9811−20. https://doi.org/10.1007/s11033-020-06000-6.
[59] Zhu W, Tan C, Zhang J. Alveolar epithelial type 2 cell dysfunction in idiopathic pulmonary fibrosis. Lung 2022;200:539−47. https://doi.org/10.1007/s00408-022-00571-w.
[60] Lehmann M, Korfei M, Mutze K, Klee S, Skronka-Wasek W, Alsafadi HN, et al. Senolytic drugs target alveolar epithelial cell function and attenuate experimental lung fibrosis ex vivo. Eur Respir J 2017;50(2).
[61] Nambiar A, Kellogg 3rd D, Justice J, Goros M, Gelfond J, Pascual R, et al. Senolytics dasatinib and quercetin in idiopathic pulmonary fibrosis: results of a phase I, single-blind, single-center, randomized, placebo-controlled pilot trial on feasibility and tolerability. EBioMedicine April 2023;90:104481. https://doi.org/10.1016/j.ebiom.2023.104481.
[62] Hecker L, Logsdon NJ, Kurundkar D, Kurundkar A, Bernard K, Hock T, et al. Reversal of persistent fibrosis in aging by targeting Nox4-Nrf2 redox imbalance. Sci Transl Med April 9, 2014;6(231):231ra47. https://doi.org/10.1126/scitranslmed.3008182.
[63] Yosef R, Pilpel N, Tokarsky-Amiel R, Biran A, Ovadya Y, Cohen S, et al. Directed elimination of senescent cells by inhibition of BCL-W and BCL-XL. Nat Commun 2016;7:11190. https://doi.org/10.1038/ncomms11190.
[64] Lagares D, Santos A, Grasberger PE, Liu F, Probst CK, Rahimi RA, et al. Targeted apoptosis of myofibroblasts with the BH3 mimetic ABT-263 reverses established fibrosis. Sci Transl Med December 13, 2017;9(420):eaal3765. https://doi.org/10.1126/scitranslmed.aal3765.
[65] Kuehl T, Lagares D. BH3 mimetics as anti-fibrotic therapy: unleashing the mitochondrial pathway of apoptosis in myofibroblasts. Matrix Biol 2018;68−69:94−105.
[66] Merkt W, Bueno M, Mora AL, Lagares D. Senotherapeutics: targeting senescence in idiopathic pulmonary fibrosis. Semin Cell Dev Biol 2020;101:104−10.
[67] Povedano JM, Martinez P, Serrano R, Tejera A, Gomez-Lopez G, Bobadilla M, et al. Therapeutic effects of telomerase in mice with pulmonary fibrosis induced by damage to the lungs and short telomerase. Elife 2018;7:e31299.
[68] Glassberg MK, Csete I, Simonet E, Elliot SJ. Stem cell therapy for COPD. Chestert Rev 2021;160(4):1271−81.

CHAPTER 10

Age-related disease: Diabetes

Allyson K. Palmer[1,2] and James L. Kirkland[3,4]

[1]Robert and Arlene Kogod Center on Aging, Mayo Clinic, Rochester, MN, United States [2]Division of Hospital Internal Medicine, Department of Medicine, Mayo Clinic, Rochester, MN, United States [3]Department of Biomedical Engineering, Mayo Clinic, Rochester, MN, United States [4]Division of General Internal Medicine, Department of Medicine, Mayo Clinic, Rochester, MN, United States

1. Introduction: Type 2 diabetes is an age-related disease

Diabetes mellitus is a chronic metabolic disease characterized by the body's inability to produce enough insulin, utilize insulin appropriately, or both, that affects approximately 10% of the world's population [1]. Diabetes poses an enormous risk to the human healthspan given its contribution to multimorbidity and detrimental impact on most organ systems in the body. Type 2 diabetes (T2D), which is primarily driven by insulin resistance, accounts for over 90% of diabetes cases [1]. The prevalence of T2D increases with age: an estimated 24% of adults over age 75 worldwide have diabetes, and the majority of individuals with diabetes are over 65 years of age [1]. Therefore, the number of individuals with diabetes is expected to rise along with aging of the world's population and will be further exacerbated by a growing obesity epidemic. Impaired glucose tolerance, which is a precursor of T2D, is a consequence of both aging and obesity that affects 1 in 10 adults across the globe, putting these individuals at higher risk of developing T2D [1].

T2D arises in the setting of two factors: (1) progressive loss of insulin sensitivity (in other words, insulin resistance) in peripheral tissues including adipose tissue, liver, and skeletal muscle and (2) impaired insulin secretion by pancreatic β cells. Insulin resistance increases with age independently of body composition in nondiabetic humans [2]. Insulin secretion also declines with aging independently of BMI or insulin resistance, due to reduced beta cell function [2,3]. Progression to overt diabetes occurs when insulin production can no longer compensate for insulin resistance to maintain euglycemia. This is clinically defined as elevated fasting glucose (>126 mg/dL or 7.0 mmol/L), severely elevated random glucose (>200 mg/dL or 11.1 mmol/L) in an individual with classic symptoms of hyperglycemia, plasma glucose over 200 mg/dL (11.1 mmol/L) at the 2 h point during an oral glucose tolerance test, or elevated hemoglobin A1c (>6.5%) [4]. The extent to which insulin resistance versus diminished insulin secretion accounts for progression to diabetes in each individual varies and is impacted by characteristics including age, obesity, genetics, and comorbidities.

Obesity, which precipitates insulin resistance, also increases in middle age and early old age. Increased adiposity with aging is largely due to inactivity, however other mechanisms including reduced resting energy expenditure, changes in hormone secretion, and impaired dietary fat storage in subcutaneous adipose tissue, also contribute [5]. Distribution of body fat shifts toward visceral depots rather than subcutaneous depots with aging, and this increased visceral adiposity is associated with age-related insulin resistance [6,7]. Conversely, gluteofemoral adipose tissue is associated with increased insulin sensitivity [8,9]. In very old age or in the setting of frailty, profound loss of peripheral subcutaneous adipose tissue can be seen, possibly related to greater rates of free fatty acid (FFA) release [5]. Increased FFA release itself can also directly cause insulin resistance [10]. Dysdifferentiation of mesenchymal progenitor cells in various tissues including adipose depots, bone marrow, and muscle into insulin-sensitive fat cells occurs with aging. This leads to accumulation of lipid into extra-adipose sites, such as the liver, which may contribute to the age-related reduction in adipose depot size and is associated with insulin resistance [11]. Taken together, changes in body composition with aging are strongly associated with the development of insulin resistance and T2D.

There is a growing awareness that social determinants of health impact an individual's risk of developing disease, including diabetes, and that this effect is likely mediated through chronic activation of hormonal stress response pathways and inflammation [12]. There is evidence to suggest that social factors influence aging mechanisms, even in early age [13]. For example, children experiencing socioeconomic disadvantage had methylation patterns consistent with accelerated aging [14]. Accelerated aging mechanisms likely play a role in the increased risk of disease in individuals experiencing adverse social factors. Premature aging syndromes with genetic underpinnings are also associated with insulin resistance and diabetes, further strengthening the notion that aging mechanisms contribute to diabetes pathogenesis. For example, Hutchinson–Gilford syndrome (related to mutation in the A type lamin gene, *LMNA*) and other progeria syndromes are recognized as monogenetic diabetes syndromes associated with premature aging [15].

2. Cellular senescence in the pathogenesis of type 2 diabetes

Fundamental aging processes, which have also been referred to as "pillars" or "hallmarks" of aging, likely play a role in the pathogenesis of T2D [16–19]. These processes generally fall into four broad categories: (1) macromolecular dysfunction (such as loss of proteostasis, aberrant mRNA processing, faulty DNA damage repair, intracellular organelle dysfunction); (2) sterile inflammation (inflammation in the absence of a specific pathogen) with fibrosis; (3) progenitor cell dysfunction (including reduced or aberrant differentiation and depletion of progenitor pools); and (4) cellular senescence [20]. Fundamental aging processes are associated with the development of age-related phenotypes and are often found at the sites of pathogenesis of age-related diseases [21]. In the context of age- and obesity-related metabolic dysfunction, progenitor cell dysfunction and cellular senescence have been encountered extensively in adipose tissue. We will also discuss what is currently known regarding how cellular senescence in other metabolic tissues, including the pancreas, liver, and skeletal muscle, affects the development of T2D.

2.1 Cellular senescence

Cellular senescence is a cell fate involving cell cycle arrest in response to a variety of stressors including but not limited to metabolic signals, DNA damage, telomere shortening, mitochondrial dysfunction, and oncogene activation [22,23]. Cellular senescence is a potent tumor-suppressive mechanism and in the acute setting may have additional beneficial functions including antifibrotic and wound-healing properties [24–26]. However, the chronic presence of senescent cells and senescent cell accumulation in tissues are drivers of organ aging and multiple diseases [27,28]. Senescent cells undergo reorganization of heterochromatin and changes in gene expression to produce the senescence-associated secretory phenotype, or SASP [29]. Increased protein synthesis is in part responsible for the enlarged morphology typically seen in senescent cells. Cellular senescence was initially described in mitotic cells, but more recent investigation has shown that terminally differentiated, postmitotic cells also develop senescent-like features [30–32]. More investigation is needed to better characterize post-mitotic senescence.

Senescent cells exert their effects through the SASP (Fig. 10.1), which can include inflammatory factors, chemokines and cytokines, matrix remodeling enzymes, extracellular vesicles, noncoding nucleotides (microRNAs and mitochondrial DNA), bioactive lipids (including ceramides, bradykines, and prostanoids), protein aggregates, hemostatic factors (such as PAI-1), and factors that impact stem cell function and fibrosis such as activin A [33–37]. The SASP varies among senescent cells and is influenced by cell type, the signal that initially drove a cell to become senescent, how long cells have been senescent, and the microenvironment [38–41]. The SASP acts as a signal to attract and activate natural killer (NK) cells, cytotoxic T cells, and macrophages that recognize surface antigens expressed by senescent cells and destroy them [42–44]. In addition, the SASP acts to spread senescence to previously normal cells, and can do so in a paracrine or endocrine manner [45–48].

Although the SASP includes pro-apoptotic factors, senescent cells are often able to evade apoptosis by upregulation of senescent cell anti-apoptotic pathways, or SCAPs [49]. These pathways include networks related to BCL-2/BCL-X$_L$, PI3K/AKT, p21/p53, dependence receptors, tyrosine kinases, and HIF-1α. Investigation into targeting these SCAPs led to the identification of the first senescence-targeting, or senolytic, drugs, which will be discussed in detail later [49]. In healthy tissues, senescent cells are removed by immune cells including natural killer (NK) cells, T-cells, and macrophages [42]. Other immune cell types may also be involved in senescent cell clearance, however more studies are needed in this area.

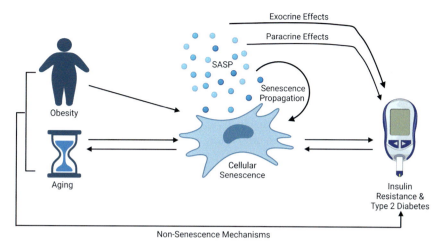

FIGURE 10.1 Cellular senescence in type 2 diabetes pathogenesis. Obesity and aging lead to the accumulation of senescent cells, which produce and secrete the senescence-associated secretory phenotype (SASP). The SASP acts through paracrine and endocrine mechanisms to promote insulin resistance and type 2 diabetes, and also causes propagation of senescence in neighboring cells. In turn, the diabetic microenvironment promotes cellular senescence, forming a pathogenic loop.

Most tissues and organs can accumulate senescent cells with aging. Senescent cells make up only a small percentage of the total cells in any tissue, however they can exert potent effects locally and systemically through their SASP [50]. For example, transplantation of a relatively small number of senescent cells into young mice induces physical dysfunction including reduced walking speed, muscle strength, activity, and food intake [48]. Fewer senescent cells are required to induce similar effects in older mice, possibly due to an existing population of senescent cells that have already accumulated with aging [48]. Similar dysfunction is seen at young age in progeroid mice with defects in DNA repair, which accumulate larger numbers of senescent cells than wild-type mice, or in mice that have undergone ionizing radiation causing senescence [49]. Taken together, these studies suggest that a certain "threshold" of senescent cells is needed to induce tissue dysfunction or cause age-related disease and frailty (Fig. 10.2).

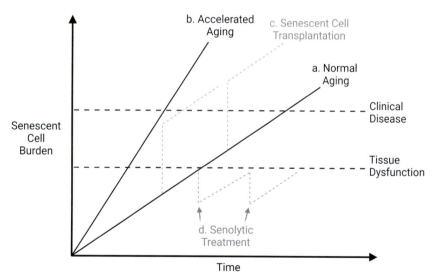

FIGURE 10.2 Threshold theory of senescent cell accumulation. Senescent cell burden increases with aging in most tissues and organs (a), eventually reaching a "threshold" at which tissue dysfunction occurs, or disease is clinically evident. In accelerated aging states, including progeroid syndromes or diseases such as type 2 diabetes, the threshold of tissue dysfunction and disease may be reached much sooner due to an increased trajectory of senescent cell accumulation (b). In experiments of senescent cell transplantation, fewer senescent cells are needed to cause physical dysfunction in older animals than younger animals, likely due to a larger number of existing senescent cells due to aging (c). Senolytic treatment, which is typically administered intermittently, has the potential to prevent tissue dysfunction and disease related to senescent cell burden by causing targeted removal of senescent cells (d).

2.2 Adipose tissue

Adipose tissue is a dynamic organ that plays a major role in energy storage and nutrient sensing, in addition to having endocrine, immune, and regenerative functions [51]. It is often the largest organ in the body, for example making up over 40% of total body mass in women whose body mass index (BMI) is greater than 35 [52,53]. Adipose tissue has a systemic influence on insulin sensitivity and inflammatory signaling even in lean individuals, through the secretion of adipokines, adipose-derived hormones, and inflammatory factors [54]. Through these mechanisms, adipose tissue can affect other organs including muscle, liver, bone, and brain in an endocrine fashion [55,56]. Many interventions that extend lifespan in model organisms impact adipose tissue, typically by changing nutrient availability [57–59]. Additionally, restricting single-gene mutations that extend lifespan in model organisms just to adipose tissue can yield similar lifespan extension to the global mutation [60,61].

White adipose tissue (WAT) is found in both subcutaneous and visceral depots and expands in response to nutrient excess through both hypertrophy (enlargement of cell size) and hyperplasia (cell division) [62]. In addition to changes in adipose tissue abundance and distribution discussed earlier, aging also leads to changes in adipose tissue signaling, miRNA processing, progenitor cell function, and cellular composition including accumulation of senescent cells [51,59]. WAT contains a large number of senescent cells even in lean individuals [63]. However, senescent cell abundance in adipose tissue is increased with both aging and obesity, primarily hypertrophic obesity [59,64,65]. Cell types that undergo senescence in adipose tissue include but are not limited to adipose progenitor cells (APCs, also called preadipocytes), endothelial cells, and post-mitotic adipocytes [59,66]. Interestingly, the extent of senescent cell accumulation is increased in adults who first became obese during childhood compared to those who first became obese later in life, supporting the theory that obesity is a state of accelerated aging (Fig. 10.2) [67].

APCs can undergo senescence due to increased proliferation in the setting of nutrient excess or aging, or due to activation of p53 by oxidative stress, which is increased in obesity [68]. Through their SASP, senescent APCs induce macrophage infiltration, promote inflammation, and directly impede insulin sensitivity by producing factors including IL-6 and TNFα. Through secretion of Activin A, senescent APCs also limit the adipogenic capacity of surrounding preadipocytes [18,68–70]. Reduced adipogenic potential then leads to hypertrophic expansion of adipose tissue, itself associated with increased lipolysis and accumulation of ectopic fat in non-adipose organs [71]. This hypertrophic adipose phenotype, which is associated with insulin resistance, can be seen prior to the development of T2D in individuals with a genetic predisposition for the disease [68]. For example, APCs isolated from healthy first-degree relatives of individuals with T2D showed reduced adipogenic potential and features of senescence [72].

Endothelial cells are abundant in the vascular network of adipose tissue and facilitate expansion of WAT by secreting angiogenic factors [73]. Lipids are also transported between the circulation and tissue across endothelial cells, a process that is mediated by PPARγ. Endothelial cell senescence has been described in the setting of a high-fat diet in model organisms and in human obesity, and may be more pronounced in visceral than subcutaneous adipose tissue [74]. VEGF within the obese adipose microenvironment likely plays a role in inducing senescence in endothelial cells [74]. PPARγ activity declines in senescent endothelial cells, potentially explaining, in part, impaired lipid handling ability [75]. In aged mice, endothelial cells comprise the highest percentage of highly p21-expressing cells in visceral adipose tissue [76]. In experiments that expose healthy mice to the circulation of mice that are genetically engineered to have senescent endothelial cells, the healthy mice develop insulin resistance [77]. In vitro experiments also show that the SASP of endothelial cells can directly cause insulin resistance in adipocytes [77].

Mature, differentiated adipose cells—adipocytes—can also develop a senescent-like phenotype, characterized by profound upregulation of p53 and p21, ROS production, and accumulation of DNA damage markers [65,78–81]. In obesity, adipocytes may be particularly susceptible to senescence due to decreased expression of SREBP1c, which is partially responsible for maintaining genome stability through enhancement of PARP1-mediated DNA repair [80]. Adipocytes in subcutaneous adipose tissue are more likely to undergo senescence than adipocytes in omental adipose tissue [81]. Senescent adipocytes seem to increase inflammation and infiltration of macrophages and other immune cells into adipose tissue, further promoting insulin resistance [80].

2.3 Pancreas

Pancreatic islets are comprised of networks of endocrine cells including alpha cells which produce glucagon, beta cells which produce insulin, delta cells which produce somatostatin, and PP cells, which produce pancreatic polypeptide. β-Cell mass seems to remain stable with aging, however human adult β-cell turnover is very low, even in response to stressors such as pancreatectomy [82–84]. This reduction in proliferation has been associated with age-related upregulation of p16^{Ink4a} in β-cells. Experimental overexpression of β-cell-specific p16 in mice also leads to features of senescence [85].

Obesity is typically accompanied by hypersecretion of insulin related to an approximately 50% increase in β-cell mass in obese vs. lean individuals [82,86]. Given the low replicative capacity of adult human β-cells, this is achieved by expansion of β-cell mass through hyperplasia and hypertrophy. In individuals with T2D, however, β-cell mass is reduced compared to nondiabetic individuals, due to increased rates of β-cell apoptosis over time in the setting of metabolic stress, specifically glucotoxicity and lipotoxicity [87–90].

The role of β-cell senescence in glucose metabolism is not completely understood. In some experiments involving in vitro induction of senescence by p16^{Ink4a} in human β-cells, senescence was associated with increased glucose-stimulated insulin secretion [85]. However, the significance of this finding for whole-body metabolic function in the setting of obesity or T2D is not understood. For example, unadjusted basal insulin secretion increases with age, however when effects of BMI and insulin sensitivity are removed, basal insulin release is actually found to decline with aging in nondiabetic individuals [91]. Additional work is needed to determine the contribution of β-cell senescence to age-related hyperinsulinemia, which in some studies has been suggested to be the initial precipitating factor of insulin resistance and T2D, rather than being a downstream effect of insulin resistance or hyperglycemia [92–94].

2.4 Muscle

Skeletal muscle is responsible for a majority of glucose disposal, is important for fatty acid metabolism, and plays a crucial role in determining energy expenditure. Muscle mass declines with aging as does muscle strength, largely due to reduced physical activity, and intramuscular adipose tissue increases [5]. Satellite cells, which are muscle progenitor cells that are located beneath the basal lamina of muscle fibers, exist in a quiescent state until activated in response to damage or stress, at which time they proliferate and fuse to form myofibers [95]. Some satellite cells re-enter a quiescent state and maintain the progenitor pool. With aging, satellite cells seem to lose their ability to stay in quiescence and instead undergo senescence, which may be in part triggered by autophagy failure or derepression of p16^{INK4a} [95,96]. In this way, activation of senescence programs in aged satellite cells likely limits regenerative capacity [97]. It is not clear whether cellular senescence in muscle is directly involved in age-related sarcopenia, although elimination of senescent cells from progeroid mice did prevent age-related sarcopenia [98]. The role that muscle senescent cells play in age- and obesity-related metabolic dysfunction is an area that requires further study.

2.5 Liver

The liver has a key role in glucose metabolism and is responsible for 80%–90% of the body's endogenous glucose production through glycogenolysis and gluconeogenesis [99]. Given that the liver vascular supply is through the portal venous circulation, the liver is also partially responsible for preventing hyperinsulinemia by removing 50% of secreted insulin from the circulation [99]. Four basic cell types are in the liver including hepatocytes, hepatic stellate cells, Kupffer cells, and liver sinusoidal endothelial cells.

The liver is the main site of ectopic lipid deposition in aging and obesity, leading to hepatic steatosis and nonalcoholic fatty liver disease (NAFLD). Two subtypes of NAFLD exist, including the nonprogressive nonalcoholic fatty liver (NAFL), which does not progress or cause complications, and nonalcoholic steatohepatitis (NASH) which is associated with inflammation and hepatocyte injury and can progress to fibrosis and cirrhosis [100]. NASH is considered to be an accelerated aging-like phenotype, given that it is associated with reduced regenerative capacity following partial liver resection, increased risk of hepatocellular carcinoma (HCC), and increased morbidity and mortality [101].

Senescent cells accumulate in the liver with aging and obesity and are a driver of NAFLD. Changes in mitochondrial function in senescent hepatocytes, namely a shift in glucose metabolism from oxidative phosphorylation to glycolysis, leads to reduced efficiency of fatty acid metabolism, which culminates in excess fat accumulation [102–104]. Liver senescent cell burden correlates positively with the degree of steatosis [104–106]. Increased levels of hepatic senescence markers were also associated with higher fasting insulin levels and insulin resistance as measured by euglycemic glucose clamp, suggesting that hepatic senescence may also promote insulin resistance [106].

3. The diabetic microenvironment drives senescent cell accumulation

T2D leads to dysfunction of multiple organ systems, much like in chronological aging, but often at a younger age [107,108]. Therefore, T2D represents a form of accelerated aging (Fig. 10.2). Comorbidities precipitated by diabetes include cardiovascular disease, cognitive impairment and dementia, renal dysfunction, peripheral neuropathy, retinopathy, osteoporosis, and frailty among others [18]. Not only does prolonged hyperglycemia directly affect cell structure and function, but fundamental aging mechanisms are also increased in individuals with diabetes and are likely to contribute significantly to this accelerated aging-like phenotype. Given that cellular senescence is implicated in the pathogenesis of insulin resistance and T2D, further promotion of cellular senescence by the diabetic milieu begets a pathogenic loop (Fig. 10.1) [19].

3.1 Hyperglycemia and advanced glycation endproducts

Prolonged exposure to hyperglycemia can induce cellular senescence and the SASP through various mechanisms, as has been observed in APCs, fibroblasts, renal mesangial cells, skin keratinocytes, macrophages, and endothelial cells [109–114]. Non-insulin-sensitive cells, including mesangial, endothelial, and immune cells, are unable to regulate glucose flux in the setting of hyperglycemia. In these cells, increased intracellular glucose concentrations activate detrimental mechanisms including nonenzymatic glycation, enhanced reactive oxygen production and signaling, renin–angiotensin system activation, endoplasmic reticulum stress, and pathogenic activation of molecular pathways including the polyol, diacylglycerol-protein kinase C, Src homology-2 domain-containing phosphatase-1, and kallikrein-bradykinin pathways [115]. Hyperglycemia also induces mitochondrial dysfunction, which involves mitochondrial fragmentation and leads to overproduction of ROS, which induces senescence [116]. Advanced glycation endproducts (AGEs), which form through the nonenzymatic Maillard reaction, can also cause premature senescence by signaling through AGE receptors (RAGE) to activate endoplasmic reticulum stress pathways, as has been observed in renal tubular cells [117].

3.2 Hyperinsulinemia

Hyperinsulinemia occurs during the natural history of insulin resistance and T2D and also in the setting of diabetes treatment with exogenous insulin. Hyperinsulinemia has been shown to drive cellular senescence in multiple cell types and tissues [81]. In adipose tissue, prolonged exposure to hyperinsulinemia causes adipocytes to undergo endoreplication, or cell cycle progression despite the inability to undergo mitosis, which can lead to activation of senescence pathways [81]. Insulin also directly promotes hydrogen peroxide generation in adipocytes by activating membrane-bound NADPH oxidase [118]. Hyperinsulinemia is a driver of hepatocyte senescence and can exacerbate the SASP in already-senescent hepatocytes in vitro [119]. In obese patients undergoing bariatric surgery, the degree of hyperinsulinemia was associated with senescence markers in liver biopsies [120].

3.3 Oxidative stress and telomere attrition

Obesity and insulin resistance are associated with increased oxidative stress. In a subanalysis of the Framingham Heart Study, hyperglycemia and indices of obesity including BMI and waist–hip ratio correlated independently with systemic oxidative stress [121]. Oxidative stress can lead to telomeric dysfunction and shortening through both direct DNA damage and increased replication. Inhibition of oxidative stress by overexpression of SOD-1 in endothelial cells blunts hyperglycemia-induced cellular senescence and SASP factor production [113]. Telomere shortening occurs in the setting of increased BMI or insulin resistance even in young individuals, indicating that this effect is distinct from age-related changes in telomere length [122]. It is also important to note that telomeric dysfunction can arise in the absence of telomere shortening through DNA damage responses to double-stranded breaks [123].

3.4 Chronic inflammation and impaired immune function

Systemic inflammation is a well-described feature of T2D characterized by elevated levels of circulating inflammatory markers such as IL-6, TNFα, and IL-1β [124,125]. This is akin to low-grade inflammation seen with aging, also known as "inflammaging" [126,127]. Inflammaging is thought to play a major role in increasing the risk of cardiovascular disease in obesity and insulin resistance, even in nondiabetic individuals [128]. Chronic activation of inflammatory pathways can trigger the production of ROS that can cause DNA damage leading to senescence.

Immune cell quantity and quality are affected by aging, both due to changes to the microenvironment of immune organs including the thymus, spleen, and lymph nodes, and also due to inflammaging given that inflammatory factors can directly inhibit immune cell function [42,129]. The pro-inflammatory microenvironment in T2D begets immune cell dysfunction, which likely contributes to the increased susceptibility to infection and malignancy that is seen in this population [130]. An impaired or overwhelmed immune system is also permissive of increased senescent cell burden, since immune cells including natural killer (NK) cells, T-cells, and macrophages that are responsible for removing senescent cells are less able to clear them efficiently [42]. Senescence of primary immune cells themselves also occurs and can impact not only accumulation of senescent cells, but also metabolic function. For example, senescent T cells have been implicated in the dysregulation of hepatic glucose homeostasis [131].

4. Cellular senescence and diabetic complications

T2D is associated with widespread tissue dysfunction and onset of similar chronic diseases as in non-diabetic individuals, but at an earlier age (Fig. 10.2). [18] Therefore not only is T2D an age-related disease, but T2D itself may represent an accelerated aging-like state. Complications of diabetes are classically organized into two groups: microvascular and macrovascular complications. However, diabetes also increases the risk of infection, cancer, osteoporosis, and cognitive dysfunction, which can additionally be classified as diabetic complications.

4.1 Microvascular complications

Microvascular dysfunction involves the blood vessels between first-order arterioles and first-order venules that are present in virtually all tissues. In diabetes, microvascular dysfunction most commonly leads to retinopathy, nephropathy, and neuropathy. Other tissues that are affected by microvascular pathology include the brain, myocardium, and skin [115]. Microvascular damage is thought to begin much earlier than clinical signs of disease. For example, retinopathy can be seen as early as 7 years prior to a formal diagnosis of diabetes in individuals with T2D [132].

Hyperglycemia, chronic inflammation, and oxidative stress are implicated in endothelial dysfunction and microvascular complications of T2D [133]. These mechanisms can also drive cellular senescence. Intensive blood glucose control has been shown to delay the onset of microvascular complications or impede their progression [134,135]. The impact of glycemic control on existing senescent cell burden, in other words whether senescent cell accumulation can be reversed in the setting of improved glycemic control, has not been reported.

Senescent cell burden is increased in the retinas of individuals with T2D. In preclinical models of retinopathy, inhibition of the SASP using intravitreal injections of metformin, or removal of senescent cells through genetic means or with senolytic treatment reduced pathological angiogenesis [136,137]. Mice with streptozotocin-induced diabetes have increased renal endothelial cell senescence, which is related to M1 macrophage accumulation in the glomerulus [138]. Podocytes, specialized glomerular cells in the kidney, also undergo senescence with aging, however this has not been shown in a diabetes model [139]. In humans with diabetic kidney disease, circulating levels of the SASP factor Activin A are increased and correlate with kidney dysfunction and kidney injury markers [140].

4.2 Macrovascular complications

Macrovascular complications of diabetes involve atherosclerotic or thrombotic obstructions of vessels, including coronary, cerebral, and peripheral arteries. Senescence is known to occur in many cell types found in the cardiovascular system including endothelial cells, vascular smooth muscle cells, fibroblasts and myofibroblasts, and cardiomyocytes [141]. Senescent cells are implicated in the pathogenesis of atherosclerosis through multiple pathways including replicative senescence, ROS, and epigenetic changes, although the specific contribution of diabetes-induced senescent cells has not been well delineated [142,143]. T-cell senescence related to chronic inflammation is also implicated in aging-related cardiovascular disease, and accelerated telomere shortening is found in CD8$^+$ T cells isolated from individuals with coronary heart disease [141,144].

4.3 Cognition

Obesity and diabetes are risk factors for cognitive impairment and dementia. Vascular senescence has been proposed as one potential mechanism underlying this association, and evidence of cellular senescence or senescent-like phenotypes has been reported in diverse brain cells including microglia, oligodendrocytes, oligodendrocyte progenitor cells, astrocytes, neurons, and neural stem cells [31]. Senolytic therapy restored neurogenesis, reduced neuroinflammation, and partially reversed brain atrophy associated with cellular senescence in aged mice with Tau overexpression [145,146].

In addition to local effects in their resident tissue, senescent cells exert systemic effects through circulating SASP factors, and there is early evidence to suggest that peripheral senescent cells can have effects on the CNS. For example, anxiety associated with obesity in mice was abrogated after senescent cell clearance [147]. In humans who had undergone total knee arthroplasty, senescent cell burden was associated with reduction of a CNS geroprotector called alpha Klotho [148].

4.4 Cancer

Cellular senescence is in general a tumor-suppressive process, however once cells undergo senescence, they may contribute to a pro-tumorigenic microenvironment through secretion of the SASP [149]. T2D is associated with increased risk of hepatocellular, pancreatic, colorectal, breast, endometrial, and ovarian cancers [150]. In preclinical models, selective elimination of senescent cells delays cancer development in aged mice and DNA repair-deficient, cancer-prone, $Ercc1^{-/\Delta}$ mice [48,151]. However, further study is needed to determine whether and how cellular senescence contributes to the elevated cancer risk seen in diabetes, and whether elimination of senescent cells or their effects is sufficient to lower risk.

4.5 Osteoporosis

Fracture risk is increased in individuals with T2D despite normal bone mass. Hyperglycemia induces upregulation of sclerostin, which inhibits Wnt signaling that is important for osteogenic capacity of osteoblasts. Accumulation of AGEs may also lead to stiffening of collagen and less resistance to fractures [152]. The role of cellular senescence in diabetes-associated osteoporosis is not known, however removal of senescent cells in a mouse model of age-related osteoporosis did reduce sclerostin levels, suggesting that similar mechanisms may underlie age- and diabetes-induced bone fragility [153].

4.6 Impaired wound healing

Cellular senescence plays an important and dichotomous role in wound healing. In acute wounds, senescent cells act to limit fibrosis through ROS-p16 signaling [154]. However, this beneficial role of senescent cells seems to be limited to acute wounds. In chronic wounds, more commonly seen in diabetes, increased oxidative stress causes senescence in fibroblasts through caveolin-1 or polymerase I and transcript release factor (PTRF)-induced activation of the p53/p21 pathway. Inhibition of this pathway accelerated tissue repair, indicating that senescent fibroblasts are involved in impaired wound healing in diabetes [155,156].

5. Current diabetes therapeutics: Impact on cellular senescence

Although therapeutic options for diabetes have expanded in the last decade, behavioral interventions, or so-called lifestyle changes, remain the first-line treatment for impaired fasting glucose and T2D. When behavioral interventions, which typically focus on modifications in diet and physical activity, are insufficient to improve glucose control, oral antihyperglycemic medications are typically introduced next. These therapies have historically focused on reduction of hyperglycemia by increasing insulin production (sulfonylureas, incretin mimetics) or increasing tissue insulin sensitivity (metformin, thiazolinediones) rather than targeting the underlying fundamental mechanisms leading to insulin resistance. If hyperglycemia is profound or refractory to oral hyperglycemics, supplemental insulin, injected subcutaneously, may be introduced to improve glycemic control. Here, we discuss what is known regarding the impacts of currently available interventions on cellular senescence.

5.1 Diet and exercise

High-fat feeding or a Western diet in preclinical models induces accumulation of senescent cells, however the contributions of different dietary components to senescent cell formation is not well understood. Caloric restriction has been widely studied as a strategy to mitigate fundamental mechanisms of aging and to prolong healthspan and lifespan in preclinical models. Caloric restriction has not been shown to reduce the number of senescent cells already present at the time of intervention, however it may serve as a strategy to prevent the accumulation of senescent cells [157].

In a mouse model of diet-induced obesity, exercise prevented the accumulation of senescent cells with a resultant reduction in the SASP and has also been shown to reduce senescent cells in livers of mice with a genetic background that promotes spontaneous liver steatosis [158,159]. In humans, aerobic exercise has been associated with fewer senescent peripheral T cells [160,161]. However, more data are needed to understand whether the positive effects of exercise on senescent cell abundance or the SASP is a direct or indirect effect. A causal association of senescent cell burden as a mediator of the positive effects of exercise on metabolic health also remains to be shown experimentally.

In humans, weight loss by a combination of caloric restriction and aerobic exercise reduces DNA damage as measured by γH2AX in preadipocytes isolated from the femoral and abdominal adipose depots [67]. Weight loss through bariatric surgery may reduce some SASP factors; however, its impact on senescent cell burden has not been definitively reported [162].

5.2 Metformin

Metformin, a biguanide, is an oral antihyperglycemic drug that improves hepatic and muscle insulin sensitivity and is the first-line pharmacologic therapy for T2D. Metformin is the most commonly prescribed medication for T2D worldwide given its ease of use, low side effect burden, and low cost. Studies in model organisms including *C. elegans* and mice have demonstrated an ability of metformin to impact several fundamental aging mechanisms and to extend healthspan [163]. Because of this, metformin is the subject of a large clinical trial in planning to study the effects of a therapy to target aging on healthspan in humans (Targeting Aging by MEtformin, or TAME) [163].

Metformin inhibits the SASP by inhibiting NF-κB activation [164]. In preclinical studies, metformin extended the healthspan and lifespan in aged mice in part through adenosine monophosphate-activated protein kinase (AMPK) pathway activation leading to reduced oxidative damage and chronic inflammation [165]. Several studies have indicated that treatment with metformin in individuals with diabetes confers lower risk of osteoarthritis development or progression [166,167]. The mechanism of this effect is also postulated to be through AMPK activation, which theoretically could also affect senescent cell burden. However, metformin's effects on cellular aging mechanisms and cellular senescence have not been reported in this context.

5.3 SGLT-2 inhibitors

SGLT-2 inhibitors promote the excretion of glucose in the urine by inhibiting a sodium-glucose cotransporter in the proximal tubule of the kidney. Early evidence exists to suggest that SGLT-2 inhibitors may impact senescent cell formation or elimination. Administration of the SGLT-2 inhibitor empagliflozin to a rat model of hypertriglyceridemia led to reduced p21 expression in WAT and liver [168]. The potential independent impact of differences due to empagliflozin in weight and adipose tissue mass (both reduced in the empagliflozin group) and food intake (increased in the empagliflozin group) on p21 in that study is unclear. In obese *db/db* mice, dapagliflozin treatment reduced the development of senescence markers in the kidney, mediated through activation of NRF2 by increased concentrations of plasma ketone [169]. In vitro, treatment with empagliflozin reduced markers of senescence including p21 expression and senescence-associated beta galactosidase staining in 3T3-L1 adipocytes and HepG2 hepatocytes [168]. SGLT-2 inhibitors have cardioprotective and renoprotective properties that are not well understood. More work is needed to determine whether the positive multi-organ impacts seen with SGLT-2 treatment may be in part due to senolytic or senomorphic properties of these agents.

5.4 GLP-1R agonists and DPP-4 inhibitors

The incretin glucagon-like peptide-1 (GLP-1), which is released after meals and signals through its receptor GLP-1R, functions to stimulate insulin secretion, prevent glucagon release, suppress appetite, and increase β-cell growth and survival. GLP-1 has also been shown to have cardiovascular protective effects. GLP-1R agonists were initially developed as antihyperglycemic therapies, however they were noted to cause weight loss and are now FDA approved for this indication. Additionally, GLP-1R agonist therapy confers reduced cardiovascular and all-cause mortality [170]. However, it is not yet known whether these effects are mediated through impacts on cellular senescence. Similarly, the impact of DPP-4 inhibitors, which block the main enzyme responsible for degrading GLP-1, on cellular senescence has not been reported.

5.5 Insulin

Although insulin is highly effective in lowering plasma glucose, hyperinsulinemia is associated with the promotion of senescence across multiple tissues as discussed previously. It is not clear whether the negative effects of insulin-induced senescent cells outweigh the benefit of improving glycemic control, or whether insulin-induced senescent cells hinder the long-term preventative effects of maintaining euglycemia using insulin therapy. Emerging evidence suggests that SGLT-2i and GLP-1R agonists may have greater impact than exogenous insulin in preventing long-term complications of diabetes, however this requires formal study. The effect of combination therapy, such as with metformin, on preventing insulin-induced senescent cell formation is also unknown. These concepts require further study, and considerations regarding cellular senescence should not inform current treatment decisions for patients with diabetes as there are no formal data to evaluate the efficacy, risks, or benefits.

Particularly in older adults, insulin is associated with adverse effects related to hypoglycemia, including falls, confusion, dizziness, and delirium. In addition, use of insulin has been associated with accelerated cognitive decline, whereas this association has not been observed with other antidiabetic therapies such as metformin [16]. It is not known whether insulin-induced senescent cells play a role in these long-term effects of insulin therapy.

5.6 Sulfonylureas

Little is known regarding the impact of sulfonylureas, which stimulate insulin production by pancreatic β-cells, on cellular senescence. In a study of metformin's impact on the micro-RNA processing protein DICER1, sulfonylurea treatment was not as effective as metformin at reducing senescence markers and the SASP [171]. Similarly, sulfonylurea treatment was utilized as a control for SGLT2i therapy in *db/db* mice and did not show effects on senescent cell markers that were observed in the dapagliflozin group [169].

5.7 Acarbose

Acarbose inhibits absorption of carbohydrates into the gut to improve glycemic control. It is seldom used clinically due to gastrointestinal side effects and hypersensitivity reactions, but has been shown to extend the lifespan in mice. In a rabbit model of atherosclerosis, acarbose did reduce neointimal senescence in blood vessels, however additional evidence that acarbose impacts cellular senescence is lacking [172].

6. Targeting cellular senescence in diabetes

The Geroscience Hypothesis holds that aging is the major risk factor for most chronic diseases and predicts that by targeting fundamental aging processes, it may be possible to prevent, alleviate, or treat age-related dysfunction across tissues and disease states all at once. This would lead to a greater impact on the healthspan than targeting any one organ system or disease individually [16,21,173]. Many fundamental aging processes have been implicated in the pathogenesis of T2D, and given that these are interrelated, an intervention targeting one fundamental aging process may reasonably be expected to affect the others. This concept has been put forward as the Unitary Theory of Targeting Fundamental Aging Mechanisms [174]. The realization that SGLT-2 inhibitors have profoundly positive impacts on heart and kidney health despite being developed as antihyperglycemic agents is an illustration of the Unitary Theory.

A growing understanding of the widespread effects of senescent cells has led to ardent attempts to mitigate their effects through strategies to eliminate senescent cells or prevent their formation. Many strategies to target senescent cells are in development both in academic research labs and in biotechnology startups [175]. Interventions that prevent the generation of senescent cells, such as knocking down $p16^{Ink4a}$ or *p53* expression, have been shown in preclinical models to promote tumor development. Therefore, it is safer and more favorable to eliminate senescent cells that have already formed, rather than preventing senescence or restoring the ability of senescent cells to replicate [27,176]. Elimination of senescent cells through genetic targeting of highly p16- or p21-expressing cells alleviates or prevents age-related diseases, further supporting a potentially causal role for senescent cells in age-related dysfunction. Specifically in obese or diabetic model organisms, elimination of senescent cells alleviated insulin resistance, reduced proteinuria, and improved diastolic function [69].

6.1 Senolytic agents

In the mid-2010s, drugs that decrease senescent cell burden, called senolytics, were identified [49,177–179]. By targeting senescence-associated apoptotic pathways (SCAPs), senolytics allow those senescent cells that have a pro-inflammatory, pro-apoptotic SASP to undergo apoptosis in response to their own SASP, leading to the clearance of senescent cells expressing a tissue-damaging SASP [27,49]. Senolytics are typically administered on an intermittent dosing schedule, much like antibiotics or chemotherapeutics, in order to clear senescent cells that have accumulated over time or since the last round of senolytic therapy (Fig. 10.2). This intermittent dosing regimen mitigates the risk of adverse or off-target effects of a senolytic drug and may promote adherence.

Senolytic drugs affect other fundamental aging mechanisms, including reversal of age-related CD38 decline and restoration of NAD^+ levels, reduction of age-related inflammation and fibrosis, improvement in progenitor cell function, and reduction of DNA damage [70,153,180–184]. Systemic administration of senolytic drugs in humans in a small, open-label trial also indicated that senolytics may be effective in clearing senescent cells in humans, as measured in adipose tissue and blood SASP factors [185].

In addition to drugs currently in development as senolytics, it is becoming clear that several additional medications already in use in humans for various conditions exhibit senolytic effects. These include zoledronic acid and digoxin [186,187]. Tailoring medication regimens to substitute senolytic for non-senolytic members of a particular drug class may be a strategy implemented in the future to impact global senescent cell burden and alleviate age-related dysfunction across different organ systems.

6.2 SASP inhibitors ("senomorphic" drugs)

Several drugs that mitigate the effects of the SASP, rather than directly eliminating senescent cells, have been investigated for their possible role in reducing the downstream adverse effects of senescent cells. This could, for example, potentially prevent the spread of senescent cells to neighboring or distant cells. These agents, which are termed "senomorphic" drugs and include metformin, glucocorticoids, rapamycin, ruxolitinib, and p28MAPK inhibitors, reduce some but not all SASP factors [70,188–192]. Therefore, certain senomorphic drugs may be more beneficial than others in a particular disease state. It has also been observed that some drugs can be both senolytic and senomorphic, for example zoledronic acid, or have senolytic effects in certain cell types but senomorphic effects in others [193,194].

6.3 Immune strategies

As previously discussed, the impaired ability of immune cells to clear senescent cells leads to increased senescent cell burden. The SASP itself is one cause of immune cell impairment and could be targeted with senomorphics as a strategy to allow immune cell clearance of senescent cells [42]. More fine-tuned strategies to hone the immune system in on senescent cells are also in development. For example, immunization of mice against GPNMB, a transmembrane protein involved in endothelial cell adhesion that was found to be enriched in senescent endothelial cells, led to decreased senescent cell burden in mice [195]. This immunization approach was also noted to improve glucose tolerance and extend the lifespan of progeroid mice [195]. The long-term effects of vaccination against senescent cells are unknown.

6.4 Impact of senescent cell elimination on metabolic tissues and phenotypes

In preclinical models of aging and obesity, senolytic treatment has been shown to improve insulin sensitivity [69,70]. Additional therapeutic effects of senolytic treatment in mouse models have included improvements in kidney function, anxiety, frailty, cardiac fibrosis, liver steatosis, recovery after coronavirus infection, pulmonary fibrosis, and intervertebral disc degeneration, among others [39,48,69,104,147,151,153,182,196—199]. Senescent cell removal leads to an improvement in several metabolic parameters in adipose tissue including a decrease in adipocyte size, improved adipogenic potential, and reduced infiltration of macrophages [69,70,200]. In experiments involving transplantation of adipose tissue from obese to lean mice, which typically induces metabolic dysfunction in the lean mice, this effect was abrogated when p21^{Cip1} highly expressing cells were eliminated from adipose tissue prior to transplantation [200]. In vitro, senolytic drugs improved insulin sensitivity in explants of human adipose tissue isolated from obese individuals and inhibited hyperinsulinemia-induced senescence in human hepatocytes [70,119]. Similarly, treating adipose tissue explants from obese humans with senolytic drugs in vitro prior to transplantation into immunodeficient mice prevented the development of insulin resistance seen in mice that received transplants without senolytic treatment [200]. However, it is not yet known what effects senolytic therapy will have on metabolism and insulin sensitivity in humans.

7. Clinical trials

Preclinical studies indicate that senescence-targeting therapies hold great promise for promotion of the healthspan, however significant effort is needed to determine whether these therapies will be safe and effective in humans. Senolytic treatment was successful in clearing senescent cells from human adipose tissue in a small open-label trial of patients with diabetic kidney disease [185]. In another open-label trial, patients with idiopathic pulmonary fibrosis had improvements in physical function including walk distance, gait speed, and chair-stand time, following senolytic treatment [201].

A retrospective study of individuals who had been prescribed dasatinib, a senolytic Src tyrosine kinase inhibitor, compared to weakly senolytic imatinib, showed a 28.8 unit reduction in insulin requirement, accompanied by a nearly 5 kg weight loss and reduction in serum glucose concentrations [202]. No prospective study of senolytic therapy for diabetes outcomes has been reported. A trial to prevent development of T2D in healthy first-degree relatives of individuals with T2D is planned. Studying the effects of senolytic drugs in T2D is reasonable given its prevalence and impact on multiple organ systems. Complications of diabetes present a particularly attractive target for clinical trials of senolytics, given that few disease-modifying interventions are available for these concerns [19].

8. Challenges and future directions

8.1 Defining senescence and biomarkers of senescence (gerodiagnostics)

A major challenge in the study of therapeutics that target senescent cells is the very definition of cellular senescence, the properties of which may vary across cell types and tissues. Large-scale efforts are underway to map, define, and characterize senescent cells across tissues in both mice and humans [203]. Similarly, biomarkers of senescence that are agreed upon by the scientific community at large have not been identified. Several groups have proposed panels of SASP factors that might represent sensitive and specific measurements of senescence, the first of which was reported in 2019, however consensus has not yet been achieved [204—207]. Given the variation in the SASP among cell types and precipitants, the definition of target engagement as measured through the SASP might differ somewhat across senescence-associated disease states and target tissues. Gerodiagnostic composite scores of SASP factors, senescence markers, and other analytes reflecting fundamental aging processes in body fluids may hold more promise than assays of any one individual factor, especially for selecting the best intervention options and predicting and following responses to treatments.

8.2 Limited knowledge of long-term effects

Chronic senolytic treatment in mice and non-human primates has been reported to have enhanced impact on target phenotypes without adverse effects [199,208]. However, long-term effects of clearing senescent cells in humans are not yet known, and judicious safety monitoring is needed when performing clinical trials of senolytics. Gaps also exist in our knowledge regarding the length of trials needed to fully test senolytic therapies, the rate at which senescent cells reappear in different individuals with aging and in different disease states, and the efficacy of senolytic drugs in early versus advanced disease.

8.3 Considerations for trials in older adults

Hypoglycemia is of particular concern in older adults with diabetes, especially those who are treated with insulin, due to its potential to cause falls, confusion, and other adverse effects [209]. Older adults are at increased risk of hypoglycemia and are less likely to become symptomatic upon becoming hypoglycemic, which can lead to underrecognition and underdiagnosis [209–211].

Testing of new interventions in the older adult population with diabetes can be challenging due to comorbidities, renal dysfunction, and drug–drug interactions that must be considered. It is important to note that heterogeneity increases with aging, and this may necessitate the use of individualized therapeutic strategies based, for example, on assessment of a particular individual's predisposition to a particular age-related mechanism and future gerodiagnostic panels. In this way, the application of individualized medicine approaches may be needed to optimize the potential and reduce side effects of therapeutics that target aging mechanisms.

9. Conclusions

Aging and obesity are inextricably linked on a population level and can both individually lead to insulin resistance and T2D, in part due to induction and exacerbation of fundamental aging mechanisms including cellular senescence. Senescent cells likely play a significant role in diabetes pathogenesis by causing dysfunction of several cell types and organs. In turn, the diabetic microenvironment and pro-inflammatory state begets additional cellular senescence, forming a pathogenic loop. Cellular senescence represents an opportunity to devise novel therapeutic strategies that prevent or alleviate insulin resistance, T2D, and their downstream effects. Therapies to target cellular senescence and other age-related cellular disorders are being tested in early clinical trials and hold promise for the prevention and treatment of metabolic syndrome, diabetes, and their complications.

References

[1] International Diabetes Federation. IDF diabetes atlas. Brussels, Belgium. 10th ed. 2021. Available at: https://www.diabetesatlas.org.
[2] Ehrhardt N, Cui J, Dagdeviren S, Saengnipanthkul S, Goodridge HS, Kim JK, et al. Adiposity-independent effects of aging on insulin sensitivity and clearance in mice and humans. Obesity 2019;27(3):434–43.
[3] Aguayo-Mazzucato C. Functional changes in beta cells during ageing and senescence. Diabetologia 2020;63(10):2022–9.
[4] American Diabetes Association Professional Practice C. 2. Classification and diagnosis of diabetes: standards of medical care in diabetes-2022. Diabetes Care 2022;45(Suppl. 1):S17–38.
[5] Palmer AK, Jensen MD. Metabolic changes in aging humans: current evidence and therapeutic strategies. J Clin Invest 2022;132(16).
[6] Kuk JL, Saunders TJ, Davidson LE, Ross R. Age-related changes in total and regional fat distribution. Ageing Res Rev 2009;8(4):339–48.
[7] Karakelides H, Irving BA, Short KR, O'Brien P, Nair KS. Age, obesity, and sex effects on insulin sensitivity and skeletal muscle mitochondrial function. Diabetes 2010;59(1):89–97.
[8] Snijder MB, Dekker JM, Visser M, Yudkin JS, Stehouwer CD, Bouter LM, et al. Larger thigh and hip circumferences are associated with better glucose tolerance: the Hoorn study. Obes Res 2003;11(1):104–11.
[9] Manolopoulos KN, Karpe F, Frayn KN. Gluteofemoral body fat as a determinant of metabolic health. Int J Obes 2010;34(6):949–59.
[10] Boden G, Jadali F, White J, Liang Y, Mozzoli M, Chen X, et al. Effects of fat on insulin-stimulated carbohydrate metabolism in normal men. J Clin Invest 1991;88(3):960–6.
[11] Kirkland JL, Tchkonia T, Pirtskhalava T, Han J, Karagiannides I. Adipogenesis and aging: does aging make fat go MAD? Exp Gerontol 2002;37:757–67.
[12] Baumer Y, Pita MA, Baez AS, Ortiz-Whittingham LR, Cintron MA, Rose RR, et al. By what molecular mechanisms do social determinants impact cardiometabolic risk? Clin Sci (Lond) 2023;137(6):469–94.
[13] Belsky DW, Caspi A, Cohen HJ, Kraus WE, Ramrakha S, Poulton R, et al. Impact of early personal-history characteristics on the pace of aging: implications for clinical trials of therapies to slow aging and extend healthspan. Aging Cell 2017;16(4):644–51.
[14] Raffington L, Belsky DW, Kothari M, Malanchini M, Tucker-Drob EM, Harden KP. Socioeconomic disadvantage and the pace of biological aging in children. Pediatrics 2021;147(6).
[15] Bonnefond A, Semple RK. Achievements, prospects and challenges in precision care for monogenic insulin-deficient and insulin-resistant diabetes. Diabetologia 2022;65(11):1782–95.
[16] Sierra F. The emergence of geroscience as an interdisciplinary approach to the enhancement of health span and life span. Cold Spring Harb Perspect Med 2016;6(4):a025163.
[17] Lopez-Otin C, Blasco MA, Partridge L, Serrano M, Kroemer G. The hallmarks of aging. Cell 2013;153(6):1194–217.
[18] Palmer AK, Gustafson B, Kirkland JL, Smith U. Cellular senescence: at the nexus between ageing and diabetes. Diabetologia 2019;62(10):1835–41.
[19] Palmer AK, Tchkonia T, LeBrasseur NK, Chini EN, Xu M, Kirkland JL. Cellular senescence in type 2 diabetes: a therapeutic opportunity. Diabetes 2015;64(7):2289–98.
[20] Tchkonia T, Kirkland JL. Aging, cell senescence, and chronic disease: emerging therapeutic strategies. JAMA 2018;320(13):1319–20.
[21] Kirkland JL. Translating the science of aging into therapeuticiInterventions. Cold Spring Harb Perspect Med 2016;6(3):a025908.

[22] Hayflick L, Moorehead P. The serial cultivation of human diploid strains. Exp Cell Res 1961;25:585–621.
[23] Munoz-Espin D, Serrano M. Cellular senescence: from physiology to pathology. Nat Rev Mol Cell Biol 2014;15(7):482–96.
[24] Campisi J. Senescent cells, tumor suppression, and organismal aging: good citizens, bad neighbors. Cell 2005;120:513–22.
[25] Krizhanovsky V, Yon M, Dickins RA, Hearn S, Simon J, Miething C, et al. Senescence of activated stellate cells limits liver fibrosis. Cell 2008; 134:657–67.
[26] Demaria M, Ohtani N, Youssef SA, Rodier F, Toussaint W, Mitchell JR, et al. An essential role for senescent cells in optimal wound healing through secretion of PDGF-AA. Dev Cell 2014;31(6):722–33.
[27] Kirkland JL, Tchkonia T. Cellular senescence: a translational perspective. EBioMedicine 2017;21:21–8.
[28] Zhu Y, Armstrong JL, Tchkonia T, Kirkland JL. Cellular senescence and the senescent secretory phenotype in age-related chronic diseases. Curr Opin Clin Nutr Metab Care 2014;17(4):324–8.
[29] Criscione SW, Teo YV, Neretti N. The chromatin landscape of cellular senescence. Trends Genet 2016;32(11):751–61.
[30] von Zglinicki T, Wan T, Miwa S. Senescence in post-mitotic cells: a driver of aging? Antioxid Redox Signal 2021;34(4):308–23.
[31] Rachmian N, Krizhanovsky V. Senescent cells in the brain and where to find them. FEBS J 2023;290(5):1256–66.
[32] Sapieha P, Mallette FA. Cellular senescence in postmitotic cells: beyond growth arrest. Trends Cell Biol 2018;28(8):595–607.
[33] Coppe JP, Desprez PY, Krtolica A, Campisi J. The senescence-associated secretory phenotype: the dark side of tumor suppression. Annu Rev Pathol 2010;5:99–118.
[34] Iske J, Seyda M, Heinbokel T, Maenosono R, Minami K, Nian Y, et al. Senolytics prevent mt-DNA-induced inflammation and promote the survival of aged organs following transplantation. Nat Commun 2020;11(1):4289.
[35] Wallis R, Mizen H, Bishop CL. The bright and dark side of extracellular vesicles in the senescence-associated secretory phenotype. Mech Ageing Dev 2020;189:111263.
[36] Tchkonia T, Zhu Y, van Deursen J, Campisi J, Kirkland JL. Cellular senescence and the senescent secretory phenotype: therapeutic opportunities. J Clin Invest 2013;123(3):966–72.
[37] Chaib S, Tchkonia T, Kirkland JL. Cellular senescence and senolytics: the path to the clinic. Nat Med 2022;28(8):1556–68.
[38] Hernandez-Segura A, de Jong TV, Melov S, Guryev V, Campisi J, Demaria M. Unmasking transcriptional heterogeneity in senescent cells. Curr Biol 2017;27(17):2652–60.
[39] Camell CD, Yousefzadeh MJ, Zhu Y, Prata L, Huggins MA, Pierson M, et al. Senolytics reduce coronavirus-related mortality in old mice. Science 2021;373(6552).
[40] Tripathi U, Nchioua R, Prata L, Zhu Y, Gerdes EOW, Giorgadze N, et al. SARS-CoV-2 causes senescence in human cells and exacerbates the senescence-associated secretory phenotype through TLR-3. Aging (Albany NY) 2021;13(18):21838–54.
[41] De Cecco M, Ito T, Petrashen AP, Elias AE, Skvir NJ, Criscione SW, et al. L1 drives IFN in senescent cells and promotes age-associated inflammation. Nature 2019;566(7742):73–8.
[42] Prata L, Ovsyannikova IG, Tchkonia T, Kirkland JL. Senescent cell clearance by the immune system: emerging therapeutic opportunities. Semin Immunol 2018;40:101275.
[43] Eggert T, Wolter K, Ji J, Ma C, Yevsa T, Klotz S, et al. Distinct functions of senescence-associated immune responses in liver tumor surveillance and tumor progression. Cancer Cell 2016;30(4):533–47.
[44] Song P, An J, Zou MH. Immune clearance of senescent cells to combat ageing and chronic diseases. Cells 2020;9(3).
[45] Acosta JC, Banito A, Wuestefeld T, Georgilis A, Janich P, Morton JP, et al. A complex secretory program orchestrated by the inflammasome controls paracrine senescence. Nat Cell Biol 2013;15(8):978–90.
[46] Wang B, Liu Z, Chen VP, Wang L, Inman CL, Zhou Y, et al. Transplanting cells from old but not young donors causes physical dysfunction in older recipients. Aging Cell 2020;329(14):e13106.
[47] Kim SR, Jiang K, Ferguson CM, Tang H, Chen X, Zhu X, et al. Transplanted senescent renal scattered tubular-like cells induce injury in the mouse kidney. Am J Physiol Ren Physiol 2020;318(5):F1167–76.
[48] Xu M, Pirtskhalava T, Farr JN, Weigand BM, Palmer AK, Weivoda MM, et al. Senolytics improve physical function and increase lifespan in old age. Nat Med 2018;24(8):1246–56.
[49] Zhu Y, Tchkonia T, Pirtskhalava T, Gower AC, Ding H, Giorgadze N, et al. The Achilles' heel of senescent cells: from transcriptome to senolytic drugs. Aging Cell 2015;14(4):644–58.
[50] Biran A, Zada L, Abou Karam P, Vadai E, Roitman L, Ovadya Y, et al. Quantitative identification of senescent cells in aging and disease. Aging Cell 2017;16(4):661–71.
[51] Palmer AK, Kirkland JL. Aging and adipose tissue: potential interventions for diabetes and regenerative medicine. Exp Gerontol 2016;86(16): 30054–7. https://doi.org/10.1016/j.exger.2016.02.013.
[52] Bonora E, Del Prato S, Bonadonna RC, Gulli G, Solini A, Shank ML, et al. Total body fat content and fat topography are associated differently with in vivo glucose metabolism in nonobese and obese nondiabetic women. Diabetes 1992;41(9):1151–9.
[53] Romero-Corral A, Somers VK, Sierra-Johnson J, Thomas RJ, Collazo-Clavell ML, Korinek J, et al. Accuracy of body mass index in diagnosing obesity in the adult general population. Int J Obes 2008;32(6):959–66.
[54] Tilg H, Moschen AR. Adipocytokines: mediators linking adipose tissue, inflammation and immunity. Nat Rev Immunol 2006;6(10):772–83.
[55] Kershaw EE, Flier JS. Adipose tissue as an endocrine organ. J Clin Endocrinol Metab 2004;89:2548–56.
[56] Fontana L, Eagon JC, Trujillo ME, Scherer PE, Klein S. Visceral fat adipokine secretion is associated with systemic inflammation in obese humans. Diabetes 2007;56(4):1010–3.
[57] Huffman DM, Barzilai N. Contribution of adipose tissue to health span and longevity. Interdiscipl Top Gerontol 2010;37:1–19.
[58] Picard F, Guarente L. Molecular links between aging and adipose tissue. Int J Obes 2005;29(Suppl. 1):S36–9.
[59] Tchkonia T, Morbeck DE, von Zglinicki T, van Deursen J, Lustgarten J, Scrable H, et al. Fat tissue, aging, and cellular senescence. Aging Cell 2010;9:667–84.
[60] Bluher M, Kahn BB, Kahn CR. Extended longevity in mice lacking the insulin receptor in adipose tissue. Science 2003;299(5606):572–4.
[61] Giannakou ME, Goss M, Junger MA, Hafen E, Leevers SJ, Partridge L. Long-lived Drosophila with overexpressed dFOXO in adult fat body. Science 2004;305(5682):361.

References

[62] Longo M, Zatterale F, Naderi J, Parrillo L, Formisano P, Raciti GA, et al. Adipose tissue dysfunction as determinant of obesity-associated metabolic complications. Int J Mol Sci 2019;20(9).

[63] Espinosa De Ycaza AE, Sondergaard E, Morgan-Bathke M, Carranza Leon BG, Lytle KA, Ramos P, et al. Senescent cells in human adipose tissue: a cross-sectional study. Obesity 2021;29(8):1320—7.

[64] Kirkland JL, Hollenberg CH, Gillon WS. Age, anatomic site, and the replication and differentiation of adipocyte precursors. Am J Physiol 1990;258:C206—10.

[65] Minamino T, Orimo M, Shimizu I, Kunieda T, Yokoyama M, Ito T, et al. A crucial role for adipose tissue p53 in the regulation of insulin resistance. Nat Med 2009;15(9):1082—7.

[66] Smith U, Li Q, Ryden M, Spalding KL. Cellular senescence and its role in white adipose tissue. Int J Obes 2021;45(5):934—43.

[67] Murphy J, Tam BT, Kirkland JL, Tchkonia T, Giorgadze N, Pirtskhalava T, et al. Senescence markers in subcutaneous preadipocytes differ in childhood- versus adult-onset obesity before and after weight loss. Obesity 2023;31(6).

[68] Gustafson B, Nerstedt A, Smith U. Reduced subcutaneous adipogenesis in human hypertrophic obesity is linked to senescent precursor cells. Nat Commun 2019;10(1):2757.

[69] Palmer AK, Xu M, Zhu Y, Pirtskhalava T, Weivoda MM, Hachfeld CM, et al. Targeting senescent cells alleviates obesity-induced metabolic dysfunction. Aging Cell 2019;18(3):e12950.

[70] Xu M, Palmer AK, Ding H, Weivoda MM, Pirtskhalava T, White TA, et al. Targeting senescent cells enhances adipogenesis and metabolic function in old age. Elife 2015;4. https://doi.org/10.7554/eLife.12997.

[71] Hammarstedt A, Gogg S, Hedjazifar S, Nerstedt A, Smith U. Impaired adipogenesis and dysfunctional adipose tissue in human hypertrophic obesity. Physiol Rev 2018;98(4):1911—41.

[72] Spinelli R, Florese P, Parrillo L, Zatterale F, Longo M, D'Esposito V, et al. ZMAT3 hypomethylation contributes to early senescence of preadipocytes from healthy first-degree relatives of type 2 diabetics. Aging Cell 2022;21(3):e13557.

[73] Cao Y. Angiogenesis modulates adipogenesis and obesity. J Clin Invest 2007;117(9):2362—8.

[74] Villaret A, Galitzky J, Decaunes P, Esteve D, Marques MA, Sengenes C, et al. Adipose tissue endothelial cells from obese human subjects: differences among depots in angiogenic, metabolic, and inflammatory gene expression and cellular senescence. Diabetes 2010;59:2755—63.

[75] Briot A, Decaunes P, Volat F, Belles C, Coupaye M, Ledoux S, et al. Senescence alters PPARgamma (peroxisome proliferator-activated receptor gamma)-dependent fatty acid handling in human adipose tissue microvascular endothelial cells and favors inflammation. Arterioscler Thromb Vasc Biol 2018;38(5):1134—46.

[76] Wang B, Wang L, Gasek NS, Zhou Y, Kim T, Guo C, et al. An inducible p21-Cre mouse model to monitor and manipulate p21-highly-expressing senescent cells in vivo. Nat Aging 2021;1(10):962—73.

[77] Barinda AJ, Ikeda K, Nugroho DB, Wardhana DA, Sasaki N, Honda S, et al. Endothelial progeria induces adipose tissue senescence and impairs insulin sensitivity through senescence associated secretory phenotype. Nat Commun 2020;11(1):481.

[78] Chen YW, Harris RA, Hatahet Z, Chou KM. Ablation of XP-V gene causes adipose tissue senescence and metabolic abnormalities. Proc Natl Acad Sci U S A 2015;112(33):E4556—64.

[79] Vergoni B, Cornejo PJ, Gilleron J, Djedaini M, Ceppo F, Jacquel A, et al. DNA damage and the activation of the p53 pathway mediate alterations in metabolic and secretory functions of adipocytes. Diabetes 2016;65(10):3062—74.

[80] Lee G, Kim YY, Jang H, Han JS, Nahmgoong H, Park YJ, et al. SREBP1c-PARP1 axis tunes anti-senescence activity of adipocytes and ameliorates metabolic imbalance in obesity. Cell Metabol 2022;34(5):702—18.

[81] Li Q, Hagberg CE, Silva Cascales H, Lang S, Hyvonen MT, Salehzadeh F, et al. Obesity and hyperinsulinemia drive adipocytes to activate a cell cycle program and senesce. Nat Med 2021;27(11):1941—53.

[82] Saisho Y, Butler AE, Manesso E, Elashoff D, Rizza RA, Butler PC. beta-cell mass and turnover in humans: effects of obesity and aging. Diabetes Care 2013;36(1):111—7.

[83] Perl S, Kushner JA, Buchholz BA, Meeker AK, Stein GM, Hsieh M, et al. Significant human beta-cell turnover is limited to the first three decades of life as determined by in vivo thymidine analog incorporation and radiocarbon dating. J Clin Endocrinol Metab 2010;95(10): E234—9.

[84] Menge BA, Tannapfel A, Belyaev O, Drescher R, Muller C, Uhl W, et al. Partial pancreatectomy in adult humans does not provoke beta-cell regeneration. Diabetes 2008;57(1):142—9.

[85] Helman A, Klochendler A, Azazmeh N, Gabai Y, Horwitz E, Anzi S, et al. p16(Ink4a)-induced senescence of pancreatic beta cells enhances insulin secretion. Nat Med 2016;22(4):412—20.

[86] Weir GC, Bonner-Weir S. Five stages of evolving beta-cell dysfunction during progression to diabetes. Diabetes 2004;53(Suppl. 3):S16—21.

[87] Rahier J, Guiot Y, Goebbels RM, Sempoux C, Henquin JC. Pancreatic beta-cell mass in European subjects with type 2 diabetes. Diabetes Obes Metabol 2008;10(Suppl. 4):32—42.

[88] Kaneto H. Pancreatic beta-cell glucose toxicity in type 2 diabetes mellitus. Curr Diabetes Rev 2015;11(1):2—6.

[89] Liang C, Hao F, Yao X, Qiu Y, Liu L, Wang S, et al. Hypericin maintians PDX1 expression via the Erk pathway and protects islet beta-cells against glucotoxicity and lipotoxicity. Int J Biol Sci 2019;15(7):1472—87.

[90] Zhu M, Liu X, Liu W, Lu Y, Cheng J, Chen Y. Beta cell aging and age-related diabetes. Aging (Albany NY) 2021;13(5):7691—706.

[91] Iozzo P, Beck-Nielsen H, Laakso M, Smith U, Yki-Jarvinen H, Ferrannini E. Independent influence of age on basal insulin secretion in nondiabetic humans. European group for the study of insulin resistance. J Clin Endocrinol Metab 1999;84(3):863—8.

[92] Shanik MH, Xu Y, Skrha J, Dankner R, Zick Y, Roth J. Insulin resistance and hyperinsulinemia: is hyperinsulinemia the cart or the horse? Diabetes Care 2008;31(Suppl. 2):S262—8.

[93] Ferrannini E, Gastaldelli A, Miyazaki Y, Matsuda M, Mari A, DeFronzo RA. beta-Cell function in subjects spanning the range from normal glucose tolerance to overt diabetes: a new analysis. J Clin Endocrinol Metab 2005;90(1):493—500.

[94] Pories WJ, Dohm GL. Diabetes: have we got it all wrong? Hyperinsulinism as the culprit: surgery provides the evidence. Diabetes Care 2012; 35(12):2438—42.

[95] Sousa-Victor P, Gutarra S, Garcia-Prat L, Rodriguez-Ubreva J, Ortet L, Ruiz-Bonilla V, et al. Geriatric muscle stem cells switch reversible quiescence into senescence. Nature 2014;506(7488):316—21.

[96] Garcia-Prat L, Martinez-Vicente M, Perdiguero E, Ortet L, Rodriguez-Ubreva J, Rebollo E, et al. Autophagy maintains stemness by preventing senescence. Nature 2016;529(7584):37–42.
[97] Schafer MJ, Miller JD, LeBrasseur NK. Cellular senescence: implications for metabolic disease. Mol Cell Endocrinol 2016;455:93–102.
[98] Baker DJ, Wijshake T, Tchkonia T, LeBrasseur NK, Childs BG, van de Sluis B, et al. Clearance of p16Ink4a-positive senescent cells delays ageing-associated disorders. Nature 2011;479(7372):232–6.
[99] Norton L, Shannon C, Gastaldelli A, DeFronzo RA. Insulin: the master regulator of glucose metabolism. Metabolism 2022;129:155142.
[100] Wang M, Li L, Xu Y, Du J, Ling C. Roles of hepatic stellate cells in NAFLD: from the perspective of inflammation and fibrosis. Front Pharmacol 2022;13:958428.
[101] Molla NW, Hassanain MM, Fadel Z, Boucher LM, Madkhali A, Altahan RM, et al. Effect of non-alcoholic liver disease on recurrence rate and liver regeneration after liver resection for colorectal liver metastases. Curr Oncol 2017;24(3):e233–43.
[102] Dorr JR, Yu Y, Milanovic M, Beuster G, Zasada C, Dabritz JH, et al. Synthetic lethal metabolic targeting of cellular senescence in cancer therapy. Nature 2013;501(7467):421–5.
[103] Kaplon J, Zheng L, Meissl K, Chaneton B, Selivanov VA, Mackay G, et al. A key role for mitochondrial gatekeeper pyruvate dehydrogenase in oncogene-induced senescence. Nature 2013;498(7452):109–12.
[104] Ogrodnik M, Miwa S, Tchkonia T, Tiniakos D, Wilson CL, Lahat A, et al. Cellular senescence drives age-dependent hepatic steatosis. Nat Commun 2017;8:15691.
[105] Ogrodnik M, Jurk D. Senescence explains age- and obesity-related liver steatosis. Cell Stress 2017;1(1):70–2.
[106] Baboota RK, Rawshani A, Bonnet L, Li X, Yang H, Mardinoglu A, et al. BMP4 and Gremlin 1 regulate hepatic cell senescence during clinical progression of NAFLD/NASH. Nat Metab 2022;4(8):1007–21.
[107] Morley JE. Diabetes and aging: epidemiologic overview. Clin Geriatr Med 2008;24(3):395–405.
[108] Aronson D, Edelman ER. Coronary artery disease and diabetes mellitus. Cardiol Clin 2014;32(3):439–55.
[109] Yokoi T, Fukuo K, Yasuda O, Hotta M, Miyazaki J, Takemura Y, et al. Apoptosis signal-regulating kinase 1 mediates cellular senescence induced by high glucose in endothelial cells. Diabetes 2006;55(6):1660–5.
[110] Blazer S, Khankin E, Segev Y, Ofir R, Yalon-Hacohen M, Kra-Oz Z, et al. High glucose-induced replicative senescence: point of no return and effect of telomerase. Biochem Biophys Res Commun 2002;296(1):93–101.
[111] Cramer C, Freisinger E, Jones RK, Slakey DP, Dupin CL, Newsome ER, et al. Persistent high glucose concentrations alter the regenerative potential of mesenchymal stem cells. Stem Cell Dev 2010;19(12):1875–84.
[112] Spravchikov N, Sizyakov G, Gartsbein M, Accili D, Tennenbaum T, Wertheimer E. Glucose effects on skin keratinocytes: implications for diabetes skin complications. Diabetes 2001;50(7):1627–35.
[113] Prattichizzo F, De Nigris V, Mancuso E, Spiga R, Giuliani A, Matacchione G, et al. Short-term sustained hyperglycaemia fosters an archetypal senescence-associated secretory phenotype in endothelial cells and macrophages. Redox Biol 2018;15:170–81.
[114] Maeda M, Hayashi T, Mizuno N, Hattori Y, Kuzuya M. Intermittent high glucose implements stress-induced senescence in human vascular endothelial cells: role of superoxide production by NADPH oxidase. PLoS One 2015;10(4):e0123169.
[115] Barrett EJ, Liu Z, Khamaisi M, King GL, Klein R, Klein BEK, et al. Diabetic microvascular disease: an endocrine society scientific statement. J Clin Endocrinol Metab 2017;102(12):4343–410.
[116] Yu T, Robotham JL, Yoon Y. Increased production of reactive oxygen species in hyperglycemic conditions requires dynamic change of mitochondrial morphology. Proc Natl Acad Sci U S A 2006;103(8):2653–8.
[117] Liu J, Huang K, Cai GY, Chen XM, Yang JR, Lin LR, et al. Receptor for advanced glycation end-products promotes premature senescence of proximal tubular epithelial cells via activation of endoplasmic reticulum stress-dependent p21 signaling. Cell Signal 2014;26(1):110–21.
[118] Krieger-Brauer HI, Kather H. Human fat cells possess a plasma membrane-bound H2O2-generating system that is activated by insulin via a mechanism bypassing the receptor kinase. J Clin Invest 1992;89(3):1006–13.
[119] Baboota RK, Spinelli R, Erlandsson MC, Brandao BB, Lino M, Yang H, et al. Chronic hyperinsulinemia promotes human hepatocyte senescence. Mol Metabol 2022;64:101558.
[120] Meijnikman AS, van Olden CC, Aydin O, Herrema H, Kaminska D, Lappa D, et al. Hyperinsulinemia is highly associated with markers of hepatocytic senescence in two independent cohorts. Diabetes 2022;71(9):1929–36.
[121] Keaney Jr JF, Larson MG, Vasan RS, Wilson PW, Lipinska I, Corey D, et al. Obesity and systemic oxidative stress: clinical correlates of oxidative stress in the Framingham Study. Arterioscler Thromb Vasc Biol 2003;23(3):434–9.
[122] Gardner JP, Li S, Srinivasan SR, Chen W, Kimura M, Lu X, et al. Rise in insulin resistance is associated with escalated telomere attrition. Circulation 2005;111(17):2171–7.
[123] Victorelli S, Passos JF. Telomeres and cell senescence—size matters not. EBioMedicine 2017;21:14–20.
[124] Festa A, D'Agostino Jr R, Howard G, Mykkanen L, Tracy RP, Haffner SM. Chronic subclinical inflammation as part of the insulin resistance syndrome: the Insulin Resistance Atherosclerosis Study (IRAS). Circulation 2000;102(1):42–7.
[125] Vozarova B, Weyer C, Hanson K, Tataranni PA, Bogardus C, Pratley RE. Circulating interleukin-6 in relation to adiposity, insulin action, and insulin secretion. Obes Res 2001;9(7):414–7.
[126] Franceschi C, Campisi J. Chronic inflammation (inflammaging) and its potential contribution to age-associated diseases. J Gerontol A Biol Sci Med Sci 2014;69(Suppl. 1):S4–9.
[127] Franceschi C, Bonafe M, Valensin S, Olivieri F, De Luca M, Ottaviani E, et al. Inflamm-aging. An evolutionary perspective on immunosenescence. Ann N Y Acad Sci 2000;908:244–54.
[128] Bardini G, Dicembrini I, Cresci B, Rotella CM. Inflammation markers and metabolic characteristics of subjects with 1-h plasma glucose levels. Diabetes Care 2010;33(2):411–3.
[129] Chinn IK, Blackburn CC, Manley NR, Sempowski GD. Changes in primary lymphoid organs with aging. Semin Immunol 2012;24(5):309–20.
[130] Berbudi A, Rahmadika N, Tjahjadi AI, Ruslami R. Type 2 diabetes and its impact on the immune system. Curr Diabetes Rev 2020;16(5):442–9.
[131] Yi HS, Kim SY, Kim JT, Lee YS, Moon JS, Kim M, et al. T-cell senescence contributes to abnormal glucose homeostasis in humans and mice. Cell Death Dis 2019;10(3):249.
[132] Fong DS, Aiello LP, Ferris 3rd FL, Klein R. Diabetic retinopathy. Diabetes Care 2004;27(10):2540–53.

[133] Horton WB, Barrett EJ. Microvascular dysfunction in diabetes mellitus and cardiometabolic disease. Endocr Rev 2021;42(1):29–55.
[134] Nathan DM, Genuth S, Lachin J, Cleary P, Crofford O, Davis M, et al. The effect of intensive treatment of diabetes on the development and progression of long-term complications in insulin-dependent diabetes mellitus. N Engl J Med 1993;329(14):977–86.
[135] Effect of intensive blood-glucose control with metformin on complications in overweight patients with type 2 diabetes (UKPDS 34). UK Prospective Diabetes Study (UKPDS) Group. Lancet 1998;352(9131):854–65.
[136] Crespo-Garcia S, Tsuruda PR, Dejda A, Ryan RD, Fournier F, Chaney SY, et al. Pathological angiogenesis in retinopathy engages cellular senescence and is amenable to therapeutic elimination via BCL-xL inhibition. Cell Metabol 2021;33(4):818–32.
[137] Oubaha M, Miloudi K, Dejda A, Guber V, Mawambo G, Germain MA, et al. Senescence-associated secretory phenotype contributes to pathological angiogenesis in retinopathy. Sci Transl Med 2016;8(362):362ra144.
[138] Yu S, Cheng Y, Li B, Xue J, Yin Y, Gao J, et al. M1 macrophages accelerate renal glomerular endothelial cell senescence through reactive oxygen species accumulation in streptozotocin-induced diabetic mice. Int Immunopharm 2020;81:106294.
[139] Fang Y, Chen B, Liu Z, Gong AY, Gunning WT, Ge Y, et al. Age-related GSK3beta overexpression drives podocyte senescence and glomerular aging. J Clin Invest 2022;132(4).
[140] Bian X, Griffin TP, Zhu X, Islam MN, Conley SM, Eirin A, et al. Senescence marker activin A is increased in human diabetic kidney disease: association with kidney function and potential implications for therapy. BMJ Open Diabetes Res Care 2019;7(1):e000720.
[141] Hu C, Zhang X, Teng T, Ma ZG, Tang QZ. Cellular senescence in cardiovascular diseases: a systematic review. Aging Dis 2022;13(1):103–28.
[142] Wang JC, Bennett M. Aging and atherosclerosis: mechanisms, functional consequences, and potential therapeutics for cellular senescence. Circ Res 2012;111(2):245–59.
[143] Roos CM, Zhang B, Palmer AK, Ogrodnik MB, Pirtskhalava T, Thalji NM, et al. Chronic senolytic treatment alleviates established vasomotor dysfunction in aged or atherosclerotic mice. Aging Cell 2016;15(5):973–7.
[144] Spyridopoulos I, Hoffmann J, Aicher A, Brummendorf TH, Doerr HW, Zeiher AM, et al. Accelerated telomere shortening in leukocyte subpopulations of patients with coronary heart disease: role of cytomegalovirus seropositivity. Circulation 2009;120(14):1364–72.
[145] Musi N, Valentine JM, Sickora KR, Baeuerle E, Thompson CS, Shen Q, et al. Tau protein aggregation is associated with cellular senescence in the brain. Aging Cell 2018;17(6):e12840.
[146] Zhang P, Kishimoto Y, Grammatikakis I, Gottimukkala K, Cutler RG, Zhang S, et al. Senolytic therapy alleviates Abeta-associated oligodendrocyte progenitor cell senescence and cognitive deficits in an Alzheimer's disease model. Nat Neurosci 2019;22(5):719–28.
[147] Ogrodnik M, Zhu Y, Langhi LGP, Tchkonia T, Kruger P, Fielder E, et al. Obesity-induced cellular senescence drives anxiety and impairs neurogenesis. Cell Metabol 2019;29(5):1061–77.
[148] Zhu Y, Prata L, Gerdes EOW, Netto JME, Pirtskhalava T, Giorgadze N, et al. Orally-active, clinically-translatable senolytics restore alpha-Klotho in mice and humans. EBioMedicine 2022;77:103912.
[149] Laberge RM, Sun Y, Orjalo AV, Patil CK, Freund A, Zhou L, et al. MTOR regulates the pro-tumorigenic senescence-associated secretory phenotype by promoting IL1A translation. Nat Cell Biol 2015;17(8):1049–61.
[150] Tomic D, Shaw JE, Magliano DJ. The burden and risks of emerging complications of diabetes mellitus. Nat Rev Endocrinol 2022;18(9):525–39.
[151] Yousefzadeh MJ, Zhu Y, McGowan SJ, Angelini L, Fuhrmann-Stroissnigg H, Xu M, et al. Fisetin is a senotherapeutic that extends health and lifespan. EBioMedicine 2018;36:18–28.
[152] Hofbauer LC, Busse B, Eastell R, Ferrari S, Frost M, Muller R, et al. Bone fragility in diabetes: novel concepts and clinical implications. Lancet Diabetes Endocrinol 2022;10(3):207–20.
[153] Farr JN, Xu M, Weivoda MM, Monroe DG, Fraser DG, Onken JL, et al. Targeting cellular senescence prevents age-related bone loss in mice. Nat Med 2017;23(9):1072–9.
[154] Andrade AM, Sun M, Gasek NS, Hargis GR, Sharafieh R, Xu M. Role of senescent cells in cutaneous wound healing. Biology 2022;11(12).
[155] Bitar MS, Abdel-Halim SM, Al-Mulla F. Caveolin-1/PTRF upregulation constitutes a mechanism for mediating p53-induced cellular senescence: implications for evidence-based therapy of delayed wound healing in diabetes. Am J Physiol Endocrinol Metab 2013;305(8):E951–63.
[156] Wyles SP, Dashti P, Pirtskhalava T, Tekin B, Inman C, Gomez LS, et al. A chronic wound model to investigate skin cellular senescence. Aging (Albany NY) 2023;15(8):2852–62.
[157] Mitterberger MC, Mattesich M, Zwerschke W. Bariatric surgery and diet-induced long-term caloric restriction protect subcutaneous adipose-derived stromal/progenitor cells and prolong their life span in formerly obese humans. Exp Gerontol 2014;56:106–13.
[158] Schafer MJ, White TA, Evans G, Tonne JM, Verzosa GC, Stout MB, et al. Exercise prevents diet-induced cellular senescence in adipose tissue. Diabetes 2016;65(6):1606–15.
[159] Bianchi A, Marchetti L, Hall Z, Lemos H, Vacca M, Paish H, et al. Moderate exercise inhibits age-related inflammation, liver steatosis, senescence, and tumorigenesis. J Immunol 2021;206(4):904–16.
[160] Spielmann G, McFarlin BK, O'Connor DP, Smith PJ, Pircher H, Simpson RJ. Aerobic fitness is associated with lower proportions of senescent blood T-cells in man. Brain Behav Immun 2011;25(8):1521–9.
[161] Liu Y, Sanoff HK, Cho H, Burd CE, Torrice C, Ibrahim JG, et al. Expression of p16(INK4a) in peripheral blood T-cells is a biomarker of human aging. Aging Cell 2009;8(4):439–48.
[162] Hohensinner PJ, Kaun C, Ebenbauer B, Hackl M, Demyanets S, Richter D, et al. Reduction of premature aging markers after gastric bypass surgery in morbidly obese patients. Obes Surg 2018;28(9):2804–10.
[163] Barzilai N, Crandall JP, Kritchevsky SB, Espeland MA. Metformin as a tool to target aging. Cell Metabol 2016;23(6):1060–5.
[164] Hirsch HA, Iliopoulos D, Struhl K. Metformin inhibits the inflammatory response associated with cellular transformation and cancer stem cell growth. Proc Natl Acad Sci U S A 2013;110(3):972–7.
[165] Martin-Montalvo A, Mercken EM, Mitchell SJ, Palacios HH, Mote PL, Scheibye-Knudsen M, et al. Metformin improves healthspan and lifespan in mice. Nat Commun 2013;4:2192.
[166] Lim YZ, Wang Y, Estee M, Abidi J, Udaya Kumar M, Hussain SM, et al. Metformin as a potential disease-modifying drug in osteoarthritis: a systematic review of pre-clinical and human studies. Osteoarthr Cartil 2022;30(11):1434–42.
[167] Baker MC, Sheth K, Liu Y, Lu D, Lu R, Robinson WH. Development of osteoarthritis in adults with type 2 diabetes treated with metformin vs a sulfonylurea. JAMA Netw Open 2023;6(3):e233646.

[168] Trnovska J, Svoboda P, Pelantova H, Kuzma M, Kratochvilova H, Kasperova BJ, et al. Complex positive effects of SGLT-2 inhibitor empagliflozin in the liver, kidney and adipose tissue of hereditary hypertriglyceridemic rats: possible contribution of attenuation of cell senescence and oxidative stress. Int J Mol Sci 2021;22(19):10606.

[169] Kim MN, Moon JH, Cho YM. Sodium-glucose cotransporter-2 inhibition reduces cellular senescence in the diabetic kidney by promoting ketone body-induced NRF2 activation. Diabetes Obes Metab 2021;23(11):2561—71.

[170] Marso SP, Daniels GH, Brown-Frandsen K, Kristensen P, Mann JF, Nauck MA, et al. Liraglutide and cardiovascular outcomes in type 2 diabetes. N Engl J Med 2016;375(4):311—22.

[171] Noren Hooten N, Martin-Montalvo A, Dluzen DF, Zhang Y, Bernier M, Zonderman AB, et al. Metformin-mediated increase in DICER1 regulates microRNA expression and cellular senescence. Aging Cell 2016;15(3):572—81.

[172] Chan KC, Yu MH, Lin MC, Huang CN, Chung DJ, Lee YJ, et al. Pleiotropic effects of acarbose on atherosclerosis development in rabbits are mediated via upregulating AMPK signals. Sci Rep 2016;6:38642.

[173] Kennedy BK, Berger SL, Brunet A, Campisi J, Cuervo AM, Epel ES, et al. Geroscience: linking aging to chronic disease. Cell 2014;159(4):709—13.

[174] Kirkland JL, Tchkonia T. Senolytic drugs: from discovery to translation. J Intern Med 2020;288(5):518—36.

[175] Dolgin E. Send in the senolytics. Nat Biotechnol 2020;38(12):1371—7.

[176] Milanovic M, Fan DNY, Belenki D, Dabritz JHM, Zhao Z, Yu Y, et al. Senescence-associated reprogramming promotes cancer stemness. Nature 2018;553(7686):96—100.

[177] Zhu Y, Tchkonia T, Fuhrmann-Stroissnigg H, Dai HM, Ling YY, Stout MB, et al. Identification of a novel senolytic agent, navitoclax, targeting the Bcl-2 family of anti-apoptotic factors. Aging Cell 2016;15(3):428—35.

[178] Chang J, Wang Y, Shao L, Laberge RM, Demaria M, Campisi J, et al. Clearance of senescent cells by ABT263 rejuvenates aged hematopoietic stem cells in mice. Nat Med 2016;22(1):78—83.

[179] Yosef R, Pilpel N, Tokarsky-Amiel R, Biran A, Ovadya Y, Cohen S, et al. Directed elimination of senescent cells by inhibition of BCL-W and BCL-XL. Nat Commun 2016;7:11190.

[180] Chini C, Hogan KA, Warner GM, Tarrago MG, Peclat TR, Tchkonia T, et al. The NADase CD38 is induced by factors secreted from senescent cells providing a potential link between senescence and age-related cellular NAD(+) decline. Biochem Biophys Res Commun 2019;513(2):486—93.

[181] Moncsek A, Al-Suraih MS, Trussoni CE, O'Hara SP, Splinter PL, Zuber C, et al. Targeting senescent cholangiocytes and activated fibroblasts with B-cell lymphoma-extra large inhibitors ameliorates fibrosis in multidrug resistance 2 gene knockout (Mdr2(-/-)) mice. Hepatology 2018;67(1):247—59.

[182] Saccon TD, Nagpal R, Yadav H, Cavalcante MB, Nunes ADC, Schneider A, et al. Senolytic combination of dasatinib and quercetin alleviates intestinal senescence and inflammation and modulates the gut microbiome in aged mice. J Gerontol A Biol Sci Med Sci 2021;76(11):1895—905.

[183] Schafer MJ, White TA, Iijima K, Haak AJ, Ligresti G, Atkinson EJ, et al. Cellular senescence mediates fibrotic pulmonary disease. Nat Commun 2017;8:14532.

[184] Lewis-McDougall FC, Ruchaya PJ, Domenjo-Vila E, Shin Teoh T, Prata L, Cottle BJ, et al. Aged-senescent cells contribute to impaired heart regeneration. Aging Cell 2019;18(3):e12931.

[185] Hickson LJ, Langhi Prata LGP, Bobart SA, Evans TK, Giorgadze N, Hashmi SK, et al. Corrigendum to 'Senolytics decrease senescent cells in humans: preliminary report from a clinical trial of Dasatinib plus Quercetin in individuals with diabetic kidney disease' EBioMedicine 47 (2019) 446—456. EBioMedicine 2020;52:102595.

[186] Lee H, Wilson D, Bunting KV, Kotecha D, Jackson T. Repurposing digoxin for geroprotection in patients with frailty and multimorbidity. Ageing Res Rev 2023;86:101860.

[187] Triana-Martinez F, Picallos-Rabina P, Da Silva-Alvarez S, Pietrocola F, Llanos S, Rodilla V, et al. Identification and characterization of Cardiac Glycosides as senolytic compounds. Nat Commun 2019;10(1):4731.

[188] Xu M, Tchkonia T, Ding H, Ogrodnik M, Lubbers ER, Pirtskhalava T, et al. JAK inhibition alleviates the cellular senescence-associated secretory phenotype and frailty in old age. Proc Natl Acad Sci USA 2015;112(46):E6301—10.

[189] Moiseeva O, Deschenes-Simard X, St-Germain E, Igelmann S, Huot G, Cadar AE, et al. Metformin inhibits the senescence-associated secretory phenotype by interfering with IKK/NF-kappaB activation. Aging Cell 2013;12(3):489—98.

[190] Laberge RM, Zhou L, Sarantos MR, Rodier F, Freund A, de Keizer PL, et al. Glucocorticoids suppress selected components of the senescence-associated secretory phenotype. Aging Cell 2012;11(4):569—78.

[191] Herranz N, Gallage S, Mellone M, Wuestefeld T, Klotz S, Hanley CJ, et al. mTOR regulates MAPKAPK2 translation to control the senescence-associated secretory phenotype. Nat Cell Biol 2015;17(9):1205—17.

[192] Singh M, Jensen MD, Lerman A, Kushwaha S, Rihal CS, Gersh BJ, et al. Effect of low-dose rapamycin on senescence markers and physical functioning in older adults with coronary artery disease: results of a pilot study. J Frailty Aging 2016;5(4):204—7.

[193] Samakkarnthai P, Saul D, Zhang L, Aversa Z, Doolittle ML, Sfeir JG, et al. In vitro and in vivo effects of zoledronic acid on senescence and senescence-associated secretory phenotype markers. Aging (Albany NY) 2023;15:3331.

[194] Zhu Y, Doornebal EJ, Pirtskhalava T, Giorgadze N, Wentworth M, Fuhrmann-Stroissnigg H, et al. New agents that target senescent cells: the flavone, fisetin, and the BCL-XL inhibitors, A1331852 and A1155463. Aging 2017;9(3):955.

[195] Suda M, Shimizu I, Katsuumi G, Yoshida Y, Hayashi Y, Ikegami R, et al. Senolytic vaccination improves normal and pathological age-related phenotypes and increases lifespan in progeroid mice. Nat Aging 2021;1(12):1117—26.

[196] Anderson R, Lagnado A, Maggiorani D, Walaszczyk A, Dookun E, Chapman J, et al. Length-independent telomere damage drives postmitotic cardiomyocyte senescence. EMBO J 2019;38(5):e100492.

[197] Kim SR, Jiang K, Ogrodnik M, Chen X, Zhu XY, Lohmeier H, et al. Increased renal cellular senescence in murine high-fat diet: effect of the senolytic drug quercetin. Transl Res J Lab Clin Med 2019;213:112—23.

[198] Schafer MJ, Haak AJ, Tschumperlin DJ, LeBrasseur NK. Targeting senescent cells in fibrosis: pathology, paradox, and practical considerations. Curr Rheumatol Rep 2018;20(1):3.

[199] Novais EJ, Tran VA, Johnston SN, Darris KR, Roupas AJ, Sessions GA, et al. Long-term treatment with senolytic drugs Dasatinib and Quercetin ameliorates age-dependent intervertebral disc degeneration in mice. Nat Commun 2021;12(1):5213.

[200] Wang L, Wang B, Gasek NS, Zhou Y, Cohn RL, Martin DE, et al. Targeting p21(Cip1) highly expressing cells in adipose tissue alleviates insulin resistance in obesity. Cell Metabol 2022;34(1):186.

[201] Justice JN, Nambiar AM, Tchkonia T, LeBrasseur NK, Pascual R, Hashmi SK, et al. Senolytics in idiopathic pulmonary fibrosis: results from a first-in-human, open-label, pilot study. EBioMedicine 2019;40:554–63.

[202] Salaami O, Kuo CL, Drake MT, Kuchel GA, Kirkland JL, Pignolo RJ. Antidiabetic effects of the senolytic agent dasatinib. Mayo Clin Proc 2021;96(12):3021–9.

[203] SenNet C. NIH SenNet Consortium to map senescent cells throughout the human lifespan to understand physiological health. Nat Aging 2022;2(12):1090–100.

[204] Hickson LJ, Langhi Prata LGP, Bobart SA, Evans TK, Giorgadze N, Hashmi SK, et al. Senolytics decrease senescent cells in humans: preliminary report from a clinical trial of Dasatinib plus Quercetin in individuals with diabetic kidney disease. EBioMedicine 2019;47:446–56.

[205] Saul D, Kosinsky RL, Atkinson EJ, Doolittle ML, Zhang X, LeBrasseur NK, et al. A new gene set identifies senescent cells and predicts senescence-associated pathways across tissues. Nat Commun 2022;13(1):4827.

[206] Schafer MJ, Zhang X, Kumar A, Atkinson EJ, Zhu Y, Jachim S, et al. The senescence-associated secretome as an indicator of age and medical risk. JCI Insight 2020;5(12).

[207] Basisty N, Kale A, Jeon OH, Kuehnemann C, Payne T, Rao C, et al. A proteomic atlas of senescence-associated secretomes for aging biomarker development. PLoS Biol 2020;18(1):e3000599.

[208] Ruggiero AD, Vemuri R, Blawas M, Long M, DeStephanis D, Williams AG, et al. Long-term dasatinib plus quercetin effects on aging outcomes and inflammation in nonhuman primates: implications for senolytic clinical trial design. Geroscience 2023:1–19.

[209] Morley JE. The elderly Type 2 diabetic patient: special considerations. Diabet Med 1998;15(Suppl. 4):S41–6.

[210] Kezerle L, Shalev L, Barski L. Treating the elderly diabetic patient: special considerations. Diabetes Metab Syndr Obes 2014;7:391–400.

[211] Boureau AS, Guyomarch B, Gourdy P, Allix I, Annweiler C, Cervantes N, et al. Nocturnal hypoglycemia is underdiagnosed in older people with insulin-treated type 2 diabetes: the HYPOAGE observational study. J Am Geriatr Soc 2023;71:2107–19.

CHAPTER

11

Age-related disease: Eyes

Coad Thomas Dow

Department of Ophthalmology and Visual Sciences, McPherson Eye Research Institute, University of Wisconsin, School of Medicine and Public Health, Madison, WI, United States

1. Introduction

Age-related diseases are burgeoning, paralleling increases in life expectancy [1]. This chapter assesses the vicissitudes of aging as they relate to the eyes and vision. It addresses the range of age-related changes from the universal difficulty with near vision that comes in middle age, the common lens opacification of cataract, the common causes of irreversible age-related blindness, glaucoma, and macular degeneration. The components of the eye are illustrated in Fig. 11.1. The retina (Fig. 11.1, #4) lines the inner aspect of the back of the eye; its primary features are the specialized center, the macula (Fig. 11.1, #3); the collected neurons that become the optic nerve; and the retinal blood vessels as seen in Fig. 11.2. While "life-style" changes in diet, exercise, sleep, and stress control are regularly encouraged to offset aging or, at least to increase "healthspan," these interventions will not be reviewed in this chapter. Rather contemporary interventions for age-related diseases of the eye are presented; moreover, prospects for curing and/or reversing age-related eye disease with "upstream" targeting the mechanics of aging are presented; these include the potential ocular benefit from incretin receptor agonists, the anti-aging Klotho protein, and telomerase activation.

2. Presbyopia

The common age-related refractive change that reduces the ability to see at near is called presbyopia (etymology—Greek presbus "old man" + ōps, ōp- "eye"). This is due to a progressive inflexibility of the natural lens (Fig. 11.1, #8). The eyes of children have a large "amplitude of accommodation." This is commonly observed with young children gleefully content to look at objects near the end of their noses. The ability of the lens to add this accommodative power steadily decreases and, typically, after 40 years of age, an individual will start removing their glasses for reading if they are near-sighted, require reading glasses if they had previously not needed glasses, or require a bifocal to add a refractive power for near reading. (A shout out to Benjamin Franklin, who on May 23, 1785, announced his invention of the "double spectacle" [2]).

The ciliary body (Fig. 11.1, #7) is the ocular anatomic structure that controls accommodation. This muscle is attached to the lens by a host of tiny cords called zonules (Fig. 11.1, #13). In presbyopia the muscle contracts in the effort to see at near but the lens, having become progressively rigid, is unable to change its shape, smaller and fatter, to "zoom" up close vision. The result is blurring at near that requires increasing bifocal power; typically, from approximately age 45 to 65 an individual will have several increases in their bifocal power.

Aside from glasses, solutions for presbyopia have included alterations to the cornea and the lens. An individual could have corneal contact lenses, one for distance and one for near, i.e., "monofocal" lenses. The same can be accomplished with LASIK, the laser-based sculpting of the cornea. The natural lens of the eye can be replaced at the time of cataract surgery (discussed later) or done strictly as a refractive procedure. Options for improving near vision with lens surgery would include the mono-focal state accomplished by calculating the new lens power—one for distance,

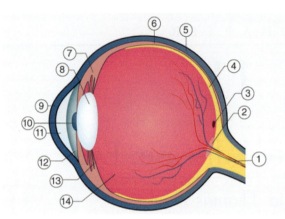

FIGURE 11.1 Basic eye anatomy.

1. Optic Nerve
2. Optic Nerve Head
3. Fovea - the Center of Macula
4. Retina
5. Sclera
6. Choroid
7. Ciliary Body
8. Lens
9. Cornea
10. Pupil
11. Aqueous in Anterior Chamber
12. Iris
13. Zonules
14. Vitreous

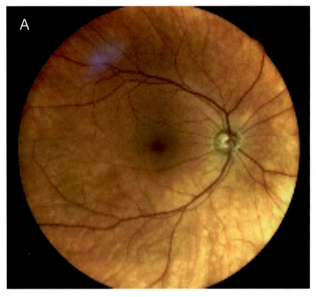
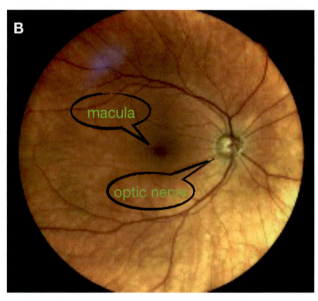

FIGURE 11.2 (A) A normal "posterior pole" of the eye. (B) The critical center of the retina, the macula, and the other major landmark, the optic nerve.

one for near. Another option is for a "premium" replacement lens—one that has rings of power to accomplish distance, intermediate, and near vision without glasses.

Lastly, and a new option, is an eye drop to help near vision. Vuity is a prescription medication and is a repurposed, eye drop used for glaucoma, pilocarpine [3]. This eye drop has a side effect of making the pupil smaller (Fig. 11.1, #10) and it is this effect that imparts benefit for presbyopia. As any photographer would know, when the aperture of the lens is made smaller, the resulting image will have a greater depth of field. It is this side effect that allows some improvement for presbyopia. However, a side effect of the smaller aperture is decreased vision, particularly in low light.

3. Glaucoma

Glaucoma is one of the most common eye diseases and, worldwide, is the leading cause of irreversible blindness; it is estimated that glaucoma will affect over 100 million people by 2040 [4]. There are three primary risk factors for glaucoma: increasing age, genetic risk, and elevated intraocular pressure (IOP). Elevated IOP is strongly associated with disease progression in 60% of patients, yet available antiglaucomatous therapies (topical medications and surgery) solely target IOP [5,6]. A unique challenge with glaucoma is that it is very difficult to diagnose early in its

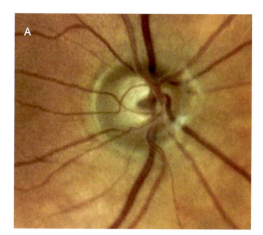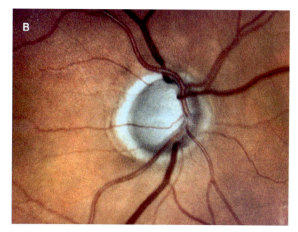

FIGURE 11.3 (A) A normal optic nerve with a normal "cupping" of the central aspect of the optic nerve. (B) Extensive "cupping" due to loss of nerve tissue from glaucoma.

course and, subsequently, there can be substantial and irreversible optic nerve damage before discernible vision loss. There are three parameters used to classify glaucoma: primary (idiopathic) or secondary (associated with other ocular or systemic diseases), the degree of opening of the anterior chamber angle (open or closed) (Fig. 11.1, #11), and the acute or chronic state. The most common type of glaucoma associated with age is chronic, primary, open-angle glaucoma (POAG). While increased IOP is strongly associated with POAG, it is neither necessary nor sufficient to cause it. Glaucoma is the collection of a variety of diseases that have a common finding, the progressive loss of retinal ganglion cells. These cells are neurons that collect as the optic nerve (Fig. 11.1, #1 and 2) to carry visual receptor-generated information to the brain. Ganglion cell damage is reflected in the loss of optic nerve tissue as can be seen as a typical excavation of the optic nerve head (Fig. 11.3).

Historic treatments for glaucoma have included topical medications, laser treatment, and a variety of surgical procedures (reviewed by Wagner [7]).

However, due to the knowledge that a number of glaucoma patients will progress to blindness despite "normal" IOP [8], there has been a continued pursuit of neuroprotective interventions to act as an adjunct to IOP management. Broadly, neuroprotection refers to efforts to preserve neuronal structure and function [9].

Despite numerous laboratory-based studies showing neuroprotection in animal models of glaucoma, there has been very limited translation to the clinic [10,11]. Notable exceptions to this are brimonidine, Betaxolol, and latanoprost. In addition to lowering IOP, brimonidine protects the optic nerve in animal models of glaucoma; and, in clinical trials, it reduced the incidence of glaucomatous nerve progression [10].

A collection of drugs that may offer exciting prospects for glaucoma neuroprotection are the pleotropic incretin receptor agonists [12]. In addition to treating insulin resistance of type 2 diabetes (T2D) [13], these are also being fast-tracked for treatment of obesity [14]. Moreover, these drugs may have a central place in the treatment of neurodegenerative diseases [15−18]. The basis of the benefit of these drugs is brain energy rescue [19] and may extend to neuroprotection in glaucoma [12].

4. Lens

Cataract is clouding of the natural lens of the eye (Fig. 11.1, #8). While cataracts can be acquired congenitally, secondary to certain medications, or traumatically induced, far and away the most common cause of cataract is age. With aging, lens proteins accumulate fluorescent chromophores which increases susceptibility to oxidation that increases scattering of light. Because the human lens grows throughout life, the center of the lens is exposed for a longer period to such influences and the risk of oxidative damage increases into the fourth decade when a barrier to the transport of glutathione forms around the lens nucleus. The steady accumulation of chromophores and complex, insoluble crystallin aggregates in the lens nucleus leads to the formation of the common brown nuclear cataract [20].

Cataract is the leading cause of treatable blindness in the world. Although cataract is treatable with a straightforward surgery, there remains an unacceptably high backlog of operable blindness from cataract, particularly in the developing world [21,22]. Cataract surgery not only improves vision but, with presurgery planning, a precise postoperative refraction target is commonly achieved, reducing the need for spectacles. This is all via preoperative measurements for the artificial lens implant that replaces the cloudy natural lens. While claims have been made that topically given medication will reduce cataract, no such medication has emerged as an effective treatment.

The pathologic pathways resulting in cataract may be tied to the Klotho genes (discussed later in this chapter). In humans, the Klotho protein decreases across organ systems during normal aging, making it an age-modulating protein. The Klotho protein is present in the lens and beneficially modulates oxidative stress; this has the potential to protect against age-related cataract [23]. This oxidative stress benefit against age-related cataract may be controlled by telomerase activity [24].

5. Retina

In the Western world, macular degeneration is the most common cause of age-related severe vision loss [25,26]. The macula is the very specialized center of the retina (Fig. 11.1, #3) and progressive changes in macular degeneration affect activities of daily living ranging from blurred and distorted reading vision to loss of central vision. A severely affected individual would be able to move about a familiar environment without difficulty but would not be able to identify another individual due to loss of central vision.

While there is great variety in the manifestation of AMD, it is classified as "early," "intermediate," and late AMD. Of the late AMD, there is atrophic ("dry") and neovascular ("wet") degeneration [27] (Fig. 11.4). Although there are newer treatments for wet degeneration, dry AMD—accounting for 80%—90% of the disease—still lacks effective therapy.

Extensive research has focused on the ability of nutrients to prevent and/or mitigate AMD. Seminal work regarding nutritional antioxidants and AMD started in the late 1980s and led to the large Age Related Eye Disease Study (AREDS). The results of the AREDS study showed that supplementation with antioxidants and zinc reduced the risk of progression to advanced AMD by approximately 25% in those with intermediate AMD or advanced AMD in one eye. Although encouraging, the study showed no benefit of antioxidants and zinc in early AMD [28]. A subsequent study with a reformulation (AREDS2) demonstrated an additional effect, again for those with moderate-stage AMD. However, as with the original AREDS study, the AREDS2 study showed no benefit (or harm) for individuals with early AMD [29]. While existing literature has suggested that lutein, zeaxanthin, and di-docosahexaenoic acid (DHA) consumption tend to protect against the development of AMD, there is little improvement to the previously involved retina [30].

The retinal pigment epithelium (RPE) plays a pivotal role in vision and in the pathogenesis of AMD. Of the multiple functions of the RPE one of the more important is the consumption capacity to support the maintenance of the retina's photoreceptors. Phagocytosis of "used" outer-segment membranes (discs) shed by the photoreceptors occurs at the apical surface of RPE. There are approximately 150 million photoreceptors in the human retina. Photoreceptors shed as many as 100 of their outer discs per day. Disc shedding is the process by which photoreceptors in

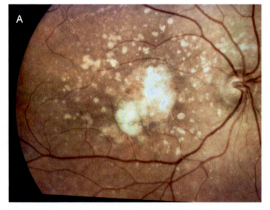
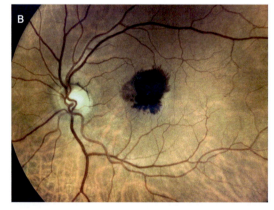

FIGURE 11.4 (A) Extensive macular degeneration, it is termed "dry" as there are no leaking blood vessels. (B) "Wet" macular degeneration with a hemorrhage in the critical macula. Both patients had poor vision due to AMD.

the eye are renewed [31]. These are continually replaced by the addition of new discs at the base of the stack. For homeostasis, each RPE cell needs to ingest as many as 4000 outer-segment discs per day [32].

The early stage of AMD is characterized by the presence of drusen. Drusen, readily seen at the time of an eye exam, are yellow deposits between the RPE and Bruch's membrane of the aged retina. Most agree that these deposits occur when the RPE can no longer provide their "recycling" function [33]. Aging of the RPE, RPE senescence, is associated with telomere shortening [34,35].

Telomeres are protective nucleotide repeats (TTAGGG) at the ends of chromosomes and due to the "end replication problem" [36] are sequentially shortened with each cell division.

Loss of telomeric DNA occurs during cell division because the end of the lagging strand of the chromosome cannot be fully replicated [37]. Cellular senescence is generally considered to result from critical telomere attrition and that telomere attrition plays an important role in a variety of age-related diseases [38,39] as well as AMD [40]. The erosion of telomeres can be restored by the enzyme telomerase conceptually offering to rescue cells from senescence and, in the case of AMD, improving RPE functioning [41,42].

In one study in early AMD participants, an oral telomerase activator showed improvement in macular function as shown by macular micro-perimetry; the benefit was seen at 6 months and maintained at 1 year compared to placebo controls (Fig. 11.5) [40].

A bold initiative in the treatment of late AMD is transplantation of pluripotential stem cells that have been trained to become RPE [43]. This ocular application is one of many potential stem cell therapy applications to be utilized in various organ systems. While only used in clinical trials thus far, ocular applications for stem cell transplant include AMD as well as blinding inherited diseases [44].

Another promising intervention for AMD may be the incretin receptor agonists (mentioned also in the glaucoma section). There are pleiotropic benefits of the cellular energy usage and antiinflammation imparted by these medications that also extend to the retina [45]. While these drugs were originally intended for metabolic syndrome, they are finding increasing applications [46] including in the RPE [47].

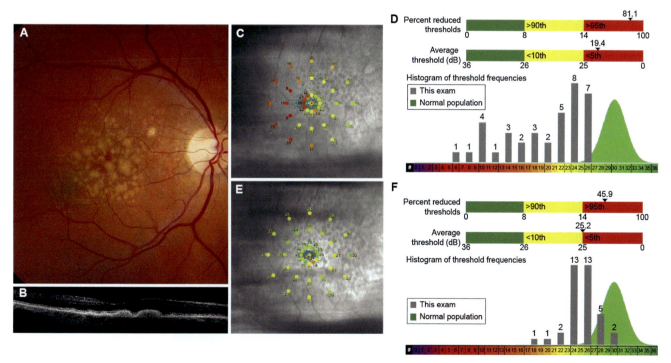

FIGURE 11.5 This montage is from one participant in a trial of oral telomerase activator for early age-related macular degeneration. (A) A color photograph of the posterior pole showing multiple drusen, deposits related to dry macular degeneration. (B) Optical coherence tomography showing the drusen. (C) Micro-perimetry prior to the study. (D) A graphic representation of the testing from (C). (E) Micro-perimetry after 1 year of taking the telomerase activator. (F) Improvement in macular function after 1 year of an oral telomerase activator. *From Dow and Harley [40]. Originally published by and used with permission from Dove Medical Press Ltd.*

The Klotho protein mentioned in the cataract section plays a protective role in the RPE, thus it may have application for AMD [48,49].

6. The eye as a window to other age-related diseases

The unique anatomy of the eye allows a glimpse into the overall health of the individual; this is particularly the case for neurologic and vascular diseases. The retina is available for direct, noninvasive imaging. It is increasingly understood that the pathologies of Alzheimer's disease start years, or even decades, prior to cognitive impairment. As the retina is an outgrowth of the brain there have been efforts to image amyloid plaques in the retina. Moreover, the "amyloid burden" in the retina is reflective of the same burden in the brain [50].

While this technology has not advanced to the clinical setting, an example is presented in Fig. 11.6. The first color photo is a customary retinal image showing a portion of the back of the right eye, the image is normal. The second photo is of the same individual taken in upward gaze (the structures have moved downwards as the individual looks upwards). There is a preferential deposition of amyloid in the superior retina. Special processing programs analyze for amyloid via autofluorescence. A "mask" is placed over the area to be assessed and the aggregated amyloid is identified (yellow-colored "dots") and scored. Scoring is tied to the gold standard for amyloid assessment,

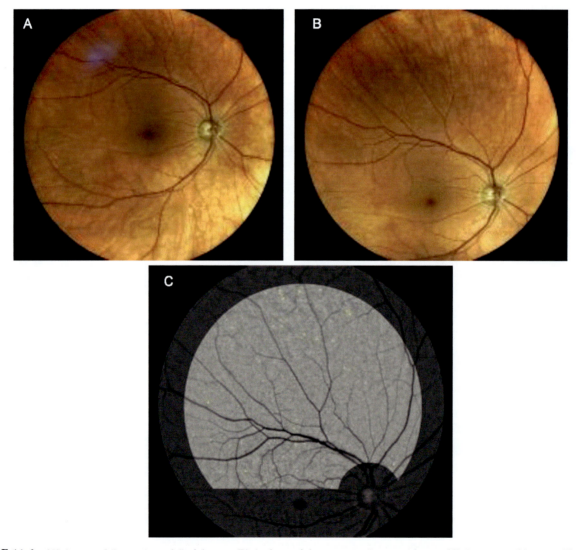

FIGURE 11.6 (A) A normal "posterior pole" of the eye. (B) A photo of the same eye in upward gaze. (C) A processed image with amyloid deposits as produced by the Retia camera.

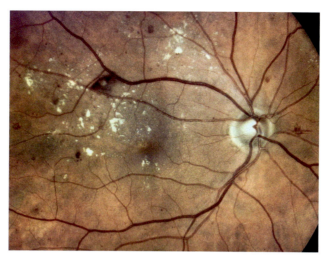

FIGURE 11.7 This image features a range of diabetic changes to the retina. The vascular changes include exudates from the retinal blood vessels and also a variety of small hemorrhages.

PET scan for amyloid. This individual's scoring was relatively low and would not connote a degree of amyloid burden associated with a positive amyloid-PET scan.

Although retinal imaging of amyloid for Alzheimer's risk has seemed promising due to its relative ease and cost, that promise has been replaced by an even more readily done testing—a blood test for amyloid. A mass-spectrometry-based plasma amyloid test has been validated [51,52] and commercialized—the PrecivityAD test (www.precivityAD.com). The benefits of this test include distinguishing between amyloid-related and nonamyloid-related neurocognitive disease as well as enhancing the enrollment of participants into Alzheimer's trials [53].

The easy visualization of the blood vessels of the retina has long been employed to monitor the vascular changes manifest as vascular changes of diabetes. Both autoimmune (type 1) and nonautoimmune (type 2) diabetics are prone to a significant range of retinal blood vessel changes—those changes associated with the duration of disease (Fig. 11.7). Some of the diabetic vascular changes, even moderate in severity, can be asymptomatic for the individual; not infrequently, the initial diabetes diagnosis is made at the time of an eye exam.

7. Upstream opportunities for life and health extension

There are characteristic hallmarks of aging, these are: genomic instability, telomere attrition, epigenetic alterations, loss of proteostasis, deregulated nutrient sensing, mitochondrial dysfunction, cellular senescence, stem cell exhaustion, and interrupted intercellular signaling [54]. While the results of interventions such as telomerase activation and the use of incretin receptor agonists are being seen in assorted age-related ocular pathologies, other lifespan extension strategies have yet to make their mark on the eye. While more broadly covered elsewhere in this book, this author would like to enumerate some of these interventions that will rightfully be expected to favorably impact age-related eye disease. Known approaches for extension of lifespan are caloric restriction (SIRT activation), inhibition of the mechanistic target of rapamycin (mTOR), use of metformin, and increasing the anti-aging protein, Klotho.

A growing health practice worldwide is intermittent fasting consisting of alternating periods of eating and fasting done on a regular basis. This simulates the life that ancestral hunter-gatherers faced with periods of food shortage. In this context, fasting promotes adaptations for weight loss, decreasing insulin resistance, and preventing age-related diseases [55]. Caloric restriction causes an increase in sirtuin (SIRT) proteins. SIRT activation can be useful against aging-related disorders [56]. Several SIRT activators currently exist and are in use for their health benefits such as green tea, turmeric, kale, and resveratrol [57].

Metformin is a well-known medication for type 2 diabetes (T2D). There is increasing realization of the potential of metformin to prevent age-related disease in non-diabetics as well as diabetics [58]. Epidemiological studies have suggested that metformin may prolong lifespan and have beneficial impacts on health beyond its antidiabetic effect

[59]. In a recent study assessing the effect of metformin on epigenetic age biomarkers, metformin was found to slow epigenetic aging [60].

The mechanistic target of rapamycin (mTOR) is an integral modulator of aging and age-related disease [61]. Inhibition of this nutrient response pathway mimics the benefit of caloric restriction resulting in lifespan extension and protection against age-related diseases. Deregulation of mTOR signaling is implicated in several age-related diseases. Acting as an energy rheostat mTOR influences cellular transcription, translation, autophagy, and metabolism [62]. Currently, FDA-approved "rapamycins" or "rapalogs" are used for assorted cancers and inflammatory diseases. These drugs have significant adverse effects and efforts continue to find analogs that have a better safety profile [63].

The Klotho protein was discovered in 1997; its name comes from Greek mythology—Clotho was one of the three Fates and she spun the thread of life. In the absence of Klotho, mice had an accelerated aging phenotype; conversely, overexpression of Klotho extended both healthspan and lifespan by 30% [64]. Klotho inhibits several aging pathways; moreover, several currently used drugs increase circulating Klotho levels. Included in these Kotho promoters are losartan, fluvastatin, various rapalogs, and vitamin D [65]. In a recent study of American adults, low serum levels of Klotho correlated with increased all-cause death rate [66]. Another recent study suggests that Klotho's role in life-extension is due to regulation of telomerase activity and telomere length [24].

8. Summary

While several of the interventions against aging discussed in this chapter are future considerations, telomerase activation with an oral agent is currently available. The knowledge that decreased visual function due to one of the most onerous age-related eye diseases, AMD, can be improved is truly exciting and available today. Moreover, telomerase activation for AMD can be achieved with the oral preparation sourced from the natural world. For both age-related eye disease and age-related disease in general, this chapter has outlined exciting contemporary efforts to extend both lifespan and healthspan. Success in these efforts will surely give support to the adage "I hope to die young—as old as I can."

References

[1] Kehler DS. Age-related disease burden as a measure of population ageing. Lancet Public Health March 2019;4(3):e123—4. https://doi.org/10.1016/S2468-2667(19)30026-X. PMID: 30851865.

[2] https://www.ushistory.org/franklin/science/bifocals.htm (accessed November 22, 2022).

[3] Meghpara BB, Lee JK, Rapuano CJ, Mian SI, Ho AC. Pilocarpine 1.25% and the changing landscape of presbyopia treatment. Curr Opin Ophthalmol July 1, 2022;33(4):269—74. https://doi.org/10.1097/ICU.0000000000000864. PMID: 35779051.

[4] Tham YC, Li X, Wong TY, Quigley HA, Aung T, Cheng CY. Global prevalence of glaucoma and projections of glaucoma burden through 2040: a systematic review and meta-analysis. Ophthalmology November 2014;121(11):2081—90. https://doi.org/10.1016/j.ophtha.2014.05.013. Epub 2014 Jun 26. PMID: 24974815.

[5] Quigley HA. 21st century glaucoma care. Eye February 2019;33(2):254—60. https://doi.org/10.1038/s41433-018-0227-8.

[6] Kolko M. Present and new treatment strategies in the management of glaucoma. Open Ophthalmol J May 15, 2015;9:89—100. https://doi.org/10.2174/1874364101509010089. PMID: 26069521; PMCID: PMC4460216.

[7] Wagner IV, Stewart MW, Dorairaj SK. Updates on the diagnosis and management of glaucoma. Mayo Clin Proc Innov Qual Outcomes November 16, 2022;6(6):618—35. https://doi.org/10.1016/j.mayocpiqo.2022.09.007. PMID: 36405987; PMCID: PMC9673042.

[8] Peters D, Bengtsson B, Heijl A. Lifetime risk of blindness in open-angle glaucoma. Am J Ophthalmol October 2013;156(4):724—30. https://doi.org/10.1016/j.ajo.2013.05.027. Epub 2013 Aug 7. PMID: 23932216.

[9] Casson RJ, Chidlow G, Ebneter A, Wood JP, Crowston J, Goldberg I. Translational neuroprotection research in glaucoma: a review of definitions and principles. Clin Exp Ophthalmol 2012 May—June;40(4):350—7. https://doi.org/10.1111/j.1442-9071.2011.02563.x. Epub 2011 Apr 27. PMID: 22697056.

[10] Guymer C, Wood JP, Chidlow G, Casson RJ. Neuroprotection in glaucoma: recent advances and clinical translation. Clin Exp Ophthalmol January 2019;47(1):88—105. https://doi.org/10.1111/ceo.13336. Epub 2018 Jul 1. PMID: 29900639.

[11] Tamm ER, Schmetterer L, Grehn F. Status and perspectives of neuroprotective therapies in glaucoma: the European glaucoma society white paper. Cell Tissue Res August 2013;353(2):347—54. https://doi.org/10.1007/s00441-013-1637-3. Epub 2013 May 28. PMID: 23712457.

[12] Mouhammad ZA, Vohra R, Horwitz A, Thein AS, Rovelt J, Cvenkel B, et al. Glucagon-like peptide 1 receptor agonists—potential game changers in the treatment of glaucoma? Front Neurosci February 21, 2022;16:824054. https://doi.org/10.3389/fnins.2022.824054. PMID: 35264926; PMCID: PMC8899005.

[13] Nauck MA, D'Alessio DA. Tirzepatide, a dual GIP/GLP-1 receptor co-agonist for the treatment of type 2 diabetes with unmatched effectiveness regrading glycaemic control and body weight reduction. Cardiovasc Diabetol September 1, 2022;21(1):169. https://doi.org/10.1186/s12933-022-01604-7. PMID: 36050763; PMCID: PMC9438179.

[14] Jastreboff AM, Aronne LJ, Ahmad NN, Wharton S, Connery L, Alves B, et al. SURMOUNT-1 investigators. Tirzepatide once weekly for the treatment of obesity. N Engl J Med July 21, 2022;387(3):205—16. https://doi.org/10.1056/NEJMoa2206038. Epub 2022 Jun 4. PMID: 35658024.

[15] Ferrari F, Moretti A, Villa RF. Incretin-based drugs as potential therapy for neurodegenerative diseases: current status and perspectives. Pharmacol Ther November 2022;239:108277. https://doi.org/10.1016/j.pharmthera.2022.108277. Epub 2022 Sep 3. PMID: 36064147.

[16] Reich N, Hölscher C. The neuroprotective effects of glucagon-like peptide 1 in Alzheimer's and Parkinson's disease: an in-depth review. Front Neurosci September 1, 2022;16:970925. https://doi.org/10.3389/fnins.2022.970925. PMID: 36117625; PMCID: PMC9475012.

[17] Hölscher C. Glucagon-like peptide 1 and glucose-dependent insulinotropic peptide hormones and novel receptor agonists protect synapses in Alzheimer's and Parkinson's diseases. Front Synaptic Neurosci July 27, 2022;14:955258. https://doi.org/10.3389/fnsyn.2022.955258. PMID: 35965783; PMCID: PMC9363704.

[18] Salameh TS, Rhea EM, Talbot K, Banks WA. Brain uptake pharmacokinetics of incretin receptor agonists showing promise as Alzheimer's and Parkinson's disease therapeutics. Biochem Pharmacol October 2020;180:114187. https://doi.org/10.1016/j.bcp.2020.114187. Epub 2020 Aug 2. PMID: 32755557; PMCID: PMC7606641.

[19] Cunnane SC, Trushina E, Morland C, Prigione A, Casadesus G, Andrews ZB, et al. Brain energy rescue: an emerging therapeutic concept for neurodegenerative disorders of ageing. Nat Rev Drug Discov September 2020;19(9):609—33. https://doi.org/10.1038/s41573-020-0072-x. Epub 2020 Jul 24. PMID: 32709961; PMCID: PMC7948516.

[20] Michael R, Bron AJ. The ageing lens and cataract: a model of normal and pathological ageing. Philos Trans R Soc Lond B Biol Sci April 27, 2011;366(1568):1278—92. https://doi.org/10.1098/rstb.2010.0300. PMID: 21402586; PMCID: PMC3061107.

[21] Hashemi H, Pakzad R, Yekta A, Aghamirsalim M, Pakbin M, Ramin S, et al. Global and regional prevalence of age-related cataract: a comprehensive systematic review and meta-analysis. Eye August 2020;34(8):1357—70. https://doi.org/10.1038/s41433-020-0806-3. Epub 2020 Feb 13. PMID: 32055021; PMCID: PMC7376226.

[22] Flaxman SR, Bourne RRA, Resnikoff S, Ackland P, Braithwaite T, Cicinelli MV, et al. Vision Loss Expert Group of the Global Burden of Disease Study. Global causes of blindness and distance vision impairment 1990—2020: a systematic review and meta-analysis. Lancet Glob Health December 2017;5(12):e1221—34. https://doi.org/10.1016/S2214-109X(17)30393-5. Epub 2017 Oct 11. PMID: 29032195.

[23] Zhang Y, Wang L, Wu Z, Yu X, Du X, Li X. The expressions of klotho family genes in human ocular tissues and in anterior lens capsules of age-related cataract. Curr Eye Res June 2017;42(6):871—5. https://doi.org/10.1080/02713683.2016.1259421. Epub 2017 Jan 17. PMID: 28095050.

[24] Ullah M, Sun Z. Klotho deficiency accelerates stem cells aging by impairing telomerase activity. J Gerontol A Biol Sci Med Sci August 16, 2019;74(9):1396—407. https://doi.org/10.1093/gerona/gly261. PMID: 30452555; PMCID: PMC6696722.

[25] Friedman DS, O'Colmain BJ, Muñoz B, Tomany SC, McCarty C, de Jong PT, et al. Eye diseases prevalence research group. Prevalence of age-related macular degeneration in the United States. Arch Ophthalmol April 2004;122(4):564—72. https://doi.org/10.1001/archopht.122.4.564. Erratum in: Arch Ophthalmol. 2011 Sep;129(9):1188. PMID: 15078675.

[26] Smith W, Assink J, Klein R, Mitchell P, Klaver CC, Klein BE, et al. Risk factors for age-related macular degeneration: pooled findings from three continents. Ophthalmology April 2001;108(4):697—704. https://doi.org/10.1016/s0161-6420(00)00580-7. PMID: 11297486.

[27] Ferris 3rd FL, Wilkinson CP, Bird A, Chakravarthy U, Chew E, Csaky K, et al. Initiative for macular research classification committee. Clinical classification of age-related macular degeneration. Ophthalmology April 2013;120(4):844—51. https://doi.org/10.1016/j.ophtha.2012.10.036. Epub 2013 Jan 16. PMID: 23332590.

[28] Age-Related Eye Disease Study Research Group. A randomized, placebo-controlled, clinical trial of high-dose supplementation with vitamins C and E, beta carotene, and zinc for age-related macular degeneration and vision loss: AREDS report no. 8. Arch Ophthalmol October 2001;119(10):1417—36. https://doi.org/10.1001/archopht.119.10.1417. Erratum in: Arch Ophthalmol. 2008 Sep;126(9):1251. PMID: 11594942; PMCID: PMC1462955.

[29] Chew EY, Clemons TE, Agrón E, Launer LJ, Grodstein F, Bernstein PS. Age-related eye disease study 2 (AREDS2) research group. Effect of omega-3 fatty acids, lutein/zeaxanthin, or other nutrient supplementation on cognitive function: the AREDS2 randomized clinical trial. JAMA August 25, 2015;314(8):791—801. https://doi.org/10.1001/jama.2015.9677. PMID: 26305649; PMCID: PMC5369607.

[30] Walchuk C, Suh M. Nutrition and the aging retina: a comprehensive review of the relationship between nutrients and their role in age-related macular degeneration and retina disease prevention. Adv Food Nutr Res 2020;93:293—332. https://doi.org/10.1016/bs.afnr.2020.04.003. Epub 2020 May 4. PMID: 32711865.

[31] Besharse JC, Pfenninger KH. Membrane assembly in retinal photoreceptors I. Freeze-fracture analysis of cytoplasmic vesicles in relationship to disc assembly. J Cell Biol November 1980;87(2 Pt 1):451—63. https://doi.org/10.1083/jcb.87.2.451. PMID: 7430251; PMCID: PMC2110759.

[32] Thumann G, Bartz-Schmidt KU, Kociok N, Heimann K, Schraemeyer U. Ultimate fate of rod outer segments in the retinal pigment epithelium. Pigm Cell Res October 1999;12(5):311—5. https://doi.org/10.1111/j.1600-0749.1999.tb00764.x. PMID: 10541040.

[33] Kinnunen K, Petrovski G, Moe MC, Berta A, Kaarniranta K. Molecular mechanisms of retinal pigment epithelium damage and development of age-related macular degeneration. Acta Ophthalmol June 2012;90(4):299—309. https://doi.org/10.1111/j.1755-3768.2011.02179.x. Epub 2011 Nov 23. PMID: 22112056.

[34] Armanios M. Telomeres and age-related disease: how telomere biology informs clinical paradigms. J Clin Invest March 2013;123(3):996—1002. https://doi.org/10.1172/JCI66370. Epub 2013 Mar 1. PMID: 23454763; PMCID: PMC3673231.

[35] Rodier F, Campisi J. Four faces of cellular senescence. J Cell Biol February 21, 2011;192(4):547—56. https://doi.org/10.1083/jcb.201009094. Epub 2011 Feb 14. PMID: 21321098; PMCID: PMC3044123.

[36] Levy MZ, Allsopp RC, Futcher AB, Greider CW, Harley CB. Telomere end-replication problem and cell aging. J Mol Biol June 20, 1992;225(4):951—60. https://doi.org/10.1016/0022-2836(92)90096-3. PMID: 1613801.

[37] Harley CB, Futcher AB, Greider CW. Telomeres shorten during ageing of human fibroblasts. Nature May 31, 1990;345(6274):458—60. https://doi.org/10.1038/345458a0. PMID: 2342578.

[38] Ruiz A, Flores-Gonzalez J, Buendia-Roldan I, Chavez-Galan L. Telomere shortening and its association with cell dysfunction in lung diseases. Int J Mol Sci December 31, 2021;23(1):425. https://doi.org/10.3390/ijms23010425. PMID: 35008850; PMCID: PMC8745057.

[39] Rattan P, Penrice DD, Ahn JC, Ferrer A, Patnaik M, Shah VH, et al. Inverse association of telomere length with liver disease and mortality in the US population. Hepatol Commun February 2022;6(2):399—410. https://doi.org/10.1002/hep4.1803. Epub 2021 Aug 28. PMID: 34558851; PMCID: PMC8793996.

[40] Dow CT, Harley CB. Evaluation of an oral telomerase activator for early age-related macular degeneration—a pilot study. Clin Ophthalmol January 28, 2016;10:243–9. https://doi.org/10.2147/OPTH.S100042. PMID: 26869760; PMCID: PMC4734847.

[41] Rowe-Rendleman C, Glickman RD. Possible therapy for age-related macular degeneration using human telomerase. Brain Res Bull February 15, 2004;62(6):549–53. https://doi.org/10.1016/S0361-9230(03)00072-8. PMID: 15036570.

[42] Blasiak J, Szczepanska J, Fila M, Pawlowska E, Kaarniranta K. Potential of telomerase in age-related macular degeneration-involvement of senescence, DNA damage response and autophagy and a key role of PGC-1α. Int J Mol Sci July 3, 2021;22(13):7194. https://doi.org/10.3390/ijms22137194. PMID: 34281248; PMCID: PMC8268995.

[43] Ahmed I, Johnston Jr RJ, Singh MS. Pluripotent stem cell therapy for retinal diseases. Ann Transl Med August 2021;9(15):1279. https://doi.org/10.21037/atm-20-4747. PMID: 34532416; PMCID: PMC8421932.

[44] Nair DSR, Thomas BB. Stem cell-based treatment strategies for degenerative diseases of the retina. Curr Stem Cell Res Ther 2022;17(3):214–25. https://doi.org/10.2174/1574888X16666210804112104. PMID: 34348629; PMCID: PMC9129886.

[45] Puddu A, Maggi D. Anti-inflammatory effects of GLP-1R activation in the retina. Int J Mol Sci October 17, 2022;23(20):12428. https://doi.org/10.3390/ijms232012428. PMID: 36293281; PMCID: PMC9604172.

[46] Laurindo LF, Barbalho SM, Guiguer EL, da Silva Soares de Souza M, de Souza GA, Fidalgo TM, et al. GLP-1a: going beyond traditional use. Int J Mol Sci January 10, 2022;23(2):739. https://doi.org/10.3390/ijms23020739. PMID: 35054924; PMCID: PMC8775408.

[47] Cui R, Tian L, Lu D, Li H, Cui J. Exendin-4 protects human retinal pigment epithelial cells from H2O2-induced oxidative damage via activation of NRF2 signaling. Ophthalmic Res 2020;63(4):404–12. https://doi.org/10.1159/000504891. Epub 2019 Dec 20. PMID: 31865348.

[48] Kokkinaki M, Abu-Asab M, Gunawardena N, Ahern G, Javidnia M, Young J, et al. Klotho regulates retinal pigment epithelial functions and protects against oxidative stress. J Neurosci October 9, 2013;33(41):16346–59. https://doi.org/10.1523/JNEUROSCI.0402-13.2013. Erratum in: J Neurosci. 2016 Jun 15;36(24):6597. PMID: 24107965; PMCID: PMC3810551.

[49] Reish NJ, Maltare A, McKeown AS, Laszczyk AM, Kraft TW, Gross AK, et al. The age-regulating protein klotho is vital to sustain retinal function. Invest Ophthalmol Vis Sci October 11, 2013;54(10):6675–85. https://doi.org/10.1167/iovs.13-12550. PMID: 24045987; PMCID: PMC3796940.

[50] Snyder PJ, Alber J, Alt C, Bain LJ, Bouma BE, Bouwman FH, et al. Retinal imaging in Alzheimer's and neurodegenerative diseases. Alzheimers Dement January 2021;17(1):103–11. https://doi.org/10.1002/alz.12179. Epub 2020 Oct 8. PMID: 33090722; PMCID: PMC8062064.

[51] Kirmess KM, Meyer MR, Holubasch MS, Knapik SS, Hu Y, Jackson EN, et al. The PrecivityAD test: accurate and reliable LC-MS/MS assays for quantifying plasma amyloid beta 40 and 42 and apolipoprotein E proteotype for the assessment of brain amyloidosis. Clin Chim Acta August 2021;519:267–75. https://doi.org/10.1016/j.cca.2021.05.011. Epub 2021 May 17. PMID: 34015303.

[52] Janelidze S, Teunissen CE, Zetterberg H, Allué JA, Sarasa L, Eichenlaub U, et al. Head-to-Head comparison of 8 plasma amyloid-β 42/40 assays in alzheimer disease. JAMA Neurol November 1, 2021;78(11):1375–82. https://doi.org/10.1001/jamaneurol.2021.3180. PMID: 34542571; PMCID: PMC8453354.

[53] West T, Kirmess KM, Meyer MR, Holubasch MS, Knapik SS, Hu Y, et al. A blood-based diagnostic test incorporating plasma Aβ42/40 ratio, ApoE proteotype, and age accurately identifies brain amyloid status: findings from a multi cohort validity analysis. Mol Neurodegener May 1, 2021;16(1):30. https://doi.org/10.1186/s13024-021-00451-6. PMID: 33933117; PMCID: PMC8088704.

[54] Mishra SK, Balendra V, Esposto J, Obaid AA, Maccioni RB, Jha NK, et al. Therapeutic antiaging strategies. Biomedicines October 8, 2022;10(10):2515. https://doi.org/10.3390/biomedicines10102515. PMID: 36289777; PMCID: PMC9599338.

[55] Mattson MP, Longo VD, Harvie M. Impact of intermittent fasting on health and disease processes. Ageing Res Rev October 2017;39:46–58. https://doi.org/10.1016/j.arr.2016.10.005. Epub 2016 Oct 31. PMID: 27810402; PMCID: PMC5411330.

[56] Mellini P, Valente S, Mai A. Sirtuin modulators: an updated patent review (2012–2014). Expert Opin Ther Pat January 2015;25(1):5–15. https://doi.org/10.1517/13543776.2014.982532. Epub 2014 Nov 29. PMID: 25435179.

[57] Lingappa N, Mayrovitz HN. Role of sirtuins in diabetes and age-related processes. Cureus September 4, 2022;14(9):e28774. https://doi.org/10.7759/cureus.28774. PMID: 36225477; PMCID: PMC9531907.

[58] Kulkarni AS, Gubbi S, Barzilai N. Benefits of metformin in attenuating the hallmarks of aging. Cell Metabol July 7, 2020;32(1):15–30. https://doi.org/10.1016/j.cmet.2020.04.001. Epub 2020 Apr 24. PMID: 32333835; PMCID: PMC7347426.

[59] Campbell JM, Bellman SM, Stephenson MD, Lisy K. Metformin reduces all-cause mortality and diseases of ageing independent of its effect on diabetes control: a systematic review and meta-analysis. Ageing Res Rev November 2017;40:31–44. https://doi.org/10.1016/j.arr.2017.08.003. Epub 2017 Aug 10. PMID: 28802803.

[60] Li M, Bao L, Zhu P, Wang S. Effect of metformin on the epigenetic age of peripheral blood in patients with diabetes mellitus. Front Genet September 26, 2022;13:955835. https://doi.org/10.3389/fgene.2022.955835. PMID: 36226195; PMCID: PMC9548538.

[61] Johnson SC, Rabinovitch PS, Kaeberlein M. mTOR is a key modulator of ageing and age-related disease. Nature January 17, 2013;493(7432):338–45. https://doi.org/10.1038/nature11861. PMID: 23325216; PMCID: PMC3687363.

[62] Bjedov I, Rallis C. The target of rapamycin signalling pathway in ageing and lifespan regulation. Genes September 3, 2020;11(9):1043. https://doi.org/10.3390/genes11091043. PMID: 32899412; PMCID: PMC7565554.

[63] Walters HE, Cox LS. mTORC inhibitors as broad-spectrum therapeutics for age-related diseases. Int J Mol Sci August 8, 2018;19(8):2325. https://doi.org/10.3390/ijms19082325. PMID: 30096787; PMCID: PMC6121351.

[64] Cheikhi A, Barchowsky A, Sahu A, Shinde SN, Pius A, Clemens ZJ, et al. Klotho: an elephant in aging research. J Gerontol A Biol Sci Med Sci June 18, 2019;74(7):1031–42. https://doi.org/10.1093/gerona/glz061. PMID: 30843026; PMCID: PMC7330474.

[65] Prud'homme GJ, Kurt M, Wang Q. Pathobiology of the klotho antiaging protein and therapeutic considerations. Front Aging July 12, 2022;3:931331. https://doi.org/10.3389/fragi.2022.931331. PMID: 35903083; PMCID: PMC9314780.

[66] Kresovich JK, Bulka CM. Low serum klotho associated with all-cause mortality among a nationally representative sample of American adults. J Gerontol A Biol Sci Med Sci March 3, 2022;77(3):452–6. https://doi.org/10.1093/gerona/glab308. PMID: 34628493; PMCID: PMC9122743.

CHAPTER

12

Age-related disease: Cancer, telomerase, and cell aging

Kurt Whittemore

Novo Nordisk, Watertown, MA, United States

1. Introduction

In humans, the thing is that as we mature, our telomeres slowly wear down. So the question has always been: "Did that matter?" Well, more and more, it seems like it matters. —*Elizabeth Blackburn.*

Nature wants 5 of your 7 children dead. It wants you dead by 50. Everything better than that is brought to you by science and technology. —*David Frum.*

The relationship between telomerase and cancer is complex. Many who have heard of telomerase associate it with cancer and may even have the notion that telomerase causes cancer. This is actually not true, however. While telomere lengthening by telomerase or an alternative lengthening of telomeres (ALT) mechanism is required for cancer to survive, such mechanisms are not sufficient to cause cancer as there are several other requirements that must be met for an aberrant cell to begin to proliferate out of control. Put simply, telomerase is not an oncogene [1]. In this chapter, we explore the complex relationships between telomerase, cancer, and aging, and we start with a brief look at the history of the discoveries which led to the knowledge we have today.

2. A short historical perspective

In the late 1800s, the German evolutionary biologist August Weismann speculated that aging may be caused by cells that cannot replicate forever [2]. In other words, he hypothesized that aging did not occur by the waste of single cells, but rather by the inability of cells to renew tissue because their capacity for cell division was not everlasting, but finite [2]. However, at that time, the experimental tools to prove or disprove this theory did not exist. A few years before Weismann died, Alexis Carrel seemingly disproved this theory by showing that mammalian cells actually have an infinite capacity for replication [3], and this became the dominant view in the early 20th century. Alexis Carrel was quite famous at the time, winning the Nobel Prize in Physiology or Medicine in 1912 for pioneering vascular suturing techniques, and even appearing on the TIME magazine cover in 1938. Charles Lindbergh (yes, the same Charles Lindbergh who became famous for completing the first nonstop flight from New York City to Paris) even developed the first perfusion pump with Dr. Carrel [4]. The mammalian cells that Dr. Carrel grew were chicken heart cells that seemed to divide and continue to live in culture longer than Dr. Carrel himself [4]. The care of the chicken cells was transferred to Carrel's associate, Albert Ebeling in 1912 [4]. These chicken cells were then grown in culture until 1946, 2 years after Carrel's death, and following 34 years of proliferation [4]. This was an interesting finding indeed, considering that the average lifespan of an actual chicken is only 5–10 years.

However, the concept that normal mammalian cells were immortal was challenged in 1961 when Leonard Hayflick showed that human cells have a limited replication capacity in culture [5], a concept coined the Hayflick limit [6]. This finding was confirmed by other scientists, and the data showed that human cells will generally divide between 40 and 60 times in cell culture before entering a state of senescence in which they can no longer divide [5,7–10]. A limited number of population doublings has also been observed for other species such as mice, horses, humans, and tortoises, and the number of population doublings before reaching senescence correlates with the lifespan of these species [11,12]. The number of cellular population doublings before reaching senescence also decreases with increasing donor age [11–15]. The reason Alexis Carrel's experiments did not reveal the limited replicative capacity of mammalian cells is unclear, but some have speculated that fresh cells were being inadvertently introduced when fresh components were being added to the culture medium [4].

In the early 1970s, the Russian theoretical biologist Alexey Olovnikov took note of Hayflick's experiments, combined them with the knowledge of telomeres at the time, and proposed a telomere hypothesis of aging as well as a relationship between telomeres and cancer [16–18]. Telomeres were first observed in the 1930s. Barbara McClintock was studying chromosomes in maize (corn), when she observed distinct knobs at the ends of these chromosomes [19]. Through further studies, Herman Muller and Barbara McClintock found that the ends of fragments of chromosomes broken by X-rays or during meiosis behaved very differently from natural chromosome ends with telomeres, showing that natural chromosome ends served a protective role [20–22]. Muller called these ends "telomeres" from the Greek words for "end" (telos) and "part" (meros). In 1953, the structure of DNA was described by James Watson and Francis Crick. Later, Arthur Kornberg discovered the DNA polymerase used to replicate DNA and characterized its mechanism [23,24]. Around this time James Watson identified the end-replication problem and realized that DNA polymerase would not be capable of replicating the entire DNA strand, leaving some of the ends of DNA lost with each replication [25]. If the ends of DNA cannot be completely replicated, then the telomeres at the ends of chromosomes must become shorter with each cell division. If the cell did not have some means of lengthening the telomeres, then the DNA would be completely lost with successive generations as it continued to shorten. The means for lengthening telomeres turned out to be an enzyme called telomerase which was discovered by Elizabeth H. Blackburn, Carol W. Greider, and Jack Szostak in the 1980s [26–28], and they were awarded the 2009 Nobel Prize in Physiology or Medicine for this discovery.

Despite having a mechanism for lengthening telomeres, most cells in the human body do not express telomerase, and in 1986 the discovery was made that telomeres shorten during the aging of humans [29]. However, there are a few cell types that express high levels of telomerase: the germline cells [29] and cancer cells [30]. This makes sense because both of these cell types are immortal, replicating without ever reaching a state of senescence. Stem cells can also express telomerase, but generally do not express enough to prevent the gradual shortening of telomeres with age [31]. The germline (the reproductive line of cells) has been replicating for about 3.5 to 4 billion years since there has been life on Earth [32,33]. These cells have been able to continue to replicate and mutate throughout the course of evolution up until the present day. Cancer cells can also replicate indefinitely. The HeLa cell line grown in labs around the world was derived from a tumor in 1951 from a woman named Henrietta Lacks who died that same year, but these cells continue to replicate to this day [34]. Researchers discovered that most cancers express the telomerase enzyme in 1994 [30], and further studies found that telomerase is significantly expressed in 85%–95% of human cancers [35]. Ironically, the first noncancer human cells to be immortalized by the artificial introduction of telomerase were cells from Dr. Hayflick himself who had first shown that human cells would senesce after a limited number of divisions, thus proving that this one enzyme, telomerase, was sufficient to allow human cells to divide indefinitely [36–39].

3. Current cancer treatments

The current methods for treating cancers over recent decades have primarily involved surgery, radiation, and chemotherapy (or to word it more bluntly, to cut, burn, and poison) [40–43]. These methods have often harmed the cancer as well as the patient in the hopes that the patient would come out ahead, and this has shown to be more effective than no treatment at all [44–47]. However, the harsh side effects and collateral damage of such treatments have left us searching for better cures for cancer. There are much more targeted approaches to cancer treatment on the horizon such as mRNA vaccines [48,49], gene therapy [50], and others [51–55], but many of these have not passed through clinical trials and are not currently in widespread use for treating humans. The use of CAR T cells is an exciting approach, but so far this method has only been approved for a limited number of antigens in six different cancer types, all of which are blood cancers [56,57].

One of the interesting aspects of cancer is that cancer rates increase with age, with age being the biggest risk factor for cancer for most people [58,59]. We have already observed that telomeres shorten with age as well. Theoretically, you could prevent telomere shortening with age by expressing telomerase. However, we have also observed that most cancers ($\approx 85\%-95\%$) express telomerase [35,60]. Clearly there is a relationship between telomeres, telomerase, aging, and cancer. However, precisely what these relationships are, and the best way to take advantage of them to enhance human health, has been somewhat difficult to untangle. Is it possible to lengthen telomeres, potentially slowing the aging process, and also avoid cancer?

4. Relationship between cancer, aging, and telomerase

4.1 Short telomeres and aging

The process of telomere shortening appears to play a causative role in aging. The Hayflick limit for the cells from a particular species correlates with the lifespan of that species, and the Hayflick limit is affected by the length of the telomeres [61]. For example, mice have a lifespan of about 2 years and their cells have a Hayflick limit of about 15 divisions. Chickens have a lifespan of about 30 years with a Hayflick limit of 15–35 divisions. Humans can live around 100 years and have a Hayflick limit of about 40–60 divisions, and Galapagos tortoises have a lifespan of about 175 years with a Hayflick limit of about 72–114 divisions [62]. However, telomere lengths cannot be compared between species in a simple way. In fact, the telomeres of lab mice are much longer than the telomeres of humans. Mouse telomeres range from 10 to 150 kb, while human telomeres range from 5 to 15 kb [63]. Therefore, the mean telomere length compared between species cannot be used to estimate lifespan. There is some research that indicates that the rate of telomere shortening rather than the telomere length alone predicts mortality [64], and the rate of telomere shortening of different species has correlated with species lifespan [65]. Other research indicates that it is the length of the shortest telomere, rather than the mean telomere length, in a cell which is the most important [63,66,67]. Regardless of which telomere measurement is most important, the number of cell divisions that can occur before a critically short telomere length is reached for a given species does appear to have a correlation with lifespan.

Telomere shortening and telomere uncapping activate senescence [68]. Short telomeres in the cell are recognized as DNA double-stranded breaks [69]. The DNA damage response of the cell can then fuse different chromosomes together, resulting in genome instability [70,71]. In order to prevent further complications, the cell uses the tumor suppressor protein p53 to activate a program of senescence in which the cell will no longer divide and continue to shorten telomeres [72]. Once the stem cells in an organism reach senescence, they can no longer proliferate to renew damaged tissue.

Short and long telomeres have been correlated with lifespan, and short telomeres are considered a hallmark of aging [73]. Human centenarians and their offspring have longer telomeres compared to the telomeres of the whole population on average [74–76]. Moreover, within twin pairs, the twin with the greater telomere length often looks younger and lives longer [77]. Additionally, knockout mice deficient for *Tert* (the protein component of the telomerase enzyme) have more complications and live shorter lifespans with subsequent generations of breeding without the telomerase enzyme [78], although substantial reductions in lifespan are typically not observed until the third or fourth generations, which may be due to the very long telomeres of laboratory mice. These *Tert* knockout mice are also physically smaller in size [79].

4.2 Telomere stability and aging diseases

Mutations in genes for proteins necessary for maintaining telomeres can cause diseases which result in short lifespans. One of the main diseases caused by mutations in genes responsible for the maintenance of telomeres is dyskeratosis congenita [80,81]. Mutations in genes such as DKC1, TERC, TERT, TINF2, NOP10, and NHP2 can lead to the condition [80]. Patients with dyskeratosis congenita have abnormal skin pigmentation, nail dystrophy, leukoplakia of the oral mucosa, short telomeres, premature aging, and progressive bone marrow failure, often resulting in a significantly shortened lifespan [80,82]. Patients with dyskeratosis congenita even have, somewhat counterintuitively, a predisposition for cancer [82]. Hoyeraal-Hreidarsson syndrome is a clinically severe variant of dyskeratosis congenita which is also caused by aberrant telomere biology with mutations found in TINF2, TERT, TPP1, and RTEL1 [83]. Revesz syndrome is another severe variant of dyskeratosis congenita, and mutations in CTC1 (CST telomere replication complex component 1) and TINF2 (part of the telomere shelterin complex) can lead to the disease [84]. Other diseases such as familial idiopathic pulmonary fibrosis and aplastic anemia have been associated with mutations in telomere maintenance machinery [85,86]. Werner syndrome is characterized by short stature, bilateral cataracts, early graying and loss of hair, and scleroderma-like skin changes [86]. This disorder is caused by mutations in the WRN gene involved in the maintenance and repair of DNA, and this protein may also play a role in telomere maintenance [86]. Hutchinson–Gilford progeria is the most severe premature aging syndrome, and patients often only live into their teenage years [87]. This disorder is caused by mutations for lamin A, a protein that is a key component of the nucleus membrane protecting DNA [87]. A defect in lamin A results in poor DNA repair and genomic instability [88]. In this case, the original cause of the disease does not appear to be due to telomere maintenance, but the defective DNA repair and genomic instability does lead to shorter telomeres in Hutchinson—Gilford progeria patients [61,88]. A groundbreaking recent study was able to cure the disease in a Hutchinson—Gilford progeria mouse model by correcting the mutation in the lamin A gene with an adenine base editor, and the mutation was also corrected in human fibroblasts in vitro [89]. Not only do the bodies of the patients of such disorders age rapidly, but cells from humans with hereditary premature aging syndromes senesce prematurely in vitro, providing further evidence that senescence plays a causal role in the aging process [90].

Patients with aging syndromes often have cells with short or dysfunctional telomeres and more senescent cells, and they can also have higher rates of cancer, which contradicts the notion that cellular senescence always acts as a tumor suppression mechanism. The predisposition for cancer in patients with telomere syndromes leads to an 11-fold increase of cancer relative to the normal population, resulting in cancers such as squamous cell carcinomas of the skin, upper aerodigestive and anogenital tract cancers, and hematological malignancies [85]. Although many of these patients have defective telomere maintenance, and cancers require some form of telomere maintenance, cancer cells can use non-traditional methods such as ALT for extending the length of telomeres.

4.3 Outcome of increased telomerase expression

Artificially increasing telomerase expression results in increases in healthspan and lifespan. In a mouse model in which mice would increase telomerase expression when fed tamoxifen, the mice exhibited numerous health improvements under the telomerase expression conditions [91]. The telomerase reactivation extended telomeres, reduced DNA damage signaling, allowed quiescent cultures to resume proliferation, and eliminated degenerative phenotypes in the testes, spleen, intestine, and brain. These results were observed after only 4 weeks of treatment. In another experiment, mice were injected with an adenovirus containing the gene for *Tert* [92]. The median lifespan of the mice was increased, and the mice were healthier in a variety of ways such as increased insulin sensitivity, less osteoporosis, enhanced neuromuscular coordination, and improvement of several biomarkers of aging. Mice treated with the virus at 1 year of age also had a 24% increase in median lifespan. Importantly, the mice did not have an increased level of cancer incidence. Another group used a high-capacity cytomegalovirus vector (CMV) to deliver *Tert* to mice and observed an increase in lifespan of 41.4% without any increased cancer incidence [93]. This treatment also resulted in hair and weight loss prevention, improved activity and motor coordination, increased glucose tolerance, and improved mitochondrial integrity [93].

Telomerase gene therapy has also been used in mouse models for specific diseases. Telomerase gene therapy has been used to provide protection for the heart under the conditions of an acute myocardial infarction [94], increase survival in mice with aplastic anemia [95], and evidence suggests that telomerase gene therapy could be used to reduce symptoms of pulmonary fibrosis driven by telomere dysfunction [96]. In the myocardial infarction study, mice were treated with telomerase gene therapy and exhibited longer telomeres, attenuated cardiac dilation, improved ventricular function, and smaller infarct scars. These improved symptoms led to an increased survival of 17% of the treated group compared with the control group which received an empty virus. An increased number of Ki-67- and pH3-positive cardiomyocytes were observed, indicating higher levels of proliferation. A gene expression signature also revealed that the treated group had a signature more similar to the signature of neonatal mice. These results indicate that telomerase gene therapy could be a viable treatment for myocardial infarction. Importantly, no increase in cancer was observed.

Aplastic anemia is a fatal bone marrow disorder characterized by marrow hypoplasia. As a result, there is replicative impairment of hematopoietic stem and progenitor cells. There is a subgroup of the inherited disease caused by mutations in telomerase and other telomere components. Telomerase gene therapy was used to treat two mouse models of aplastic anemia: one mouse model deficient for the Trf1 shelterin protein, and one mouse model deficient for the Tert telomerase protein [95]. The research study found that the AAV9-*Tert* virus was able to target the bone marrow compartment, including hematopoietic stem cells. The mice treated with the telomerase gene therapy exhibited increased telomere length, improved blood counts, and increased survival, with 87% of *Tert*-treated mice surviving and only 55% of the empty virus mice surviving in the study. No increase in cancer was observed in the *Tert*-treated mice.

Idiopathic pulmonary fibrosis (IPF) is a degenerative disease of the lungs with an average survival post-diagnosis of 2—3 years [97]. IPF is an age-associated disease, and environmental factors such as smoking are known to increase the risk of developing this disease. A study with IPF and telomeres was performed with two mouse models of IPF:

TRF1-deficient mice, and short telomere mice treated with low-dose bleomycin [96]. The study found that *Trf1* deletion induces pulmonary fibrosis and decreased mouse survival. This effect is observed without changing the telomere length, but rather by inducing dysfunctional telomeres. In late-generation telomerase knockout mice, a low bleomycin dose synergizes with short telomeres to trigger pulmonary fibrosis. The bleomycin triggered the DNA damage response in the lungs of telomerase-deficient mice. The results from the study indicated that DNA damage above a certain threshold (5%–10%), combined with defective stem cell regeneration ability, may lead to the development of IPF. These results suggest that it may be possible to treat IPF using telomerase gene therapy to prevent DNA damage, prevent telomere uncapping, and increase stem cell regeneration.

One important feature of increased telomerase expression is that it appears to even benefit tissues with low proliferation rates such as tissues of the heart and brain. Although these tissues have low proliferation rates, they do contain specific stem cells and undergo certain levels of regeneration. Additionally, although some specific cell types never proliferate, these cells depend on other cells that do proliferate in order to function well. These cells depend on other cells through intercellular signaling and the level of inflammation in the surrounding environment. These cells also depend on immune cells to clean up plaques and other harmful substances in the surrounding environment. For example, microglia can remove or prevent the plaques that form in the brain which are associated with Alzheimer's disease. Since the telomeres of the immune cells will shorten with each cell division, the quiescent cells would naturally benefit from immune cell pools with longer telomeres. Therefore, telomerase gene therapy has the potential to treat diseases which affect the cardiovascular system and the central nervous system as well. In one study that examined the effects of telomerase gene therapy on the brain, there was less DNA damage, more neurogenesis, less inflammation, and improved performance in a Barnes maze memory test [98].

4.4 Telomerase and cancer

In general, when many people think of telomerase, they often think of cancer. This makes sense since telomerase is expressed in about 90% of all tumors, and telomerase allows the cancer to continue to proliferate and avoid reaching senescence [99]. However, telomerase does not cause cancer [1,100,101]. Some kind of telomere maintenance, whether through telomerase or alternative lengthening of telomeres (ALT) [102], is necessary for cancer cells to survive, but this does not indicate that telomerase caused the cancer. In fact, there are multiple cellular pathways that must become deregulated in a cell to allow for aggressive tumor growth [103]: proliferative signaling, evasion of growth suppressors, activating invasion and metastasis, replicative immortality (through telomere elongation or ALT), induction of angiogenesis, and resistance to cell death. When researchers treated cell lines with TERT alone, the cell lines did become immortalized, but the cells retained normal growth control and did not exhibit a transformed phenotype [100,101]. More factors such as SV40 large T antigen and oncogenic Ras in some cases are required to obtain a transformed phenotype [104]. The exact factors and mutations which cause transformation will depend on the specific context. An interesting note is that although cancer cells have high telomerase expression, they can often actually have shorter telomeres than the cells around them due to their high proliferation rates [105,106].

Telomerase does not cause cancer, but the lack of it might. Short telomeres may actually be the cause of some cancers [107,108]. In some ways, short telomeres and senescence are a double-edged sword since senescence can serve as a tumor suppression mechanism and prevent uncontrolled growth, but at the same time senescent cells have short telomeres which can initiate the chromosome instability found in cancer cells [90,109]. Senescent cells can also secrete inflammatory factors which result in an SASP (senescence-associated secretory phenotype) and produce a microenvironment conducive to cancer initiation [110]. Here is a brief summary about how short telomeres could cause cancer (Fig. 12.1). The cells in the body continue to proliferate with age, and the telomeres shorten with each cell division. Eventually the telomeres become critically short, and the cells enter into a state of crisis. At this point, the cell may enter into a state of senescence and never divide again; the cell would not become a cancer cell, but the cell could also never proliferate to renew damaged tissue. This state of senescence would result in tissue atrophy and aging. On the other hand, the short telomeres are unstable and can lead to genome instability as the DNA damage response and repair pathways recognize the chromosome ends as DNA double-strand breaks and fuse chromosomes together. This chromosome instability could lead to the necessary epimutations, mutations, and gene fusions for a cancer to arise. Measurements of telomere length support this model since short telomere length can be associated with a greater risk of cancer [107,108].

Note also that regardless of the lifespan of the organism, organisms experience increased cancer incidence with age [111,112], with a few rare exceptions such as in the naked mole rat species [113]. Mice experience increased cancer rates as their age increases even though their lifespans are much shorter than human lifespans. Part of the reason for

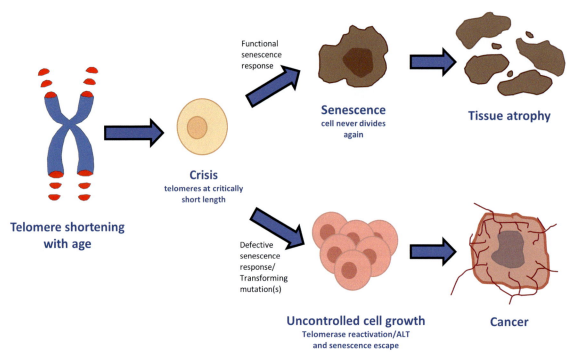

FIGURE 12.1 Model for how telomere shortening can lead to both the tissue atrophy of old age and cancer depending on the path a cell takes once telomeres become critically short and a state of crisis is reached.

this phenomenon is certainly due to the fact that more time has allowed more mutations throughout the genome to occur. Another factor though could be that the telomeres shorten more with age, resulting in more genome instability, DNA mutations, and microenvironments containing senescent cells as well as the resulting inflammation that accompanies senescent cells. Therefore, although the atrophy that occurs with aging and the cancer that occurs with aging are opposite processes, both processes correlate with an increase in age and can be linked to the process of telomere shortening.

There are circumstances when telomerase overexpression has led to increases in cancer in mice and humans. For example, in the K5-mTert mice, higher levels of spontaneous and induced cancer incidence were observed in the mice at young ages [114]. However, the mice which did not succumb to cancer at a young age ended up living longer than normal mice, with a 10% increase in mean lifespan. The increased cancer incidence at a young age may have been the result of higher proliferation rates present in young ages compared with older ages. The higher proliferation rates may have led to more mutations due to mistakes in DNA replication. Normally, the cells may have made a mutation without the presence of telomerase; therefore, the cell could not become immortal without telomerase and no cancer was able to develop. In general though, mice have higher cancer rates than humans do in a short period of time [111]. If telomerase were expressed highly in the same manner in human cells, it is possible that increases in cancer would not be observed. In another strain of mice, *Tert* was constitutively expressed under the control of the beta-actin promoter and CMV enhancer elements and increased spontaneous breast cancer was observed [115]. In humans, a German family was found with a greatly increased risk of cancer due to a mutation in the Tert promoter which increased telomerase expression and resulted in longer telomeres [116,117]. One feature in common with all of these examples in which increased telomerase expression resulted in increased rates of cancer is that they all involved increased telomerase expression from birth. In order to avoid complications, if telomerase gene therapy is applied to humans, it may be most effective and safe when given at later ages.

4.5 Safely applying telomerase therapy

There could be great benefits, but also potential risks, to therapies which lengthen telomeres through telomerase. Current data suggest that telomerase activators can slow telomere shortening and improve health in humans [118,119]. Work from the group of Maria Blasco and also Bioviva have shown that gene therapy with telomerase increases mouse health and lifespan without increasing cancer incidence [92,93]. Mouse health was also increased in a genetically engineered mouse model with inducible telomerase expression [91]. Nevertheless, we have seen that increased telomerase expression can result in increased cancer in some circumstances, particularly when there is increased telomerase expression from birth rather than started later in life [114,115]. If the cell does not have a method to lengthen telomeres, the telomeres will eventually reach a critically short length and the cell will not be able to continue to divide. To prevent increased cancer risk with increased telomerase expression, we may want to enhance tumor suppressor mechanisms as well. For example, p53, the "guardian angel of the genome" is mutated in about 50%–60% of all cancers [120]. The p53 gene is often inactivated by alternative mechanisms in cancers with low mutation rates [121]. Elephants actually have 20 paired copies of genes for p53 isoforms, whereas humans only have one paired copy [122,123]. One might think that larger animals such as elephants would have increased risk of cancer since they have more cells which could possibly become mutated and cancerous. However, such increased cancer rates are not observed, and this is known as Peto's paradox [124]. This paradox may be explained by the numerous copies of the tumor suppressor p53 gene that elephants possess. Although not practical or cost-effective today, perhaps humans will one day be able to apply gene therapy to give ourselves extra copies of p53. Under some conditions, high expression of p53 can be harmful [125], but extra copies of p53 under proper regulatory control could be beneficial. In one research project, super p53 mice with an extra copy of p53 and increased telomerase expression lived 40% longer than the controls [126].

Another important factor of telomerase gene therapy that has been demonstrated by Maria Blasco's group is that with the AAV-*Tert* therapy, the *Tert* gene does not integrate into the genome, or integrates at a very low frequency [127]. Therefore, once the cell divides, the *Tert* gene is not copied into the new cell. Therefore, a cell which does become malignant would not have the benefit of increased telomerase expression from the telomerase gene therapy in the progeny from that cell. Additionally, the AAV-*Tert* virus does not replicate since it requires a helper plasmid to do so [92]. AAV viruses generally have low immunogenicity and have been used in numerous clinical trials worldwide [128]. One could imagine, however, other systems which use alternative viruses or gene-delivery methods (e.g., one study delivered *Tert* with the CMV virus [93]), or even integrate into the genome, and still prove safe.

Although there may not be much cancer risk associated with telomerase gene therapy, the ideal therapy, when the technology becomes available, may involve telomere lengthening with increased telomerase expression combined with enhanced tumor suppressor mechanisms.

4.6 Inhibiting telomerase to fight cancer

Due to the fact that telomerase is over-expressed in 85%–95% of cancers [35,60], telomerase has been the subject of much cancer research, and many companies and research projects have viewed telomerase as an attractive target to inhibit in order to fight cancer [129–133]. One telomerase inhibitor known as imetelstat is currently in phase III clinical trials for myelofibrosis [134]. Quite a few other telomerase inhibitors are being researched [131], and new approaches such as using siRNAs to inhibit telomerase are also being explored [135]. This telomerase inhibition approach, like so many other cancer therapies, may have adverse side effects, however, since inhibiting telomerase non-discriminately may inhibit healthy cells that express telomerase such as stem cells needed to regenerate tissues and body systems. If one could specifically target the telomerase-inhibiting therapies to the cancer cells and avoid normal cells or restrict the telomerase inhibition to a specific window of time, however, this approach could greatly inhibit the ability of cancers to benefit from replicative immortality.

4.7 Mechanisms through which short telomeres affect cellular aging

There are still many discoveries that must be made in order to fully understand how short telomeres cause cellular aging apart from inducing cellular senescence, but shelterin proteins, TERRA, and gene expression changes may play a role. Shelterin is a protein complex which binds to and protects telomeres. The shelterin complex is composed of six sub-unit proteins: TRF1, TRF2, POT1, RAP1, TIN2, and TPP1 [136]. Of these six proteins, TRF1 appears to be particularly important because it is highly expressed in stem cell compartments [137], and therefore TRF1 would be important for regeneration and repair. The structure of TRF1 favors its interaction with DNA [138]. However, an investigation was performed to deduce whether TRF1 binds to other parts of the genome, but TRF1 was found to exclusively bind to telomeric repeats [139]. TRF1 is an essential protein for an organism, but some initial experiments also indicate that inhibiting TRF1 may prove to be a viable strategy for suppressing cancer [140]. POT1 is another shelterin that is important for telomere structure and function, and mutations of POT1 have been found in patients with chronic lymphocytic leukemia and TP53-negative Li-Fraumeni-like syndrome

[141,142]. TRF2 is another important shelterin protein which is essential for adult skin homeostasis [143]. The RAP1 protein is particularly interesting because it interacts with other parts of the genome in addition to telomeres, and regulates genes related to metabolism [144]. This finding could be particularly important because numerous studies have already demonstrated that food intake and metabolism are related to health and longevity [145–148]. RAP1 provides a tangible link between the telomeres and metabolism genes, and scientists may uncover more interesting phenomena related to this connection in the future. The shelterin proteins play an important role in protecting the telomeres, and they also provide a connection between the telomeres and the rest of the genome.

The TERRA molecule is another important molecule for the telomere structure. TERRA is actually RNA that is transcribed from the subtelomere region of the chromosome, and it then binds to the telomere region. The transcription region for TERRA seems to be located on chromosome 18 and chromosome 9 in mice, but then this TERRA binds to nearly all of the chromosome ends in the cell [149]. The TERRA seem to provide a protective role to the telomeres. Downregulation of TERRA resulted in telomere dysfunction-induced foci [149]. In humans, the TERRA locus is located at the 20q and Xp subtelomeres [150]. Deletion of the 20q locus results in decreased TERRA levels, and a dramatic loss of telomere sequences as well as the induction of a large DNA damage response [150]. There are still many more mysteries to uncover about TERRA and its role in telomere maintenance and health.

There are also several other possible methods through which telomere length could affect the cell. Some genes which are located close to the telomere are expressed at low levels, but as the telomeres shorten with subsequent cell divisions, access to these genes becomes easier and the expression of these genes increases [151]. This phenomenon is known as the "telomere position effect." Newly discovered miRNAs may also affect telomeres and the genome [152]. Also, as previously mentioned, short telomeres can induce a DNA damage response as well as the activation of p53. The resulting senescence results in atrophy and aging, and this senescence can also provide an environment that is conducive to developing cancer.

5. Conclusion

To summarize, I am reproducing a series of questions and answers that Michael Fossel and I put into a recent publication [153]. Several points should be stressed in summarizing our current knowledge of the relationship between telomere length, telomerase, genomic stability, and cancer.

(1) Does telomerase activity increase the risk of cancer?

No. Malignant transformation requires other mutations or disrupted pathways. Telomerase itself is not sufficient to cause growth deregulation, mutation, or malignant transformation. Once a cell becomes malignant, however, then either telomerase or ALT are necessary for such cells to proliferate and result in clinical cancer.

(2) Does telomerase benefit the normal cell?

Yes. Telomerase prevents genomic instability and decreases the risk of transformation from a normal to a malignant state. Once such transformation occurs, however, then the absence of telomerase (or the presence of telomerase inhibition) may prevent further cell division and proliferation of the malignant cell.

(3) Is the telomerase/cancer relationship simple?

No. Its effects are complex and are dependent upon whether or not a cell has already incurred genetic damage and consequent growth dysregulation. In normal cells, telomerase can prevent mutation by preventing genomic instability and malignant transformation. Once such a transition occurs, however, telomerase activity (or ALT) may permit such proliferation of malignant cells, often culminating in overt clinical cancer.

(4) Does telomerase protect genomic stability?

Yes. Telomere shortening, on the other hand, increases genomic instability and thereby increases cancer risk. Long telomeres lower the risk of some clinical cancers, while short telomeres are part of a causal cascade of intracellular events that results in oncogenesis and, ultimately, clinical cancer. Telomerase therapy is not only unlikely to result in an increased risk of cancer, but is likely to lower the risk of cancer compared to age-matched patients not treated with telomerase therapy.

At the end of the day though, to address whether increased telomerase expression through gene therapy will be beneficial and safe, the actual evidence from experiments and trials will be more important than logical arguments. We will not know for sure whether or not telomerase gene therapy is safe in humans until the data have been collected. Even if such therapy proves safe, it may be wise to pursue enhanced tumor suppression strategies as well.

We have come a long way since the days of August Weismann and his philosophical theories in the late 1800s. Although the path has often been confusing and counter-intuitive, we have learned a great deal in a century and genetically engineered mouse model with inducible telomerase expression [91]. Nevertheless, we have seen that increased telomerase expression can result in increased cancer in some circumstances, particularly when there is increased telomerase expression from birth rather than started later in life [114,115]. If the cell does not have a method to lengthen telomeres, the telomeres will eventually reach a critically short length and the cell will not be able to continue to divide. To prevent increased cancer risk with increased telomerase expression, we may want to enhance tumor suppressor mechanisms as well. For example, p53, the "guardian angel of the genome" is mutated in about 50%—60% of all cancers [120]. The p53 gene is often inactivated by alternative mechanisms in cancers with low mutation rates [121]. Elephants actually have 20 paired copies of genes for p53 isoforms, whereas humans only have one paired copy [122,123]. One might think that larger animals such as elephants would have increased risk of cancer since they have more cells which could possibly become mutated and cancerous. However, such increased cancer rates are not observed, and this is known as Peto's paradox [124]. This paradox may be explained by the numerous copies of the tumor suppressor p53 gene that elephants possess. Although not practical or cost-effective today, perhaps humans will one day be able to apply gene therapy to give ourselves extra copies of p53. Under some conditions, high expression of p53 can be harmful [125], but extra copies of p53 under proper regulatory control could be beneficial. In one research project, super p53 mice with an extra copy of p53 and increased telomerase expression lived 40% longer than the controls [126].

Another important factor of telomerase gene therapy that has been demonstrated by Maria Blasco's group is that with the AAV-*Tert* therapy, the *Tert* gene does not integrate into the genome, or integrates at a very low frequency [127]. Therefore, once the cell divides, the *Tert* gene is not copied into the new cell. Therefore, a cell which does become malignant would not have the benefit of increased telomerase expression from the telomerase gene therapy in the progeny from that cell. Additionally, the AAV-*Tert* virus does not replicate since it requires a helper plasmid to do so [92]. AAV viruses generally have low immunogenicity and have been used in numerous clinical trials worldwide [128]. One could imagine, however, other systems which use alternative viruses or gene-delivery methods (e.g., one study delivered *Tert* with the CMV virus [93]), or even integrate into the genome, and still prove safe.

Although there may not be much cancer risk associated with telomerase gene therapy, the ideal therapy, when the technology becomes available, may involve telomere lengthening with increased telomerase expression combined with enhanced tumor suppressor mechanisms.

4.6 Inhibiting telomerase to fight cancer

Due to the fact that telomerase is over-expressed in 85%—95% of cancers [35,60], telomerase has been the subject of much cancer research, and many companies and research projects have viewed telomerase as an attractive target to inhibit in order to fight cancer [129—133]. One telomerase inhibitor known as imetelstat is currently in phase III clinical trials for myelofibrosis [134]. Quite a few other telomerase inhibitors are being researched [131], and new approaches such as using siRNAs to inhibit telomerase are also being explored [135]. This telomerase inhibition approach, like so many other cancer therapies, may have adverse side effects, however, since inhibiting telomerase non-discriminately may inhibit healthy cells that express telomerase such as stem cells needed to regenerate tissues and body systems. If one could specifically target the telomerase-inhibiting therapies to the cancer cells and avoid normal cells or restrict the telomerase inhibition to a specific window of time, however, this approach could greatly inhibit the ability of cancers to benefit from replicative immortality.

4.7 Mechanisms through which short telomeres affect cellular aging

There are still many discoveries that must be made in order to fully understand how short telomeres cause cellular aging apart from inducing cellular senescence, but shelterin proteins, TERRA, and gene expression changes may play a role. Shelterin is a protein complex which binds to and protects telomeres. The shelterin complex is composed of six sub-unit proteins: TRF1, TRF2, POT1, RAP1, TIN2, and TPP1 [136]. Of these six proteins, TRF1 appears to be particularly important because it is highly expressed in stem cell compartments [137], and therefore TRF1 would be important for regeneration and repair. The structure of TRF1 favors its interaction with DNA [138]. However, an investigation was performed to deduce whether TRF1 binds to other parts of the genome, but TRF1 was found to exclusively bind to telomeric repeats [139]. TRF1 is an essential protein for an organism, but some initial experiments also indicate that inhibiting TRF1 may prove to be a viable strategy for suppressing cancer [140]. POT1 is another shelterin that is important for telomere structure and function, and mutations of POT1 have been found in patients with chronic lymphocytic leukemia and TP53-negative Li-Fraumeni-like syndrome

[141,142]. TRF2 is another important shelterin protein which is essential for adult skin homeostasis [143]. The RAP1 protein is particularly interesting because it interacts with other parts of the genome in addition to telomeres, and regulates genes related to metabolism [144]. This finding could be particularly important because numerous studies have already demonstrated that food intake and metabolism are related to health and longevity [145–148]. RAP1 provides a tangible link between the telomeres and metabolism genes, and scientists may uncover more interesting phenomena related to this connection in the future. The shelterin proteins play an important role in protecting the telomeres, and they also provide a connection between the telomeres and the rest of the genome.

The TERRA molecule is another important molecule for the telomere structure. TERRA is actually RNA that is transcribed from the subtelomere region of the chromosome, and it then binds to the telomere region. The transcription region for TERRA seems to be located on chromosome 18 and chromosome 9 in mice, but then this TERRA binds to nearly all of the chromosome ends in the cell [149]. The TERRA seem to provide a protective role to the telomeres. Downregulation of TERRA resulted in telomere dysfunction-induced foci [149]. In humans, the TERRA locus is located at the 20q and Xp subtelomeres [150]. Deletion of the 20q locus results in decreased TERRA levels, and a dramatic loss of telomere sequences as well as the induction of a large DNA damage response [150]. There are still many more mysteries to uncover about TERRA and its role in telomere maintenance and health.

There are also several other possible methods through which telomere length could affect the cell. Some genes which are located close to the telomere are expressed at low levels, but as the telomeres shorten with subsequent cell divisions, access to these genes becomes easier and the expression of these genes increases [151]. This phenomenon is known as the "telomere position effect." Newly discovered miRNAs may also affect telomeres and the genome [152]. Also, as previously mentioned, short telomeres can induce a DNA damage response as well as the activation of p53. The resulting senescence results in atrophy and aging, and this senescence can also provide an environment that is conducive to developing cancer.

5. Conclusion

To summarize, I am reproducing a series of questions and answers that Michael Fossel and I put into a recent publication [153]. Several points should be stressed in summarizing our current knowledge of the relationship between telomere length, telomerase, genomic stability, and cancer.

(1) Does telomerase activity increase the risk of cancer?
   No. Malignant transformation requires other mutations or disrupted pathways. Telomerase itself is not sufficient to cause growth deregulation, mutation, or malignant transformation. Once a cell becomes malignant, however, then either telomerase or ALT are necessary for such cells to proliferate and result in clinical cancer.

(2) Does telomerase benefit the normal cell?
   Yes. Telomerase prevents genomic instability and decreases the risk of transformation from a normal to a malignant state. Once such transformation occurs, however, then the absence of telomerase (or the presence of telomerase inhibition) may prevent further cell division and proliferation of the malignant cell.

(3) Is the telomerase/cancer relationship simple?
   No. Its effects are complex and are dependent upon whether or not a cell has already incurred genetic damage and consequent growth dysregulation. In normal cells, telomerase can prevent mutation by preventing genomic instability and malignant transformation. Once such a transition occurs, however, telomerase activity (or ALT) may permit such proliferation of malignant cells, often culminating in overt clinical cancer.

(4) Does telomerase protect genomic stability?
   Yes. Telomere shortening, on the other hand, increases genomic instability and thereby increases cancer risk. Long telomeres lower the risk of some clinical cancers, while short telomeres are part of a causal cascade of intracellular events that results in oncogenesis and, ultimately, clinical cancer. Telomerase therapy is not only unlikely to result in an increased risk of cancer, but is likely to lower the risk of cancer compared to age-matched patients not treated with telomerase therapy.

At the end of the day though, to address whether increased telomerase expression through gene therapy will be beneficial and safe, the actual evidence from experiments and trials will be more important than logical arguments. We will not know for sure whether or not telomerase gene therapy is safe in humans until the data have been collected. Even if such therapy proves safe, it may be wise to pursue enhanced tumor suppression strategies as well.

We have come a long way since the days of August Weismann and his philosophical theories in the late 1800s. Although the path has often been confusing and counter-intuitive, we have learned a great deal in a century and a half. We now know that normal human cells cannot divide indefinitely and will reach a state of senescence after enough cell divisions. We have discovered DNA and understand the molecular basis of life. Through the study of the replication of DNA, it became clear that there was an end-replication problem, and that the genetic code would be lost over generations unless cells had a way to lengthen the chromosome ends (the telomeres). The method by which cells lengthened chromosome ends was through the telomerase enzyme. Through numerous studies, we have begun to understand the close connection between telomeres, telomerase, cancer, and aging. With this foundation, we may be close to harnessing this knowledge to both fight cancer and prevent or slow down the atrophy and decay that occur with natural aging.

References

[1] Harley CB. Telomerase is not an oncogene. Oncogene 2002;21(4):494−502. https://doi.org/10.1038/sj.onc.1205076.
[2] Strehler BL. Understanding aging. Methods Mol Med 2000;38:1−19. https://doi.org/10.1385/1-59259-070-5:1.
[3] Carrel A. On the permanent life of tissues outside of the organism. J Exp Med 1912;15(5):516. https://doi.org/10.1084/JEM.15.5.516.
[4] Witkowski JA. Dr. Carrel's immortal cells. Med Hist 1980;24(2):129. https://doi.org/10.1017/S0025727300040126.
[5] Hayflick L, Moorhead PS. The serial cultivation of human diploid cell strains. Exp Cell Res 1961;25(3):585−621. https://doi.org/10.1016/0014-4827(61)90192-6.
[6] Shay JW, Wright WE. Hayflick, his limit, and cellular ageing. Nat Rev Mol Cell Biol 2000;1(1):72−6. https://doi.org/10.1038/35036093.
[7] Todaro GJ, Wolman SR, Green H. Rapid transformation of human fibroblasts with low growth potential into established cell lines by SV40. J Cell Comp Physiol 1963;62(3):257−65. https://doi.org/10.1002/jcp.1030620305.
[8] Yoshida MC, Makino S. A chromosome study of non-treated and an irradiated human in vitro cell line. Jpn J Hum Genet 1963;8:39−45.
[9] Pilyugin S, Mittler J, Antia R. Modeling T-cell proliferation: an investigation of the consequences of the Hayflick limit. J Theor Biol 1997;186(1):117−29. https://doi.org/10.1006/jtbi.1996.0319.
[10] Chan M, Yuan H, Soifer I, Maile TM, Wang RY, Ireland A, et al. Novel insights from a multiomics dissection of the Hayflick limit. Elife 2022;11. https://doi.org/10.7554/eLife.70283.
[11] Juckett DA. Cellular aging (The hayflick limit) and species longevity: a unification model based on clonal succession. Mech Ageing Dev 1987;38(1):49−71. https://doi.org/10.1016/0047-6374(87)90110-2.
[12] Goldstein S. Aging in vitro: growth of cultured cells from the Galapagos tortoise. Exp Cell Res 1974;83(2):297−302. https://doi.org/10.1016/0014-4827(74)90342-5.
[13] Bierman EL. The effect of donor age on the in vitro life span of cultured human arterial smooth-muscle cells. In Vitro 1978;14(11):951−5. https://doi.org/10.1007/BF02616126.
[14] Choudhery MS, Badowski M, Muise A, Pierce J, Harris DT. Donor age negatively impacts adipose tissue-derived mesenchymal stem cell expansion and differentiation. J Transl Med 2014;12(1):8. https://doi.org/10.1186/1479-5876-12-8.
[15] Schneider EL, Mitsui Y. The relationship between in vitro cellular aging and in vivo human age. Proc Natl Acad Sci U S A 1976;73(10):3584−8. https://doi.org/10.1073/pnas.73.10.3584.
[16] Olovnikov AM. Principle of marginotomy in template synthesis of polynucleotides. Dokl Akad Nauk SSSR 1971;201(6):1496−9.
[17] Olovnikov AM. A theory of marginotomy. The incomplete copying of template margin in enzymic synthesis of polynucleotides and biological significance of the phenomenon. J Theor Biol 1973;41(1):181−90.
[18] Greider CW. Telomeres and senescence: the history, the experiment, the future. Curr Biol 1998;8(5):R178−81. https://doi.org/10.1016/S0960-9822(98)70105-8.
[19] Creighton HB, McClintock B. A correlation of cytological and genetical crossing-over in Zea mays. Proc Natl Acad Sci U S A 1931;17(8):492−7. https://doi.org/10.1073/pnas.17.8.492.
[20] Muller HJ. The remaking of chromosomes. Collecting net 1938;13:181−98.
[21] McClintock B. The stability of broken ends of chromosomes in Zea mays. Genetics 1941;26(2):234−82. https://doi.org/10.1093/genetics/26.2.234.
[22] McClintock B. The behavior in successive nuclear divisions of a chromosome broken at meiosis. Proc Natl Acad Sci U S A 1939;25(8):405−16. https://doi.org/10.1073/pnas.25.8.405.
[23] Schekman R, Weiner A, Kornberg A. Multienzyme systems of DNA replication. Science 1974;186(4168):987−93. https://doi.org/10.1126/science.186.4168.987.
[24] Kornberg T, Gefter ML. Purification and DNA synthesis in cell-free extracts: properties of DNA polymerase II. Proc Natl Acad Sci U S A 1971;68(4):761−4. https://doi.org/10.1073/pnas.68.4.761.
[25] Watson JD. Origin of concatemeric T7 DNA. Nat New Biol 1972;239(94):197−201.
[26] Szostak JW, Blackburn EH. Cloning yeast telomeres on linear plasmid vectors. Cell 1982;29(1):245−55. https://doi.org/10.1016/0092-8674(82)90109-X.
[27] Greider CW, Blackburn EH. Identification of a specific telomere terminal transferase activity in Tetrahymena extracts. Cell 1985;43(2 Pt 1):405−13.
[28] Greider CW, Blackburn EH. A telomeric sequence in the RNA of Tetrahymena telomerase required for telomere repeat synthesis. Nature 1989;337(6205):331−7. https://doi.org/10.1038/337331a0.
[29] Cooke HJ, Smith BA. Variability at the telomeres of the human X/Y pseudoautosomal region. Cold Spring Harb Symp Quant Biol 1986;51(0):213−9. https://doi.org/10.1101/SQB.1986.051.01.026.
[30] Kim N, Piatyszek M, Prowse K, Harley C, West M, Ho P, et al. Specific association of human telomerase activity with immortal cells and cancer. Science 1994;266(5193):2011−5. https://doi.org/10.1126/science.7605428.
[31] Hiyama E, Hiyama K. Telomere and telomerase in stem cells. Br J Cancer 2007;96(7):1020−4. https://doi.org/10.1038/sj.bjc.6603671.

[32] Betts HC, Puttick MN, Clark JW, Williams TA, Donoghue PCJ, Pisani D. Integrated genomic and fossil evidence illuminates life's early evolution and eukaryote origin. Nat Ecol Evol 2018;2(10):1556−62. https://doi.org/10.1038/s41559-018-0644-x.

[33] Dodd MS, Papineau D, Grenne T, Slack JF, Rittner M, Pirajno F, et al. Evidence for early life in Earth's oldest hydrothermal vent precipitates. Nature 2017;543(7643):60−4. https://doi.org/10.1038/nature21377.

[34] Lucey BP, Nelson-Rees WA, Hutchins GM. Henrietta lacks, HeLa cells, and cell culture contamination. Arch Pathol Lab Med 2009;133(9): 1463−7. https://doi.org/10.5858/133.9.1463.

[35] Shay JW, Wright WE. Telomeres and telomerase: three decades of progress. Nat Rev Genet 2019;20(5):299−309. https://doi.org/10.1038/s41576-019-0099-1.

[36] Bodnar AG, Ouellette M, Frolkis M, Holt SE, Chiu CP, Morin GB, et al. Extension of life-span by introduction of telomerase into normal human cells. Science 1998;279(5349):349−52.

[37] Hall S, Kingsland J. Merchants of immortality: chasing the dream of human life extension. New Sci 2003.

[38] West M, Lessons I've learned, with Michael West. Transl Sci. Published online November 23, 2006.

[39] West M. The immortal cell: one scientist's quest to solve the mystery of human aging. Doubleday; 2003.

[40] Baskar R, Lee KA, Yeo R, Yeoh KW. Cancer and radiation therapy: current advances and future directions. Int J Med Sci 2012;9(3):193−9. https://doi.org/10.7150/ijms.3635.

[41] Schirrmacher V. From chemotherapy to biological therapy: a review of novel concepts to reduce the side effects of systemic cancer treatment (Review). Int J Oncol 2019;54(2):407−19. https://doi.org/10.3892/ijo.2018.4661.

[42] Loughlin KR. William B. Coley: his hypothesis, his toxin, and the birth of immunotherapy. Urol Clin 2020;47(4):413−7. https://doi.org/10.1016/j.ucl.2020.07.001.

[43] Yaqub F. Cut poison burn. Lancet Oncol 2012;13(6):578. https://doi.org/10.1016/S1470-2045(12)70251-6.

[44] Slater JM. From X-rays to ion beams: a short history of radiation therapy. In: Ion beam therapy: fundamentals, technology, clinical applications; 2012. p. 3−16. https://doi.org/10.1007/978-3-642-21414-1_1.

[45] Chabner BA, Roberts TG. Chemotherapy and the war on cancer. Nat Rev Cancer 2005;5(1):65−72. https://doi.org/10.1038/nrc1529.

[46] Galmarini D, Galmarini CM, Galmarini FC. Cancer chemotherapy: a critical analysis of its 60 years of history. Crit Rev Oncol Hematol 2012; 84(2):181−99. https://doi.org/10.1016/j.critrevonc.2012.03.002.

[47] DeVita VT, Chu E. A history of cancer chemotherapy. Cancer Res 2008;68(21):8643−53. https://doi.org/10.1158/0008-5472.CAN-07-6611.

[48] Miao L, Zhang Y, Huang L. mRNA vaccine for cancer immunotherapy. Mol Cancer 2021;20(1):41. https://doi.org/10.1186/s12943-021-01335-5.

[49] Jahanafrooz Z, Baradaran B, Mosafer J, Hashemzaei M, Rezaei T, Mokhtarzadeh A, et al. Comparison of DNA and mRNA vaccines against cancer. Drug Discov Today 2020;25(3):552−60. https://doi.org/10.1016/j.drudis.2019.12.003.

[50] Roma-Rodrigues C, Rivas-García L, Baptista PV, Fernandes AR. Gene therapy in cancer treatment: Why Go Nano? Pharmaceutics 2020;12(3): 233. https://doi.org/10.3390/pharmaceutics12030233.

[51] Pucci C, Martinelli C, Ciofani G. Innovative approaches for cancer treatment: current perspectives and new challenges. Ecancermedicalscience 2019;13. https://doi.org/10.3332/ecancer.2019.961.

[52] Jou J, Harrington KJ, Zocca MB, Ehrnrooth E, Cohen EEW. The changing landscape of therapeutic cancer vaccines—novel platforms and neoantigen identification. Clin Cancer Res 2021;27(3):689−703. https://doi.org/10.1158/1078-0432.CCR-20-0245.

[53] Ouyang X, Telli ML, Wu JC. Induced pluripotent stem cell-based cancer vaccines. Front Immunol 2019;10. https://doi.org/10.3389/fimmu.2019.01510.

[54] Pusztai L, Karn T, Safonov A, Abu-Khalaf MM, Bianchini G. New strategies in breast cancer: immunotherapy. Clin Cancer Res 2016;22(9): 2105−10. https://doi.org/10.1158/1078-0432.CCR-15-1315.

[55] Yeh JE, Toniolo PA, Frank DA. Targeting transcription factors: promising new strategies for cancer therapy. Curr Opin Oncol 2013;25(6): 652−8. https://doi.org/10.1097/01.cco.0000432528.88101.1a.

[56] Hong M, Clubb JD, Chen YY. Engineering CAR-T cells for next-generation cancer therapy. Cancer Cell 2020;38(4):473−88. https://doi.org/10.1016/j.ccell.2020.07.005.

[57] CAR T cells: engineering patients' immune cells to treat their cancers. Published March 10, 2022. Accessed December 10, 2022. https://www.cancer.gov/about-cancer/treatment/research/car-t-cells.

[58] White MC, Holman DM, Boehm JE, Peipins LA, Grossman M, Jane Henley S. Age and cancer risk. Am J Prev Med 2014;46(3):S7−15. https://doi.org/10.1016/j.amepre.2013.10.029.

[59] Laconi E, Marongiu F, DeGregori J. Cancer as a disease of old age: changing mutational and microenvironmental landscapes. Br J Cancer 2020;122(7):943−52. https://doi.org/10.1038/s41416-019-0721-1.

[60] Lansdorp PM. Telomeres, telomerase and cancer. Arch Med Res 2022;53(8):741−6. https://doi.org/10.1016/j.arcmed.2022.10.004.

[61] Allsopp RC, Vaziri H, Patterson C, Goldstein S, Younglai EV, Futcher AB, et al. Telomere length predicts replicative capacity of human fibroblasts. Proc Natl Acad Sci U S A 1992;89(21):10114−8.

[62] Hayflick L. The longevity of cultured human cells. J Am Geriatr Soc 1974;22(1):1−12. https://doi.org/10.1111/j.1532-5415.1974.tb02152.x.

[63] Zijlmans JM, Martens UM, Poon SS, Raap AK, Tanke HJ, Ward RK, et al. Telomeres in the mouse have large inter-chromosomal variations in the number of T2AG3 repeats. Proc Natl Acad Sci U S A 1997;94(14):7423−8.

[64] Epel ES, Merkin SS, Cawthon R, Blackburn EH, Adler NE, Pletcher MJ, et al. The rate of leukocyte telomere shortening predicts mortality from cardiovascular disease in elderly men. Aging 2008;1(1):81−8. https://doi.org/10.18632/aging.100007.

[65] Whittemore K, Vera E, Martínez-Nevado E, Sanpera C, Blasco MA. Telomere shortening rate predicts species life span. Proc Natl Acad Sci USA 2019;116(30):15122−7. https://doi.org/10.1073/pnas.1902452116.

[66] Hemann MT, Strong MA, Hao LY, Greider CW. The shortest telomere, not average telomere length, is critical for cell viability and chromosome stability. Cell 2001;107(1):67−77. https://doi.org/10.1016/S0092-8674(01)00504-9.

[67] Xu Z, Duc KD, Holcman D, Teixeira MT. The length of the shortest telomere as the major determinant of the onset of replicative senescence. Genetics 2013;194(4).

[68] Herbig U, Jobling WA, Chen BPC, Chen DJ, Sedivy JM. Telomere shortening triggers senescence of human cells through a pathway involving ATM, p53, and p21CIP1, but not p16(INK4a). Mol Cell 2004;14(4):501−13. https://doi.org/10.1016/S1097-2765(04)00256-4.

[69] Bailey SM, Cornforth MN. Telomeres and DNA double-strand breaks: ever the twain shall meet? Cell Mol Life Sci 2007;64(22):2956–64. https://doi.org/10.1007/s00018-007-7242-4.
[70] Artandi SE, Attardi LD. Pathways connecting telomeres and p53 in senescence, apoptosis, and cancer. Biochem Biophys Res Commun 2005; 331(3):881–90. https://doi.org/10.1016/j.bbrc.2005.03.211.
[71] Martínez P, Blasco MA. Replicating through telomeres: a means to an end. Trends Biochem Sci 2015;40(9):504–15. https://doi.org/10.1016/j.tibs.2015.06.003.
[72] Rufini A, Tucci P, Celardo I, Melino G. Senescence and aging: the critical roles of p53. Oncogene 2013;32(43):5129–43. https://doi.org/10.1038/onc.2012.640.
[73] López-Otín C, Blasco MA, Partridge L, Serrano M, Kroemer G. The hallmarks of aging. Cell 2013;153(6):1194–217. https://doi.org/10.1016/j.cell.2013.05.039.
[74] Honig LS, Kang MS, Cheng R, Eckfeldt JH, Thyagarajan B, Leiendecker-Foster C, et al. Heritability of telomere length in a study of long-lived families. Neurobiol Aging 2015;36(10):2785–90. https://doi.org/10.1016/j.neurobiolaging.2015.06.017.
[75] Deelen J, Beekman M, Codd V, Trompet S, Broer L, Hägg S, et al. Leukocyte telomere length associates with prospective mortality independent of immune-related parameters and known genetic markers. Int J Epidemiol 2014;43(3):878–86. https://doi.org/10.1093/ije/dyt267.
[76] Terry DF, Nolan VG, Andersen SL, Perls TT, Cawthon R. Association of longer telomeres with better health in centenarians. J Gerontol A Biol Sci Med Sci 2008;63(8):809–12.
[77] Christensen K, Thinggaard M, McGue M, Rexbye H, Hjelmborg JVB, Aviv A, et al. Perceived age as clinically useful biomarker of ageing: cohort study. BMJ 2009;339.
[78] Rudolph KL, Chang S, Lee HW, Blasco M, Gottlieb GJ, Greider C, et al. Longevity, stress response, and cancer in aging telomerase-deficient mice. Cell 1999;96(5):701–12.
[79] Raval A, Behbehani GK, Nguyen LXT, Thomas D, Kusler B, Garbuzov A, et al. Reversibility of defective hematopoiesis caused by telomere shortening in telomerase knockout mice. PLoS One 2015;10(7). https://doi.org/10.1371/journal.pone.0131722.
[80] AlSabbagh MM. Dyskeratosis congenita: a literature review. JDDG J Dtsch Dermatol Ges 2020;18(9):943–67. https://doi.org/10.1111/ddg.14268.
[81] Nelson ND, Bertuch AA. Dyskeratosis congenita as a disorder of telomere maintenance. Mutat Res Fund Mol Mech Mutagen 2012;730(1–2):43–51. https://doi.org/10.1016/j.mrfmmm.2011.06.008.
[82] Batista LFZ, Pech MF, Zhong FL, Nguyen HN, Xie KT, Zaug AJ, et al. Telomere shortening and loss of self-renewal in dyskeratosis congenita induced pluripotent stem cells. Nature 2011;474(7351):399–402. https://doi.org/10.1038/nature10084.
[83] Glousker G, Touzot F, Revy P, Tzfati Y, Savage SA. Unraveling the pathogenesis of Hoyeraal-Hreidarsson syndrome, a complex telomere biology disorder. Br J Haematol 2015;170(4):457–71. https://doi.org/10.1111/bjh.13442.
[84] Karremann M, Neumaier-Probst E, Schlichtenbrede F, Beier F, Brümmendorf TH, Cremer FW, et al. Revesz syndrome revisited. Orphanet J Rare Dis 2020;15(1):299. https://doi.org/10.1186/s13023-020-01553-y.
[85] Armanios M, Blackburn EH. The telomere syndromes. Nat Rev Genet 2012;13(10):693–704. https://doi.org/10.1038/nrg3246.
[86] Kudlow BA, Kennedy BK, Monnat RJ. Werner and Hutchinson–Gilford progeria syndromes: mechanistic basis of human progeroid diseases. Nat Rev Mol Cell Biol 2007;8(5):394–404. https://doi.org/10.1038/nrm2161.
[87] Lai W, Wong W. Progress and trends in the development of therapies for Hutchinson–Gilford progeria syndrome. Aging Cell 2020;19(7). https://doi.org/10.1111/acel.13175.
[88] Gonzalo S, Kreienkamp R. DNA repair defects and genome instability in Hutchinson–Gilford Progeria Syndrome. Curr Opin Cell Biol 2015; 34:75–83. https://doi.org/10.1016/j.ceb.2015.05.007.
[89] Koblan LW, Erdos MR, Wilson C, Cabral WA, Levy JM, Xiong Z-M, et al. In vivo base editing rescues Hutchinson–Gilford progeria syndrome in mice. Nature 2021;589(7843):608–14. https://doi.org/10.1038/s41586-020-03086-7.
[90] Campisi J. Aging and cancer: the double-edged sword of replicative senescence. J Am Geriatr Soc 1997;45(4):482–8. https://doi.org/10.1111/j.1532-5415.1997.tb05175.x.
[91] Jaskelioff M, Muller FL, Paik JH, Thomas E, Jiang S, Adams AC, et al. Telomerase reactivation reverses tissue degeneration in aged telomerase-deficient mice. Nature 2011;469(7328):102–6. https://doi.org/10.1038/nature09603.
[92] Bernardes de Jesus B, Vera E, Schneeberger K, Tejera AM, Ayuso E, Bosch F, et al. Telomerase gene therapy in adult and old mice delays aging and increases longevity without increasing cancer. EMBO Mol Med 2012;4(8):691–704. https://doi.org/10.1002/emmm.201200245.
[93] Jaijyan DK, Selariu A, Cruz-Cosme R, Tong M, Yang S, Stefa A, et al. New intranasal and injectable gene therapy for healthy life extension. Proc Natl Acad Sci USA 2022;119(20). https://doi.org/10.1073/pnas.2121499119.
[94] Bär C, de Jesus BB, Serrano R, Tejera A, Ayuso E, Jimenez V, et al. Telomerase expression confers cardioprotection in the adult mouse heart after acute myocardial infarction. Nat Commun 2014;5.
[95] Bär C, Povedano JM, Serrano R, Benitez-Buelga C, Popkes M, Formentini I, et al. Telomerase gene therapy rescues telomere length, bone marrow aplasia and survival in mice with aplastic anemia. Blood 2016;127(14). https://doi.org/10.1182/blood-2015-08-667485.
[96] Povedano JM, Martinez P, Flores JM, Mulero F, Blasco MA. Mice with pulmonary fibrosis driven by telomere dysfunction. Cell Rep 2015; 12(2):286–99. https://doi.org/10.1016/j.celrep.2015.06.028.
[97] King TE, Pardo A, Selman M. Idiopathic pulmonary fibrosis. Lancet 2011;378(9807):1949–61. https://doi.org/10.1016/S0140-6736(11)60052-4.
[98] Whittemore K, Derevyanko A, Martinez P, Serrano R, Pumarola M, Bosch F, et al. Telomerase gene therapy ameliorates the effects of neurodegeneration associated to short telomeres in mice. Aging 2019;11(10):2916–48. https://doi.org/10.18632/aging.101982.
[99] Shay JW, Wright WE. Telomerase therapeutics for cancer: challenges and new directions. Nat Rev Drug Discov 2006;5(7):577–84. https://doi.org/10.1038/nrd2081.
[100] Morales CP, Holt SE, Ouellette M, Kaur KJ, Yan Y, Wilson KS, et al. Absence of cancer-associated changes in human fibroblasts immortalized with telomerase. Nat Genet 1999;21(1):115–8. https://doi.org/10.1038/5063.
[101] Jiang XR, Jimenez G, Chang E, Frolkis M, Kusler B, Sage M, et al. Telomerase expression in human somatic cells does not induce changes associated with a transformed phenotype. Nat Genet 1999;21(1):111–4. https://doi.org/10.1038/5056.

[102] Cesare AJ, Reddel RR. Alternative lengthening of telomeres: models, mechanisms and implications. Nat Rev Genet 2010;11(5):319–30. https://doi.org/10.1038/nrg2763.

[103] Hanahan D, Weinberg RA. Hallmarks of cancer: the next generation. Cell 2011;144(5):646–74.

[104] Weinberg RA, Hahn WC, Counter CM, Lundberg AS, Beijersbergen RL, Brooks MW. Creation of human tumour cells with defined genetic elements. Nature 1999;400(6743):464–8. https://doi.org/10.1038/22780.

[105] Zhan WH, Ma JP, Peng JS, Gao J-S, Cai S-R, Wang J-P, et al. Telomerase activity in gastric cancer and its clinical implications. World J Gastroenterol 1999;5(4):316–9. https://doi.org/10.3748/wjg.v5.i4.316.

[106] Counter CM, Hirte HW, Bacchetti S, Harley CB. Telomerase activity in human ovarian carcinoma. Proc Natl Acad Sci U S A 1994;91(8):2900–4. https://doi.org/10.1073/pnas.91.8.2900.

[107] Broberg K, Björk J, Paulsson K, Höglund M, Albin M. Constitutional short telomeres are strong genetic susceptibility markers for bladder cancer. Carcinogenesis 2004;26(7):1263–71. https://doi.org/10.1093/carcin/bgi063.

[108] Ma H, Zhou Z, Wei S, et al. Shortened telomere length is associated with increased risk of cancer: a meta-analysis. In: Toland AE, editor. PLoS one, vol 6; 2011. p. e20466. https://doi.org/10.1371/journal.pone.0020466.

[109] Bendix L, Kølvraa S. The role of telomeres in cancer. Ugeskr Laeger 2010;172(40):2748–51.

[110] Campisi J, Andersen JK, Kapahi P, Melov S. Cellular senescence: a link between cancer and age-related degenerative disease? Semin Cancer Biol 2011;21(6):354–9. https://doi.org/10.1016/j.semcancer.2011.09.001.

[111] Rangarajan A, Weinberg RA. Opinion: Comparative biology of mouse versus human cells: modelling human cancer in mice. Nat Rev Cancer 2003;3(12):952–9. https://doi.org/10.1038/nrc1235.

[112] DePinho RA. The age of cancer. Nature 2000;408(6809):248–54. https://doi.org/10.1038/35041694.

[113] Seluanov A, Hine C, Azpurua J, Feigenson M, Bozzella M, Mao Z, et al. Hypersensitivity to contact inhibition provides a clue to cancer resistance of naked mole-rat. Proc Natl Acad Sci U S A 2009;106(46):19352–7. https://doi.org/10.1073/pnas.0905252106.

[114] González-Suárez E, Geserick C, Flores JM, Blasco MA. Antagonistic effects of telomerase on cancer and aging in K5-mTert transgenic mice. Oncogene 2005;24(13):2256–70. https://doi.org/10.1038/sj.onc.1208413.

[115] Artandi SE, Alson S, Tietze MK, Sharpless NE, Ye S, Greenberg RA, et al. Constitutive telomerase expression promotes mammary carcinomas in aging mice. Proc Natl Acad Sci U S A 2002;99(12):8191–6. https://doi.org/10.1073/pnas.112515399.

[116] Horn S, Figl A, Rachakonda PS, Fischer C, Sucker A, Gast A, et al. TERT promoter mutations in familial and sporadic melanoma. Science 2013;339(6122):959–61. https://doi.org/10.1126/science.1230062.

[117] Rachakonda S, Kong H, Srinivas N, Garcia-Casado Z, Requena C, Fallah M, et al. Telomere length, telomerase reverse transcriptase promoter mutations, and melanoma risk. Genes Chromosomes Cancer 2018;57(11):564–72. https://doi.org/10.1002/gcc.22669.

[118] Salvador L, Singaravelu G, Harley CB, Flom P, Suram A, Raffaele JM. A natural Product telomerase activator lengthens telomeres in humans: a randomized, double blind, and placebo controlled study. Rejuvenation Res 2016;19(6):478–84. https://doi.org/10.1089/rej.2015.1793.

[119] Harley CB, Liu W, Flom PL, Raffaele JM. A natural Product telomerase activator as part of a health maintenance program: metabolic and cardiovascular response. Rejuvenation Res 2013;16(5):386–95. https://doi.org/10.1089/rej.2013.1430.

[120] Baugh EH, Ke H, Levine AJ, Bonneau RA, Chan CS. Why are there hotspot mutations in the TP53 gene in human cancers? Cell Death Differ 2018;25(1):154–60. https://doi.org/10.1038/cdd.2017.180.

[121] Olivier M, Hollstein M, Hainaut P. TP53 mutations in human cancers: origins, consequences, and clinical use. Cold Spring Harb Perspect Biol 2010;2(1). https://doi.org/10.1101/cshperspect.a001008.

[122] Padariya M, Jooste ML, Hupp T, Fåhraeus R, Vojtesek B, Vollrath F, et al. The elephant evolved p53 isoforms that escape MDM2-mediated repression and cancer. Mol Biol Evol 2022;39(7). https://doi.org/10.1093/molbev/msac149.

[123] Haupt S, Haupt Y. P53 at the start of the 21st century: lessons from elephants. F1000Res 2017;6:2041. https://doi.org/10.12688/f1000research.12682.1.

[124] Caulin AF, Maley CC. Peto's Paradox: evolution's prescription for cancer prevention. Trends Ecol Evol 2011;26(4):175–82. https://doi.org/10.1016/j.tree.2011.01.002.

[125] Bauer JH, Helfand SL. New tricks of an old molecule: lifespan regulation by p53. Aging Cell 2006;5(5):437–40. https://doi.org/10.1111/j.1474-9726.2006.00228.x.

[126] Tomás-Loba A, Flores I, Fernández-Marcos PJ, Cayuela ML, Maraver A, Tejera A, et al. Telomerase reverse transcriptase delays aging in cancer-resistant mice. Cell 2008;135(4):609–22. https://doi.org/10.1016/j.cell.2008.09.034.

[127] Daya S, Berns KI. Gene therapy using adeno-associated virus vectors. Clin Microbiol Rev 2008;21(4):583–93. https://doi.org/10.1128/CMR.00008-08.

[128] Verdera HC, Kuranda K, Mingozzi F. AAV vector immunogenicity in humans: a long journey to successful gene transfer. Mol Ther 2020;28(3):723–46. https://doi.org/10.1016/j.ymthe.2019.12.010.

[129] Guterres AN, Villanueva J. Targeting telomerase for cancer therapy. Oncogene 2020;39(36):5811–24. https://doi.org/10.1038/s41388-020-01405-w.

[130] Vertecchi E, Rizzo A, Salvati E. Telomere targeting approaches in cancer: beyond length maintenance. Int J Mol Sci 2022;23(7):3784. https://doi.org/10.3390/ijms23073784.

[131] Fragkiadaki P, Renieri E, Kalliantasi K, Kouvidi E, Apalaki E, Vakonaki E, et al. Telomerase inhibitors and activators in aging and cancer: a systematic review. Mol Med Rep 2022;25(5):158. https://doi.org/10.3892/mmr.2022.12674.

[132] Fossel M. Telomerase and the aging cell. JAMA 1998;279(21):1732. https://doi.org/10.1001/jama.279.21.1732.

[133] Fossel MB. Cells, aging, and human disease. Oxford University Press; 2004.

[134] Tremblay D, Mascarenhas J. Next generation therapeutics for the treatment of myelofibrosis. Cells 2021;10(5):1034. https://doi.org/10.3390/cells10051034.

[135] Ghareghomi S, Ahmadian S, Zarghami N, Hemmati S. hTERT-molecular targeted therapy of ovarian cancer cells via folate-functionalized PLGA nanoparticles co-loaded with MNPs/siRNA/wortmannin. Life Sci 2021;277:119621. https://doi.org/10.1016/j.lfs.2021.119621.

[136] Martínez P, Blasco MA. Telomeric and extra-telomeric roles for telomerase and the telomere-binding proteins. Nat Rev Cancer 2011;11(3):161–76. https://doi.org/10.1038/nrc3025.

[137] Schneider RP, Garrobo I, Foronda M, Palacios JA, Marión RM, Flores I, et al. TRF1 is a stem cell marker and is essential for the generation of induced pluripotent stem cells. Nat Commun 2013;4:1946. https://doi.org/10.1038/ncomms2946.

[138] Boskovic J, Martinez-Gago J, Mendez-Pertuz M, Buscato A, Martinez-Torrecuadrada JL, Blasco MA. Molecular architecture of full-length TRF1 favors its interaction with DNA. J Biol Chem 2016;291(41):21829−35. https://doi.org/10.1074/jbc.M116.744896.

[139] Garrobo I, Marión RM, Domínguez O, Pisano DG, Blasco MA. Genome-wide analysis of *in vivo* TRF1 binding to chromatin restricts its location exclusively to telomeric repeats. Cell Cycle 2014;13(23):3742−9. https://doi.org/10.4161/15384101.2014.965044.

[140] García-Beccaria M, Martínez P, Méndez-Pertuz M, Martínez S, Blanco-Aparicio C, Cañamero M, et al. Therapeutic inhibition of TRF1 impairs the growth of p53-deficient K-RasG12V-induced lung cancer by induction of telomeric DNA damage. EMBO Mol Med 2015;7(7): 930−49. https://doi.org/10.15252/emmm.201404497.

[141] Ramsay AJ, Quesada V, Foronda M, Conde L, Martínez-Trillos A, Villamor N, et al. POT1 mutations cause telomere dysfunction in chronic lymphocytic leukemia. Nat Genet 2013;45(5):526−30. https://doi.org/10.1038/ng.2584.

[142] Calvete O, Martinez P, Garcia-Pavia P, Benitez-Buelga C, Paumard-Hernández B, Fernandez V, et al. A mutation in the POT1 gene is responsible for cardiac angiosarcoma in TP53-negative Li-Fraumeni-like families. Nat Commun 2015;6:8383. https://doi.org/10.1038/ncomms9383.

[143] Martínez P, Ferrara-Romeo I, Flores JM, Blasco MA. Essential role for the TRF2 telomere protein in adult skin homeostasis. Aging Cell 2014; 13(4):656−68. https://doi.org/10.1111/acel.12221.

[144] Martínez P, Gómez-López G, García F, Mercken E, Mitchell S, Flores JM, et al. RAP1 protects from obesity through its extratelomeric role regulating gene expression. Cell Rep 2013;3(6):2059−74. https://doi.org/10.1016/j.celrep.2013.05.030.

[145] Bordone L, Guarente L. Calorie restriction, SIRT1 and metabolism: understanding longevity. Nat Rev Mol Cell Biol 2005;6(4):298−305. https://doi.org/10.1038/nrm1616.

[146] Harrison DE, Strong R, Sharp ZD, Nelson JF, Astle CM, Flurkey K, et al. Rapamycin fed late in life extends lifespan in genetically heterogeneous mice. Nature 2009;460(7253):392−5. https://doi.org/10.1038/nature08221.

[147] Weindruch R, Walford R. Dietary restriction in mice beginning at 1 year of age: effect on life-span and spontaneous cancer incidence. Science 1982;215(4538).

[148] Lee CK, Klopp RG, Weindruch R, Prolla TA. Gene expression Profile of aging and its retardation by caloric restriction. Science 1999; 285(5432).

[149] López de Silanes I, Graña O, De Bonis ML, Dominguez O, Pisano DG, Blasco MA. Identification of TERRA locus unveils a telomere protection role through association to nearly all chromosomes. Nat Commun 2014;5:4723. https://doi.org/10.1038/ncomms5723.

[150] Montero JJ, López de Silanes I, Graña O, Blasco MA. Telomeric RNAs are essential to maintain telomeres. Nat Commun 2016;7:12534. https://doi.org/10.1038/ncomms12534.

[151] Baur JA, Zou Y, Shay JW, Wright WE. Telomere position effect in human cells. Science 2001;292(5524).

[152] Slattery ML, Herrick JS, Pellatt AJ, Wolff RK, Mullany LE. Telomere length, TERT, and miRNA expression. In: Santos J, editor. PLoS one, vol 11; 2016. p. e0162077. https://doi.org/10.1371/journal.pone.0162077.

[153] Fossel M, Whittemore K. Telomerase and cancer: a complex relationship. OBM Geriat 2021;5(1). https://doi.org/10.21926/obm.geriatr.2101156.

CHAPTER

13

Age-related disease: Effective intervention

Michael Fossel[1] and John P. Cooke[2]

[1]Telocyte LLC, Ada, MI, United States [2]Houston Methodist Research Institute, Houston, TX, United States

1. Perspective

... But ignorance more frequently begets confidence than does knowledge: it is those who know little, and not those who know much, who so positively assert that this or that problem will never be solved by science. —*Charles Darwin*, The Descent of Man, and Selection in Relation to Sex, 1871.

The major barrier to reversing aging is that we think of it as a problem that will never be solved. In our current regulatory environment, aging is considered a natural process and not a target for therapeutics. These assumptions regarding the aging process impose an obstacle to medical progress and derive from a conceptual framework based less on a scientific foundation and more on cultural beliefs. Underlying this conceptual framework is the assumption that aging is a natural process, and that although we might be able to ameliorate the effects of aging, we cannot reverse the process of aging. The world abounds with myths about rejuvenation or immortality, with the two often confused for one another. Every culture has its history of attempts to reverse aging, which are almost always frustrated and intertwined with cautionary warnings and catastrophic endings. Historically, the pessimistic thread chasing an impossible immortality goes back as far as recorded human history and probably well beyond. It is one of the two threads that define our beliefs and self-imposed limits on our ability to cure and prevent age-related disease, but it is not the only thread. The more optimistic thread, that aging might be reversed, is a view that is older (at least in documented human literature) and possibly the more plausible of these two conceptual strands.

This more hopeful view regarding aging is found in the oldest recorded human story, the *Epic of Gilgamesh*, a Mesopotamian story originating more than 4000 years ago and rediscovered in the Library of Ashurbanipal in only the mid-1800s. In 11th tablet of the epic, the hero is told of a plant that grows underwater that will restore lost youth [1]. This approach to intervention, although fictional, is echoed repeatedly throughout the world's literature in every culture and every epoch of human history. The urge to slow or reverse the aging process is a practical, quotidian concern: can we alter the aging process and undo the diseases that create so much human misery?

An opposing view is reflected by the Greek legend of Tithonus, a tragic love story of a mortal, beloved of Eos, the goddess of the dawn, who is granted the gift of eternal life by Zeus, but without the more important gift of eternal youth. In its several iterations, such as Sappho's poem [2], the story ends with Tithonus always aging, never dying, until naught remains save perhaps a shade, perhaps a cricket, perhaps a cicada, old, crippled, and almost mindless, begging for death to release him and cursing the gods for his cruel fate [3]. The Tithonian view of aging, in which we might prevent or postpone death, but cannot prevent, postpone, or reverse aging, is found frequently in human history and resurfaces even today in, for example, modern surveys [4], in which people are asked whether they would like to live longer. In an echo of Tithonus, and probably fearing a similar outcome, most would rather not live an extra several decades. Behind this opinion, however, lies the horror and fear that a longer life—like that of Tithonus—would not be healthier, just grimmer. They fear that a longer life would not bring more joy, but merely more years of suffering, more decades of disease. Many fear ending their life in a nursing home; few would willingly extend misery.

As we will suggest, this view is at odds with biological possibility, but it is perhaps understandable. After all, over the past few centuries we have extended the mean human life span without altering the aging process.

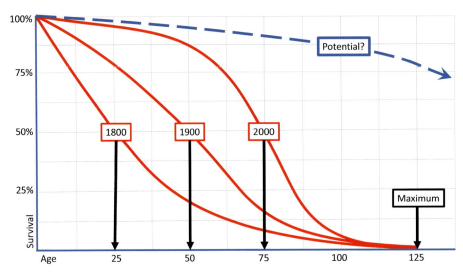

FIGURE 13.1 Over the past 200 years, the *mean* life span has (in many countries) risen from roughly 25 years to 50, and to 75 years. However, the *maximum* life span has not changed from roughly 125 years during that time. We may soon alter the aging process, preventing age-related disease and extending the healthy maximum life span an indeterminate amount.

Although the mean human life span has tripled over the past few centuries (largely the result of better nutrition, sanitation, and antibiotics), the maximum life span has remained unaffected (Fig. 13.1). On the other hand, the conclusion that medical science can extend life span indefinitely (either mean or maximum) without simultaneously extending biological health is erroneous. Further extension of the mean life span will be constrained by the limits of the human life span imposed by the aging process. At this point, to extend the mean human life span significantly further, we will need to intervene directly in the aging process itself. Put simply, our goal should be to reset physiologic age and directly restore youthful function. We need not resign ourselves to the fate of Tithonus. There is a substantial understanding of the mechanisms of aging that may facilitate therapies to provide a longer and healthier life.

The most important advances in medicine that led to mean life span extension were not related to high technology. Although robotic surgery, transaortic valve replacement, and four-dimensional imaging enhanced by artificial intelligence have extended and improved human life, their public health benefit has been incremental and at great financial cost. None of these technical improvements has equaled the inflection point achieved by the single key conceptual revolution in medical history: microbial theory. This conceptual advance allowed many changes—hand washing during deliveries, sterile surgeries, immunizations, antibiotics, and other advances—that saved lives and made those lives better, and reduced the costs of medical care.

Consider the example of polio in 1950. Treatment consisted of rehabilitation, braces, nursing care, and iron lungs. Estimates were that the United States and other countries might become medically bankrupt by 2000 because of the increasing costs of those treatments. By 2000, however, the cost of preventing polio was less than 10¢ per patient. The key to both better medical care and lower medical costs was not a better iron lung, but a polio vaccine. Lives were saved and children grew to become adults not because of better braces, but because of a better understanding of polio itself. Polio is but a small example of 2 centuries of improved lives, owing to a conceptual revolution that allowed for the myriad technical changes.

The single greatest medical advance in history came about because we began to understand microbial disease. We are at the next inflection point in human medical history, as we finally begin to understand aging and implement rational therapies to reverse aging. In doing so, the critical question is not whether we can extend the life span, but whether we can do so while improving human health. It is not Tithonus, but Gilgamesh who will direct our path forward.

Any human endeavor requires both understanding and ability. We need an understanding of reality and the ability to take that understanding and translate it into tangible results. To adopt our lexicon to an ancient seafarer's world, we need both a map and a ship. The map tells us where we are going; the ship allows us to get there. In curing smallpox 2 centuries ago, the greatest hurdle was the map: our lack of understanding blocked progress. The ship was relatively easy: once we understood the concept of vaccination, the technology was no more than a small, sharp instrument to scratch the skin, allowing entry of a vaccinia virus. We have improved since our technology, but it was the *understanding* that held us back, or we might have conquered smallpox thousands of years earlier [5–7]. Getting humans to the moon was the opposite: we had the map, but we lacked the ship. There has never been a question of where the moon was (the map), but getting there (the ship) could not have been accomplished until two-thirds of the way through the 20th century. The concept was painfully simple, but the technology was fiendishly complicated.

Reversing aging requires both: we need to improve our understanding and to enhance our technology. Without a fundamental reappraisal of how aging works and how the process results in age-related disease (an appraisal that the previous chapters addressed), we could not make progress. We needed to understand not only the full, complicated process of aging, but also to realize that intervention is feasible. However, it is also true that our technical ability is only now barely able to reset the aging process. In short, we now have a better map and are rapidly improving our ship. As the previous chapters have brought us a better understanding of the aging process, let us now focus on our ability to intervene in that process effectively and safely. We need to reset cell aging at the fundamental level to address fundamental processes in aging.

2. Current approaches

Many current interventions purporting to target aging often use small molecular approaches that target *biomarkers* of aging rather than aging per se. This is much like targeting symptoms (e.g., a cough) of a disease (e.g., a COVID-19 infection) with a small-molecular medication (e.g., an antitussive agent) rather than targeting the cause (COVID-19 virus) with an effective agent (a vaccine). When we target biomarkers, we are not targeting the process, only the outcomes. Approaches that target downstream biomarkers, signs, or symptoms of a disease may be effective against specific downstream outcomes of aging, but the fact that they are targeting downstream outcomes rather than the underlying biological process accounts for their lack of significant clinical effect. To put this problem into a day-to-day clinical context, we can dye gray hair, inject botulinum toxin for wrinkles, and replace arthritic joints, but the process of aging continues apace, unabated and unaffected. Targeting biomarkers of aging is often a cosmetic approach that may not affect aging at all (Fig. 13.2).

At the clinical level, the common downstream targets include the known histologic or biochemical outcomes of specific age-related diseases. In the case of the age-related neurodegenerative diseases, downstream targets (see Chapter 2) include β-amyloid plaque, tau tangles, and α-synuclein. In the case of age-related cardiovascular disease (see Chapter 3), downstream targets include endothelial dysfunction, vascular inflammation, calcification, fibrosis, atherosclerotic plaque, and thrombosis. Attempts to intervene solely at downstream clinical levels have at best been only partially effective. Thus, in cardiovascular medicine, we lower cholesterol levels to reduce morbidity and mortality of the relevant age-related cardiovascular diseases. In neurology, we use monoclonal antibodies to remove

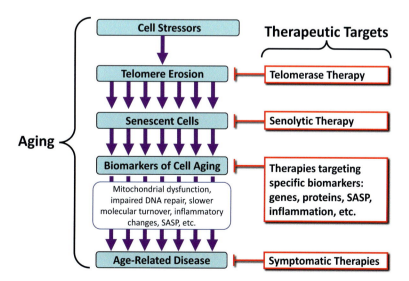

FIGURE 13.2 Targets in age-related disease. Most current and proposed therapies target downstream outcomes rather than critical upstream processes. Cell stressors include multiple factors that may cause telomere erosion, including cell division, inflammatory signaling, radiation or chemical DNA damage, and oxidative stress. *SASP*, senescence-associated secretory phenotype. *Adapted from Fossel [36].*

amyloid plaque, in hopes of slowing the relentless progress of cognitive decline in age-related neurodegenerative diseases such as Alzheimer's disease [8,9]. The problems with these approaches are that (1) in some cases we have a limited ability to normalize the targeted biomarker; (2) there are other downstream biomarkers in addition to the one we target; and most important, (3) none of the biomarkers is the disease process. Biomarkers are not a disease.

Such biomarker targets may include reactive oxygen species (free radicals), lipofuscin, and β-galactosidase. Consider parallel clinical hallmarks of aging that we find on physical examination, including wrinkles (as well as thin, dry skin), gray hair, swollen joints, gait instability, cataracts, presbyopia, poor healing, presbycusis, slow reaction times, and sarcopenia. As discussed in Chapter 1, these are generally the consequence rather than the drivers of aging, or are at best components of the aging process without being the entire aging process. Biomarkers may have a significant role within the aging process, but they are individual components rather than the entire, complex system. These components offer us convenient measures of the progress of aging without being the underlying pathology, much as vital signs may allow us to monitor a microbial disease without being the infectious process. One cannot expect to intervene effectively in aging using an antioxidant any more than we might intervene effectively in a microbial disease using antipyretics.

The classic hallmarks of aging (genomic instability, telomere attrition, epigenetic alterations, loss of proteostasis, deregulated nutrient-sensing, mitochondrial dysfunction, cellular senescence, stem cell exhaustion, dysbiosis, altered mechanobiology, aberrant messenger RNA (mRNA) splicing, and altered intercellular communication [10]) are useful in terms of categorizing processes involved in aging. However, these processes are not likely to be operative to the same degree in different individuals (e.g., there are different senotypes). Furthermore, some processes may be more important clinically than others. It is also possible that in the same individual, different aging processes may dominate in different tissues.

Understandably, and in default of a deeper understanding, our current clinical approach to aging is to treat the signs and symptoms as best we are able: nonsteroidal medications, glasses, hearing aids, walkers, exercise, surgical replacements, and so forth. Our current biochemical approach to aging takes a parallel track: we target identifiable biochemical biomarkers without regard to the aging process. Such approaches include the use of antioxidants [11,12], nicotinamide adenine dinucleotide (NAD$^+$) [13,14], rapamycin [15,16], resveratrol [17,18], and so on. Although these approaches (clinically or biochemically) are understandable and may be justifiable if they achieve some degree of clinical improvement, they do not target the underlying processes in aging.

3. Targeting aging processes

Inflammatory signaling of senescent cells is a well-defined molecular process involved in aging. Small molecules have been used to target senescent cells (senolytics) or the biochemical characteristics of such cells (senomorphics) [19–21]. Senescent cells are associated with aging tissue. Senescent cells may actively interfere with the function of nearby nonsenescent cells [22]. Senescent cells may exhibit a senescence-associated secretory phenotype (SASP) [23–25] that results in dysfunction in surrounding cells, particularly through proinflammatory changes [26].

However, senescent cells are also associated with wound healing and stem cell regeneration [27]. Cells characterized by senescent markers can be found in association with stem cells in the lung, colon, small intestine, and skin. Senolytic drugs destroy these cells and impair regeneration of the basement membrane of the lung. Thus, there may be subsets of senescent cells that are beneficial and/or contexts in which senescent cells are physiologically adaptive.

In the case of senomorphics, the approach is to target the abnormal secretory profile of the senescent cells, such as by using a small-molecule agent to suppress the phenotype [28,29]. The goal is not to change or remove the senescent cell, let alone reverse the aging process in any fundamental way, only to address the downstream outcome of cell senescence: the abnormal secretory profile. Like many such partial approaches to aging processes, this may result in some clinical benefits, but it does not address the issue of cell aging directly. Conceptually, this approach parallels the use of small molecules (e.g., pressors, antipyretics, blood products) to address severe infectious disease (e.g., an Ebola infection) in an attempt to ameliorate symptoms (e.g., hypotension, fever, bleeding) rather than using a definitive molecular approach (e.g., a vaccine) in an attempt to prevent or cure the fundamental disease process. Senomorphics target the secretory biomarkers of cell aging rather than the cell aging process that causes such secretory biomarkers. The approach exposes two problems: (1) it does not address the upstream cascade of processes that result in an SASP, and (2) it employs an approach to target a limited subset within a large set of biochemical abnormalities resulting from the upstream cascade of processes that define aging.

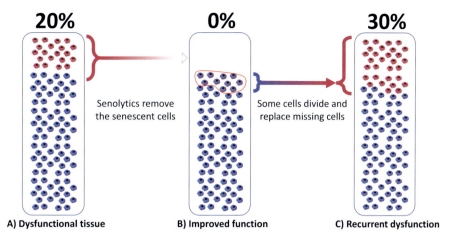

FIGURE 13.3 Potential adverse effect of senolytics. (A) Tissue starts with 20% senescent cells. (B) Senolytic therapy removes the senescent cells, initially improving tissue function. (C) Some remaining cells divide to replace cells that have been removed, which accelerates senescence in the dividing cells, resulting in 30% senescent cells and even poorer tissue function than before the intervention. Repeated senolytic intervention may ultimately accelerate age-related pathology.

In the case of senolytics, the approach has been to remove senescent cells rather than address their secretory outcomes. Specifically, the focus is to remove rather than restore the function of aging cells. The senolytic approach relies on the inference that if senescent cells create problems, removing such cells may result in therapeutic benefits. At least initially, this appears to be the case [30]. Unsurprisingly, the potential clinical use of senolytics has prompted considerable enthusiasm [31–33], with significant interest among basic researchers, startup biotechnology firms [34], and investors [35]. However, concerns have been raised that senolytics might accelerate the underlying pathology [36]. This concern is supported by animal data suggesting that despite initial extensions of survival, the slope of the survival curves shows an accelerated mortality slope afterward [30]. This begs the practical question of the balance between the initial benefit and the accelerated mortality rate.

To consider a trivially simplistic model (Fig. 13.3), imagine that we have 100 cells, 20 of which are senescent. After using senolytics to remove the senescent cells, the remaining cells divide to replace the missing 20 cells, which accelerates their own cell aging. As a result, we again have 100 cells, but now 30 are senescent. The result of removing senescent cells is to improve the function of the tissue transiently, followed by an even greater degree of senescence and worsening pathology. Furthermore, as described earlier, there may be a physiologic role for senescent cells, or subsets of these cells, which would be adversely affected by senolytics.

The issue of clinical efficacy is complicated. Both theory and data support the notion that clearing senescent cells from a tissue will effectuate a short-term improvement in the function of the remaining nonsenescent cells and therefore overall tissue function. This assumes that the senescent cells can be removed efficiently and without direct effects on nonsenescent cells and/or useful senescent cells. As before, the more critical question is that of the long-term consequences of removing such cells. In particular, how quickly will removing senescent cells result in the acceleration of cell aging in the remaining cells, and will that acceleration result in near-term adverse effects, specifically a notable acceleration in age-related disease (Fig. 13.4)?

Therefore, senolytics have three intrinsic problems that may limit their usefulness: (1) they may prove ineffective except transiently; (2) they do not address other aging processes that cause cells to become senescent in the first place; and (3) in some contexts, senescent cells may have important physiologic functions in stem cell maintenance and wound healing.

4. Targeting a dominant process in cell aging

The optimal intervention in age-related disease is one that will beneficially affect most if not all hallmarks of aging. As health care providers, we have generally focused on downstream disease and have targeted known biomarkers (e.g., monoclonal antibodies to target β-amyloid), anatomic pathology (e.g., cardiac bypass grafts, stents, joint replacement) and symptoms (e.g., using antiinflammatories, botulinum toxin, etc.), none of which has an effect on the underlying processes of cell aging. Because upstream or downstream interventions have marginal efficacy

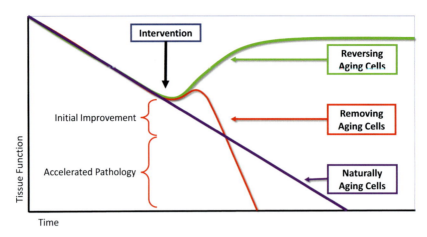

FIGURE 13.4 Telomerase versus senolytics. The intention of telomerase therapy (reversing aging cells) is to reset cell function in aging cells. The intention of senolytic therapy (removing aging cells) is to remove senescent cells. Although evidence supports initial improvement after senolytics use, it also suggests this may accelerate cell aging in the remaining cells, resulting in accelerated (and more severe) age-related pathology. *Adapted from Fossel [36].*

(neither stopping nor reversing the underlying pathology), the question becomes assessing the potential impact of interventions in the cell aging process itself (Fig. 13.5).

Specifically, can we effectively reverse cell aging by targeting one or more of the processes underlying cell aging? Alternatively, if it is effective, targeting a dominant molecular process would offer a potentially more effective point of therapeutic intervention [37–40].

As we describe next, telomere erosion appears to be a dominant process in cell aging, because its correction leads to the reversal of multiple processes in cell aging. Telomere erosion can be reversed by inducing the expression of telomerase. The telomerase enzyme complex includes (1) telomerase reverse transcriptase (TERT) and (2) telomerase

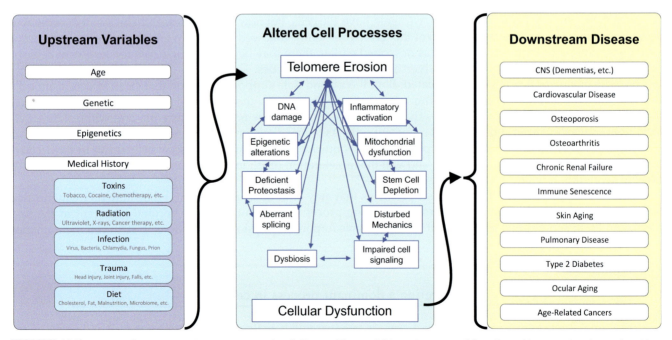

FIGURE 13.5 Points of intervention in treating age-related disease. The crucial issue is not possible points of intervention, but rather, identifying the optimal point of intervention vis-à-vis efficacy, safety, cost, and technical feasibility. Some upstream variables cannot be modified (e.g., age) whereas targeting others is only partially effective in slowing age-related disease. Treating downstream disease reduces morbidity and mortality, but it does not reverse age-related disease. To reverse age-related disease, cellular processes involved in aging must be targeted. Targeting certain dominant processes (e.g., telomere erosion) may have the greatest benefit. *CNS*, central nervous system.

RNA (TR or TERC). The former is a protein of 1132 amino acids [41] and is the active enzymatic portion of the complex. More important, it is generally not expressed in somatic cells, allowing cell aging to proceed. The latter is the template portion, a noncoding RNA of 451 nucleotides that provides the template for adding hexameric repeat sequences (TTAGGG) to the telomere. TERC is generally expressed constitutively in somatic cells. Hence, it is not considered the rate-limiting factor in attempts to reset cell aging. The hTERT gene [42,43] is located on the distal part of chromosome 5p15.33 [44–46], whereas the hTERC gene is located on chromosome 3q26.2 [47]. These two basic components have been generally known for more than 3 decades [48], but we continue to refine our knowledge, particularly with regard to multiple adjunctive molecules that constitute the telomerase complex [49].

Targeting telomere erosion led to the first demonstrated reversal of cell aging (in human cells in vitro) published in *Science* in 1998 [50]. In this example (and similar studies [51–53]), epithelial cells and fibroblasts, were transfected with telomerase, maintaining the ability to divide while exhibiting the normal characteristics of young cells. Although groundbreaking, these studies left open the question of whether reversal of cell aging could be accomplished in human tissues. This was addressed affirmatively within the following few years in human skin cells [54], vascular endothelial cells (ECs) [55], and osteocytes [56]. Whether these studies could be parlayed into an effective intervention in organisms remained open, and to some extent, it still does.

Subsequent work employed first a genetic switch to turn on the telomerase gene (mTERT) in mice [57], and then the use of an adeno-associated virus (AAV) delivery of the mTERT gene [58], again in mice. In both cases, the results were remarkable, showing dramatic effects not only on life span, but on multiple indices of age-related physiologic function, including behavior, bone mineral density, immune response, glucose response, neural stem cell proliferation and differentiation, and brain volume.

At the current juncture, there are several routes (regarding both the method of extending telomeres and methods of delivering the agent that does so) that might reasonably be employed. Given our existing knowledge, key targets for these approaches are the telomeres themselves. Nevertheless, telomerase may have noncanonical effects (e.g., on DNA damage and mitochondrial function) that may be beneficial independent of telomere extension. If the optimally effective clinical target is indeed telomere erosion, we have multiple potential approaches to restoring telomeres in vivo (Fig. 13.6). These approaches include telomerase DNA, TERC, telomerase protein, or telomerase activators, all of which have some theoretical and (in most cases) experimental support, although there are marked practical differences as we assay their use in animal studies or human trials.

An interesting set of studies [59–61] offered tantalizing if not overwhelming evidence that telomerase activators (e.g., astragenol) might be partially effective in this regard, but the effects were scarcely dramatic, and they were ambiguous. Theoretical considerations suggest that this approach may be of limited value and may encounter unacceptably higher safety risks than other approaches. Small molecules such as astragenol, are insufficiently specific to avoid unintended activation or interference with other biological processes with consequent risks of toxicity. To some extent, this concern has been borne out by the difficulty of identifying highly effective telomerase activators that simultaneously have low toxicity.

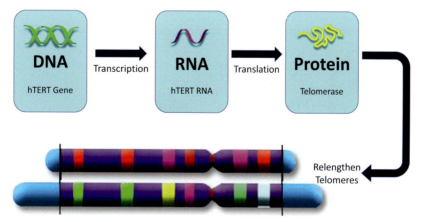

FIGURE 13.6   Potential therapeutic options for telomere therapy include (1) administration of human telomerase (hTERT) DNA that codes for the telomerase RNA, (2) administration of telomerase RNA that codes for the telomerase protein, and (3) administration of telomerase protein itself. In all cases, the goal is to relengthen telomeres and thus restore a more effective pattern of gene expression consistent with healthy cell function.

4.1 DNA and its utility for age-reversal strategies

The use of the DNA encoding TERT has proven effective in multiple animal trials. As discussed earlier, estrogen switches to turn on telomerase were inserted into the germ cell line of mice, with results that were remarkable and significant [57], but not directly applicable for use in human trials. A more pertinent example used AAV vectors (Fig. 13.7) to deliver mTERT to older mice [58], with results which were again remarkable but this time using an AAV vector. Such vectors have previously been used successfully in human trials for spinal muscular atrophy [62]. This gene therapy technique was cleared for commercial use by the US Food and Drug Administration and has provided support for its wider use in other genetic therapies [63,64]. Nevertheless, the industry continues to be cognizant of the side effects and risks of AAV therapy, as well as dosage uncertainties, immunogenicity [65], quality control (i.e., chemistry, manufacturing, and control) issues in production [66], and cost concerns [67]. Newer serotypes [68] and capsids [69], tailored promoters [70] (as well as serotype–promoter interactions), and tailoring of the genes themselves have made the AAV approach prone to rapid changes and almost equally rapid improvements. One major obstacle has been that AAV vectors tend to be cell-type specific [71]. Depending on the serotype, capsid, and promoter, some target cells may be transduced at a high efficiency, whereas other cells may not be transduced at all. This has prompted considerable interest in finding optimal ways to target some specific cells (e.g., glial cells) as well as in broadening the efficiency in targeting a wide gamut of human cells simultaneously. In the case of AAV delivery, questions remain as to what percentage of relevant cells we transduce, and with what safety. In the case of age-related behavioral changes, for example, our current ability to transduce neural and glial cells is severely limited, and this appears to be the primary variable in clinical success, whether in mice or humans.

Whereas AAV vectors have received the most attention, at least in the initial human gene therapy trials, other viral vectors, including hybrid vectors, have certainly not escaped growing interest in finding optimal delivery systems [72]. A host of such vectors are under consideration, but a great deal of attention has been paid to lentiviruses in this context [73]. The major concern is the insertion of the therapeutic gene into the chromosome, disrupting native gene sequences and creating a risk of mutagenesis (i.e., cancer). This has created some trepidation in using lentivirus, but it has also prompted a search for nonintegrating lentivirus techniques [74]. In addition, the lentivirus capsid has been used as a chimera with bacteriophage in an attempt to employ the broad targeting of lentiviral particles, with the hope of lowering the risk of chromosomal insertion [75].

DNA delivery techniques are changing rapidly in regard to the serotypes and virus chosen for use as well as the use of lipid nanoparticles (LNPs) [76,77]. The potential for artificially constructed delivery agents is hard to overestimate [77,78]. Perhaps the best summation of the use of DNA delivery (such as TERT), is that although the use of viral vectors or LNPs to deliver DNA to target cells in human trials has drawbacks, limitations, and understandable safety concerns, it has proven both feasible and at least reasonably effective. Limiting variables in regard to safety are apparently not the risk of cancer [58,79], but rather immunogenicity and the side effects of gene delivery [80,81]. To some extent, our status is reminiscent of the Wright brothers in 1903, when they had unequivocally demonstrated the

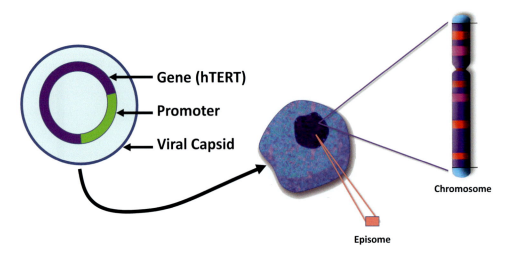

FIGURE 13.7 The adeno-associated virus (AAV) incorporates a gene and a promoter (as well as other elements) encased in the viral capsid. The capsid determines target cell specificity. Once the AAV enters the target cell, the gene human telomerase (hTERT) forms an episome, which is independent of the cell's chromosomes and expresses telomerase, resetting telomere lengths.

ability to show that powered flight was possible, but it was scarcely ready for safe commercial use, let alone able to transport hundreds of people safely between continents at a reasonable cost. Much the same pertains to our current ability to demonstrate the effectiveness of gene therapy: there is no question that it can be done, but the issues of safe, widespread, and commercially viable use remain in flux. An alternative approach is to use mRNA constructs [82].

4.2 The emergence of mRNA therapeutics and its utility for age-reversal strategies

The advent of mRNA vaccines against SARS-CoV-2 focused attention on the rapidly emerging field of mRNA therapeutics [83,84]. As dramatically manifested during the COVID-19 crisis, the development of mRNA therapeutics can be most rapid and effective. Because mRNA is essentially biological software, one can swiftly generate the code for the protein or antigen of choice. By comparison, small-molecule therapies may take a decade for development, requiring the development of assays for a particular pathway, high-throughput screening, and orthogonal confirmation of putative hits, followed by extensive testing of new compounds for their pharmacokinetics, pharmacodynamics, distribution, metabolism, excretion, and toxicities. Because mRNA is nonintegrating, there is less concern for genomic disruption as may occur with DNA-based gene therapies. Whereas the manufacture of recombinant proteins is made more complex by the need for appropriate folding and posttranslational modification of the drug product, with mRNA therapies the target cell does this work. Thus, the generation of mRNA therapies is swifter and less costly.

One may ask why mRNA therapies did not come to fruition earlier, because mRNA was discovered well over half a century ago. The answer is that two important obstacles had to be overcome [83,84]. First, mRNA is highly unstable, subject to rapid hydrolytic degradation in aqueous solution, and quickly degraded by RNases, which are ubiquitous in plasma and tissue. This problem has been solved in part by use of carriers such as LNPs. Indeed, the two mRNA COVID-19 vaccines each were composed of mRNA against the spike protein, encapsulated in LNPs. Second, mRNA made in the laboratory using standard ribonucleotides (adenosine, cytosine, guanine, and uracil triphosphates) are toxic to mammalian cells. The work of Weissman and Kariko showed that mammalian ribonucleotides are modified and that mRNA made using modified ribonucleotides is well-tolerated. This opened the gates to modern mRNA therapeutics [85]. Common modified nucleotides include pseudouridine (replacing uracil) and 5-methylcytosine (replacing cytosine).

The application of mRNA therapeutics to age reversal is well under way and includes the use of mRNA encoding the Yamanaka factors (Oct4, Klf4, Sox2, and cMyc) to induce partial reprogramming from a senescent to a juvenile form; the use of mRNA vaccines directed against senescent cells; and the use of mRNA encoding human telomerase (hTERT mRNA). Using this latter approach, we showed that a single treatment of hTERT mRNA was sufficient to extend the telomeres of replicatively aged human fibroblasts [86]. Furthermore, senescence was reversed (as shown by a decline in β-galactosidase) and the aged human fibroblasts could reenter the cell cycle. There was a dose-dependent increase in population doublings, with four treatments nearly doubling the replicative capacity of human fibroblasts.

Subsequently, we have studied the potential therapeutic effects of hTERT mRNA in cell from patients with Hutchinson-Gilford progeria syndrome (HGPS) [87–90]. A mutation in the nuclear matrix protein lamin A is responsible for generating the abnormal protein progerin. Children with this spontaneous mutation age rapidly and manifest many of the clinical signs and symptoms of aging, including loss of subcutaneous fat, sarcopenia, osteopenia, arthritis, and alopecia. In their teen years they succumb to myocardial infarction or stroke, and at autopsy, they have severe arterial occlusive disease in the coronary and carotid arteries.

Fibroblasts derived from HGPS patients grow poorly and manifest senescence markers compared with those derived from their parents (who lack the mutation). Treatment of HGPS fibroblasts with hTERT mRNA reversed senescence markers and restored cell proliferation [87,88]. To determine whether the vascular disease of HGPS might be responsive to hTERT mRNA, we generated vascular smooth muscle cells (VSMCs) and ECs from HGPS and parental induced pluripotent stem cells [89,90] (iPSCs). (The Progeria Research Foundation generates iPSCs from parental and patient-derived fibroblasts, by overexpressing the master regulators of pluripotency, the previously mentioned transcriptional factors Oct4, Sox2, KLF4, and c-Myc. The parental- and patient-derived iPSCs express the expected markers of pluripotency and can be differentiated into all three germ lines.)

The iPSC-derived ECs and VSMCs from the parental iPSCs express the expected surface markers and functions of vascular cells. By comparison, ECs and VSMCs from the HGPS patients manifested shorter telomeres, proliferated poorly, expressed senescence markers, and generated inflammatory cytokines (which is a feature of the senescence-associated secretory phenotype) [89,90]. The senescent phenotype was more severe in ECs, which exhibited a high prevalence of altered cell (fried egg appearance) and nuclear (lobulated) morphology. An aberrant transcriptional profile was revealed by RNA sequencing, and ECs made less nitric oxide and network formation in Matrigel.

Two treatments with hTERT mRNA separated by 48 h extended telomeres in the HGPS EC and VSMCs [89,90] This effect of hTERT mRNA was associated with a reduced expression of senescence markers and normalized replicative capacity. The effect of hTERT mRNA on the functions and morphology of HGPS EC was striking, with a normalization of cell shape and reduction in lobulated nuclei, and an improvement in nitric oxide production and network formation in Matrigel. The transcriptional profile was normalized. In addition, we noted a reduction in DNA damage.

These findings were highly similar to those obtained previously using a retroviral construct encoding hTERT [55] In this study, replicatively aged ECs exhibited the altered cell and nuclear morphology described earlier, had impaired replicative capacity, generated less nitric oxide, exhibited inflammatory activation, and had impaired responses to fluid shear stress. Each of these abnormalities was reversed in cells with the induced expression of telomerase. Considered together, in vitro work with human senescent vascular cells (whether senescence was induced by a genetic mutation or with replicative aging) strongly suggested that reactivating telomerase activity in senescent human cells could reverse many of the hallmark processes in aging. These include the reversal of abnormal cell and nuclear morphology and disturbed mechanobiology, inflammatory activation, impaired cell signaling, aberrant transcriptional profile, and DNA damage. Thus, one therapeutic intervention directed at one of the hallmark processes in aging (telomere erosion) resulted in the reversal of many of the other hallmarks.

Subsequently, we tested the hypothesis that TERT mRNA might reverse the vascular dysfunction and extend life span in vivo [90]. We used the mouse model of Dr. Frances Collins at the National Institutes of Health, which manifests the human mutation in lamin A. These animals grow poorly and die early, like children with HGPS. Although they have evidence of vascular senescence, with media atrophy, they do not exhibit the arterial occlusive disease seen in children with HGPS. We treated these animals with placebo or one dose of a lentiviral construct encoding telomerase. Animals treated with telomerase had a reduction in measures of vascular inflammation (e.g., vascular cell adhesion molecule) and DNA damage (gH2ax). The mTERT-treated animals also had a significantly extended life span [90]. Based on this successful preclinical study, we intend to develop an mRNA/LNP strategy for treating patients with HGPS and others with disease caused by accelerated aging.

The therapeutic use of telomerase protein has also been considered. The major drawback of using proteins delivered by the vein (or in the vascular space generally) is that proteins have too short a serum half-life (typically minutes to hours) [91], as well as unreliable cellular uptake (depending on their size and specific membrane receptors). Systemically administered telomerase needs to endure long enough in the blood to enter cell membranes, but there is little evidence that it can do so, although engineering telomerase with a cell permeation domain may be possible [92]. Moreover, because the nucleus is the canonical site of action, it needs to have a sufficiently long serum half-life as well as efficient transport across both the cell and the nuclear membranes, while having a sufficiently long half-life within the cytoplasm and the nucleoplasm to affect the telomere length. Although the use of telomerase protein (or telomerase-derived peptides) has proponents and some suggestive data [93–97], pragmatic hurdles set a high bar for the effective therapeutic use of telomerase protein in extending telomeres in vivo.

5. Where do we stand now?

Telomere extension is the optimal theoretical target in reversing cell aging as a therapeutic target for age-related disease, and telomerase is the obvious means to effect such an extension. Telomere extension has proven to be theoretically optimal, and it is effective in vitro, ex vivo, and (within our current technical limitations) in vivo. Yet, several key practical questions remain:

(1) Will the cell, tissue, and animal data translate well into human clinical trials?
(2) Can we improve the efficacy of any of several current approaches?
(3) What adverse effects will limit the clinical use of telomerase therapy?

Age-related diseases, in which Alzheimer's disease is the obvious example, are notorious for proving the depressing observation that everything works in mice, but nothing works in humans. To a large extent, the validity of this observation rests on trials that target mechanisms not shared between mice and humans, whereas the telomere mechanism in cell aging is shared broadly not only in all vertebrates, but in all eukaryotes in one manner or another. Nevertheless, credible clinical use does not rest on having a consistent, logical theory, but on data. In short, the question of whether current data indicate of human clinical effectiveness can be answered only by taking an effective means of delivery into human trials.

The second question, that of improved delivery (or, for that matter, proved delivery) haunts those involved in clinical research, who worry that although the model may reflect the biology accurately, it might fail to use a sufficiently effective delivery. Given the attractions of the model and the accelerating improvements in the gene therapy field, we can expect efficient delivery and therefore a clear test of the model in the immediate future.

The third question, that of adverse effects, will likewise require human testing. Because of the nature of age-related diseases, the risk—benefit ratio will be weighted in favor of accepting moderate risks, but even then there are obvious limits, both medically and ethically. Throughout the past several decades, the most common concern has been cancer. The common trope has been that telomerase causes cancer, although even a superficial understanding suggests that at worst, telomerase is permissive but not causal. A more careful analysis of the literature [79], and diligent attention to the nature of the data and the unexamined assumptions in them, show that not merely is the relationship between telomerase and cancer more complicated than it first appears, but that telomerase may in many cases be protective against cancer and in some cases even therapeutic. The latter potential effects may be attributable to the upregulation of DNA repair: as DNA repair returns to baseline functional levels typical of young cells, fewer DNA errors may remain, lowering mutagenic risk. The view that telomerase will in itself increase the risks of cancer is not well-supported by a diligent appraisal of currently available data, but as in all such cases, the actual outcome awaits the verdict of rigorous clinical trials.

Additional safety concerns are largely those relating to our current ability to deliver gene therapy rather than to the specific gene used, in this case hTERT. Such risks are well-identified in the literature [98] and are likely to change as we become more adept at delivery techniques, adjunctive therapy, and a basic understanding of gene therapy as a clinical field. They include issues related to production quality [99], injury to nontargeted organs, and immunogenicity [100—102].

6. Where we are going

There is an ongoing shift in the tides of medicine, barely felt by some yet obvious to others. We are on the verge of demonstrating an ability to cure and prevent most age-related diseases. That capability hinges on our ability to reverse the aging process at its most fundamental cellular, genetic, and epigenetic levels. Reversing cell aging promises to improve human health to an almost unparalleled degree, matched only by our ability to understand and intervene in microbial disease over the past 2 centuries. This shift depends on our willingness to step back and understand that aging is more than mere passive wear and tear, but also our willingness to step forward and employ biotechnological tools that are just now coming into use in our laboratories, our clinical trials, and our care of patients. To paraphrase Charles Darwin, in dealing with the challenge of aging and age-related disease, the ability to solve the problem requires both deeper knowledge and dedicated science. Together, they promise to change our lives.

References

[1] http://www.ancienttexts.org/library/mesopotamian/gilgamesh/tab11.htm.
[2] Sappho, fragment 58. See for example http://www-personal.umich.edu/~rjanko/Tithonus%20Eos%20and%20the%20cicada.pdf.
[3] Propertius, Elegies 2.18b. http://www.perseus.tufts.edu/hopper/text?doc=Perseus%3Atext%3A2008.01.0494%3Abook%3D2%3Apoem%3D18b.
[4] Living to 120 and beyond: Americans' views on aging, medical advances and radical life extension. Pew Research Center; https://www.pewresearch.org/religion/2013/08/06/living-to-120-and-beyond-americans-views-on-aging-medical-advances-and-radical-life-extension/.
[5] Shchelkunov SN. Emergence and reemergence of smallpox: the need for development of a new generation smallpox vaccine. Vaccine 2011;29(Suppl. 4):D49—53. https://doi.org/10.1016/j.vaccine.2011.05.037.
[6] Hopkins DR. The greatest killer: smallpox in history. University of Chicago Press; 2002. 978-0-226-35168-1. Originally published as Hopkins DR. Princes and Peasants: Smallpox in History. 1983. ISBN 0-226-35177-7.
[7] Needham J. "Part 6, medicine". Science and civilization in China. In: Biology and biological technology, vol. 6. Cambridge: Cambridge University Press; 1999. p. 134.
[8] Shi M, Chu F, Zhu F, Zhu J. Impact of anti-amyloid-β monoclonal antibodies on the pathology and clinical profile of Alzheimer's disease: a focus on aducanumab and lecanemab. Front Aging Neurosci 2022;14:870517. https://doi.org/10.3389/fnagi.2022.870517.
[9] Siemers E, Aisen PS, Carrillo MC. The ups and downs of amyloid in Alzheimer's. J Prev Alzheimer's Dis. 2022;9(1):92—5. https://doi.org/10.14283/jpad.2021.54.

[10] Lopez-Otin C, Blasco MA, Partridge L, Serrano M, Kroemer G. The hallmarks of aging. Cell 2013;153(6):1194−217. https://doi.org/10.1016/j.cell.2013.05.039.
[11] Steinhubl SR. Why have antioxidants failed in clinical trials? Am J Cardiol 2008;101(10):S14−9. https://doi.org/10.1016/j.amjcard.2008.02.003.
[12] Rotllan N, Camacho M, Tondo M, Diarte-Añazco EM, Canyelles M, Méndez-Lara KA, et al. Therapeutic potential of emerging NAD+-increasing strategies for cardiovascular diseases. Antioxidants 2021;10(12):1939. https://doi.org/10.3390/antiox10121939.
[13] Braidy N, Lie Y. NAD+ therapy in age-related degenerative disorders: a benefit/risk analysis. Exp Gerontol 2020;132:110831. https://doi.org/10.1016/j.exger.2020.110831.
[14] Gan E. Therapeutic potential of NAD+ precursors and aging-related diseases. Int J Biomed Sci 2022;18(2):35−40. http://www.ijbs.org/User/ContentFullText.aspx?VolumeNO=18&StartPage=35&Type=pdf.
[15] Wang R, Yu Z, Sunchu B, Shoaf J, Dang I, Zhao S, et al. Rapamycin inhibits the secretory phenotype of senescent cells by a Nrf2-independent mechanism. Aging Cell 2017;16(3):564−74. https://doi.org/10.1111/acel.12587.
[16] Hambright WS, Philippon MJ, Huard J. Rapamycin for aging stem cells. *Aging* (Albany NY) 2020;12(15):15184−5. https://doi.org/10.18632/aging.103816.
[17] Kaeberlein M. Resveratrol and rapamycin: are they anti-aging drugs? Bioessays 2010;32(2):96−9. https://doi.org/10.1002/bies.200900171.
[18] Palmer AK, Jensen MD. Metabolic changes in aging humans: current evidence and therapeutic strategies. Clin Invest 2022;132(16):e158451. https://doi.org/10.1172/JCI158451.
[19] Lagoumtzi SM, Chondrogianni N. Senolytics and senomorphics: natural and synthetic therapeutics in the treatment of aging and chronic diseases. Free Radic Biol Med 2021;171:169−90. https://doi.org/10.1016/j.freeradbiomed.2021.05.003.
[20] Martel J, Ojcius DM, Wu CY, Peng HH, Voisin L, Perfettini JL, et al. Emerging use of senolytics and senomorphics against aging and chronic diseases. Med Res Rev 2020;40(6):2114−31. https://doi.org/10.1002/med.21702.
[21] Zhang L, Pitcher LE, Prahalad V, Neidernhofer LJ, Robbins PD. Targeting cellular senescence with senotherapeutics: senolytics and senomorphics. FEBS J 2022; January 11. https://doi.org/10.1111/febs.16350.
[22] Lewis-McDougall FC, Ruchaya PJ, Domenjo-Vila E, Teoh TS, Prata L, Cottle BJ, et al. Aged-senescent cells contribute to impaired heart regeneration. Aging Cell 2019;18(3):e12931. https://doi.org/10.1111/acel.12931.
[23] Birch J, Gil J. Senescence and the SASP: many therapeutic avenues. Genes Dev 2020;34(23−24):1565−76. https://doi.org/10.1101/gad.343129.120.
[24] Chapman J, Fielder E, Passos JF. Mitochondrial dysfunction and cell senescence: deciphering a complex relationship. FEBS (Fed Eur Biochem Soc) Lett 2019;593(13):1566−79. https://doi.org/10.1002/1873-3468.13498.
[25] Kirkland JL, Tchkonia T. Cellular senescence: a translational perspective. EBioMedicine 2017;21:21−8. https://doi.org/10.1016/j.ebiom.2017.04.013.
[26] Serrano M. The InflammTORy powers of senescence. Trends Cell Biol 2015;25(11):634−6. https://doi.org/10.1016/j.tcb.2015.09.011.
[27] de Mochel NR, Cheong KN, Cassandras M, Wang C, Krasilnikov M, Matatia P, et al. Sentinel p16INK4a+ cells in the basement membrane form a reparative niche in the lung. bioRxiv January 1, 2020. https://doi.org/10.1101/2020.06.10.142893.
[28] Lim JS, Kim HS, Park SC, Park JT, Kim HS, Oh WK, et al. Identification of a novel senomorphic agent, avenanthramide C, via the suppression of the senescence-associated secretory phenotype. Mech Ageing Dev 2020;192:111355. https://doi.org/10.1016/j.mad.2020.111355.
[29] Kaur J, Farr JN. Cellular senescence in age-related disorders. Transl Res 2020;226:96−104. https://doi.org/10.1016/j.mad.2020.111355.
[30] Baker DJ, Childs BG, Durik M, Wijers ME, Sieben CJ, Zhong J, et al. Naturally occurring p16Ink4a- positive cells shorten healthy lifespan. Nature 2016;530:184−9. https://doi.org/10.1038/nature16932.
[31] Ellison-Hughes GM. First evidence that senolytics are effective at decreasing senescent cells in humans. EBioMedicine June 2020;56:102473. https://doi.org/10.1016/j.ebiom.2019.09.053.
[32] Kirkland JL, Tchkonia T, Zhu Y, Niedernhofer LJ, Robbins PD. The clinical potential of senolytic drugs. J Am Geriatr Soc 2017;65(10):2297−301. https://doi.org/10.1111/jgs.14969.
[33] Chaib S, Tchkonia T, Kirkland JL. Cellular senescence and senolytics: the path to the clinic. Nat Med (N Y, NY, U S) 2022;28(8):1556−68. https://doi.org/10.1038/s41591-022-01923-y.
[34] Dolgin E. Send in the senolytics. Nat Biotechnol 2020;38(12):1371−8. https://doi.org/10.1038/s41587-020-00750-1.
[35] De Magalhães JP. Longevity pharmacology comes of age. Drug Discov Today 2021;26(7):1559−62. https://doi.org/10.1016/j.drudis.2021.02.015.
[36] Fossel M. Cell senescence, telomerase, and senolytic therapy. OBM Geriatrics 2019;3(1):1. https://doi.org/10.21926/obm.geriatr.1901034.
[37] Fossel M. Telomerase and the aging cell: implications for human health. JAMA 1998;279:1732−5. https://doi.org/10.1001/jama.279.21.1732.
[38] Banks D, Fossel M. Telomeres, cancer, and aging: altering the human lifespan. JAMA 1997;278:1345−8. https://doi.org/10.1001/jama.1997.03550160065040.
[39] Fossel M. Reversing human aging. New York: William Morrow & Co.; 1996.
[40] Fossel M. Cells, aging, and human disease. Oxford University Press; 2004.
[41] https://www.ncbi.nlm.nih.gov/protein/109633031.
[42] https://www.ncbi.nlm.nih.gov/nuccore/AF015950.
[43] Nakamura TM, Morin GB, Chapman KB, Weinrich SL, Andrews WH, Lingner J, et al. Telomerase catalytic subunit homologs from fission yeast and human. Science 1997;277(5328):955−9. https://doi.org/10.1126/science.277.5328.955.
[44] https://www.ncbi.nlm.nih.gov/gene?Cmd=DetailsSearch&Term=7015.
[45] Bryce LA, Morrison N, Hoare SF, Muir S, Keith WN. Mapping of the gene for the human telomerase reverse transcriptase, hTERT, to chromosome 5p15.33 by fluorescence in situ hybridization. Neoplasia 2000;2(3):197−201. https://doi.org/10.1038/sj.neo.7900092.
[46] Meyerson M, Counter CM, Eaton EN, Ellisen LW, Steiner P, Caddle SD, et al. hEST2, the putative human telomerase catalytic subunit gene, is up-regulated in tumor cells and during immortalization. Cell 1997;90(4):785−95. https://doi.org/10.1016/s0092-8674(00)80538-3.
[47] https://www.ncbi.nlm.nih.gov/gtr/genes/7012/.
[48] Greider CW. Telomeres, telomerase and senescence. Bioessays 1990;12(8):363−9. https://doi.org/10.1002/bies.950120803.
[49] He Y, Feigon J. Telomerase structural biology comes of age. Curr Opin Struct Biol 2022;76:102446. https://doi.org/10.1016/j.sbi.2022.102446.

References

[50] Bodnar AG, Chiu C, Frolkis M, Harley CB, Holt SE, Lichtsteiner S, et al. Extension of life-span by introduction of telomerase into normal human cells. Science 1998;279:349—52. https://doi.org/10.1126/science.279.5349.349.

[51] Vaziri H, Benchimol S. Reconstruction of telomerase activity in normal cells leads to elongation of telomeres and extended replicative life span. Curr Biol 1998;8:279—82. https://doi.org/10.1016/s0960-9822(98)70109-5.

[52] Vaziri H. Extension of life span in normal human cells by telomerase activation: a revolution in cultural senescence. J Anti Aging Med 1998;1:125—30. https://doi.org/10.1089/rej.1.1998.1.125.

[53] Shelton DN, Chang E, Whittier PS, Choi D, Funk W. Microarray analysis of replicative senescence. Curr Biol 1999;9:939—45. https://doi.org/10.1016/S0960-9822(99)80420-5.

[54] Funk WD, Wang CK, Shelton DN, Harley CB, Pagon GD, Hoeffler WK. Telomerase expression restores dermal integrity to in vitro-aged fibroblasts in a reconstituted skin model. Exp Cell Res 2000;258:270—8. https://doi.org/10.1006/excr.2000.4945.

[55] Matsushita H, Chang E, Glassford AJ, Cooke JP, Chiu CP, Tsao PS. eNOS activity is reduced in senescent human endothelial cells. Preservation by hTERT immortalization. Circ Res 2001;89:793—8. https://doi.org/10.1161/hh2101.098443.

[56] Yudoh K, Matsuno H, Nakazawa F, Katayama R, Kimura T. Reconstituting telomerase activity using the telomerase catalytic subunit prevents the telomere shorting and replicative senescence in human osteoblasts. J Bone Miner Res 2001;16:1453—64. https://doi.org/10.1359/jbmr.2001.16.8.1453.

[57] Jaskelioff M, Muller FL, Paik JH, Thomas E, Jiang S, Adams A, et al. Telomerase reactivation reverses tissue degeneration in aged telomerase deficient mice. Nature 2011;469:102—6. https://doi.org/10.1038/nature09603.

[58] de Jesus BB, Vera E, Schneeberger K, Tejera AM, Ayuso E, Bosch F, et al. Telomerase gene therapy in adult and old mice delays aging and increases longevity without increasing cancer. EMBO Mol Med 2012;4:691—704. https://doi.org/10.1002/emmm.201200245.

[59] Harley CB, Liu WM, Blasco M, Vera E, Andrews WH, Briggs LA, et al. A natural product telomerase activator as part of a health maintenance program. Rejuvenation Res 2011;14:45—56. https://doi.org/10.1089/rej.2010.1085.

[60] Harley CB, Liu W, Flom PL, Raffaele JM. A natural product telomerase activator as part of a health maintenance program: metabolic and cardiovascular response. Rejuvenation Res 2013;16:386—95.

[61] de Jesus BB, Schneeberger K, Vera ME, Tejera A, Harley CB, Blasco MA. The telomerase activator TA-65 elongates short telomeres and increases health span of adult old mice without increasing cancer incidence. Aging Cell 2011;10:604—21. https://doi.org/10.1111/j.1474-9726.2011.00700.x.

[62] Mendell JR, Al-Zaidy S, Shell R, Arnold WD, Rodino-Klapac LR, Prior TW, et al. Single-dose gene-replacement therapy for spinal muscular atrophy. N Engl J Med November 2, 2017;377(18):1713—22. https://doi.org/10.1056/NEJMoa1706198.

[63] Chamberlain JS. A boost for muscle with gene therapy. N Engl J Med 2022;386(12):1184—6. https://doi.org/10.1056/NEJMcibr2118576.

[64] FDA approves first gene therapy to treat adults with hemophilia B. FDA.

[65] Li X, Wei X, Lin J, Ou L. A versatile toolkit for overcoming AAV immunity. Front Immunol 2022;13:991832. https://doi.org/10.3389/fimmu.2022.991832.

[66] Sivagourounadin K, Ravichandran M, Rajendran P. National guidelines for gene therapy product (2019): a road-map to gene therapy products development and clinical trials. Perspect Clin Res 2021;12(3):118. https://doi.org/10.4103/picr.PICR_189_20.

[67] Ho JK, Borle K, Dragojlovic N, Dhillon M, Kitchin V, Kopac N, et al. Economic evidence on potentially curative gene therapy products: a systematic literature review. Pharmacoeconomics 2021;39(9):995—1019. https://doi.org/10.1007/s40273-021-01051-4.

[68] Byrne BJ, Corti M, Muntoni F. Considerations for systemic use of gene therapy. Mol Ther 2021;29(2):422. https://doi.org/10.1016/j.ymthe.2021.01.016.

[69] Stanton AC, Lagerborg KA, Tellez L, Krunnfusz A, King EM, Ye S, et al. Systemic administration of novel engineered AAV capsids facilitates enhanced transgene expression in the macaque CNS. Méd 2022;4:1—20. https://doi.org/10.1016/j.medj.2022.11.002.

[70] O'Carroll SJ, Cook WH, Young D. AAV targeting of glial cell types in the central and peripheral nervous system and relevance to human gene therapy. Front Mol Neurosci 2021;13:618020. https://doi.org/10.3389/fnmol.2020.618020.

[71] Naso MF, Tomkowicz B, Perry WL, Strohl WR. Adeno-associated virus (AAV) as a vector for gene therapy. BioDrugs 2017;31(4):317—34. https://doi.org/10.1007/s40259-017-0234-5.

[72] Fakhiri J, Grimm D. Best of most possible worlds: hybrid gene therapy vectors based on parvoviruses and heterologous viruses. Mol Ther 2021;29(12):3359—82. https://doi.org/10.1016/j.ymthe.2021.04.005.

[73] Poletti V, Mavilio F. Designing lentiviral vectors for gene therapy of genetic diseases. Viruses 2021;13(8):1526. https://doi.org/10.3390/v13081526.

[74] Gurumoorthy N, Nordin F, Tye GJ, Wan Kamarul Zaman WS, Ng MH. Non-integrating lentiviral vectors in clinical applications: a glance through. Biomedicines 2022;10(1):107. https://doi.org/10.3390/biomedicines10010107.

[75] Prel A, Caval V, Gayon R, Ravassard P, Duthoit C, Payen E, et al. Highly efficient in vitro and in vivo delivery of functional RNAs using new versatile MS2-chimeric retrovirus-like particles. Mol Ther Methods Clin Dev 2015;2:15039. https://doi.org/10.1038/mtm.2015.39.

[76] Albertsen CH, Kulkarni J, Witzigmann D, Lind M, Petersson K, Simonsen JB. The role of lipid components in lipid nanoparticles for vaccines and gene therapy. Adv Drug Deliv Rev 2022;3:114416. https://doi.org/10.1016/j.addr.2022.114416.

[77] Godbout K, Tremblay JP. Delivery of RNAs to specific organs by lipid nanoparticles for gene therapy. Pharmaceutics 2022;14(10):2129. https://doi.org/10.3390/pharmaceutics14102129.

[78] Mukai H, Ogawa K, Kato N, Kawakami S. Recent advances in lipid nanoparticles for delivery of nucleic acid, mRNA, and gene editing-based therapeutics. Drug Metab Pharmacokinet 2022; February 5:100450. https://doi.org/10.1016/j.dmpk.2022.100450.

[79] Fossel M, Whittemore K. Telomerase and cancer: a complex relationship. OBM Geriatrics 2021;5:18. https://doi.org/10.21926/obm.geriatr.2101156.

[80] Kishimoto TK, Samulski RJ. Addressing high dose AAV toxicity – 'one and done' or 'slower and lower'. Expet Opin Biol Ther 2022;22(9):1067—71. https://doi.org/10.1080/14712598.2022.2060737.

[81] McElroy A, Sena-Esteves M, Arjomandnejad M, Keeler AM, Gray-Edwards HL. Redosing adeno-associated virus gene therapy to the central nervous system. Hum Gene Ther 2022;33:17—8. https://doi.org/10.1089/hum.2022.170.

[82] Meyer RA, Neshat SY, Green JJ, Santos JL, Tuesca AD. Targeting strategies for mRNA delivery. Mat Today Adv 2022;14:100240. https://doi.org/10.1016/j.mtadv.2022.100240.

[83] Damase TR, Sukhovershin R, Boada C, Taraballi F, Pettigrew RI, Cooke JP. The limitless future of RNA therapeutics. Front Bioeng Biotechnol March 18, 2021;9:628137. https://doi.org/10.3389/fbioe.2021.628137.

[84] Chanda PK, Sukhovershin R, Cooke JP. mRNA-enhanced cell therapy and cardiovascular regeneration. Cells January 19, 2021;10(1):187. https://doi.org/10.3390/cells10010187.

[85] Weissman D, Karikó K. mRNA: fulfilling the promise of gene therapy. Mol Ther September 2015;23(9):1416−7. https://doi.org/10.1038/mt.2015.138.

[86] Ramunas J, Yakubov E, Brady JJ, Corbel SY, Holbrook C, Brandt M, et al. Transient delivery of modified mRNA encoding TERT rapidly extends telomeres in human cells. Faseb J May 2015;29(5):1930−9. https://doi.org/10.1096/fj.14-259531.

[87] Li Y, Zhou G, Bruno IG, Cooke JP. Telomerase mRNA reverses senescence in progeria cells. J Am Coll Cardiol August 8, 2017;70(6):804−5. https://doi.org/10.1016/j.jacc.2017.06.017.

[88] Li Y, Zhou G, Bruno IG, Zhang N, Sho S, Tedone E, et al. Transient introduction of human telomerase mRNA improves hallmarks of progeria cells. Aging Cell August 2019;18(4):e12979. https://doi.org/10.1111/acel.12979.

[89] Xu Q, Mojiri A, Boulahouache L, Morales E, Walther BK, Cooke JP. Vascular senescence in progeria: role of endothelial dysfunction. Eur Heart J Open July 28, 2022;2(4):oeac047. https://doi.org/10.1093/ehjopen/oeac047.

[90] Mojiri A, Walther BK, Jiang C, Matrone G, Holgate R, Xu Q, et al. Telomerase therapy reverses vascular senescence and extends lifespan in progeria mice. Eur Heart J November 7, 2021;42(42):4352−69. https://doi.org/10.1093/eurheartj/ehab547.

[91] Tan H, Su W, Zhang W, Wang P, Sattler M, Zou P. Recent advances in half-life extension strategies for therapeutic peptides and proteins. Curr Pharmaceut Des 2018;24(41):4932−46. https://doi.org/10.2174/1381612825666190206105232.

[92] Yang WC, Patel KG, Lee J, Ghebremariam YT, Wong HE, Cooke JP, et al. Cell-free production of transducible transcription factors for nuclear reprogramming. Biotechnol Bioeng December 15, 2009;104(6):1047−58. https://doi.org/10.1002/bit.22517.

[93] Kim H, Seo EH, Lee SH, Kim BJ. The telomerase-derived anticancer peptide vaccine GV1001 as an extracellular heat shock protein-mediated cell-penetrating peptide. Int J Mol Sci 2016;17(12):2054. https://doi.org/10.3390/ijms17122054.

[94] Lee SA, Kim BR, Kim BK, Kim DW, Shon WJ, Lee NR, et al. Heat shock protein-mediated cell penetration and cytosolic delivery of macromolecules by a telomerase-derived peptide vaccine. Biomaterials 2013;34(30):7495−505. https://doi.org/10.1016/j.biomaterials.2013.06.015.

[95] Park HH, Lee KY, Kim S, Lee JW, Choi NY, Lee EH, et al. The novel vaccine peptide GV1001 effectively blocks β-amyloid toxicity by mimicking the extra-telomeric functions of human telomerase reverse transcriptase. Neurobiol Aging 2014;35(6):1255−74. https://doi.org/10.1016/j.neurobiolaging.2013.12.015.

[96] Park HH, Yu HJ, Kim S, Kim G, Choi NY, Lee EH, et al. Neural stem cells injured by oxidative stress can be rejuvenated by GV1001, a novel peptide, through scavenging free radicals and enhancing survival signals. Neurotoxicology 2016;55:131−41. https://doi.org/10.1016/j.neuro.2016.05.022.

[97] Koh SH, Kwon HS, Choi SH, Jeong JH, Na HR, Lee CN, et al. Efficacy and safety of GV1001 in patients with moderate-to-severe Alzheimer's disease already receiving donepezil: a phase 2 randomized, double-blind, placebo-controlled, multicenter clinical trial. Alzheimer's Res Ther 2021;13(1). https://doi.org/10.1186/s13195-021-00803-w. 1-1.

[98] Horton RH, Saade D, Markati T, Harriss E, Bönnemann CG, Muntoni F, et al. A systematic review of adeno-associated virus gene therapies in neurology: the need for consistent safety monitoring of a promising treatment. J Neurol Neurosurg Psychiatr 2022;93(12):1276−88. https://doi.org/10.1136/jnnp-2022-329431.

[99] Mohammad R. Key considerations in formulation development for gene therapy products. Drug Discov Today 2021; September 6. https://doi.org/10.1016/j.drudis.2021.08.013.

[100] Gorovits B, Azadeh M, Buchlis G, Harrison T, Havert M, Jawa V, et al. Evaluation of the humoral response to adeno-associated virus-based gene therapy modalities using total antibody assays. AAPS J 2021;23(6):1−7. https://doi.org/10.1208/s12248-021-00628-3.

[101] Muhuri M, Maeda Y, Ma H, Ram S, Fitzgerald KA, Tai PW, et al. Overcoming innate immune barriers that impede AAV gene therapy vectors. J Clin Invest 2021;131(1). https://doi.org/10.1172/JCI143780.

[102] Rapti K, Grimm D. Adeno-associated viruses (AAV) and host immunity−A race between the hare and the hedgehog. Front Immunol 2021; 12:753467. https://doi.org/10.3389/fimmu.2021.753467.

CHAPTER 14

Summary

Michael Fossel

Telocyte LLC, Ada, MI, United States

If you hunt only rabbits, Tigers and dragons go uncaught. —*Li Bai (李白), ~750*

1. Introduction

Medical school professors are fond of pointing out that as our classes adapt to advances in medical science, our final exam questions remain the same but the correct answers change year to year. Ironic and humorous, perhaps, but if that were not at least somewhat true, medicine would still be in the dark ages. The best way to treat a fever may not be willow bark or even aspirin, but an antibiotic. The best way to treat a broken hip may not be months of bed rest, but an orthopedic screw. The best way to deal with polio is not a better iron lung, but a better polio vaccine. The questions remain the same; our answers change dramatically and usually in unforeseen ways. Over the centuries, over the decades, and sometimes even from year to year, many of us have come to realize that last year's answers were merely that: last year's answers.

The correct answers, assuming we can ever actually know what is correct, vary not only over time but from expert to expert. This is not to say that there are no correct answers, but merely an acknowledgment that, all academic hubris aside, there is much that we still do not know. To identify a particular answer as final and ultimately correct is merely an example of historical provincialism. Believing our contemporary answers were final is a trap that we have fallen into repeatedly throughout history. Unfortunately, humanity's fundamental errors never change, and no matter how modern we think we are, we echo the same error, believing that, except perhaps for crossing a few trivial *t*'s and dotting a few inconsequential *i*'s, we have finally come to understand medicine and science. We may be bright, hard-working, well-educated, technically well-trained, and honest, but the result is that we are at risk (to put it humorously) of becoming some of the best 20th-century minds in a 21st-century world.

If medicine and science are to progress, we need to open minds, reexamine assumptions, open debate, and tolerate diverse opinions. This book has more than a dozen well-respected authors from around the world and multiple specialties. They not only have different perspectives and interests, their views of how aging and age-related disease work are different. Some of us agree and some of us disagree, even about how to interpret fundamental and seemingly obvious data. For that reason, not only do the chapters vary in the organ system they address, the authors interpret basic data on disease in subtly different ways, and they sometimes draw markedly different conclusions.

At least at this juncture and perhaps with no temporal limit, disagreement is the mark of a healthy scientific approach. The underlying truism is that whereas logical and consistent theories are good, data win every time. Although we have varying views of how the complex cascades of aging occur and result in clinical disease, and although we hold varying predictions of what will happen if we try to intervene in the aging process, the ultimate judge is whether such interventions succeed in human trials, in medical practice, and in improving human lives. The key question is not which is the better theory of aging, but which intervention results in a more effective clinical outcome for patients with age-related diseases. A careful reader will have noted that our authors do not always agree about precisely how cell aging occurs or how cell aging results in pathology within any specific organ system, let alone the optimal point of intervention or the best technical vehicle that might be used to target a potentially optimal point of intervention.

One example of this, typical of the literature, is the question of how mitochondrial dysfunction is related to cell aging. Historically, the fact of the age-related, progressive deterioration of mitochondrial dysfunction has been accepted for several decades, and there is an enormous body of literature to support the general observation: our mitochondria become less functional as we get older [1–3]. There remains disagreement, however, as to whether (and to what degree) mitochondrial dysfunction causes and accelerates cell aging, or whether cell aging causes and accelerates mitochondrial dysfunction. The literature can be interpreted—perhaps with a tussle—to support either view. Many would compromise, suggesting that cell aging and mitochondrial dysfunction are causally interrelated and that neither single-minded view is supportable or accurate. That the mitochondria we inherit maternally come with a provenance going back more than a billion years would, at least at first glance, make it hard to blame cell aging on mitochondrial dysfunction per se, because our billion-year-old mitochondria apparently did not have evidence of aging on arrival in the fertilized gametes from which we each derive. This might suggest that cell aging (perhaps by slowing the turnover of key aerobic enzymes, mitochondrial lipids, and scavenger molecules) should shoulder the entire blame for mitochondrial dysfunction and the increase in age-related free radicals (Fig. 14.1). Yet the precise mechanisms of energy production in the mitochondria are known to shift as we move from gamete to fertilized ovum, and to multicellular organism, so the fact that we inherit old yet fully functional mitochondria may have no relevance to which way causation runs (and perhaps both ways) as we look at aging mitochondria and aging cells.

None of this might matter if were it not for the practical clinical question of where and how effectively we might intervene (Fig. 14.2). Do we target cell aging (as assessed by telomere shortening, for example) or mitochondrial aging (as assessed by mitochondrial dysfunction)? There is some evidence (discussed earlier) that restoring lost telomere lengths is sufficient to restore much of young normal cell, tissue, and organism function, but to what extent? And might such intervention be made more effective if we were to target mitochondria as well as telomeres directly? The same argument arises with regard to DNA damage and a host of other biomarkers of aging. In every case, at the end of the day, the diversity of academic opinion becomes a question of clinical intervention. What is the optimal point of intervention, both clinically and financially? How can we most effectively treat and prevent age-related disease, and how can do so without unnecessary or unavoidable costs? In an ideal world, we would effectively treat and prevent age-related diseases while lowering the medical costs of doing so.

Theory can guide us, diversity of opinion can instruct us, but only clinical trials will provide the important answers.

Assuming we can effectively intervene, what are the limits? The question is not merely the limits of the maximum human life span, but the extent to which we can improve health and prevent age-related disease over that life span. We may find that we can effectively reverse cell aging, but would that be a panacea for all age-related diseases? Even in the best of cases, we might combine as age-related some diseases that will not prove to be amenable to the reversal of cell aging. Consider the case of silicosis, for example. After decades of exposure to crystalline silica dust (silicon

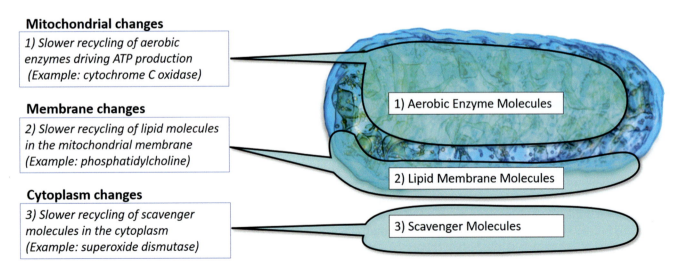

FIGURE 14.1 Age-related mitochondrial dysfunction may be driven by epigenetic changes in the nucleus that result in (1) slower turnover of aerobic enzymes (increasing the reactive oxygen species [ROS]/adenosine triphosphate ratio), (2) slower turnover of mitochondrial lipids (increasing permeability), and (3) slower turnover of scavenger molecules (increasing ROS damage throughout the cell). These processes result in a gradual but exponential increase in mitochondrial dysfunction as cell aging progresses.

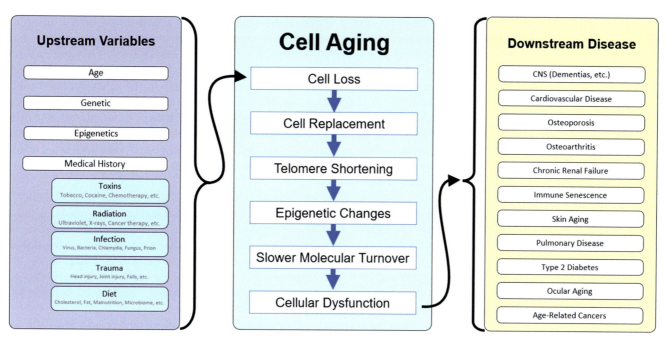

FIGURE 14.2 Points of intervention. The crucial issue is not possible points of intervention, but identifying the optimal point of intervention vis-à-vis efficacy, safety, cost, and technical feasibility. Upstream targets are only partially effective in slowing (but not stopping) age-related disease. Downstream targets are only partially effective in addressing symptoms or in postponing mortality, but neither stop nor reverse age-related disease. *CNS*, central nervous system.

dioxide exposure, a common risk for miners), the affected lung becomes progressively dysfunctional. Even young, healthy cells cannot remove all of the accumulating silicon, however, and there is a linear increase in the burden to those cells. Even the perfect restoration of young cell function would not remove the accumulated burden of a lifetime of exposure and silica deposition. The lung may function better if we can reset cell aging in the pneumocytes and other relevant cells, but it cannot be expected to function as well as a lung that has never experienced silica deposition in and around those cells. Silica deposition is likely an example of an accumulated burden that correlates with chronologic age but is not related to cell aging or (to put it bluntly) aging per se. The outcome, then, may be that we can prevent or cure age-related diseases, but only if we are precise and perhaps even pedantic in our definition of age-related.

We are left with the observation that most (but not all) of what we call aging is likely attributable to cell aging and the subsequent failure of cell maintenance with resulting disease, but that there are cases that we innocently think of as aging, which are of a different etiology. These latter cases, unrelated to cell aging, are examples of the progressive accumulation of damage, which occur regardless of and independently of the cell aging process. We may be able to reverse cell aging and address most age-related diseases, but the approach is not a clinical panacea.

2. Is aging a disease?

Until the past century or so, although a small percentage of us lived long enough to acquire age-related diseases, most of us died young, and we died of infections. The past century has seen a historic change in the most common causes of death. Most of us now live long enough to age, and we die of age-related diseases. Put differently, and striking to the core of the problem, in the 21st century we die of aging.

Yet a perennial trope in the field of gerontology is the question of whether aging is a disease [4–6]. It would remain a merely odd and purely academic question if it were not for the regulatory concerns that it raises for those attempting to find successful clinical interventions. In many respects, the question elicits a purely arbitrary, subjective definition of disease, yet the question has objective ramifications for the development of interventions, including its implications for funding, clearing regulatory hurdles, and successful commercialization. The question may be academic and elicit parallels from medieval scholasticism (how many angels can dance on the head of a pin?), but the question has tangible implications for our ability to develop novel and effective treatments.

On the one hand, aging is clearly a natural process, and we could well argue that a natural process is automatically precluded from the definition of disease. On the other hand, we clearly include an enormous number of natural processes—including microbial infections and genetically inherited problems—as diseases. They are natural, but they are also worthy clinical targets. Even if we narrow our consideration to age-related natural processes such as menopause, we treat hot flashes and related clinical symptoms as being entirely justifiable targets for medical intervention. In short, whether we define menopause as a disease, we certainly treat menopause (or its clinical symptoms) as a disease when judged from a practical perspective. Much the same is true of other age-related findings such as baldness and wrinkles. Few would define these as diseases, yet whatever we choose to define as a disease, an enormous clinical market remains for such interventions. Whether or not aging is a disease, we must admit that aging causes dis-ease, and that both patients and clinicians act on that observation.

Whether or not aging is a disease, we treat it as a disease.

From both funding and regulatory perspectives, whether aging is a disease has practical implications. In general, both funding agencies and the US Food and Drug Administration (as well as other global regulatory agencies) have enforced a distinction between aging and age-related disease. Generally, the former has not been regarded as an appropriate target for either funds or clinical trials, whereas the latter earns the imprimatur of all relevant regulatory agencies. Put simply, it is difficult to obtain funding or regulatory approval for an intervention targeting aging, but relatively easy to get funding or regulatory approval for an intervention targeting an age-related disease. Both age-related neurodegenerative diseases and age-related cardiovascular diseases, for example, are deemed suitable targets for funding and for human trials, whereas aging per se lacks the same cachet.

To a degree this is changing, because we begin to suspect that age-related diseases might best be addressed (in the sense of developing truly effective and fundamental interventions) if we address the aging process more directly. Nonetheless, it is easier to define and agree on objective biomarkers (and clinical end points) for age-related diseases than for aging, and this difficulty has retarded progress. The upshot is that most attempts to deal with aging have taken a narrower and more accepted focus, choosing to target specific age-related diseases rather than to target aging.

To paraphrase the opening poem of this chapter, although we generally find it easier to hunt rabbits, the field is changing, and there is a growing recognition that it may well be worthwhile to see whether we might hunt—and learn to slay—the dragons. Aging remains the single most predictive variable in age-related disease, and the field is beginning to address not merely age-related diseases, but the fundamental, driving cause of such clinical diseases.

From a human perspective, we act as if aging is a disease, so perhaps it is appropriate to treat it as though it were exactly that: a disease. From a practical perspective, putting aside definitional disputes, if we are both to meet regulatory needs and achieve clinical success, we need to ensure that we have objective and measurable end points to assess the efficacy of interventions in the aging process. The question is not whether aging is a disease, but rather whether we can objectively quantify success in clinical interventions that target aging. In short, whatever definitions we choose and whatever outcome measures we employ, the bottom line is whether we can improve human lives.

There is no more practical question.

3. Organs

Aging occurs in parallel in all of our organs, but the parallels are not tightly linked, and we commonly find that any one given organ may age more rapidly than other organs and may present with earlier and more compelling clinical disease. Each of us starts with different genes and different epigenetic backgrounds, but we have different medical histories and engage in different behaviors. A patient with a history of high-impact sports to the knee may have earlier osteoarthritis than another patient who is continuously exposed to ultraviolet radiation and has earlier skin aging. Trauma, infections, toxic exposures, radiation, and dozens of other factors affect which age-related diseases we experience, at what age they show themselves, and how they progress. We each vary in how quickly we age, and we see equal heterogeneity in comparing how quickly each organ ages. Our organs age at different rates, although in all cases the individual clinical outcomes hinge on a common process: cell aging. Regardless of the universality of cell aging, however, as our organs undergo aging, we generally experience—and die of—the predominant aging of a specific organ that happens to proceed a bit more quickly into age-related disease. With that in mind, it is reasonable to address aging system by system, as we have in the chapters of this book.

3.1 Age-related central nervous system disease

Age-related diseases of the central nervous system are not the most common cause of death as we grow older, but they are perhaps the most poignant. The dementias, for example, appear to erase not only our physical abilities, but our very selves. To put it poetically, dementia is a disease that steals our souls. Whatever the name used, the age-related neurodegenerative disease has been observed throughout all of human history and in all cultures. It is certainly not a disease of the 21st century, although as the mean life span increases, so does the prevalence of dementia. Over the centuries, the nomenclature has changed (we no longer talk of senile dementia, for example), gone through contradictions and arguments, and generated almost innumerable subdiagnoses. Despite the long, alphabetic litany of age-related neurodegenerative diseases (even restraining ourselves to abbreviations, Alzheimer's disease, Parkinson's disease, progressive supernuclear palsy, fronto-temporal dementia, Lewy body disease, corticobasal syndrome, medial temporal lobe atrophy, fronto-temporal lobe atrophy, Huntington's disease, primary progressive aphasia, and posterior cortical atrophy are only a few of those under consideration), when we deal with actual patients, we find that these diagnoses, although all too real, have overlaps and exceptions. Some patients involve classic cases that precisely embody descriptions in the textbook of neurology, but many patients present with aspects of two or more diagnoses, unusual findings, or the lack of typical or even key findings. Appropriately, we use diagnostic categories to help craft treatment plans or establish prognostic vectors, but the reality is that all of the age-related neurodegenerative diseases might best be viewed as sharing a common pathologic pathway: that of aging. The clinical presentations are heterogeneous because patients have heterogeneous starting points. The model of cell aging as the essential pathology is entirely consistent with all current clinical and animal data and offers a novel and feasible point of intervention [7]. The tragedy is that we have as yet been unable to intervene effectively in any of these diseases, and the reason for our failure is likely that we have targeted downstream biomarkers such as amyloid rather than the underlying pathologic cascade that drives such biomarkers. Current data suggest that if we target cell aging itself (and not merely the biomarkers of cell aging), there is considerable reason to think we may be able to prevent and, to an unknown extent, reverse much of the pathology.

3.2 Age-related cardiovascular disease

As Thomas Sydenham famously said, you are "only as old as your arteries," a sentiment which remains as accurate today as it was when he said it almost 4 centuries ago [8]. We might reasonably quibble and add your veins and capillaries, and the entire cardiovascular system, but the relationship between the age-related decline in physical function and the status of the cardiovascular system is hard to overstate. Age-related cardiovascular disease is the major killer of most of the world's population. Live long enough, and it is likely to be the diagnosis at death for most of us. Like age-related neurodegenerative diseases, the clinical outcomes vary, probably owing to differences in the medical history as well as genetic and epigenetic starting points. Also like those diseases, there is a wide gamut of diagnoses for age-related cardiovascular disease, including but not limited to such common categories as myocardial infarction, cerebrovascular accident, stroke, aneurysm, atherosclerosis, atherogenesis, coronary artery disease, congestive heart failure, venous stasis, and peripheral capillary pruning. Moreover, and again just like central nervous system diseases, most patients do not have merely one type of clinical presentation (aneurysm for example), but degrees of several other presentations as well. Although a single diagnostic finding may bring the patient to the attention of the physician, an attentive workup usually finds that age-related cardiovascular pathology is a broader phenomenon. This observation alone is enough for physicians to recognize the general clinical problem of cardiovascular disease as a tangible entity rather than a mere conglomeration of diagnostic pigeonholes. The recognition is apt and likely to represent the underlying reality: age-related cardiovascular disease is the result of the fundamental changes in cell function incumbent on cell aging itself, and, as a result it may be amenable to interventions targeting cell aging rather than the individual biomarkers we associate with the individual types of cardiovascular disease. Although it is true that we have made remarkable advances in treating and preventing many types of cardiovascular disease over the past century, the interventions generally target either risk factors (such as blood pressure, smoking, blood glucose, and exercise) or outcomes (such a replacement or stenting of significantly obstructed coronary arteries) rather than the fundamental changes of cell aging that operate between upstream risk factors and downstream histologic changes. As a result, our current cardiovascular interventions, although laudable, remarkable, and valuable, might well be surpassed if we can approach the problem of aging more directly. As in the case of age-related neurodegenerative disease, a fundamental approach using a unified, systems model may yet prove able to prevent and largely reverse much of the pathology [9].

3.3 Age-related bone disease

Although they are not fatal per se, some age-related diseases cause significant clinical problems, and when complications ensue, they may become the prelude to a fatal outcome. Osteoporosis falls into this category: it can cause pain and disability and may be the opening scene in a series of medical complications that may close with the patient's death. Although there is considerable variance among patients, in most patients, bone mass falls with age, and this is finding is accentuated in females, especially after menopause (although the menopausal etiology is intrinsically distinct from aging). The outcome is an age-associated risk of osteoporotic fractures, along with complications inherent in such fractures. Current therapies serve to address parts of the pathology, endeavoring to increase bone formation or mineralization of the matrix, for example, yet leave the underlying cell changes unaffected. The clinical success of current therapies remains arguable. Clearly, there is room for improvement in our ability to renormalize bone mineral density and the risk of fractures. To a large extent, such improvement hinges on a better understanding of the role of aging in the observed changes. The balance between osteoclast and osteoblast activity, as well as the function of each type of osteocyte individually, is a complicated cascade of events in which cell aging has a prominent role as we age and both body mass index and structural strength diminish. Even if osteoporosis is not a major cause of death (as is the case with the age-related nervous system and cardiovascular disease) the same key questions pertain: can we reverse cell aging, and if so, to what extent can we prevent and cure age-related bone disease? What few data we have suggest that age-related osteoporosis may prove an effective clinical target for therapies that reset cell aging directly.

3.4 Age-related joint disease

Osteoarthritis, the most obvious age-related joint disease, has long been the object of clinical therapies. Until 1890, such therapies remained purely symptomatic, but attempts to replace joint surfaces, starting in 1890 [10] and culminating in the advent of total joint arthroplasties [11] in the 1970s, has led to what is now a standard therapeutic option for severe age-related joint disease. Despite the widespread and effective use of surgical arthroplasty for a number of affected joints, the procedure has significant costs, requires expertise, and carries risks to the patient, such as pain and infectious complications. This has prompted a search for other approaches, although with as yet limited and arguable efficacy, especially compared with solid clinical experience that supports the use of arthroplasty. Both pathology and clinical data support the key role of cell aging of the chondrocytes in joint surfaces, which has focused attention on interventions targeting such cells in vivo. One such approach is the use of senolytics, which purport to remove senescent chondrocytes in an effort to restore, or at least improve, joint function. Although an interesting approach, it has a limited clinical history and may prove to be less than optimal based on theoretical considerations [12], as well as the initial animal data. However, human trials will provide insight into the actual clinical value, if any. Senolytic and senomorphic approaches presuppose that the consequences of cell aging are amenable to therapeutic targeting, but that the process of cell aging itself is not. A more fundamental approach would be to reset cell aging itself in the chondrocytes, a technique that deserves attention and is supported by previous work in reversing cell aging in several types of human cells and tissues. The use of gene therapy in the joint is, technically speaking, easier than similar approaches in other organs or systems. The knee joint, for example, is a relatively easy target for localized joint infusions via arthroscopy. There are still few data (at least joint-related data), but the use of gene therapy in clinical trials targeting age-related joint disease by directly resetting chondrocyte cell aging is a promising venue.

3.5 Age-related kidney disease

Chronic kidney disease (also termed chronic renal failure) is attributable to any number of factors including diabetes, hypertension, infection, toxic exposure, genetic predilections, and other upstream risks, but as in so many other organs, such upstream risk factors are expressed with increasing emphasis as we grow older. As ever, aging itself remains the most important risk factor in age-related disease, and chronic kidney disease is no exception. Age is the single most predictive risk factor and the common denominator in such a disease [13]. The other known risk factors serve to accelerate cell aging in renal tissue, specifically in renal podocytes, as well as in renal vascular endothelial cells. Before the availability of renal dialysis, there was little medicine could offer, but the technique has come a long way since the first dialysis in 1943 [14], including various peritoneal and hemodialysis techniques [15,16]. The option of renal transplantation [17] has also effected a revolution in the treatment of chronic kidney disease, although with caveats: it is available to a limited number of patients, has well-defined clinical risks, and incurs significant financial (and emotional) costs [18]. As with other age-related diseases, chronic kidney disease has a number of effective interventions, yet the current armamentarium is far from optimal from several vantage points. The central stage on which chronic kidney disease occurs—that of cell aging—suggests a potentially more effective and less costly option. The key question is whether we can effectively reset aging in podocytes (and other renal cells) by intervening directly in the central modulators of cell aging, such as telomere length. The potential benefits would be enormous, but regardless of the potential to improve health and lower costs, the actual efficacy of resetting renal cell aging remains both unknown and untested.

3.6 Age-related immune dysfunction

The fact of immunosenescence has long been acknowledged and studied by clinicians [19] as well as accepted to be common knowledge. We accept that older individuals are more likely to succumb to infections that may lack significant clinical consequences in a young adult. The global experience of COVID-19, with its increased mortality among the elderly population, emphasized this discrepancy [20–22], an effect that may be linked to cell aging and telomerase attrition in relevant immune cells [23]. Despite the link between aging and a decrease in immune competence (the immune response is typically more sluggish and less accurate as we grow older) there are few clinical options to address the problem successfully despite suggested alterations in the microbiome [24], the use of small molecular compounds [25,26], endocrinal interventions [27], interventions targeting inflammaging [28], and so forth. However, the cells of the immune system are relatively easy to access for gene therapy compared with most other tissues. The ease of intravenous access to immune cells, current chimeric antigen receptor T-cell technology, and our ability to administer viral vectors, lipid nanoparticles (LNPs), or messenger RNA (mRNA) therapies directly to the vascular system and its circulating immune cells all speak to a potentially fertile area for clinical trials in addressing cell aging in the immune system.

3.7 Age-related skin dysfunction

Cell senescence clearly underlies aging in the skin. Skin aging is ubiquitous, and although affected by genetic and behavioral factors, it is so highly affected by radiation exposure (generally ultraviolet radiation) that many authors argue that skin aging and photoaging are synonymous [29], or at least that photoaging accelerates skin aging [30], and point out the role of cell aging as central to skin aging [31,32]. Given the cosmetic importance of skin aging and the consequently large commercial market, an endless number of interventions have been suggested or become commercially available. For the most part, these have included products offering transient or purely cosmetic benefits, generally small-molecular skin creams and moisturizers. A more credible approach has been the use of various retinoids with claims of having antiaging effects [33–35]. More recent approaches, and currently in vogue in the biotechnology world, is the use of senolytic compounds to remove senescent cells [36,37] or senomorphic compounds to address some consequences of senescent cell function [38]. Compared with using senolytics in other organs and tissues, this approach has the advantage of permitting topical application. A promising but daunting (see the chapter on skin aging) approach remains the use of telomerase (e.g., via gene therapy) to reset cell aging directly. The literature is tantalizing but occasionally contradictory and critical issues remain unresolved. As my coauthor, Saranya Wyles, summarizes earlier in the chapter on skin aging, "the results of intervention in skin cell senescence are startling and clinically intriguing. Initial work suggests that our knowledge of cell senescence in aging can be translated into clinical interventions for aging and age-related disease." It remains to be seen, but the evidence that aging is clinically modifiable makes it all the more important that potential interventions be brought to rigorous clinical trials.

3.8 Age-related lung disease

Like many other organs, age-related diseases of the lungs might be thought of as having two distinct, if occasionally overlapping etiologies: those due to aging of the parenchymal cells and those to accrued damage over the life span, independent of cell aging. In the latter category, for example, we might include silicosis, in which the primary driving pathology is the degree to which the lungs have acquired a burden of silicates. Although this burden rises with exposure time, the effects do not depend upon cell aging per se, but rather simply on the silicate load. A lifetime of smoking presents a parallel example in which the primary etiology is recurrent and prolonged damage, although in the case of smoking, such damage would accelerate cell loss and hence cell aging. In cases such as these, cell aging would be expected to accelerate or exacerbate the clinical disease, but the underlying pathology increases over time regardless of aging. Exposure is not the same as aging, although the two etiologies commonly coexist.

In most cases of age-related pulmonary diseases, however, we can make a clear case for cell aging of the involved pulmonary parenchyma to drive the pathology. This is likely true for chronic obstructive pulmonary disease, emphysema, interstitial lung disease, idiopathic pulmonary fibrosis, and so on, as well as parallel aging of the immune system, which would drive the increase in susceptibility to (and mortality of) age-related pulmonary infections. Cynicism and irony notwithstanding, it is for a good reason that pneumonia has long been called the old man's friend. As the authors of the chapter on pulmonary aging point out, good data show telomere loss and cell aging in multiple types of lung cells in patients with age-related pulmonary disease, and even more so among smokers. Moreover, both telomerase activators and telomerase gene therapy have shown promising results in mice. It remains tempting to suspect that the same therapeutic target would be effective in age-related human pulmonary disease.

3.9 Age-related diabetes

Diabetes is the disease that strikes at every organ. Whereas type 1 diabetes (once termed juvenile or insulin-dependent diabetes) is typically the provenance of young patients with autoimmune disease that destroys the islets of Langerhans, type 2 diabetes (once termed adult onset diabetes) has long been viewed as an age-related disease. The latter is generally typified by aging cells that lack an appropriate response to insulin (or glucose). This is true not only of a wide array of cell types, but may include the islet cells. In short, aged cells of whatever location become increasingly at risk of being unable to manage glucose appropriately. The details of those changes are remarkably complex, and our chapter on age-related diabetes covers these changes in detail.

As with the affected cell types, the complications of type 2 diabetes extend into an equally broad range of clinical complications, encompassing obesity, cardiovascular disease, and changes in brain, gastrointestinal tract, muscle, urinary, hepatic, pulmonary, ocular, renal, and other systems. Current therapy generally aims at either behavioral issues (diet, exercise, etc.) or the consequent complications of the disease. As pointed out in the chapter on age-related diabetes, a number of novel approaches targeting cell aging and senescent cells are on the horizon. These approaches include senolytic therapies to remove senescent cells, as well as the potential for telomerase gene therapy to reset cell aging altogether. The outcomes of such approaches in rigorous clinical trials offer reasonable promise of revolutionizing the care of age-related diabetes as well as the significant complications that currently follow.

3.10 Age-related eye disease

As the eye ages, there are multiple age-related problems. These include presbyopia, cataracts, glaucoma, and macular degeneration. Although cell aging almost certainly has a role in each of these (and other) age-related ocular diseases, in some cases, the precise mechanisms are remarkably hard to pin down. How, precisely, does cell aging trigger or promote presbyopia? Multiple pathways remain possible, and the same might equally be said for many of the other age-related eye problems we see in clinical practice. The uncertainty of the role of cell aging in such diseases results in a matching uncertainty when trying to predict the effects of interventions that target cell aging in the eye.

Available data, such as they are in regard to age-related eye disease, are suggestive, particularly work on macular degeneration and telomerase activators. As in some other organ systems, however, the data suggesting a potential benefit of ophthalmologic interventions targeting cell aging remain hypothetical, if enticing. Suggestive data, yes; unambiguous data, no. Although theoretical models and extant data would strongly support the value of taking telomerase therapy into trials for macular degeneration (and probably other age-related eye problems), the outcome remains unknown. A rigorous human trial is well-worth pursuing, but which age-related diseases will benefit from cell aging therapies, and to what degree, remain to be seen.

3.11 Age-related cancer

Like many things in life, and certainly many topics in medicine, the relationship between cell aging and cancer has been recurrently oversimplified. At no point is this more true than in discussions of the relationship between cancer and telomeres. Telomeres, and more specifically telomere shortening and consequent changes in the patterns of gene expression, clearly have a central, albeit not an exclusive role in cell aging. Unfortunately, the correlation between telomerase expression and carcinogenesis has been confused with a potentially causal role and viewed

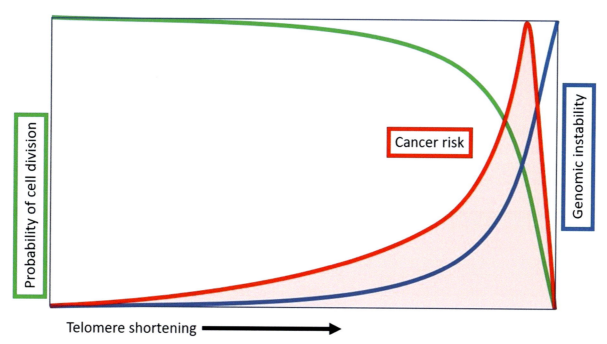

FIGURE 14.3 Telomere length and cancer risk. As telomeres shorten, genomic instability rises, as does cancer risk. Cancer risk falls as cell division is inhibited at minimum telomere lengths. Cancer risk maximizes with short, but not minimum telomeres. Both long and extremely short telomeres correlate with lower cancer risk. *Adapted from Ref. [39].*

as a linear relationship. The reality is unsurprisingly far more complex (Fig. 14.3) [39]. As is pointed out in the chapter on age-related disease and cancer, telomerase expression may be necessary, but is insufficient, for many cancer cells.

Moreover, in many cases, it may be the lack rather than the presence of telomerase that causes cancer. Although cell senescence may serve to inhibit cell division and hence inhibit carcinogenesis, cell senescence also increases chromosomal instability and hence promotes carcinogenesis. The balance between these opposing mechanisms is part of the inherent complexity of the relationship between telomerase, which extends telomeres, and cancer. Telomerase expression may permit continued cell division in chromosomally damaged cells, but it may also improve chromosomal repair. A more nuanced and critical view of the literature suggests that far from being a risk, telomerase-based therapies are more likely to lower the risk of cancer in aging patients. The role of such therapies in patients who already have known cancers is more problematic and concerning, but it remains an unknown.

Shorn of theoretical arguments, however, two practical clinical questions remain: (1) Will telomerase therapy increase cancer risk in aging patients without a known cancer history? (2) Will telomerase therapy increase cancer mortality in aging patients with a known cancer? Oddly, the data actually suggest that the answer to the first question is no, whereas the answer to the second question is anyone's guess. As the chapter on cancer points out, these are about to become practical questions that can be addressed only in human trials. Whereas the concern regarding cancer is reasonable and the argument for a beneficial (rather than a detrimental) effect of telomerase therapy on cancer risk is rational and well-supported, only careful, cautious, and rigorous human trials will provide the answer to these questions.

3.12 Mechanisms, caveats, and points of intervention

The insights and perspectives presented in this text express a variety of views and cite an astonishing landscape of clinical data. Making sense of that data (being able to provide a coherent and consistent overview of those data) is not simple, nor can it be accomplished neatly. Moreover, there is a changing set of assumptions, a changing understanding that is ongoing in the field of aging and age-related disease as this text goes to press.

For many people, and even for many clinicians and researchers, aging was viewed as a passive effect of entropy: organic systems fail over time because of wear and tear at all levels from the molecular through the cellular, and to entire organ systems. In fact, this seemed so self-evident that the assumption was not only unquestioned but often unrecognized. For many, this assumption appeared to be self-evident. The common trope was that everything ages, a view consistent with a superficial consideration of our fellow humans, our domestic animals, and most other organisms we encounter in daily life. Put simply, the provincial view was that all organisms age.

A broader view of biology suggested that organic systems are more complex than that simple statement, and that aging was not simply a passive outcome of entropic change, but the result of a failure of biological maintenance in the *face* of entropic forces. After all, biological systems have been present on our planet for several billion years, so clearly biological maintenance systems have been sufficient to ensure that organisms still exist and flourish despite entropic forces. As a more specific example, although our human somatic cells clearly demonstrate aging changes over a mere handful of decades (hence, age-related disease), we are the result of innumerable divisions of a fertilized ovum, and the provenance of that ovum goes back several billion years as an unbroken line of living material. In short, we inherit a several—billion year old unaged ovum, yet our resultant cells show aging changes almost from fertilization.

Aging is not a purely passive phenomenon, but the result of the gradual failure of maintenance mechanisms and these mechanisms cannot only be traced to changes in gene expression, they can be reset effectively in human cells, human tissues, and (within our current technical limits) in organisms. Whether we can do so effectively in human patients, whether for aging generally or with regard to specific age-related diseases, remains an open question, one that requires better techniques and rigorous testing via clinical trials.

Significant conceptual obstacles have impeded progress in this regard. The first is that alluded to earlier: if we regard aging as a purely passive entropic process, the prospect of intervention becomes problematic or perhaps not viable. A second common conceptual stumbling block has been the recurrent tendency to view biomarkers as essential clinical targets per se, rather than useful markers of clinical intervention. Aging or age-related diseases are not biomarkers; biomarkers are the (useful and often clinically convenient) outcomes of aging and age-related diseases, but not the fundamental processes themselves. Like clinical signs or symptoms, biomarkers are useful measures of a disease, but they are not the disease itself. Whether we are considering chest pain (a symptom) or the degree of coronary narrowing (a biomarker), these are correlated with, but are not identical to, age-related cardiovascular disease.

In this text, one underlying motif has been the use of cell aging to understand age-related diseases. The overall model can be clearly outlined (if overly simplified) as a progression:

- Loss of cells
- Division and replacement by remaining cells
- Telomere shortening
- Alterations in epigenetic pattern
- Decreased rates of molecular turnover
- Decreased percentage of effective molecules
- Decreased cellular function
- Decreased tissue function
- Clinically apparent age-related disease

The reality, as is always the case, is more complex and involves exceptions, additional inputs, interdependent cascades, and random occurrences, all of which make aging and age-related disease a field that will engage the best clinical minds for centuries to come. Yet the outline remains not only valid in a practical sense, but useful as well. It offers novel and feasible points of intervention that were far from obvious when we viewed age-related disease as the passive, entropic outcome of getting old (Fig. 14.4).

In addition to enormous and perhaps largely unforeseen complexities, other caveats remain if we are to make progress in curing and preventing age-related diseases. It might be tempting to view telomeres as the indispensable key to aging, whereas in reality they are merely an important part of the cast of characters as cell aging progresses. Telomeres are a key part, but they are far from being the only participant in aging and age-related disease. Moreover, most of the literature makes jejune mistakes that muddy our ability to understand. Too often, we see a focus on absolute rather than relative telomere lengths, not only between, but within species. A careful look at the data suggests that it is not the length, but the change in length of telomeres that is a critical variable [40—42], and even then, only because it serves as a reasonable indicator of (and point of intervention in) the epigenetic status of a cell. The pattern of gene expression is not a matter of absolute telomere length, but of an alteration in the relative telomere length (as well as myriad other inputs affecting gene expression). In addition, the literature is replete with irrelevant data comparing the telomere lengths of circulating leukocytes with various other variables, such as age-related diseases in nonleukocyte tissues. The result is an almost unusable literature on telomeres and aging in organisms [43,44].

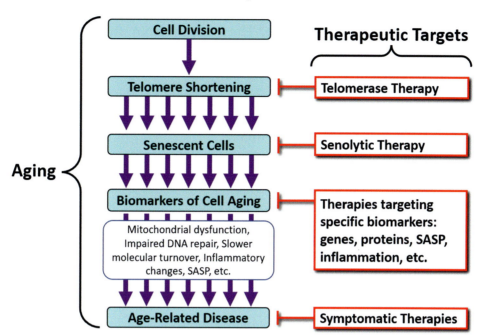

FIGURE 14.4 Targets in age-related disease. Most current and proposed therapies target downstream outcomes rather than fundamental upstream processes. *SASP*, senescence-associated secretory phenotype. *Adapted from Ref. [12].*

The critical question for the clinician, however, is how to identify the optimal point of intervention, both clinically and financially. We would like to find an intervention that not only cures and prevents age-related diseases, but is less costly than current interventions, which even at their best do not reverse the age-related pathology. In addition to identifying the optimal target of intervention (whether telomeres or other points in the cascade of aging pathology), we must also identify the optimal method of intervention. If the telomere were an optimal point of intervention, what would be the optimal method for creating such telomere changes? We might, for example (see the chapter on interventions and the subsequent discussion) use DNA (as hTERT), mRNA, or the protein itself (telomerase). To deliver the telomeric payload, we might use viral vectors (such as adeno-associated virus [AAV], lentivirus, or artificially constructed virions), LNPs, or other delivery methods. Targets are becoming clearer, but additional uncertainties continually arise in our efforts to move from a clearer understanding to effective clinical use.

3.13 What we do not know

Statistics is a measure of our ignorance. More accurately, statistics is a conceptual approach that permits us to quantify the degree to which we are uncertain. In short, statistics allow us to have an estimate of how far our ignorance extends. A dispassionate understanding of our degree of ignorance is often more enlightening (and frequently more useful) than a passionate understanding of our degree of knowledge. We have an enormous body of knowledge, often statistical, about aging, and specifically about cell aging and its role in causing age-related disease.

But although our knowledge is vast, our understanding is not.

We are only beginning to understand precisely how cell aging occurs and how it links upstream risk factors with downstream clinical outcomes. We know much about the upstream risks, and we know the downstream clinical diseases only too well, yet we lack an understanding of the unified systems process that is aging itself. This lacuna in our understanding has until now kept us from intervening successfully in age-related disease.

Nor is our lack of understanding the only gap if we are to cure and prevent age-related disease. We tend to focus narrowly on the downstream outcomes of cell aging, the effects of senescing cells both functionally and in terms of their impact on surrounding tissue. The result is an emphasis on removing (senolytics) or moderating (senomorphics) the effects of senescent cells, rather than resetting the affected cells. To what extent is the resetting of cell aging (a more fundamental approach) achievable in clinical trials? Can we not only reverse aging in cells and tissues in vitro and ex vivo (and even there, our reversal has limits), but do so effectively in vivo in aging human patients? Additional practical gaps remain in our knowledge that must be addressed if we are to proceed to effective

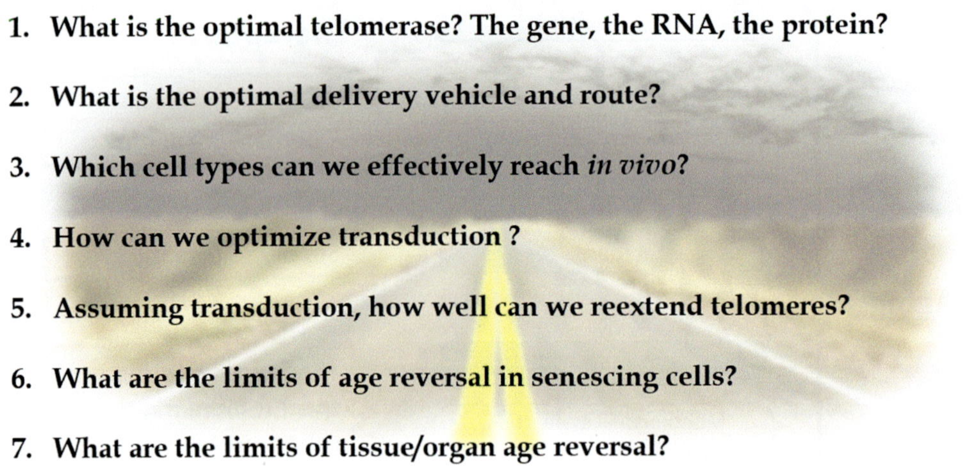

1. What is the optimal telomerase? The gene, the RNA, the protein?

2. What is the optimal delivery vehicle and route?

3. Which cell types can we effectively reach *in vivo*?

4. How can we optimize transduction?

5. Assuming transduction, how well can we reextend telomeres?

6. What are the limits of age reversal in senescing cells?

7. What are the limits of tissue/organ age reversal?

FIGURE 14.5 Uncertainties in reversing cell aging. A number of uncertainties are apparent, many of which will have a strong impact on the clinical efficacy of attempts to reset cell aging as a therapy for age-related diseases in human clinical trials.

interventions that target cell aging itself. If we simply list the known unknowns, we might find them to include a spectrum of uncertainties (Fig. 14.5).

With regard to the use of telomerase, do we use the protein, the RNA, or the gene itself? Is the optimal sequence the natural one, or is there a more effective gene sequence that would enhance therapeutic efficacy? The optimal delivery vehicle is currently in flux and likely to remain so in the near future. Although already in the midst of change, a current default is often an AAV vector, yet even here the question of optimal serotype (and promoter) creates both frustrations and opportunities. But would an alternative viral vector or perhaps an artificial construct be preferable? Or a lipid nanoparticle? The options are changing quickly, and the default choice today is unlikely to be the preferred choice tomorrow. The route of administration is a problem, with pros and cons for each choice of approach, although we can foresee a time when intravenous, intranasal, or oral will perhaps become the accepted (and optimally effective) route of delivery. Current approaches to gene therapy are characterized by a narrow match to specific cell types, while proving less effective in other cell types. This is clearly an opportunity to improve delivery. Yet, efficient delivery to preferred cells is no guarantee of efficient transduction. Again, there is clearly room for improvement (although overproduction of a gene product can itself create significant risks in some current clinical trials).

To what extent can we increase telomere lengths, and more crucially, to what extent will such increases translate into improved cell function? We expect limits, and they are likely bounded by the degree of cell aging. That is, cells that have reached replicative senescence are not expected to respond as favorably as would younger cells that, although characterized by relatively shortened telomeres, are not yet at the terminal phase of cell aging (i.e., replicative senescence). Finally, even if we can effectively reset a significant percentage of aging cells within a tissue, to what extent would this translate into improved tissue and organ function? What are the practical limits in living organisms? In all cases (cells, tissues, organs, and systems), we should reasonably anticipate diminishing returns with age. The older the patient (biologically), the lower our expectations should be as we intervene in age-related disease.

We know a great deal, but the unknowns constitute a far larger landscape.

3.14 A historical perspective

And yet, despite the unknowns and the limitations, the potential for effective interventions in age-related disease is not only unprecedented but tantalizing. Perhaps the only parallel in the history of medicine was the microbial revolution, which allowed us to understand, prevent, and cure many diseases that were previously simply accepted. Here, too, unknowns and limitations remain: we are still coming to terms with rapidly changing viruses, and we are still unable to prevent or cure many microbial diseases completely. Our ability to intervene and improve human life by understanding microbial disease was a stunning change in the history of human disease.

Perhaps in the case of age-related disease, we are on the verge of a similarly stunning change in the history of human disease.

The importance of microbial theory was the role it had not as a technical advance, but as a conceptual watershed. Handwashing, sterile surgeries, immunizations, and other changes were predicated on the understanding of how microbial diseases occur. Much the same will be true for our understanding of how aging diseases occur. Moreover, an understanding of microbial disease lowered medical costs. The cost of handwashing during a delivery is negligible, whereas avoidable maternal and neonatal deaths are both expensive and tragic. Incremental technical medical advances often raise health care costs with marginal improvements in human health, but our understanding of microbial disease was a conceptual revolution that saved lives and lowered the costs of medical care.

We now stand at a point where increasingly marginal improvements in treating age-related disease are linked to increasingly large financial costs. An ability to intervene effectively in the fundamental causes of age-related disease offers us the opportunity to drastically improve human health and also to lower current health care costs globally, nationally, and personally. An effective therapy that could cure or prevent Alzheimer's disease, for example, would be desirable clinically and is likely to come at a cost that is far less than that of nursing care for such patients as they progress inexorably downhill. The same can be said for the likely savings in the cost of coronary artery bypass grafts, stents, statins, and the panoply of current interventions for cardiovascular disease, to say nothing of the age-related diseases in other organ systems.

Li Bai's opening poem is apt. For too long, we have been content to hunt merely the rabbits of age-related disease, to treat symptoms rather than causes. However, the rabbits are the inconsequential and costly targets. They are easy to catch, but they offer us little, and the costs of hunting rabbits has become increasingly insupportable. It is time for us to understand aging and the diseases of aging, and hunt for more than rabbits.

It is time we went hunting for the dragons of aging.

References

[1] Ojaimi J, Byrne E. Mitochondrial function and Alzheimer's disease. Biol Signals Recept 2001;10:254—62. https://doi.org/10.1159/000046890.

[2] Trounce I, Byrne E, Marzuki S. Decline in skeletal muscle mitochondrial respiratory chain function: possible factor in ageing. Lancet 1989;333: 637—9. https://doi.org/10.1016/S0140-6736(89)92143-0.

[3] Ojaimi J, Masters CL, McLean C, Opeskin K, McKelvie P, Byrne E. Irregular distribution of cytochrome c oxidase protein subunits in aging and Alzheimer's disease. Ann Neurol 2001;46:656—60. https://doi.org/10.1002/1531-8249(199910)46:4<656::AID-ANA16>3.0.CO;2-Q.

[4] Novoselov VM. Is aging a disease? Adv Gerontol 2018;8:119—22. https://doi.org/10.1134/S2079057018020121.

[5] Lakatta EG. So! What's aging? Is cardiovascular aging a disease? J Mol Cellular Cardiol 2015;83:1—3. https://doi.org/10.1016/j.yjmcc.2015.04.005.

[6] Caplan A. Is aging a disease? In: Caplan A, editor. If I were a rich man could I buy a pancreas? And other essays on the ethics of healthcare. Bloomington: Indiana University Press; 1992. p. 195—209.

[7] Fossel M. A unified model of dementias and age-related neurodegeneration. Alzheimer's & dementia. J Alzheimers Assoc 2020;16:365—83. https://doi.org/10.1002/alz.12012.

[8] Weber T, Mayer CC. "Man is as old as his arteries" taken literally: in search of the best metric. Hypertension 2020;76:1425—7. https://doi.org/10.1161/HYPERTENSIONAHA.120.16128.

[9] Fossel M, Bean J, Khera N, Kolonin MG. A unified model of age-related cardiovascular disease. Biology 2022;11(12):1768. https://doi.org/10.3390/biology11121768.

[10] Gkiatas I, Sculco TP, Sculco PK. The history of total knee arthroplasty. In: Hansen E, Kühn KD, editors. Essentials of cemented knee arthroplasty. Berlin, Heidelberg: Springer; 2022. p. 3—14. https://doi.org/10.1007/978-3-662-63113-3_1.

[11] Ranawat AS, Ranawat CS. The history of total knee arthroplasty. In: Bonnin M, Amendola A, Bellemans J, MacDonald S, Ménétrey J, Ranawat AS, et al., editors. The knee joint. Paris: Springer; 2012. p. 699—707. https://doi.org/10.1007/978-2-287-99353-4_63.

[12] Fossel M. Cell senescence, telomerase, and senolytic therapy. OBM Geriatr 2019;3(1):1. https://doi.org/10.21926/obm.geriatr.1901034.

[13] Kitai Y, Nangaku M, Yanagita M. Aging-related kidney diseases. In: Nephrology and Public Health Worldwide, vol. 199; 2021. p. 266—73. https://doi.org/10.1159/000517708.

[14] Ash SR. The wearable artificial kidney created by Kolff and Jacobsen. Artif Organs 2022;46:159—61. https://doi.org/10.1111/aor.14108.

[15] Ng CH, Ong ZH, Sran HK, Wee TB. Comparison of cardiovascular mortality in hemodialysis versus peritoneal dialysis. Int Urol Nephrol July 2021;53:1363—71. https://doi.org/10.1007/s11255-020-02683-9.

[16] Purnell TS, Auguste P, Crews DC, Lamprea-Montealegre J, Olufade T, Greer R, et al. Comparison of life participation activities among adults treated by hemodialysis, peritoneal dialysis, and kidney transplantation: a systematic review. Am J Kidney Dis 2013;62:953—73. https://doi.org/10.1053/j.ajkd.2013.03.022.

[17] Fenton SS, Schaubel DE, Desmeules M, Morrison HI, Mao Y, Copleston P, et al. Hemodialysis versus peritoneal dialysis: a comparison of adjusted mortality rates. Am J Kidney Dis 1997;30:334—42. https://doi.org/10.1016/S0272-6386(97)90276-6.

[18] Axelrod DA, Schnitzler MA, Xiao H, Irish W, Tuttle-Newhall E, Chang SH, et al. An economic assessment of contemporary kidney transplant practice. Am J Transplant 2018;18:1168—76. https://doi.org/10.1111/ajt.14702.

[19] Pawalec G, editor. Immunosenescence. NY: Springer; 2008.

[20] Péterfi A, Mészáros Á, Szarvas Z, Pénzes M, Fekete M, Fehér Á, et al. Comorbidities and increased mortality of COVID-19 among the elderly: a systematic review. Physiol Int May 16, 2022. https://doi.org/10.1556/2060.2022.00206.

[21] Parra PN, Atanasov V, Whittle J, Meurer L, Luo QE, Zhang R, et al. The effect of the covid-19 pandemic on the elderly: population fatality rates, covid mortality percentage, and life expectancy loss. Elder Law J 2022;30:33.

[22] Harris E. Most COVID-19 deaths worldwide were among older people. JAMA 2023;329:704. https://doi.org/10.1001/jama.2023.1554.

[23] Sanchez-Vazquez R, Guío-Carrión A, Zapatero-Gaviria A, Martínez P, Blasco MA. Shorter telomere lengths in patients with severe COVID-19 disease. Aging 2021;13:1. https://doi.org/10.18632/aging.202463.

[24] Bosco N, Noti M. The aging gut microbiome and its impact on host immunity. Gene Immun 2021;22:289–303. https://doi.org/10.1038/s41435-021-00126-8.

[25] Lian J, Yue Y, Yu W, Zhang Y. Immunosenescence: a key player in cancer development. J Hematol Oncol 2020;13:1–8. https://doi.org/10.1186/s13045-020-00986-z.

[26] Aiello A, Farzaneh F, Candore G, Caruso C, Davinelli S, Gambino CM, et al. Immunosenescence and its hallmarks: how to oppose aging strategically? A review of potential options for therapeutic intervention. Front Immunol 2019. 25;10:2247. https://doi.org/10.3389/fimmu.2019.02247.

[27] Bauer ME, Muller GC, Correa BL, Vianna P, Turner JE, Bosch JA. Psychoneuroendocrine interventions aimed at attenuating immunosenescence: a review. Biogerontology 2013;14:9–20. https://doi.org/10.1007/s10522-012-9412-5.

[28] Teissier T, Boulanger E, Cox LS. Interconnections between inflammageing and immunosenescence during ageing. Cells 2022;11:359. https://doi.org/10.3390/cells11030359.

[29] Krutmann J, Gilchrest BA. Photoaging of skin. Skin Aging 2006:33–43.

[30] Yaar M, Gilchrest BA. Aging versus photoaging: postulated mechanisms and effectors. J Invest Dermatol Symp Proc 1998;3:47–51. https://doi.org/10.1038/jidsymp.1998.12. Elsevier.

[31] Kosmadaki MG, Gilchrest BA. The role of telomeres in skin aging/photoaging. Micron 2004;35:155–9. https://doi.org/10.1016/j.micron.2003.11.002.

[32] Jacczak B, Rubiś B, Totoń E. Potential of naturally derived compounds in telomerase and telomere modulation in skin senescence and aging. Int J Mol Sci 2021;22:6381. https://doi.org/10.3390/ijms22126381.

[33] Helfrich YR, Sachs DL, Voorhees JJ. Overview of skin aging and photoaging. Dermatol Nurs 2008;20(3).

[34] Mukherjee S, Date A, Patravale V, Korting HC, Roeder A, Weindl G. Retinoids in the treatment of skin aging: an overview of clinical efficacy and safety. Clin Interv Aging 2006;1:327–48. https://doi.org/10.2147/ciia.2006.1.4.327.

[35] Spierings NM. Evidence for the efficacy of over-the-counter vitamin A cosmetic products in the improvement of facial skin aging: a systematic review. J Clin Aesthet Dermatol 2021;14:33. https://www.ncbi.nlm.nih.gov/pmc/articles/PMC8675340/.

[36] Kim H, Jang J, Song MJ, Kim G, Park CH, Lee DH, et al. Attenuation of intrinsic ageing of the skin via elimination of senescent dermal fibroblasts with senolytic drugs. J Eur Acad Dermatol Venereol 2022;36:1125–35. https://doi.org/10.1111/jdv.18051.

[37] Pils V, Ring N, Valdivieso K, Lämmermann I, Gruber F, Schosserer M, et al. Promises and challenges in skin regeneration, pathology and ageing. Mech Ageing Dev 2021;200:111588. https://doi.org/10.1016/j.mad.2021.111588.

[38] Dańczak-Pazdrowska A, Gornowicz-Porowska J, Polańska A, Krajka-Kuźniak V, Stawny M, Gostyńska A, et al. Cellular senescence in skin-related research: targeted signaling pathways and naturally occurring therapeutic agents. Aging Cell 2023;11:e13845. https://doi.org/10.1111/acel.13845.

[39] Fossel M, Whittemore K. Telomerase and cancer: a complex relationship. OBM Geriatr 2021;5:18. https://doi.org/10.21926/obm.geriatr.2101156.

[40] Vera E, DeJesus BB, Foronda M, Flores JM, Blasco M. The rate of increase of short telomeres predicts longevity in mammals. Cell Rep 2012. https://doi.org/10.1016/j.celrep.2012.08.023.

[41] Whittemore K, Martinez-Nevado E, Blasco MA. Slower rates of accumulation of DNA damage in leukocytes correlate with longer lifespans across several species of birds and mammals. Aging 2019. https://doi.org/10.18632/aging.102430.

[42] Whittemore K, Fossel M. Editorial: telomere length and species lifespan. Front Genet 2023;14:1199667. https://doi.org/10.3389/fgene.2023.1199667.

[43] Fossel M. Use of telomere length as a biomarker for aging and age-related disease. Curr Transl Geriatr Exp Gerontol Rep 2012;1:121–7. https://link.springer.com/article/10.1007/s13670-012-0013-6.

[44] Semeraro MD, Almer G, Renner W, Gruber HJ, Herrmann M. Telomere length in leucocytes and solid tissues of young and aged rats. Aging 2022, Feb 27. https://doi.org/10.18632/aging.203922.

Index

Note: Page numbers followed by "f" indicate figures and "t" indicate tables.

A

Acarbose, 184
Acidosis, 108
Action to Control Cardiovascular Risk in Diabetes (ACCORD) research, 108
Activation (phase 1), 54–55
Acute kidney damage (AKI), 103
Acute kidney injury (AKI), 106, 108–109
Adeno-associated virus (AAV), 136, 225, 226f
Adipocytes, 62
Adipose-derived stem cells (ADSCs), 150–151
Adipose tissue, 178
Adrenoleukodystrophy (ALD), 136
Advanced glycation end-products (AGE), 81–82, 105, 180
Age-related bone disease, 238
Age-related cancer, 240–241
Age-related cardiovascular disease, 237
  arterial stiffness and early vascular aging, 36–37
  arterial stiffness and vascular aging, 36
  determinants of arterial stiffness, 35–36
  early vascular aging, 36
  healthy and supernormal vascular aging, 37
  mechanisms and clinical aspects, 40
  polypill role for cardiovascular prevention, 39–40
  treatment of cardiovascular risk and clinical manifestations, 38–39
Age-related central nervous system disease, 237
Age-related changes
  in adaptive immune system, 123
  in innate immune system, 122
Age-related decline in GFR, 96
Age-related diabetes, 240
Age-related diseases, 10f, 15f, 195
  eye as window to other, 200–201
  targets in, 221f, 243f
  type 2 diabetes, 175–176
Age-related eye disease, 240
Age Related Eye Disease Study (AREDS), 198
Age-related immune dysfunction, 239
Age-related joint disease, 238
Age-related kidney disease, 238–239
Age-related lung disease, 239–240
  role of cell aging in, 167
Age-related neurologic disease, 21

Age-related skin dysfunction, 239
Age-reversal strategies, 227–228
Aging
  diseases, 207–208
  of immune system, 119–124
  as key process, 25–27
  lung, 166–167
  and osteoarthritis, 75–76
  processes, 222–223
Aging and pulmonary disease, 165–166
  aging lung, 166–167
  role of cell aging in age-related lung disease, 167
  specific lung diseases correlated with advanced age, 167–171
Aging kidney, 92–97
  functional changes of, 95–96
  structural changes of, 92–95
Albumin to creatinine ratio (ACR), 102
Alternative lengthening of telomeres (ALT) mechanism, 205
Alveolar type II cells (ATII), 171
Alzheimer's disease, 19–20, 200, 221–222, 245
AMP-activated protein kinase (AMPK), 79, 156
"Amyloid burden", 200
Ang II type 1 (AT1) receptor, 99
Angiotensin-converting enzyme inhibitors (ACEI), 97–98, 107
Angiotensin II converting enzyme (ACE) inhibitors, 106
Angiotensin II type 1a (AT1a) receptor, 98
Angiotensin receptor blockers (ARB), 106–107
Antiresorptive and osteoclast-targeting therapeutics, 57–59
Antiresorptive drugs, 58
Aplastic anemia, 208
Arterial stiffness
  determinants of, 35–36
  and early vascular aging, 36–37
  and vascular aging, 36
Astragalus membranaceus, 136
Ataxia-telangiectasia mutated kinase (ATM), 77, 96
Atherosclerotic process, 1–2
Augmentation index (Aix), 36–37
Autoimmune thyroid disease (AITD), 125
Autoimmunity, 125
Autologous chondrocyte implantation (ACI), 75

Autophagy, 78–79
Autosomal dominant polycystic kidney disease (ADPKD), 95

B

Base excision repair (BER) fixes oxidative, 63
Basic multicellular units (BMUs), 54
Biochemical changes, 23–25
Biochemical markers of cellular senescence, 154
Bioenergetic Health Index (BHI), 135
Biology of aging bone, 62–65
Biomarkers, 6–7
Biomarkers/downstream factors, 134–135
Biomolecular mechanisms of immunosenescence, 126–130
Blood pressure monitoring, 107
Blood transfer, rejuvenation effects of, 136
Body mass index (BMI), 105, 178
Bone anabolics, 59–60
Bone anatomy and homeostasis, 53–54, 54f
Bone-forming osteoblasts (OBs), 54f
Bone marrow adipose tissue (BMAT), 62
Bone marrow monocytes, 53–54
Bone remodeling, 55f
Brain energy rescue, 197
Bronchioles, 165–166
Bubonic plague, 4

C

Calcineurin inhibitors, 106
Calorie restriction (CR), 132, 183
cAMP response element binding protein (CREB), 80
Cancer, 182
  aging and telomerase, 207–212
  and other neoplasms, 125
  treatments, 206–207
Cardiovascular diseases (CVD), 35, 101–102, 105–106
  role of cell aging in, 40–43
Cardiovascular prevention, polypill role for, 39–40
Cardiovascular risk
  assessment, 108
  and clinical manifestations, 38–39
Cardiovascular system, 35, 40, 237
Cartilage structure and function, 73–75
Caveat medicus, 13
Caveat physicus, 13

Caveats, 13−14
Cell aging
  in central nervous system, 27−30
  process, 15
  targeting dominant process in, 223−228
Cell therapeutics, 81
Cell types, 22−23
Cellular aging, 128−129
Cellular aging in bone
  biology of aging bone, 62−65
  bone anatomy and homeostasis, 53−54
  cellular components of aging bone, 60−62
  osteoporosis, 56−57
  skeletal maintenance and remodeling, 54−56
  therapeutic options, 65
  therapeutic options to treat osteoporosis, 57−60
Cellular components of aging bone, 60−62
Cellular damage, 82
Cellular replicometer, 3
Cellular senescence, 62−63, 96−97, 151−152, 176−177, 182−184
  in diabetes, 184−186
  and diabetic complications, 181−182
  in pathogenesis of type 2 diabetes, 176−179
  in renal aging, 96−97
  with senotherapies, 158−159
  in type 2 diabetes pathogenesis, 177f
Centella asiatica, 136
Centers for Disease Control (CDC) analysed, 75
Central nervous system (CNS) disease, 14, 19
  aging as key process, 25−27
  biochemical changes, 23−25
  cell types, 22−23
  dementias, 19−21
  neuroanatomy, 22
  therapeutic status, 21−22
  unified systems model, 27−30
Cerebral microbleeds (CMBs), 104
Cerebrovascular disease, 104
Checkpoint kinase-2 (CHK2), 77
Chimeric antigen receptor (CAR), 157
Chondrocyte senescence
  aging and osteoarthritis, 75−76
  autophagy, 78−79
  cartilage structure and function, 73−75
  chondrosenescence, 82−83
  circadian clocks in senescence, 79
  extracellular matrix in initiating and stabilising cellular senescence, 81−82
  mTOR, nutrient sensing, and senescence in cartilage, 80
  replicative senescence in cartilage, 76−77
  role of sirtuins in chondrosenescence, 79
  senescence-associated secretory profile in chondrocytes, 77−78
  stem cell exhaustion, 80−81
  stress-induced senescence in cartilage, 77
Chondrocytes increases type II collagen (COL2A1) gene, 79
Chondrosenescence, 82−83
  role of sirtuins in, 79
Chronic inflammation, 100, 180−181
Chronic kidney disease (CKD), 91−92, 97−101, 238−239
  care and management for the elderly, 107−109
  in elderly, prevalence of, 101−104
  patients, vulnerability of aged, 104
  progression, stabilization of, 109
  risk factors for, 104−107
  and staging, 102
  vascular consequences of, 104
Chronic myeloid leukemia (CML), 155−156
Chronic obstructive pulmonary disease (COPD), 166−167
Chronological age, 148−149
Chronological *versus* biological skin aging, 147−149
CKD with uncertain etiology (CKDu), 102−103
Clonal hematopoiesis, 125
Cognition, 182
Cognitive functioning, 104
Common alterations in cellular, 97−101
Congestive heart failure (CHF), 106
Coronary and peripheral vascular disease, 104
Coronary artery disease (CAD), 1−2, 39, 108
Correction of hemoglobin and outcomes in renal insufficiency (CHOIR), 107−108
Creutzfeldt Jakob disease (CJD), 24
Cryptochrome (CRY) genes, 79
Curcuma longa L, 60
Cyclic adenosine monophosphate (cAMP), 96−97
Cyclin-dependent kinase (CDK) 4, 62
Cyclin-dependent kinase (CDK) inhibitors, 96
Cytokine and protein production, 123−124
Cytomegalovirus (CMV), 124
Cytosine-phosphate-guanine (CpG), 169−170

D
Dementias, 19−21
Destabilised medial meniscus (DMM) model, 80
Diabetes, 105
  cellular senescence and diabetic complications, 181−182
  cellular senescence in pathogenesis of type 2 diabetes, 176−179
  clinical trials, 186
  diabetic microenvironment drives senescent cell accumulation, 180−181
  limited knowledge of long-term effects, 186
  senescence and biomarkers of senescence (gerodiagnostics), 186
  targeting cellular senescence in diabetes, 184−186
  therapeutics, 182−184
  trials in older adults, 187
  type 2 diabetes is age-related disease, 175−176
Diabetes mellitus (DM), 101, 175
Diabetes therapeutics, 182−184
Diabetic microenvironment, 180−181
Dickkopf-related protein-1 (DKK1), 80−81
Didocosahexaenoic acid (DHA) consumption, 198
Diet and exercise, 183
DNA
  and its utility for age-reversal strategies, 226−227
  methylation, 148−149
  mutations or organelle damage, 155
DNA damage, 77, 180, 183
  and genomic instability, 63, 149−150
  and repair mechanisms, 129−130
DNA damage response (DDR), 129−130
  proteins, 77
  stabilizes, 62−63
DNA methylation (DNAm), 130, 135
"Double spectacle", 195
Downstream outcomes, 11−13
DPP-4 inhibitors, 183−184
Dunkin-Hartley guinea pig develops, 84

E
Early vascular aging (EVA), 35−36
  arterial stiffness and, 36−37
Emphysema and COPD, 167−170
Endothelial cells, 62, 178
End-stage renal disease (ESRD), 92
Epigenetic aging clocks, 135
Epigenetic regulations, 64
Erythrocytes, 121
Erythropoiesis is erythropoietin (EPO), 121
Erythropoiesis-stimulating agents (ESAs), 107−108
Estimated glomerular filtration rate (eGFR), 91−92
Etiology, 102−103
Extensive extracellular matrix (ECM) composed, 73
Extracellular matrix (ECM), 36
Extracellular membrane (ECM), 168
Eyes, 195
  glaucoma, 196−197
  lens, 197−198
  presbyopia, 195−196
  retina, 198−200
  upstream opportunities for life and health extension, 201−202
  as window to other age-related diseases, 200−201

F

Fibroblast growth factor-2 (FGF2), 76—77
First- and second-generation senolytics, 155—157
Formation (phase 4), 56
Free fatty acid (FFA) release, 175—176
Free radicals, indirect and direct measures of, 134
Frontotemporal dementia (FTD), 20
Future potential interventions and emerging trials, 65

G

Ganglion cell damage, 196—197
Gastrointestinal (GI) tract, 57
General anti-aging strategies, 83—84
Genetic components, 107
Genetic factors, 130—131
Genetic therapy, 136
Genome-wide association studies (GWAS), 5, 76—77
Geriatric dermatology, 147
Germ cells and mitochondria, 5
Geron Corporation, 3
Geroscience hypothesis, 184
Glaucoma, 196—197
Globally sclerotic glomeruli (GSG), 93
Glomerular filtration rate (GFR), 92t, 95
Glucagon-like peptide-1R (GLP-1R) agonists, 183—184
Glycogen synthase kinase 3β (GSK3β), 96—97
Glycosaminoglycan carbohydrate chains, 73
Gobally sclerotic glomeruli (GSG), 94f

H

Hashimoto thyroiditis (HT), 125
Health and disease, immunosenescence role in, 125—126
Healthy and supernormal vascular aging, 37
Helix-loop-helix (HLH) proteins, 60
Hematopoietic stem cell (HSC) system, 119—121, 120f
Hepatocellular carcinoma (HCC), 134—135, 179
Histone deacetylase 3 (Hdac3), 64
Histone deacetylase (HDAC) levels, 170
Human cord blood-derived plasma (hUCP), 136
Human erythrocytes, 121
Human immunodeficiency virus (HIV), 131
Human telomerase reverse transcriptase (hTERT)
　protein, 128
　treatment, 43—44
Human vascular endothelial tissue, 3—4
Huntingtin (HTT) gene, 24
Huntingtin protein, 24
Hutchinson—Gilford progeria syndrome (HGPS), 63, 131, 227
Hyaluronic acid binding, 81—82
Hyperglycemia, 180
Hyperinsulinemia, 180
Hyperinsulinemia-induced senescence, 185—186
Hyperlipidemia, 38
Hypertension, 38, 105—106
"Hypoxia-responsive" reporter, 101

I

Idiopathic interstitial pneumonias (IIPs), 170
Idiopathic pulmonary fibrosis (IPF), 166—167, 170, 208—209
Immune risk phenotype (IRP), 124
Immune strategies, 185
Immune system, 119—120
　aging of, 119—126
　biomarkers/downstream factors related to aging and immunosenescence, 134—135
　biomolecular mechanisms of immunosenescence, 126—130
　conditions related to, 124—126
　genetic therapy and pharmacological interventions to stimulate telomerase activity, 136
　nicotinamide adenine dinucleotide (NAD), 137
　rejuvenation effects of blood transfer, 136
　stem cells from umbilical cord blood, 136
　upstream risk factors, 130—134
　use of senolytics, 137
Immunoglobulin G (IgG), 134—135
"Immunological Theory of Aging", 119
"Immunosenescence", 119
Impaired immune function, 180—181
Impaired wound healing, 182
Induced pluripotent stem cells (iPSCs), 151
Infectious diseases, 124
Inflamm-aging, 119, 123—124
Inflammatory bowel disease, 36
Innate and adaptive immune system, 122
Insulin, 184
　and glucose control with oral agents, 108
Inter-cross-link repair (ICL) resolves, 63
Interleukin-1β (IL1β), 74
Interstitial lung diseases/interstitial pulmonary fibrosis, 170—171
Interventions
　current approaches, 221—222
　perspective, 219—221
　targeting aging processes, 222—223
　targeting dominant process in cell aging, 223—228
Intracranial hemorrhage (ICH), 104
Intraocular pressure (IOP), 196—197
Ischemia-reperfusion injury (IRI), 99

K

Kelch-like ECH-associated protein-1 (KEAP1), 78—79
Kellegren-Lawrence (KL) system, 75
Kidney, 91
　aging, 92—97
　CKD
　　care and management for elderly, 107—109
　　cellular and molecular mechanisms of renal aging and, 97—101
　　in elderly, 101—107
Kidney cysts and tumors, 95
Kidney Disease Improving Global Outcomes (KDIGO), 102
Kidney disorders, 91
Kidney function, 97—99
Kidney volume, 94—95
Klotho and Wnt/β-catenin signaling, 99—100

L

Lens, 197—198
Lifestyle and environmental factors, 131—134
Lipid nanoparticles (LNPs), 226—227, 244
Lipofuscin, 153—154
Liver, 179
Long-term effects, 186
Loss of proteostasis, 64—65
Lung diseases, 167—171

M

Macro-anatomical changes, 94—95
Macrovascular complications, 181
Magnetic resonance imaging (MRI), 94—95
Mammalian target of rapamycin (mTOR), 105
Matrix metalloproteinases (MMPs), 56, 151, 154
Mechanistic target of rapamycin (mTOR), 80, 202
Megakaryocyte, 121
"Memory inflation", 131
Mental distress, 132
Mesenchymal stem cells (MSCs), 53—54, 62, 76—77, 80—81
Messenger RNA (mRNA), 222
Metabolically healthy obesity (HMO), 37
Metformin, 183, 201—202
Micro-anatomical changes, 93—94
Microvascular complications, 181
Mismatch repair (MMR), 63
Mitochondria, 128—129
Mitochondrial DNA (mtDNA), 169
Mitochondrial dysfunction, 64, 77, 171
Mitochondrial health index (MHI), 135
Modification of diet in renal disease (MDRD), 102
Monoclonal antibodies (MABs), 21f, 28f
"Monofocal" lenses, 195—196
mRNA therapeutics, 227—228
mtDNA copy number (mtDNAcn), 129
mTOR complex 1 (mTORC1), 80
Muscle, 179
Myelodysplastic syndromes (MDS), 125
Myriad biochemical abnormalities, 25f

N

National Institutes of Health (NIH), 157
Natural killer (NK) cells, 120, 122, 176
Nephron hypertrophy, 93–94
Nephrosclerosis, 93
Nephrotoxins, 106
Neuroanatomy, 22
Neurodegenerative diseases, 19, 30, 237
Neutrophil adhesive capacity, 122
N-glycosylated proteins, 134–135
Nicotinamide adenine dinucleotide (NAD), 79, 123, 137
Nicotinamide phosphoribosyl transferase (NAMPT), 79, 98
Nonalcoholic fatty liver disease (NAFLD), 179
Nonalcoholic steatohepatitis (NASH), 179
Nonautoimmune (type 2) diabetics, 201
Non-insulin-sensitive cells, 180
Non-sclerotic glomeruli (NSGs), 92, 94f
Nonsteroidal antiinflammatory drugs (NSAIDs), 106–107
Nuclear factor erythroid-2 related factor-2 (NRF2), 78–79
Nuclear factor erythroid 3-related factor 2 (Nrf2), 169
Nuclear factor kappa-B (RANK), 57
Nuclear factor-κB (NF-κB), 169
Nuclear factor κ-light-chain-enhancer of activated B cells (NF-κB), 99
Nucleotide-binding domain leucine-rich repeat (NLR), 100
Nucleotide excision repair (NER) controls, 63
Nutrient sensing, 80
Nutrition, 132–133

O

Obesity, 105
Obesity-related glomerulopathy (ORG), 105
Older adults, considerations for trials in, 187
Organs, 236–245
Osteoarthritic knee, 1–2
Osteoblasts, 60–61
Osteoclasts, 61
Osteocytes, 61
Osteoporosis, 56–57, 182
  therapeutic options to treat, 57–60
Osteoprotegerin (OPG), 54–55
Oxidative stress, 77, 100–101, 180
Oxygen-dependent cells, 121

P

Pancreas, 178–179
Parasympathetic systems, 165–166
Parathyroid hormone (PTH), 59–60
Parkinson's disease, 10–11, 20
Peripheral artery disease (PAD), 104
Peripheral blood mononuclear cells (PBMCs) telomerase, 127–128
Peritoneal dialysis (PD), 104
Peroxisome proliferator-activated receptor (PPAR)-γ, 100
Peto's paradox, 210–211
Pharmacological interventions, 136
Phosphoinositol-3-kinase (PI3K), 80
Physical activity and exercise on immune functions, 133
Physiological tissue function, 147–148
Phytohemagglutinin (PHA), 127–128
Platelets, 121
Polio vaccine, 220
Polymerase I and transcript release factor (PTRF)-induced activation, 182
Polypill/polycaps, 40
Postmenopausal Chinese women, 60
Presbyopia, 195–196
Primary, open-angle glaucoma (POAG), 196–197
Programmed death receptor 1 (PD-1), 127
Pro-inflammatory and matrix-degrading factors, 159
Prospective therapeutics, 60
Protein-expressing genes, 11–12
Protein glycosylation and its changes with age, 134–135
Protein kinase C (PKC) pathway, 105
Pulmonary hypertension (PH), 170
Pulse wave velocity (PWV), 35

Q

Quantitative polymerase chain reaction (qPCR), 126

R

Reactive nitrogen species (RNS), 129–130
Reactive oxygen species (ROS), 77, 152–154, 169
Receptors for AGEs (RAGE), 81–82
Regenerative aesthetics, 150–151, 150f
Renal aging, 96–101
Renin–angiotensin system (RAS) activity, 97–99
  in AKI-CKD progression, 99
  with Klotho in the kidney, 99
Replicative senescence in cartilage, 76–77
Resorption (phase 2), 56
Respiratory system, 166f
Restriction fragment length polymorphism (RFLP) analysis, 107
Retina, 198–200
Retinal pigment epithelium (RPE), 198–199
Reversal (phase 3), 56
Revesz syndrome, 207–208
Rheumatoid arthritis (RA), 36

S

SASP inhibitors ("senomorphic" drugs), 185
Senescence and biomarkers of senescence (gerodiagnostics), 186
Senescence-associated apoptotic pathways (SCAPs), 185
Senescence-associated heterochromatin foci (SAHF), 62, 154
Senescence-associated secretory phenotype (SASP), 63, 96, 130, 151–152, 155f
Senescence-associated secretory profile in chondrocytes, 77–78
Senescence in cartilage, 80
Senescence phenotype secretes factors (SASP), 171
Senescence resistance (SAM R) counterparts, 167
Senescent cell accumulation, 177f, 180–181
Senescent cell anti-apoptotic pathways (SCAPs), 155, 172
Senescent cell elimination on metabolic tissues and phenotypes, 185–186
Senescent cells, 152–154
  features of, 153–154
  in skin, 154–155
"Senile emphysema", 167
Senolytic agents, 185
Senolytics, 137, 156–158, 158f
Senomorphics, 158f
Senotherapeutic agents for skin, 155
Serum calcium, 57
Sirtuin (SIRT) proteins, 201
Skeletal maintenance and remodeling, 54–56
Skeletal system, 53
Skin aging, 148f
  biochemical markers of cellular senescence, 154
  cellular senescence, 151–152
  chronological *versus* biological, 147–149
  clinical manifestations of, 149–150
  features of senescent cells, 153–154
  first- and second-generation senolytics, 155–157
  function of senescent cells, 152–153
  geriatric dermatology, 147
  hallmarks of, 149f
  regenerative aesthetics, 150–151
  senescence in, 154
  senotherapeutic agents for skin, 155
  targeting cellular senescence with senotherapies, 158–159
  targeting senescent cells in skin, 154–155
  translation to clinical practice, 157–158
  youth and skin health, 147
Sleep, importance of, 133–134
Small airway remodeling (SAR), 167–168
Small interfering RNA (siRNA) molecules, 38
Smoking, 106–107
Sodium-glucose cotransporter- 2 (SGLT2) inhibitors, 109, 183
SOST gene expression, 64
Src family kinases (SFKs), 155–156
Staphylococcus aureus, 124
Stem cell exhaustion, 80–81
Stress-induced premature senescence (SIPS), 62
Stress-induced senescence in cartilage, 77

Subchondral bone cortical plate (SB-CP), 74f
Sulfonylureas, 184
Superficial zone chondrocytes, 73

T
T cell receptors (TCR), 122–123
Telomerase, 211
    activity, 127–128, 198
    and cancer, 209–210
    expression, 127f, 208–209
    gene, 4
    gene therapy, 208
    versus senolytics, 224f
    therapy, 210–211
Telomerase reverse transcriptase (TERT), 63, 76–77, 109–110, 168–169, 224–225
Telomerase RNA (TR), 168–169
Telomere attrition, 180
Telomere biology, 125–126, 128–129
    and epigenetic processes, 130
    system, 126–127
Telomere biology disorders (TBDs), 130–131
Telomere position effect (TPE), 130, 212
Telomeres, 206, 211–212
    and aging, 207
    dysfunction and shortening, 63–64
    erosion, 76
    shortening, 101
    stability, 207–208
Telomeric repeat-containing RNA (TERRA), 79
Tertiary lymphoid tissues (TLTs), 100
T-helper type 2 cells (Th2), 123
Therapeutic status, 21–22
Thrombocyte, 121
Time to reconsider evidence for anemia treatment (TREAT), 107–108
Tissue inhibitors of metalloproteinases (TIMPs), 151
Trabecular bone, 53
Transactive response DNA binding protein 43 (TDP-43 protein), 24
Transforming growth factor-α (TNFα), 74
Trichothiodystrophy (TTD), 63
Tumor necrosis factor (TNF), 100
Tumour necrosis factor-α (TNFα), 77–78
Type 1 diabetes, 240
Type 2 diabetes, 175–176, 201–202

U
Ultraviolet (UV) light exposure, 147
Umbilical cord blood as cellular reservoir, 136
Umbilical cord-MSCs (UC-MSC-EVs), 151
Unfolded protein response (UPR), 169
Unified systems model, 7–13, 27–30
Unsaturated fatty acids (UFAs), 121
Upstream opportunities for life and health extension, 201–202
Upstream risk factors, 10–11, 130–134
Uromodulin (UMOD), 107

V
Vascular aging, 36
    and brain, 37
    and its consequences, treatment of, 37
Vascular changes, 101
Vascular endothelial cells, 8
Vascular endothelial growth factor, 101
Vascular smooth muscle cells (VSMCs), 36, 227
Veterans Affairs Diabetes Trial (VADT), 108

W
White adipose tissue (WAT), 178
Wilms Tumor 1 (WT1) protein, 99

X
Xeroderma pigmentosum (XP)-type D (XPD) gene, 63

Y
Youth and skin health, 147

Z
Zingiber officinale, 60